BIOPHYSICAL CHEMISTRY

BIOPHYSICAL CHEMISTRY

PART
III

THE BEHAVIOR OF
BIOLOGICAL MACROMOLECULES

Charles R. Cantor
COLUMBIA UNIVERSITY

Paul R. Schimmel
MASSACHUSETTS INSTITUTE OF TECHNOLOGY

W. H. FREEMAN AND COMPANY
San Francisco

Cover drawing after G. G. Hammes and C.-W. Wu, "Regulation of Enzyme Activity," *Science* 172:1205–1211 at 1205. Copyright © 1971 by the American Association for the Advancement of Science.

Sponsoring Editor: Arthur C. Bartlett
Project Editor: Pearl C. Vapnek
Manuscript Editor: Lawrence W. McCombs
Designer: Robert Ishi
Production Coordinator: Linda Jupiter
Illustration Coordinator: Cheryl Nufer
Artists: Irving Geis and Eric Hieber
Compositor: Syntax International
Printer and Binder: R. R. Donnelley & Sons Company

Figures 16-11, 16-12, 16-14, 16-15b, 17-25, 17-26, 24-16 copyright © 1980 by Irving Geis.

Library of Congress Cataloging in Publication Data

Cantor, Charles R   1942–
    The behavior of biological macromolecules.

    (Their Biophysical chemistry; pt. 3)
    Includes bibliographies and index.
    1. Macromolecules. I. Schimmel, Paul Reinhard,
1940–   joint author.   II. Title
QH345.C36 pt. 3   [QP801.P64]   574.19′283s   79-27860
ISBN 0-7167-1191-5   [574.8′8]
ISBN 0-7167-1192-3 pbk.

9 8 7 6 5 4 3 2

To Louis and Ida Dianne Cantor
and Alfred and Doris Schimmel

Contents in detail of Part III

Contents in brief of Parts I and II

Preface

Biophysical Chemistry is concerned with biological macromolecules and complexes or arrays of macromolecules. The work deals with the conformation, shape, structure, conformational changes, dynamics, and interactions of such systems. Our goal is to convey the major principles and concepts that are at the heart of the field. These principles and concepts are drawn from physics, chemistry, and biology.

We have aimed at creating a multilevel textbook in three separately bound parts. The material covers a broad range of sophistication so that the text can be used in both undergraduate and graduate courses. It also should be of value to general scientific readers who simply wish to become familiar with the field, as well as to experienced research scientists in the biophysical area. For example, perhaps half of the material requires only the background provided by a one-semester undergraduate course in physical chemistry. A somewhat smaller fraction necessitates the use of concepts and mathematical techniques generally associated with a more sophisticated background, such as elementary statistical thermodynamics and quantum mechanics.

Biophysical Chemistry is organized into three parts. The first part deals with the structure of biological macromolecules and the forces that determine this structure. Chapter 1 introduces the fundamental questions of interest to biophysical chemists, Chapters 2–4 summarize the known structures of proteins, nucleic acids, and other biopolymers, and Chapters 5–6 treat noncovalent forces and conformational analysis.

Part II summarizes some of the techniques used in studying biological structure and function. The emphasis is on a detailed discussion of a few techniques rather than an attempt to describe every known technique. Chapters 7–9 cover spectroscopic methods, Chapters 10–12 deal with hydrodynamic methods, and Chapters 13–14 discuss x-ray and other scattering and diffraction techniques.

Part III demonstrates how techniques and principles are used in concert to gain an understanding of the behavior and properties of biological macromolecules. The focus is on the thermodynamics and kinetics of conformational changes and ligand interactions. New techniques are introduced as needed, and a few selected case histories or systems are discussed in considerable detail. The topics include ligand interactions (Chapters 15–17), the special theories and techniques used to study molecules that are statistical chains rather than definite folded conformations (Chapters 18–19), protein conformational changes (Chapters 20–21), nucleic acid conformational changes (Chapters 22–24), and membranes (Chapter 25).

We have made every effort to keep the chapters as independent as possible, so that the reader has a wide choice of both the material to be covered and the order in which it is to be treated. Extensive cross-references to various chapters are included to help the reader find necessary background material if the parts are not read in sequence. Where possible, examples are taken repeatedly from a small number of systems, so that the reader can have the experience of contrasting information gained about the same protein or nucleic acid from a variety of different approaches.

Within each chapter, we have attempted to maintain a uniform level of rigor or sophistication. Short digressions from this level are segregated into boxes; longer digressions are indicated by a bullet (•) preceding the section or subsection heading. Readers with a less rigorous background in physics, mathematics, and physical chemistry should find helpful the many boxes that review elementary material and make the text fairly self-contained; Appendix A provides a basic review of principles of matrix algebra. Other boxes and special subsections are aimed at advanced readers; in many cases, these discussions attempt to illuminate points that we ourselves found confusing.

In different sections, the level of mathematical sophistication varies quite significantly. We have tried to use the simplest mathematical formulation that permits a clear presentation of each subject. For example, hydrodynamic properties are treated in one dimension only. The form of a number of the fundamental equations is extracted by dimensional analysis rather than through lengthy (and not particularly instructive) solutions of hydrodynamic boundary-value problems. On the other hand, x-ray and other scattering phenomena are treated by Fourier transforms, and many problems in statistical mechanics are treated with matrix methods. These advanced mathematical techniques are used in only a few chapters, and numerous boxes are provided to assist the reader with no previous exposure to such methods. The remaining sections and chapters are self-contained and can be understood completely without this advanced mathematical formalism.

Some techniques and systems are not covered in any fair detail. This represents a biased choice by the authors, not necessarily of which techniques we feel are important, but simply of which are instructive for the beginning student in this field.

Each chapter concludes with a summary of the major ideas covered. In addition, each chapter is heavily illustrated, including some special drawings by Irving Geis. Certainly, much can be learned simply by reading the chapter summaries and by studying the illustrations. Also, we believe the illustrations convey some of the excitement of the field.

Problems are provided at the end of each chapter. These vary in difficulty from relatively simple to a few where the full answer is not known, at least to the authors. Answers to problems are provided in Appendix B.

Detailed literature citations are not included, except to acknowledge the source of published material reproduced or adapted here. However, a list of critical references for each chapter is included. In virtually all cases, these articles will provide an immediate entrée to the original papers needed for more detailed study.

The problem of notation and abbreviations in this field is a difficult one. In drawing together material from so many different types of research, we have had to adapt the notation to achieve consistency and to avoid confusion among similar symbols. Wherever possible, we have followed the recommendations of the American Chemical Society, but inevitably we have had to develop some conventions of our own. A glossary of some of the more frequently used symbols is provided.

At MIT some of this material has been used in an undergraduate course in biophysical chemistry. The course was designed to meet the needs of students wishing a second course in physical chemistry, but developed in a biochemical framework. The idea was to construct a course that covered much of the same material with the same rigor as a parallel, more traditional course. The only preparation required was a one-semester course in undergraduate physical chemistry, which at MIT is largely concerned with chemical thermodynamics.

Over the years graduate courses in biophysical chemistry at MIT and at Columbia have made use of much of the material presented here. In addition, a special-topics course in protein structure has used some of the material. Because a broad range of subjects is covered, its usefulness as a text will hopefully meet a variety of individual teaching tastes and preferences, as well as enable instructors to vary content as needs develop and change.

It is obvious that a work of this complexity cannot represent solely the efforts of its two authors. As we sought to master and explain the wide range of topics represented in biophysical chemistry, we learned why so few books have been written in this field in the past two decades. We owe a great debt to many who helped us in ways ranging from sharing their understanding to providing original research data.

We give special thanks to Irving Geis, for his effort on a number of complex illustrations and for his helpful advice on numerous other drawings; to Wilma Olson, for reading a major portion of the entire manuscript; to Robert Alberty and Gordon Hammes, for their influence, through teaching and discussions, on the material on biochemical equilibria and kinetics; to Richard Dickerson, for providing material and advice that were essential for the preparation of Chapter 13; to Paul Flory, for inspiring our treatment of conformational energies and configurational statistics of macromolecules; to Howard Schachman, whose course at Berkeley inspired parts of several chapters; to R. Wayne Oler, for bringing the authors together for this undertaking, and to Bruce Armbruster, for sealing the commitment; to the helpful people at W. H. Freeman and Company, including Ruth Allen, Arthur Bartlett, Robert Ishi, Larry McCombs, and Pearl Vapnek; to Kim Engel, Karen Haynes, Marie Ludwig, Joanne Meshna, Peggy Nelson, Cathy Putland, and Judy Schimmel, for typing and related work associated with the manuscript; and to Cassandra Smith and to Judy, Kathy, and Kirsten Schimmel, for their patience with the intrusion this work has made on the authors' lives.

Many people read and commented on specific chapters, provided figures, notes and materials, and spent much time with us in helpful discussions. We gratefully thank these people: Robert Alberty, Arthur Arnone, Struther Arnott, P. W. Atkins, Robert Baldwin, Larry Berliner, Bruce Berne, Richard Bersohn, Sherman Beychok, Victor Bloomfield, David Brandt, John Brandts, John Chambers, Sunney Chan, Patricia Cole, Robert Crichton, Francis Crick, Donald Crothers, Norman Davidson, Richard Dickerson, David Eisenberg, Robert Fairclough, Gerry Fasman, George Flynn, David Freifelder, Ronald Gamble, Robert Gennis, Murray Goodman, Jonathan Greer, O. Hayes Griffith, Gordon Hammes, John Hearst, Ellen Henderson, James Hildebrandt, Wray Huestis, Sung Hou Kim, Aaron Klug, Nelson Leonard, H. J. Li, Stephen Lippard, Richard Lord, Brian Matthews, Harden McConnell, Peter Moore, Garth Nicolson, Leonard Peller, Richard Perham, Michael Raftery, Alexander Rich, Frederick Richards, David Richardson, Wolfram Saenger, Howard Schachman, Harold Scheraga, Benno Schoenborn, Verne Schumaker, Nadrian Seeman, Robert Shulman, Mavis Shure, Louise Slade, Cassandra Smith, Hank Sobell, Thomas Steitz, Robert Stroud, Lubert Stryer, Serge Timasheff, Ignacio Tinoco, Jr., Richard Vandlen, Jerome Vinograd, Peter von Hippel, Christopher Walsh, James Wang, Gregorio Weber, Peter Wellauer, Barbara Wells, Robert Wells, William Winter, Harold Wyckoff, Jeffries Wyman, and Bruno Zimm.

November 1979   *Charles R. Cantor*
                                                         Paul R. Schimmel

Glossary of symbols

This glossary includes some of the symbols used extensively throughout the text. In many cases, the same or very similar symbols are used in certain contexts with other meanings; the meaning of a symbol is explained in the text where it is introduced.

Symbol	Meaning
A	Absorbance.
A_{ij}	Amplitude of kinetic decay.
Å	Angstroms.
a	Hyperfine splitting constant. Long semi-axis of ellipse. Persistence length.
a	Unit cell basis vector.
a*	Reciprocal cell basis vector.
a_{ij}	Parameters composed of rate constants.
a_s	Exponent relating sedimentation to chain length.
a_η	Exponent relating viscosity to chain length.
b	Short semiaxis of ellipse.
b	Unit cell basis vector.
b*	Reciprocal cell basis vector.
C	Molar concentration.
C_n	Rotational symmetry group element. Characteristic ratio.
C_∞	Limiting characteristic ratio.
ΔC_p^0	Standard constant pressure heat capacity change per mole.
c	Velocity of light in vacuum. Ratio of k_R/k_T. Weight concentration.
c_p	Plateau weight concentration.
$\hat{c}_i$	Weight concentration of ith species or component.
c	Unit cell basis vector.
c*	Reciprocal cell basis vector.
D	Debye.
D	Translational diffusion constant.
D_n	Dihedral symmetry group element.
D_{rot}	Rotational diffusion constant.
$D_{20,w}$	D extrapolated to 20° C, water.
E_a	Activation energy.
E_d	Interaction energy between two dipoles.
E_{kl}	Nonbonded pair interaction potential.
E_{tor}	Torsional potential energy.
$E(\Phi_i,\Psi_i),$ E_i	Total rotational potential for residue i.
E	Electric field.
e	Exponential function. Unit of charge on electron.
F	Frictional coefficient ratio.
$F(\mathbf{S})$	Structure factor.
$F_H(\mathbf{S})$	Structure factor, heavy-atom contribution.

Symbol	Meaning
$F_{Tot}(S)$	Structure factor for an array.
$F_m(S)$	Molecular structure factor.
$\mathbf{F}$	Force.
$\mathscr{F}$	The Faraday.
f	Translational frictional coefficient.
f_{app}	Apparent fractional denaturation.
f_D	Fraction in denatured state.
f_N	Fraction in native state.
f_{min}	Translational friction coefficient of anhydrous sphere.
f_{rot}	Rotational friction coefficient for sphere.
f_{sph}	Translational friction coefficient for sphere.
f_a, f_b	Rotational friction coefficient around a, b axis of ellipse.
G	Gibbs free energy.
ΔG^0	Standard Gibbs free energy change per mole.
$\Delta \bar{G}^0$	Intrinsic standard free energy change (with statistical component removed).
$\Delta G_{I,ij}$	Free energy of interaction between two ligands.
ΔG_r	ΔG per residue.
ΔG_{Tot}	Total free energy change per mole.
ΔG_{el}	Change in electrostatic free energy.
ΔG_T	Total free energy of formation of configuration.
$\Delta \Delta G_T$	Difference in ΔG_T between two configurations.
$\Delta \bar{G}_{gr}$	Average helix growth free energy change per residue pair.
g	g value for free electron, 2.00232.
g_x, etc.	Component of g-factor tensor.
H	Enthalpy.
H_{xy}	Magnetic field in xy plane.
ΔH	Enthalpy change per mole.
ΔH^0	Standard enthalpy change per mole.
ΔH_r	ΔH per residue.
ΔH_D	Enthalpy change for conversion from fully native to fully denatured state.
ΔH_{app}	Apparent enthalpy change for conversion from fully native to fully denatured state.
$\mathbf{H}$	Magnetic field.

Symbol	Meaning
$\mathbf{H}_{res}$	Magnetic field at which resonance occurs.
$\underline{\mathbf{H}}$	Hamiltonian operator.
$\Delta \mathbf{H}_{loc}$	Magnetic field generated by local environment.
h	Planck's constant.
$\hbar$	$h/2\pi$
I	Intensity of radiation. Nuclear spin quantum number. Ionic strength.
$I(S)$	Scattering intensity relative to a single electron at the origin.
i	$\sqrt{-1}$
$\hat{\mathbf{i}}$	Cartesian unit vector.
J	NMR coupling constant.
$\mathbf{J}_2$	Solute flux.
$\hat{\mathbf{j}}$	Cartesian unit vector.
K_D	True equilibrium constant for conversion from fully native to fully denatured state.
K_p	Michaelis constant for product.
K_S	Michaelis constant for substrate.
K_η	Coefficient relating viscosity to chain length.
K_s	Coefficient relating sedimentation to chain length.
K_{app}	Apparent equilibrium constant for conversion from fully native to fully denatured state.
K_i	Macroscopic equilibrium constant. Equilibrium constant for forming ith configuration. Equilibrium constant for transition from native state to intermediate state i.
$\tilde{K}_i$	Apparent dissociation constant, one-ligand system.
$\tilde{K}_{ij}$	Apparent dissociation constant, two-ligand system.
k	Boltzmann's constant. Microscopic equilibrium dissociation constant.
k_R	Microscopic dissociation constant for R state.
k_T	Microscopic dissociation constant for T state.
k_i	Microscopic equilibrium constant.
$\hat{\mathbf{k}}$	Cartesian unit vector.

Symbol	Meaning
L_c	Contour length.
L, L'	Equilibrium constant for $R_0 \quad T_0$.
$\mathbf{L}$	Angular momentum.
l	Length of one polymer bond.
l_e	Length of statistical segment.
M	Molecular weight.
$\bar{M}_n$	Number-average molecular weight.
$\bar{M}_w$	Weight-average molecular weight.
$\bar{M}_i$	Molecular weight of ith macromolecular species.
M_{ij}	Species with i bound L_1 and j bound L_2.
$M^{(j)}$	Set of all species with j bound L_2.
$\mathbf{M}$	Magnetization.
$\mathbf{M}_{xy}$	Magnetization in xy plane.
$\underline{\mathbf{M}}$	Statistical weight matrix.
m	Colligative molality. Mass of object.
m_e	Mass of electron.
m_i	Molality of ith species.
m_s	Quantum number of electron spin along z axis.
m_I	Quantum number of nuclear spin along z axis.
m'	Total molality.
$\underline{\mathbf{m}}$	Magnetic dipole operator.
N_0	Avogadro's number.
N_C	Number of carbons in amphiphile R chain.
N'_C	Number of carbons in amphiphile that are imbedded in hydrocarbon core of micelle.
N_e	Number of statistical segments.
N_{Ch}	Number of chains in micelle.
N_{hg}	Number of head groups in micelle.
n	Refractive index. Number of sites. Number of bonds in polymer.
n_i	Number of moles of component i. Number of sites of certain type.
n_w	Weight-average degree of polymerization.
P	Pitch of helix. Pressure. Patterson function.
P_0	Solvent vapor pressure.
P_v	Solvent vapor pressure in presence of solute.

Symbol	Meaning
P_r	Axial ratio.
pK_a	$-\log_{10} K_a$
pO_2	Partial pressure of oxygen.
$(pO_2)_{1/2}$	Partial pressure of oxygen at half saturation.
$\underline{\mathbf{p}}$	Momentum operator.
Q	Configurational partition function.
R	Gas constant.
R_G	Radius of gyration.
$\langle R_G^2 \rangle_0$	Unperturbed mean square radius of gyration.
$\bar{R}$	Fraction of molecules in R state.
$\underline{\mathbf{R}}$	Nuclear position operator.
$\underline{\mathbf{R}}(\alpha,\beta)$	Coordinate transformation matrix.
r	Distance of separation.
r_D	Donnan ratio.
r_e	Radius of equivalent sphere.
$\langle r^2 \rangle_0$	Unperturbed mean square end-to-end distance.
$\mathbf{r}$	Polymer end-to-end vector.
$\underline{\mathbf{r}}$	Electron position operator.
S	Svedberg (unit of sedimentation coefficient).
S_A	Partial molal entropy.
S'_A	Unitary part of S_A.
ΔS_r	ΔS per residue.
ΔS^0	Standard entropy change.
ΔS_u^0	Unitary standard entropy change.
$\mathbf{S}$	Scattering vector.
s	Sedimentation coefficient. Statistical weight. Equilibrium constant for helix growth. Equilibrium constant for base-pair formation.
$s_{20,w}$	Sedimentation coefficient corrected to $20°$ C, water
$\hat{\mathbf{s}}$	Unit vector along scattered radiation.
$\hat{\mathbf{s}}_0$	Unit vector along incident radiation.
T	Temperature (in degrees Kelvin usually).
T_m	Melting temperature.
T_1	Longitudinal relaxation time.
T_2	Transverse relaxation time.
$\underline{\mathbf{T}}_i$	Transformation matrix.
t	Time.

Symbol	Meaning
U_{mic}^0	Attractive part of μ_{mic}^0.
u	Component of $\mathbf{M}_{xy}$ in phase with $\mathbf{H}_{xy}$. Electrophoretic mobility.
V	Volume.
V_{h}	Hydrated volume.
$\bar{V}_i$	Partial specific volume of component i.
V_{p}	Maximum reaction velocity in reverse direction.
V_{s}	Maximum reaction velocity in forward direction.
v	Speed (also called velocity). Component of $\mathbf{M}_{xy}$ out of phase with $\mathbf{H}_{xy}$.
v_i	Initial reaction velocity.
$\langle v_2 \rangle$	Effective average solute velocity.
$\bar{v}$	Partial molar volume.
$\bar{v}_{\text{s}}$	Partial molar volume of pure solvent.
$\mathbf{v}$	Velocity.
$W(r)$	Radial distribution function of end-to-end distance.
$W(x,y,z)$	End-to-end distance distribution function.
W_{mic}^0	Repulsive part of μ_{mic}^0.
$(\bar{X}_i)$	Equilibrium concentration.
$\Delta(X_i)$	Difference between temporal and equilibrium concentration.
x_{b}	Bottom of cell.
x_{m}	Meniscus position.
y	General physical property.
y_{D}	Physical property of denatured state.
y_{N}	Physical property of native state.
$\bar{y}$	Fractional saturation of site.
$\bar{y}_{\text{F}}$	Fractional saturation with ligand F.
z	Charge on macromolecule or ion in units of e.
z_i	Ionic valence of ith ion.
α	Degree of association. Dimensionless binding parameter like $(F)/k_{\text{R}}$.
α_{H}	Hill constant.
β	Dimensionless binding parameter.
β_{e}	Bohr magneton.
β_{n}	Nuclear magneton.
β_{s}	Mandelkern–Flory–Scheraga parameter.
β'	Scheraga–Mandelkern parameter.

Symbol	Meaning
Γ	Parameter affecting relaxation amplitudes.
γ	Magnetogyric ratio. $(A)/K_{\text{AR}}$ binding parameter. Velocity gradient dv_x/dz.
λ_1, λ_2	Parameters composed of rate constants.
δ	Chemical shift parameter. Phase shift.
$\delta(x)$	Dirac delta function of argument x.
δ_1	Hydration (in grams per gram).
δ_{ij}	Kronecker delta.
ε	Dielectric constant. Molar decadic or residue extinction coefficient.
$\Delta\varepsilon$	Circular dichroism ($\varepsilon_{\text{L}} - \varepsilon_{\text{R}}$).
η	Solution viscosity.
η_0	Solvent viscosity.
η_{rel}	Relative viscosity.
η_{sp}	Specific viscosity.
$[\eta]$	Intrinsic viscosity.
Θ_i	Fractional saturation of ith site.
θ	Scattering angle. Fractional helicity.
$[\theta]$	Molar ellipticity.
$\underset{\sim}{\Lambda}$	Matrix of λ_i's.
$\underset{\sim}{\lambda}$	Eigenvalue. Wavelength. Kinetic decay time.
λ_j	jth kinetic decay time of jth eigenvalue.
μ_i	Chemical potential per mole.
μ_i^0	Standard chemical potential per mole.
$\hat{\mu}_i$	Chemical potential per gram.
$\hat{\mu}_i^0$	Standard chemical potential per gram.
μ_{mic}^0	Standard chemical potential of amphiphile in micelle.
μ_{w}^0	Standard chemical potential of amphiphile in aqueous phase.
$\boldsymbol{\mu}_{\text{m}}$	Magnetic moment.
$\underset{\sim}{\boldsymbol{\mu}}$	Electric dipole moment operator.
$\underset{\sim}{v}$	Frequency. Simha factor in viscosity. Moles of ligand bound per mole of macromolecule.
v_N	Saturation density for lattice with N units.
π	Osmotic pressure.
ρ	Mass density (in grams per cm^3).
$\rho(\mathbf{r})$	Electron density.
σ	Nucleation constant.

Symbol	Meaning
σ_h	Superhelix density.
τ	Number of supercoils.
τ_F	Fluorescence decay time.
τ_a, τ_b	Rotational relaxation time for a-, b-axis orientation.
τ_c	Rotational correlation time.
τ_r	Rotational relaxation time of sphere.
τ, τ_j	Reaction relaxation times.
Φ	Electrical potential. Voltage difference.
Φ_c	Universal constant for random coils $2 \cdot 1 \times 10^{23}$.
ϕ	N–C' torsional angle. Phase of complex number.
$\phi_{1a}, \phi_{20},$ etc	Monomer wave functions.
ϕ_F	Fluorescence quantum yield.
ϕ_p	Practical osmotic coefficient.
ϕ', ϕ''	Nucleic acid backbone torsional angles.
$[\phi]$	Molar rotation per residue.
χ	Mole fraction of all solute species.
χ_i	Mole fraction of ith component.
χ_A	Mole fraction of Ath component.
χ_{gc}	Mole fraction G + C.
χ	Glycosidic bond torsional angle.
ψ	C'–C torsional angle.
ψ', ψ''	Nucleic acid backbone torsional angles.
Ω_{jk}	Number of ways of putting k helical units into j separated sequences.
Ω_k	$(n - k + 1)$ number of ways of placing k helical units in one sequence within chain of n residues.
$\Omega_{n,i}$	Number of ways of assorting i items (ligands) in n boxes (sites).

Symbol	Meaning
ω	Circular frequency or angular velocity.
ω_0	Larmor frequency.
ω', ω''	Nucleic acid backbone torsional angles.
$\Delta\omega_{1/2}$	Line width.
$\boldsymbol{\omega}$	Angular velocity.
imag	Imaginary part of.
$\langle\rangle$	Average.
$\langle|\rangle$	Overlap integral.
$\langle\|\rangle$	Expectation value integral.
$*$	Superscript, complex conjugate, as in F^*.
$|\ $	Amplitude of complex number or length of vector, as in $|F|$.
$\mathbf{V}$	Vector differential.
$(\ )$	Molar concentration, as in (A).
$\dagger$	Superscript, transpose of matrix, as in $A^\dagger$.
$\frown$	Superscript, convolution product, as in $\widehat{AB}$.

General Rules

K	Macroscopic equilibrium constant.
k	Microscopic equilibrium constant or rate constant.
C	Molar concentration.
c	Weight concentration.
$\mathbf{M}$	All matrices and operators.
$\hat{\imath}$	All unit vectors.
R_G	Radius of gyration.
χ	Mole fraction.
Φ	Voltage or electrical potential.

THE BEHAVIOR OF BIOLOGICAL MACROMOLECULES

The first two Parts emphasize the structure of biological molecules (Part I) and the techniques available for determining their structure and behavior (Part II). Part III focuses on the relationship between structure and behavior, and on the mechanisms that lie at the heart of biological function. The physiological function of proteins and nucleic acids commonly is manifested through interactions with specific ligands; Chapters 15, 16, and 17 treat equilibrium and kinetic aspects of ligand interactions, including (in Chapter 17) a discussion of control mechanisms that modulate biological activity. Much of the treatment is applicable to interactions of biological macromolecules in general, although most examples are drawn from systems of proteins.

After this treatment, attention shifts to the conformational behavior of biological polymers. For example, Chapter 18 takes up polymer configuration statistics, with emphasis on the statistics of polypeptides in relation to chain conformation. Chapter 19 outlines how the hydrodynamic properties of double-stranded DNA in solution reveal that it is a wormlike coil. In Chapters 20 and 21, we consider conformational changes of proteins and polypeptides, including the well-studied polypeptide helix–coil transition (Chapter 20) and the problem of the reversible folding of proteins (Chapter 21).

The emphasis shifts to nucleic acids in Chapters 22 through 24, where we take up the ligand interactions, conformational changes, and tertiary structural organization of certain nucleic acids. Some of these issues parallel those of the preceding chapters, where more emphasis was placed on proteins and polypeptides. The book concludes in Chapter 25 with a shift to the important subject of membrane equilibria and the structure and behavior of lipid bilayers.

Of course, much of the material in Part III is tied closely to the discussion in Parts I and II. In appropriate places, reference is made to the earlier chapters. However, many readers will find it possible to read a good portion of this Part without having read the other Parts, particularly if the appropriate section of an earlier chapter is consulted when needed.

15

Ligand interactions at equilibrium

15-1 IMPORTANCE OF LIGAND INTERACTIONS

A wide variety of physiological processes are the reflection of ligand interactions with macromolecules, especially with proteins. The most common are interactions between enzymes and their substrates and with other molecules that influence activity. In addition, there are interactions between hormones and hormone receptors, between small molecules and proteins involved in the active transport of the small molecules, between ions and both nucleic acids and proteins, and so on. Upon reflection, it is clear that virtually all biological phenomena depend on one or more ligand interactions. It is not surprising, therefore, that a large amount of biochemical and biophysical research has been directed at exploring these interactions in depth.

Our first consideration is to develop the statistical framework that enables us to treat (and to gain insight into) the principal features of an equilibrium ligand association system. These are general considerations that apply to any equilibrium system. In addition, there are many special features, such as site–site interactions and cooperativity, linkage relationships between two different ligands binding to the same macromolecule, and statistical complications associated with linear, latticelike chains.

General features of ligand interactions at equilibrium are developed in this chapter; Chapter 16 treats kinetic phenomena. In addition, Chapters 16 and 17 deal with some of the special areas that play a prominent role in biochemistry. These include enzymatic systems (Chapter 16) and regulation phenomena (Chapter 17) commonly known as allosteric interactions. The treatment of some of these issues draws on the general framework laid down in this chapter.

15-2   LIGAND EQUILIBRIA

Macroscopic and microscopic constants

Before discussing the association of ligands with multiple sites on macromolecules, it is useful to discuss briefly the distinction between microscopic and macroscopic equilibrium constants. A concrete example is provided by the titration of the amino acid glycine. This can be viewed as a dibasic acid. We define GH_2^+, GH, and G^- as the forms bearing two, one, and no protons, respectively. The *macroscopic* equilibria are

$$GH_2^+ \rightleftharpoons GH + H^+ \tag{15-1a}$$

$$GH \rightleftharpoons G^- + H^+ \tag{15-1b}$$

and the two macroscopic dissociation constants are given by

$$K_1 = (GH)(H^+)/(GH_2^+) \tag{15-2a}$$

$$K_2 = (G^-)(H^+)/(GH) \tag{15-2b}$$

The two pK values can be obtained from a titration; at 25°C, extrapolated to zero ionic strength, they are $pK_1 = 2.35$, and $pK_2 = 9.78$.

We now examine the *microscopic* states of glycine during the titration. Altogether there are four forms, where

$$GH_2^+ = {}^+H_3NCH_2COOH \tag{15-3a}$$

$$GH = {}^+H_3NCH_2COO^- + H_2NCH_2COOH \tag{15-3b}$$

$$G^- = H_2NCH_2COO^- \tag{15-3c}$$

and the microscopic ionization equilibria are

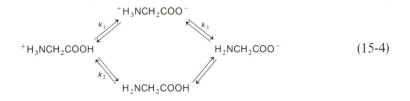

$$\tag{15-4}$$

where the k_i values are microscopic dissociation constants. According to Equations 15-2 and 15-3,

$$K_1 = [({}^+H_3NCH_2COO^-) + (H_2NCH_2COOH)](H^+)/({}^+H_3NCH_2COOH)$$

$$= k_1 + k_2 \tag{15-5a}$$

$$K_2 = (H_2NCH_2COO^-)(H^+)/[(^+H_3NCH_2COO^-) + (H_2NCH_2COOH)]$$

$$= 1/(k_3^{-1} + k_4^{-1}) \tag{15-5b}$$

Equation 15-5 shows the relationships between the microscopic and macroscopic dissociation constants.

The four microscopic constants are not independent. In particular,

$$\boxed{k_1 k_3 = k_2 k_4} \tag{15-6}$$

Equation 15-6 is easy to verify; it is a direct consequence of detailed balancing. Equations 15-5 and 15-6 give three relationships involving the four microscopic constants. A fourth relationship may be obtained by assuming that k_2 has the same value as the single dissociation constant for the methyl ester of glycine ($^+H_3NCH_2CO_2CH_3 \rightleftharpoons H_2NCH_2CO_2CH_3 + H^+$). This assumption gives $pk_2 = 7.7$. With the values of pK_1 and pK_2 given earlier, it then is easy to calculate from Equations 15-5 and 15-6 that $pk_1 = 2.35$, $pk_3 = 9.78$, and $pk_4 = 4.43$. From these values, the reader should be able to deduce whether dissociation from $^+H_3NCH_2COOH$ to neutral glycine proceeds predominantly by the top path or the bottom path in Equation 15-4.

This simple illustration serves as a concrete example of the meanings of microscopic and macroscopic constants, and of their interrelationships. As a second example, we treat a situation in which *statistical effects* come into play. Consider a molecule A, which has two equivalent sites for a specific ligand. For instance, A might be a long-chain aliphatic dicarboxylic acid in which the microscopic dissociation constant is the same for each carboxylic group, regardless of the ionization state of the other group (this condition can be fulfilled if the aliphatic chain is long enough that electrostatic interactions between the two carboxyl groups are negligible). The macroscopic equilibria are

$$A + H^+ \rightleftharpoons A(H^+)_1 \tag{15-7a}$$

$$A(H^+)_1 + H^+ \rightleftharpoons A(H^+)_2 \tag{15-7b}$$

and the macroscopic dissociation constants are given by

$$K_1 = (A)(H^+)/(A(H^+)_1) \tag{15-8a}$$

$$K_2 = (A(H^+)_1)(H^+)/(A(H^+)_2) \tag{15-8b}$$

The microscopic equilibria can be written schematically as

$$A + H^+ \rightleftharpoons AH^+ \tag{15-9a}$$

$$A + H^+ \rightleftharpoons {}^+HA \tag{15-9b}$$

$$AH^+ + H^+ \rightleftarrows {}^+HAH^+ \tag{15-9c}$$

$$^+HA + H^+ \rightleftarrows {}^+HAH^+ \tag{15-9d}$$

where the microscopic dissociation constant k is the same for each step. In Equation 15-9 we have distinguished between microspecies by assigning one ionization site to the left and the other to the right side of A. In terms of microscopic species, the macroscopic forms are defined as

$$A = A \tag{15-10a}$$

$$A(H^+)_1 = AH^+ + {}^+HA \tag{15-10b}$$

$$A(H^+)_2 = {}^+HAH^+ \tag{15-10c}$$

From Equations 15-8 to 15-10, we conclude that

$$K_1 = k/2 \tag{15-11a}$$

$$K_2 = 2k \tag{15-11b}$$

$$K_1/K_2 = 1/4 \tag{15-11c}$$

Thus, even though the microscopic dissociation constant is the same for each ionization, statistical effects make the first apparent macroscopic dissociation constant *four times smaller* than that of the second one.

In this chapter and in Chapter 17, we frequently use the concepts of microscopic and macroscopic constants, and it will be important to keep firmly in mind the distinctions between them that are illustrated in the preceding examples.

15-3   IDENTICAL INDEPENDENT SITES

Calculating the number of microscopic species

We first consider a macromolecule M, which contains n sites for the ligand L. Each site has the same microscopic ligand dissociation constant k. The sites also are assumed to be independent—that is, the microscopic dissociation constant k for a particular site is the same regardless of the state of occupancy of the other sites. The equilibria that characterize the interaction may be written as

$$M_0 + L \rightleftarrows M_1$$

$$M_1 + L \rightleftarrows M_2$$

$$\vdots \qquad \vdots \qquad \vdots$$

$$M_{n-1} + L \rightleftarrows M_n \tag{15-12}$$

where the index on M denotes the total number of molecules of L that are bound. Thus, M_i is taken to mean the *total set of microscopic species that have i bound molecules of L*. For example, if $n = 4$, and we schematically represent our macromolecule as a square with four sites,

$$M_2 = \begin{array}{|c|c|}\hline & L \\\hline & L \\\hline\end{array} + \begin{array}{|c|c|}\hline & L \\\hline L & \\\hline\end{array} + \begin{array}{|c|c|}\hline L & L \\\hline & \\\hline\end{array} + \begin{array}{|c|c|}\hline & \\\hline L & L \\\hline\end{array} + \begin{array}{|c|c|}\hline & L \\\hline & L \\\hline\end{array} + \begin{array}{|c|c|}\hline L & \\\hline L & \\\hline\end{array} \qquad (15\text{-}13)$$

where each microscopic form is present in equal amounts. Thus, with $n = 4$, there are six microscopic species that comprise M_2. In general, there are $\Omega_{n,i}$ distinct ways to put i ligands on n sites, where[§]

$$\Omega_{n,i} = \frac{n \times (n-1) \times (n-2) \times \cdots \times (n-i+1)}{i!} = \frac{n!}{(n-i)!\,i!} \qquad (15\text{-}14)$$

Consequently, there are $\Omega_{n,i}$ microscopic forms that make up M_i.

Calculation of v

Equilibrium measurements of ligand binding typically yield the moles of ligand bound per mole of macromolecule. This parameter generally is designated v; it is given by

$$v = \sum_{i=0}^{n} i(M_i) \Big/ \sum_{i=0}^{n} (M_i) \qquad (15\text{-}15)$$

Our goal is to express v in terms of the free ligand concentration, (L).

In general, we can express the concentration of any form M_i in terms of any

---

[§] Equation 15-14 is easy to derive. There are n different sites in which to place the first ligand; after it has been placed, there are $n - 1$ sites available for the second, then $n - 2$ for the third, and so on, with $n - i + 1$ sites available for the ith ligand. The product $n \times (n-1) \times \cdots \times (n-i+1)$ would give the total arrangements possible except that there is a redundancy; this arises because we have counted each distinct arrangement of i ligands in n sites more than once. For example, if we place the first ligand in site 2 and the second in site 4, this gives the same end result as if we had placed the first in site 4 and the second in site 2. In the product $n \times (n-1) \times \cdots \times (n-i+1)$, we have counted each distinct arrangement $i!$ times, so a correction must be made.

Note also that $\Omega_{n,i}$ is the binomial coefficient of x^i in the expansion of $(1 + x)^n$.

other form by making use of the *macroscopic* dissociation constants. For example,

$$K_1 = (M_0)(L)/(M_1) \tag{15-16a}$$

$$\vdots \qquad \vdots$$

$$K_i = (M_{i-1})(L)/(M_i) \tag{15-16b}$$

$$\vdots \qquad \vdots$$

$$K_n = (M_{n-1})(L)/(M_n) \tag{15-16c}$$

and

$$(M_i) = (M_{i-1})(L)/K_i = (M_0)(L)^i \Big/ \prod_{j=1}^{i} K_j \tag{15-17}$$

The macroscopic constant K_i is to be distinguished from the single microscopic constant k that characterizes all of the sites. The dissociation constant k refers to the equilibrium with respect to particular microscopic species, whereas the macroscopic constant K_i involves the entire ensemble of species represented by M_i and M_{i-1}. For example, with $n = 4$, and again using the format of the schematic illustration from Equation 15-13,

$$k = \frac{\left(\boxed{\ }\right)(L)}{\left(\boxed{L}\right)} = \frac{\left(\boxed{L\,L}\right)(L)}{\left(\boxed{L\,L}\right)} = \frac{\left(\boxed{L\,L / L}\right)(L)}{\left(\boxed{L\,L / L\,L}\right)} = \cdots \tag{15-18}$$

whereas K_1, for example, is

$$K_1 = \frac{\left(\boxed{\ }\right)(L)}{\left(\boxed{L}\right) + \left(\boxed{L}\right) + \left(\boxed{L}\right) + \left(\boxed{L}\right)} \tag{15-19}$$

The relationship between K_i and k is governed by the simple statistical factors $\Omega_{n,i}$. In particular, it is easy to show (Problem 15-1) that

$$K_i = (\Omega_{n,i-1}/\Omega_{n,i})k \tag{15-20}$$

Therefore, we can rewrite Equation 15-17 as

$$(M_i) = (M_{i-1})(L)/K_i = (M_{i-1})[(n - i + 1)/i][(L)/k] \tag{15-21}$$

With similar expressions for (M_{i-1}), (M_{i-2}), etc., we obtain

$$(M_i) = (M_0)\left\{\prod_{j=1}^{i} [(n - j + 1)/j]\right\}[(L)/k]^i \tag{15-22}$$

Substitution of Equation 15-22 into Equation 15-15 gives

$$v = \frac{\sum\limits_{i=1}^{n} i\left\{\prod\limits_{j=1}^{i} [(n-j+1)/j]\right\}[(L)/k]^i}{1 + \sum\limits_{i=1}^{n} \left\{\prod\limits_{j=1}^{i} [(n-j+1)/j]\right\}[(L)/k]^i} \tag{15-23}$$

Although Equation 15-23 appears algebraically complex, it readily simplifies. The product term is identical to $\Omega_{n,i}$ (Eqn. 15-14):

$$\prod_{j=1}^{i} [(n-j+1)/j] = n!/(n-i)!i! \tag{15-24}$$

Substituting Equation 15-24 into Equation 15-23, we obtain

$$v = \frac{\sum\limits_{i=1}^{n} i[n!/(n-i)!i!][(L)/k]^i}{1 + \sum\limits_{i=1}^{n} [n!/(n-i)!i!][(L)/k]^i} \tag{15-25}$$

The denominator of Equation 15-25 is simply the binomial expansion of $[1+(L)/k]^n$:

$$[1 + (L)/k]^n = 1 + \sum_{i=1}^{n} [n!/(n-i)!i!][(L)/k]^i \tag{15-26}$$

Differentiation of Equation 15–26 with respect to $(L)/k$, followed by multiplication by $(L)/k$, gives

$$n[(L)/k][1 + (L)/k]^{n-1} = \sum_{i=1}^{n} i[n!/(n-i)!i!][(L)/k]^i \tag{15-27}$$

The right-hand side of Equation 15-27 corresponds to the numerator of Equation 15-25. Substituting Equations 15-26 and 15-27 into Equation 15-25, we obtain

$$v = \frac{n(L)/k}{1 + (L)/k} \tag{15-28}$$

or

$$v/(L) = n/k - v/k \tag{15-29}$$

The simple forms of Equations 15-28 and 15-29 suggest that these expressions can be derived without recourse to the statistical framework we have generated. This is indeed the case, although the derivation just given is useful in that it gives good insight into the statistical features of the binding equilibria.

A simple derivation

An easy way to derive Equation 15-28 is to focus on the binding equilibrium of site i only. Let Θ_i be the fractional saturation of site i. Then,

$$\Theta_i = (\text{Bound site } i)/[(\text{Free site } i) + (\text{Bound site } i)]$$

$$= \frac{(\text{Free site } i)[(\text{Bound site } i)/(\text{Free site } i)]}{(\text{Free site } i)[1 + (\text{Bound site } i)/(\text{Free site } i)]} \tag{15-30}$$

Because $(\text{Bound site } i)/(\text{Free site } i) = (\text{L})/k$, we have

$$\Theta_i = \frac{(\text{L})/k}{1 + (\text{L})/k} \tag{15-31}$$

A similar expression may be written for each of the n identical sites. Adding these n expressions together, we obtain Equation 15-28 (note that $\sum_i \Theta_i = v$).

Scatchard plot

Equation 15-29 is a useful representation of the relationship between v and (L) for the simple case of identical independent sites. A plot of $v/(\text{L})$ versus v is sometimes known as a Scatchard plot (see Scatchard, 1949). This plot is linear with an ordinate intercept of n/k, an abscissa intercept of n, and a slope of $-k^{-1}$ (Fig. 15-1). Clearly, this plot provides a simple and convenient way to obtain the two parameters that characterize the binding equilibria.

15-4   MULTIPLE CLASSES OF INDEPENDENT SITES

Curved Scatchard plots

In many cases, a Scatchard plot of $v/(\text{L})$ versus v proves to be curved rather than linear. This may mean that more than one class of sites are present. If there are n_1 independent sites with the intrinsic microscopic dissociation constant k_1, and n_2

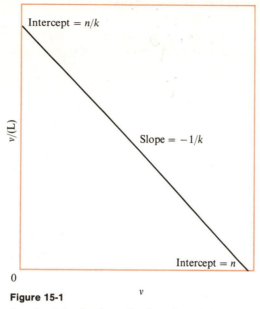

Figure 15-1

Scatchard plot for identical, independent binding sites.

sites with dissociation constant k_2, and so on, then an equation analogous to Equation 15-28 holds for each class of sites. Thus we obtain

$$v = \sum_i \frac{n_i(L)/k_i}{1 + (L)/k_i} \tag{15-32}$$

and

$$v/(L) = \sum_i \frac{n_i/k_i}{1 + (L)/k_i} \tag{15-33}$$

Equations 15-32 and 15-33 are parametric forms that may be used to obtain the parameters n_i and k_i from a Scatchard plot. Figure 15-2 is an illustration of a biphasic plot for the case of two classes of independent sites.

Decomposition of a biphasic Scatchard plot

The plot in Figure 15-2 may be decomposed as follows. A tangent line is drawn to the plot around $v = 0$. The $v/(L)$ intercept of this line[§] is $n_1/k_1 + n_2/k_2$. As a first

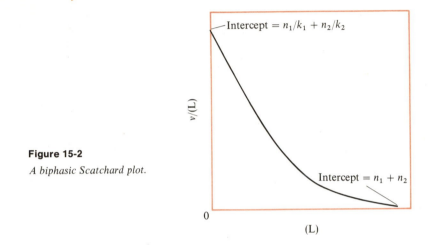

Figure 15-2

A biphasic Scatchard plot.

approximation, we can assume that it is dominated by the smallest k (defined as k_1) and estimate that the intercept is equal to n_1/k_1. Likewise, the v intercept of the tangent line is taken to be a first estimate of n_1. With estimates of n_1 and k_1, we can subtract from the data the contribution of the strongest-binding (smallest-k) sites. We then can construct a new plot that can be analyzed according to Equation 15-29 in order to obtain estimates of n_2 and k_2.

The first estimate of all the parameters may be improved by a refinement process. For example, a new estimate of n_1/k_1 may be obtained by subtracting the approximate values of n_2/k_2 from the $v/(L)$ intercept of the tangent line mentioned above. After this, the process can be continued to obtain a new estimate of n_2 and k_2. Throughout the procedure, the constraint is used that $n_1 + n_2$ equals the observed v intercept. The refinement procedure is continued until $\sum_i (n_i/k_i)$ equals the observed $v/(L)$ intercept.

Figure 15-3 gives data for the binding of Mn^{2+} to the 5'-(three-fifths molecule) of a specific transfer RNA in 0.1 M triethanolamine. Based on the tRNA cloverleaf

---

[§] The ratio $v/(L)$ appears to go to $0/0$ when $(L) \to 0$. The value of this indeterminate form can be obtained from l'Hôpital's rule, which says that the limiting ratio is given by the limit of the derivative of the numerator (v) divided by the derivative of the denominator, (L). From Equation 15-32, $[dv/d(L)]_{(L)\to 0} = \sum_i n_i/k_i$, and $d(L)/d(L) = 1$; therefore,

$$\lim_{(L)\to 0} [v/(L)] = \sum_i n_i/k_i$$

This result also is obtained by letting $(L) \to 0$ on the right-hand side of Equation 15-33.

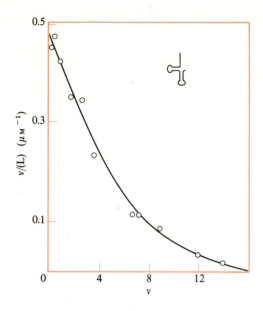

Figure 15-3

Biphasic Scatchard plot of Mn^{2+} binding to the 5'-(three-fifths molecule) of a specific tRNA. [After A. A. Schreier and P. R. Schimmel, *J. Mol. Biol.* 86:601 (1974).]

structure, this nucleic acid fragment contains single-stranded regions and a double-helical hairpin stem and hairpin loop. The data were analyzed as just described to give two classes of sites with $n_1 = 6$ and $n_2 = 10$; the dissociation constants are $k_1 = 14$ μM and $k_2 = 200$ μM. The curve is constructed from these calculated parameters, whereas the points are experimental. Good agreement is achieved between the calculated and observed behaviors.

Are the parameters obtained from a multiphasic Scatchard plot unique? For example, could other n_i and k_i values equally well fit the data in Figure 15-3? With the constraint that $n_1 + n_2 =$ constant, variations of ± 1 in n_i give relatively small (less than $\pm 50\%$) changes in the k_i values for this particular example. This suggests that the k_i values are reliable. A related question is whether the data might also be described well by positing more than two classes of sites. Of course, the greater the number of parameters available to fit any data to a model, the better will be the agreement between theory and experiment. The best procedure is to account for data, within the limits of experimental error, with the fewest possible assumptions and parameters. This approach gives a picture of the minimal (and presumably dominant) features of the system.

15-5 INTERACTION BETWEEN SITES

Some general considerations

Now we must ask whether the assumption of separate classes of sites, and the representation of Equations 15-32 and 15-33, is the only way to account for curved

Scatchard plots. It clearly is not. It is possible, for example, that binding of one ligand alters the affinity of the macromolecule for the successive one, and so on, effectively producing a continuous variation in the microscopic dissociation constant.

For the simple case of one class of identical sites, we can define k_0 as the microscopic dissociation constant at $v = 0$. As v increases, interactions between sites cause a change in k. Let ΔG^0 be the standard free energy change for dissociation of a bound ligand. This is given by

$$\Delta G^0 = \Delta G_0^0 + RT\phi(v) \tag{15-34}$$

where $\Delta G_0^0 = -RT \ln k_0$, and $\phi(v)$ is a function that, by definition, takes into account the effects of interactions between sites that vary with the degree of saturation. From Equation 15-34, and the relationship $k(v) = e^{-\Delta G^0/RT}$, we obtain

$$k(v) = k_0 e^{-\phi(v)} \tag{15-35}$$

where $\phi(v)$ is zero at $v = 0$.

Equations 15-28 and 15-29 can now be used, but with k replaced by $k(v)$ from Equation 15-35. If $\phi(v)$ is a decreasing function of v, then $k(v)$ increases as saturation proceeds. In this case, the Scatchard plot according to Equation 15-29 will be curved, concave upwards. On the other hand, if $\phi(v)$ increases as v increases, then the Scatchard plot can be "humped," or concave downwards. As binding proceeds, successive ligands are bound more strongly (smaller dissociation constants). This situation corresponds to one in which a cooperative interaction between sites occurs as v increases. Figure 15-4 illustrates the two cases.

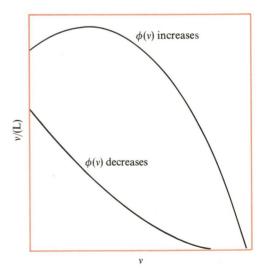

Figure 15-4

Hypothetical Scatchard plots for cases where $\phi(v)$ decreases or increases with increasing v.

Because $\phi(v)$ is a completely arbitrary function, it can always be defined so as to explain any data according to Equations 15-28 and 15-35. In an ideal situation, enough is known about the system under investigation that one can reasonably choose between the description of Equation 15-35 or the assumption of independent classes of sites as the best way to account for a curved Scatchard plot. If Equation 15-35 is believed to be the best description, then it is desirable to have a model for the system that permits a theoretical derivation of the functional form of $\phi(v)$. For example, simple electrostatic theory can be used to estimate $\phi(v)$ for the association of ions with a charged macromolecule (see Tanford, 1961). With a definite form assigned to $\phi(v)$, one then can test whether the data actually do conform to Equation 15-35.

However, in many situations it is not possible to derive an expression for $\phi(v)$. Enough information about the system simply is not available.

In the absence of accurate information on (or evidence for) negatively interacting sites as described by Equation 15-35, it is best to treat a concave-up Scatchard plot in terms of independent classes of sites (according to Eqns. 15-32 and 15-33). This, at the least, provides a useful phenomenological description of the system. Moreover, in any given situation, the likelihood of genuinely distinct classes of sites may be self-evident. In the example of Figure 15-3, the macromolecule under investigation presumably contains both single-stranded and double-stranded sections. Because the two types of sections are known to have significantly different ligand (Mn^{2+}) affinities, a model with at least two classes of sites is physically reasonable. If, as in the example, the data can be quantitatively accounted for by the different classes of sites known to exist in the macromolecule, then there is no reason to invoke possible effects due to $\phi(v)$.

However, in the event of a concave-down Scatchard plot (Fig. 15-4), separate classes of noninteracting sites cannot be assumed. This is because the description of Equations 15-32 and 15-33 gives only concave-up plots. Therefore, a concave-down plot is definitive evidence for interactions between sites: $\phi(v)$ decreasing with increasing v. We treat such systems in following subsections.

In the general case where there are several classes of interacting sites, then Equations 15-32 and 15-33 apply, with each k_i replaced by $k_i(v)$ where, by analogy with Equation 15-35,

$$k_i(v) = k_{0i}e^{-\phi_i(v)} \tag{15-36}$$

where k_{0i} is the intrinsic microscopic dissociation constant at $v = 0$ for sites in class i, and $\phi_i(v)$ is the interaction function for sites in class i. The interaction function for each class of sites may be unique, so that $\phi_i(v)$ can be different from $\phi_j(v)$. Of course, Equations 15-32 and 15-33, in conjunction with Equation 15-36, are useful only if enough information is available that $\phi_i(v)$ is known for each of the various classes of sites.

The preceding discussion serves to sketch the general issues that must be considered in treating interacting sites. In practice, cooperative interactions are probably the most commonly encountered examples of interacting sites. These are considered next.

Prevalence of cooperative interactions

There are many examples in biology of the association of ligands with a macromolecule being a cooperative process. One of the best-studied examples is the association of oxygen with hemoglobin, discussed in greater depth in Chapter 17. In addition, many multisubunit enzymes bind substrates or other molecules in a cooperative fashion. The enzyme aspartate transcarbamoylase (also considered in Chapter 17) exhibits this kind of behavior. And, in some instances, certain nucleic acids cooperatively bind particular ligands. Thus, *cooperative interactions are widespread in biological systems.*

The cooperative association of ligands with macromolecules has been treated by many authors. Some aspects of these treatments, and some of the models for cooperativity put forth, are discussed in Chapter 17. At this point, however, it is worth considering some of the elementary features of these kind of interactions.

Statistical effects and interaction energy

For the sake of illustration, consider a macromolecule that combines with four ligands, L. If all of the sites are identical and independent and bind L with a microscopic dissociation constant k, then, according to Equation 15-20, the four macroscopic constants are

$$K_1 = (1/4)k \tag{15-37a}$$

$$K_2 = (2/3)k \tag{15-37b}$$

$$K_3 = (3/2)k \tag{15-37c}$$

$$K_4 = 4\,k \tag{15-37d}$$

Therefore, in this case, $K_1 < K_2 < K_3 < K_4$; that is, viewed from the standpoint of the macroscopic constants, the binding appears to become progressively weaker as saturation proceeds, even though the same microscopic constant holds for each site. Thus, from the standpoint of the macroscopic dissociation constants, statistical effects introduce some apparent *anticooperativity* into the binding equilibria.

In a cooperative system, when corrected for statistical effects, the apparent dissociation constant for one or more of the successive steps decreases as saturation progresses. In the example of a macromolecule with four sites, this means that, if cooperativity occurs between the first and second step, then (as a consequence of Eqn. 15-37) $4K_1 > (3/2)K_2$; if all four steps involve progressively stronger binding, then $4K_1 > (3/2)K_2 > (2/3)K_3 > (1/4)K_4$.

The magnitude of the cooperativity involved in binding two ligands can be cast into units of energy by a simple procedure. Let $\Delta G_i^0 = RT \ln K_i$ be the apparent standard free energy change for binding the ith ligand in a series. (Recall that K_i is a *dissociation* constant, so that $-RT \ln K_i$ is the free energy change associated with *dissociation*; therefore, $+RT \ln K_i$ is that associated with *association*.) This free energy change contains a pure statistical component given by $RT \ln (\Omega_{n,i-1}/\Omega_{n,i})$ (cf. Eqn. 15-20). To correct for this, we define the intrinsic standard free energy change associated with binding of the ith ligand in a series as $\Delta \bar{G}_i^0$, which is

$$\Delta \bar{G}_i^0 = +RT \ln K_i - RT \ln(\Omega_{n,i-1}/\Omega_{n,i}) \tag{15-38}$$

We define the interaction energy $\Delta G_{I,ij}$ per site as the difference in the intrinsic free energies of association of the ith and jth ligands. This interaction energy is

$$\Delta G_{I,ij} = \Delta \bar{G}_j^0 - \Delta \bar{G}_i^0$$

$$= -RT \ln(K_i/K_j) + RT \ln\left(\frac{\Omega_{n,i-1}/\Omega_{n,i}}{\Omega_{n,j-1}/\Omega_{n,j}}\right) \tag{15-39}$$

With this definition of $\Delta G_{I,ij}$, if the jth ligand binds more strongly than the ith ($j > i$), then as in a cooperative system, $\Delta G_{I,ij} < 0$. Note also that, if each site has the same intrinsic dissociation constant, then the two terms on the right-hand side of Equation 15-39 cancel, and $\Delta G_{I,ij} = 0$.

In the case of oxygen binding to human hemoglobin, Equation 15-39 gives $\Delta G_{I,ij} \cong -2$ kcal mole^{-1} site^{-1} for $i = 1$ and $j = 4$. This means that site–site interactions stabilize a bound oxygen molecule in the saturated hemoglobin tetramer by approximately 2 kcal mole^{-1} over an oxygen molecule bound to a hemoglobin species that has three vacant sites.

A semiempirical approach: the Hill constant

For the purpose of treating and characterizing data on the cooperative association of ligands, it is common practice to use a semiempirical approach and then to interpret the physical significance of the empirical parameters that are obtained. This approach is based on the assumption that the binding over part of the saturation range can be described by equations phenomenologically resembling those for an infinitely cooperative system. In the extreme case of infinite cooperativity, the binding can be represented as an "all-or-none" reaction:

$$M_0 + nL \rightleftarrows M_n \tag{15-40}$$

$$K^n = (M_0)(L)^n/(M_n) \tag{15-41}$$

where K is the apparent dissociation constant for the interacting sites. For this case, the parameter v is given by

$$v = n(M_n)/[(M_0) + (M_n)]$$

$$= [n(L)^n/K^n]/[1 + (L)^n/K^n] \tag{15-42a}$$

$$v/(L) = [n(L)^{n-1}/K^n]/[1 + (L)^n/K^n] \tag{15-42b}$$

whereas the fractional saturation $\bar{y} = v/n$ is

$$\bar{y} = [(L)^n/K^n]/[1 + (L)^n/K^n] \tag{15-43}$$

Equations 15-40 through 15-43 are based on the assumption that binding is infinitely cooperative for all n ligands. In practice, infinite cooperatively is not observed. Instead, data on cooperative interactions commonly are described over part of the saturation range (typically 25% to 75%) by semiempirical relationships analogous to Equations 15-40 through 15-43. These semiempirical relationships are

$$v = [n(L)^{\alpha_H}/K^{\alpha_H}]/[1 + (L)^{\alpha_H}/K^{\alpha_H}] \tag{15-44a}$$

$$v/(L) = [n(L)^{\alpha_H-1}/K^{\alpha_H}]/[1 + (L)^{\alpha_H}/K^{\alpha_H}] \tag{15-44b}$$

$$\bar{y} = [(L)^{\alpha_H}/K^{\alpha_H}]/[1 + (L)^{\alpha_H}/K^{\alpha_H}] \tag{15-45}$$

where $1 \leqslant \alpha_H \leqslant n$. *The parameter α_H commonly is known as the Hill constant* (see Hill, 1910); it is an index to the cooperativity. When $\alpha_H = n$, the system behaves as perfectly cooperative, whereas $\alpha_H = 1$ indicates no cooperativity. Figure 15-5 shows several plots of $\bar{y}$ versus $(L)/K$ for various values of α_H. It is clear that the steepness of the curve is very sensitive to α_H.

From Equation 15-45, the parameter α_H is given by

$$\frac{d\{\ln[\bar{y}/(1 - \bar{y})]\}}{d[\ln(L)]} = \alpha_H \tag{15-46}$$

Equation 15-46 serves as a convenient definition of the Hill constant. In general, Equation 15-45 does not hold over the entire range of values of $\bar{y}$, so that α_H is a

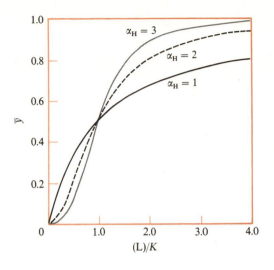

Figure 15-5

Effect of α_H on fractional saturation curves.

function of the degree of saturation. Often the parameter α_H is evaluated at $\bar{y} = 1 - \bar{y} = 1/2$. In this case, Equation 15-46 becomes (note that $d \ln x = dx/x$)

$$\left(\frac{d(\bar{y}/(1 - \bar{y}))]}{d(L)}\right)_{\bar{y} = 1/2} = \frac{\alpha_{H,1/2}}{(L)_{1/2}} \tag{15-47}$$

where $(L)_{1/2}$ is the concentration of L at half-saturation, and $\alpha_{H,1/2}$ is the value of α_H when $\bar{y} = 1/2$. Equations 15-46 and 15-47 are useful relationships; they show that the Hill constant can be obtained from the slope of a plot of $\ln[\bar{y}/(1 - \bar{y})]$ versus $\ln(L)$, which is called a Hill plot.

Equation 15-44 gives parametric relationships that can be used to analyze Scatchard plots of cooperative associations, sometimes over a broad range of values of v and (L). These plots are markedly different from those discussed earlier for independent, noninteracting sites. According to Equation 15-44b, for $\alpha_H > 1$ the plot actually passes through the origin—as when $v = 0$ [or $(L) = 0$] and $v/(L) = 0$. At low values of v or (L), the curve rises and reaches a maximum at $v_{max} = n(\alpha_H - 1)/\alpha_H$, and then descends to intercept the v axis at $v = n$.

Figure 15-6 shows an example of this kind of Scatchard plot. This figure gives data on the cooperative association of Mn^{2+} to transfer RNA. The concave-down character of the plot is clearly evident. Parameters that characterize the interaction may be obtained by defining K^{α_H} in terms of experimentally determined variables as follows:

$$K^{\alpha_H} = (L)^{\alpha_H}[n(M)_0 - (L)_b]/(L)_b \tag{15-48}$$

and rearranging Equation 15-48 to give

$$\ln(L) = -(1/\alpha_H) \ln[(n/v) - 1] + \ln K \tag{15-49}$$

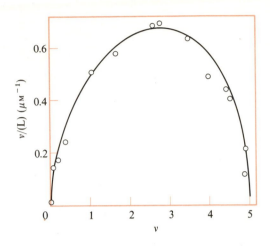

Figure 15-6

Scatchard plot of Mn^{2+} binding to a specific tRNA. [After A. A. Schreier and P. R. Schimmel, *J. Mol. Biol.* 86:601 (1974).]

where $(L)_b$ is the concentration of bound ligand, $(M)_0$ is the total macromolecule concentration, and $v = (L)_b/(M)_0$. It is clear from Equation 15-49 that a plot of $\ln(L)$ versus $\ln[(n/v) - 1]$ should give a straight line with slope of $-1/\alpha_H$ and intercept of $-\ln K$. The linearity is very sensitive to the value of n assumed, so that a good check is obtained on the value of n determined from the Scatchard plot. When the data in Figure 15-6 are plotted according to Equation 15-49, the parameters $n = 5$, $\alpha_H = 2.3$, and $K = 3.7$ μM are found to describe the data accurately. These data show that Mn^{2+} binding is cooperative, and that about five sites are involved; however, they do not act in a wholly "all-or-none" fashion, because $\alpha_H < n$. This is analogous to oxygen binding to hemoglobin, where α_H is 2.5 to 3.0 and $n = 4$.

Although not considered further here, these data and other results are particularly useful in constructing a model for the association of cations with tRNA (see Schimmel, 1976).

15-6 BINDING OF TWO DIFFERENT LIGANDS: LINKED FUNCTIONS

Thus far we have considered only situations in which one kind of ligand binds to a macromolecule. However, there are situations encountered in practice where the simultaneous binding of two different ligands must be considered. One example is the effect of pH on ligand binding. In this instance, the binding of H^+ to one or more critical sites is closely linked to the association of ligand.

When two different ligands bind to a macromolecule, there is an interesting and useful set of interrelationships that describe the binding equilibria. Our treatment follows lines developed by J. Wyman (1964). We start by considering again the binding of a single ligand L to n sites on a macromolecule M. We use a somewhat different formalism than that developed earlier; this one is particularly useful for treating linked equilibria.

● **Formalism**

The macroscopic equilibria may be written as follows:

$$M_0 + iL \rightleftharpoons M_i \qquad \text{for } i = 1, 2, \ldots, n \tag{15-50}$$

and the apparent dissociation constant $\tilde{K}_i$ is

$$\tilde{K}_i = (M_0)(L)^i/(M_i) \qquad \text{for } i = 1, 2, \ldots, n \tag{15-51}$$

where (M_i) is the concentration of macromolecules that have i bound ligand molecules. Note that our definition of $\tilde{K}_i$ differs from that of the macroscopic constant K_i used earlier (Eqn. 15-16). The number of moles of L bound per mole of M is $n\bar{y}$ and is given by

$$n\bar{y} = \frac{\displaystyle\sum_{i=0}^{n} i(M_i)}{\displaystyle\sum_{i=0}^{n} (M_i)} = \frac{\displaystyle\sum_{i=0}^{n} [i(L)^i/\tilde{K}_i]}{\displaystyle\sum_{i=0}^{n} [(L)^i/\tilde{K}_i]} \tag{15-52}$$

$$n\bar{y} = \frac{d\left(\ln \displaystyle\sum_{i=0}^{n} [(L)^i/\tilde{K}_i]\right)}{d[\ln(L)]} \tag{15-53}$$

where $\tilde{K}_0$ is equal to unity. Equation 15-53 is a particularly useful form.

Each $\tilde{K}_i$ is related to the microscopic constants that characterize the binding to each of the n sites on M. Letting the microscopic constants be distinct and designated as $k_1, k_2, \ldots, k_n$, we can easily show (Problem 15-5) that

$$\sum_{i=0}^{n} [(L)^i/\tilde{K}_i] = [1 + (L)/k_1][1 + (L)/k_2] \cdots [1 + (L)/k_n] \tag{15-54}$$

If every $k_i = k$, then

$$\sum_{i=0}^{n} [(L)^i/\tilde{K}_i] = [1 + (L)/k]^n \tag{15-55}$$

and, from Equation 15-53,

$$\bar{y} = [(L)/k]/[1 + (L)/k] \tag{15-56}$$

Equation 15-56 is identical to Equation 15-28, with $\bar{y} = v/n$.

● **Two ligands and a basic linkage relationship**

Assume now that there are two ligands, L_1 and L_2. There are n sites for L_1 and m sites for L_2. Let $\bar{y}_1$ be the fractional saturation with respect to L_1, and $\bar{y}_2$ be that with respect to L_2. We define the macroscopic equilibrium constant $\tilde{K}_{ij}$ for the equilibrium

$$iL_1 + jL_2 + M_0 \rightleftharpoons M_{ij} \tag{15-57}$$

as

$$\tilde{K}_{ij} = (L_1)^i (L_2)^j (M_0)/(M_{ij}) \tag{15-58}$$

where M_{ij} has i molecules of L_1 and j molecules of L_2 bound to it, and M_0 has neither ligand bound. Note that $K_{00} = 1$. The parameter $n\bar{y}_1$ thus is given by

$$n\bar{y}_1 = \frac{\displaystyle\sum_{i=0}^{n}\sum_{j=0}^{m} i(M_{ij})}{\displaystyle\sum_{i=0}^{n}\sum_{j=0}^{m}(M_{ij})} \tag{15-59}$$

Using Equation 15-58, we obtain

$$n\bar{y}_1 = \frac{\displaystyle\sum_{i=0}^{n}\sum_{j=0}^{m}[i(L_1)^i(L_2)^j/\tilde{K}_{ij}]}{\displaystyle\sum_{i=0}^{n}\sum_{j=0}^{m}[(L_1)^i(L_2)^j/\tilde{K}_{ij}]} \tag{15-60}$$

or

$$n\bar{y}_1 = \left(\frac{\partial\left\{\ln\displaystyle\sum_{i}\sum_{j}[(L_1)^i(L_2)^j/\tilde{K}_{ij}]\right\}}{\partial[\ln(L_1)]}\right)_{(L_2)} \tag{15-61}$$

Likewise, for $m\bar{y}_2$ we have

$$m\bar{y}_2 = \left(\frac{\partial\left\{\ln\displaystyle\sum_{i}\sum_{j}[(L_1)^i(L_2)^j/\tilde{K}_{ij}]\right\}}{\partial[\ln(L_2)]}\right)_{(L_1)} \tag{15-62}$$

For given $\tilde{K}_{ij}$ values, the double sum in Equation 15-61 or 15-62 is a function only of (L_1) and (L_2). Therefore, we have

$$d(\ln\textstyle\sum\sum) = \left(\frac{\partial\sum\sum}{\partial[\ln(L_1)]}\right)_{(L_2)} d[\ln(L_1)] + \left(\frac{\partial\sum\sum}{\partial[\ln(L_2)]}\right)_{(L_1)} d[\ln(L_2)] \tag{15-63}$$

$$d(\ln {\textstyle \sum\sum}) = n\bar{y}_1\, d[\ln(L_1)] + m\bar{y}_2\, d[\ln(L_2)] \tag{15-64}$$

where $\sum\sum$ denotes the double sum in the numerator of Equation 15-61 or 15-62. By cross-differentiation (Box 15-1) of Equation 15-64, we obtain

$$n\left(\frac{\partial \bar{y}_1}{\partial[\ln(L_2)]}\right)_{(L_1)} = m\left(\frac{\partial \bar{y}_2}{\partial[\ln(L_1)]}\right)_{(L_2)} \tag{15-65}$$

Equation 15-65 is a basic linkage relationship. It shows that, at constant (L_1), a rise in (L_2) changes the number of moles of bound L_1 by an amount exactly equal to the change in the number of moles of bound L_2 accompanying an increase in (L_1) at constant (L_2).

● Another equation for the linkage effect

Of the four variables $\bar{y}_1$, $\bar{y}_2$, (L_1), and (L_2), only two are independent. Therefore, it is possible to transform Equation 15-65 into alternative forms. A particularly useful one involves the derivative $(\partial \bar{y}_1/\partial \bar{y}_2)_{(L_1)}$, which by the chain rule (Box 15-2) is

$$(\partial \bar{y}_1/\partial \bar{y}_2)_{(L_1)} = (\partial \bar{y}_1/\partial[\ln(L_2)])_{(L_1)}(\partial[\ln(L_2)]/\partial \bar{y}_2)_{(L_1)} \tag{15-66}$$

Box 15-1   CROSS-DIFFERENTIATION

For a function $f(x, y)$ of two independent variables x and y,

$$df = (\partial f/\partial x)_y\, dx + (\partial f/\partial y)_x\, dy$$
$$= f_x\, dx + f_y\, dy$$

where $f_x = (\partial f/\partial x)_y$, and $f_y = (\partial f/\partial y)_x$. The cross-differentiation relationship says that

$$(\partial f_x/\partial y)_x = (\partial f_y/\partial x)_y$$

This follows from the second-derivative relationships:

$$(\partial f_x/\partial y)_x = \frac{\partial}{\partial y}(\partial f/\partial x)_y = \frac{\partial^2}{\partial y \partial x}f = \frac{\partial^2}{\partial x \partial y}f = \frac{\partial}{\partial x}(\partial f/\partial y)_x = (\partial f_y/\partial x)_y$$

Substituting Equation 15-65 into Equation 15-66, we obtain

$$(\partial \bar{y}_1 / \partial \bar{y}_2)_{(L_1)} = (m/n)(\partial \bar{y}_2 / \partial [\ln(L_1)])_{(L_2)}(\partial [\ln(L_2)] / \partial \bar{y}_2)_{(L_1)} \qquad (15\text{-}67)$$

The derivative with respect to $\bar{y}_2$ on the right-hand side of Equation 15-67 can be evaluated from the total differential for $d\bar{y}_2$:

$$d\bar{y}_2 = (\partial \bar{y}_2 / \partial [\ln(L_1)])_{(L_2)} d[\ln(L_1)] + (\partial \bar{y}_2 / \partial [\ln(L_2)])_{(L_1)} d[\ln(L_2)] \qquad (15\text{-}68)$$

At constant $\bar{y}_2$, we have $d\bar{y}_2 = 0$, and Equation 15-68 gives

$$(\partial [\ln(L_2)] / \partial \bar{y}_2)_{(L_1)} = -(\partial [\ln(L_1)] / \partial \bar{y}_2)_{(L_2)}(\partial [\ln(L_2)] / \partial [\ln(L_1)])_{\bar{y}_2} \qquad (15\text{-}69)$$

Substituting Equation 15-69 into Equation 15-67, we obtain

$$\boxed{\left(\frac{\partial [\ln(L_2)]}{\partial [\ln(L_1)]} \right)_{\bar{y}_2} = -\frac{n}{m} \left(\frac{\partial \bar{y}_1}{\partial \bar{y}_2} \right)_{(L_1)}} \qquad (15\text{-}70)$$

The quantity $n \, d\bar{y}_1$ is the change in the number of occupied L_1 sites, and $-n \, d\bar{y}_1 = d[n(1 - \bar{y}_1)]$ is the change in the number of free L_1 sites. Therefore, Equation 15-70

Box 15-2   THE CHAIN RULE

Consider a system of four variables (f, g, x, and y) of which only two can be independent. Choose x and y as independent variables. We then have

$$df = (\partial f / \partial x)_y \, dx + (\partial f / \partial y)_x \, dy \qquad (A)$$

$$dg = (\partial g / \partial x)_y \, dx + (\partial g / \partial y)_x \, dy \qquad (B)$$

Because x and y are independent, they can be varied at will. Choose y to be fixed, so that $dy = 0$; then divide Equation A by Equation B to obtain

$$(\partial f / \partial g)_y = (\partial f / \partial x)_y / (\partial g / \partial x)_y \qquad (C)$$

$$(\partial f / \partial g)_y = (\partial f / \partial x)_y (\partial x / \partial g)_y \qquad (D)$$

Equation D is the chain-rule relationship, so named because the two derivatives on the right-hand side of Equation D appear to form a chain.

says that the expression on the left-hand side of the equation is the change in the number of free L_1 sites (or the number of moles of L_1 released) upon binding of a mole of L_2. Of course, the number released can be positive or negative (negative release corresponds to binding).

The derivative on the left-hand side of Equation 15-70 is experimentally accessible. Figure 15-7 shows plots of $\bar{y}_2$ versus $\ln(L_2)$ for different values of $\ln(L_1)$.

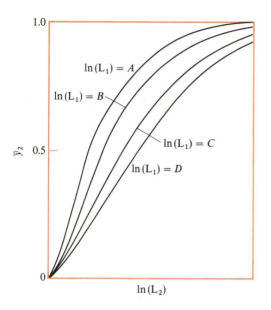

$\ln(L_1) = A$

$\ln(L_1) = B$

$\ln(L_1) = C$

$\ln(L_1) = D$

$\bar{y}_2$ 0.5

$\ln(L_2)$

Figure 15-7

Hypothetical plots of $\bar{y}_2$ *versus* $\ln(L_2)$ *for different values of* $\ln(L_1)$.

The ligand L_1 could be H^+, for example, and the curves then would simply illustrate the familiar observation that ligand binding (L_2 in this case) is affected by pH. Clearly, the horizontal difference between two adjacent curves divided by $\Delta[\ln(L_1)]$ gives the desired derivative. An analogous plot for the linked binding of oxygen and diphosphoglycerate to hemoglobin is given in Chapter 17 (Fig. 17-21). In that case, L_2 = oxygen and L_1 = diphosphoglycerate.

● An additional relationship

In order to grasp concretely how the coupling between ligands takes place, it is helpful to extend our analysis in a somewhat different way. Let us denote by $M^{(j)}$ all species that have j bound L_2 molecules. Therefore,

$$(M^{(j)}) = \sum_{i=0}^{n} (M_{ij}) \tag{15-71}$$

and, using Equation 15-58,

$$(M^{(j)}) = (M_0)(L_2)^j \sum_{i=0}^{n} [(L_1)^i/\tilde{K}_{ij}] \qquad (15\text{-}72)$$

We define the equilibrium constant $\bar{K}_j$ as

$$\bar{K}_j = (M^{(0)})(L_2)^j/(M^{(j)}) \qquad (15\text{-}73)$$

which corresponds to the reaction

$$M^{(0)} + jL_2 \rightleftarrows M^{(j)} \qquad (15\text{-}74)$$

Using Equation 15-72, we can write the equilibrium constant $\bar{K}_j$ in Equation 15-73 as

$$\bar{K}_j = \frac{\{(M_0) \sum_{i=0}^{n} [(L_1)^i/\tilde{K}_{i0}]\}(L_2)^j}{(M_0)(L_2)^j \sum_{i=0}^{n} [(L_1)^i/\tilde{K}_{ij}]} \qquad (15\text{-}75)$$

After cancellations, we obtain

$$\bar{K}_j = \frac{\sum_{i=0}^{n} [(L_1)^i/\tilde{K}_{i0}]}{\sum_{i=0}^{n} [(L_1)^i/\tilde{K}_{ij}]} \qquad (15\text{-}76)$$

If each microscopic constant for the binding of L_2 to M is distinct, then (by analogy with Eqn. 15-54)

$$\sum_{i=0}^{n} [(L_1)^i/\tilde{K}_{ij}] = (1/\tilde{K}_{0j})([1 + (L_1)/k_{11}][1 + (L_1)/k_{12}] \cdots [1 + (L_1)/k_{1n}])_{M^{(J)}}$$
$$(15\text{-}77)$$

and

$$\sum_{i=0}^{n} [(L_1)^i/\tilde{K}_{i0}] = ([1 + (L_1)/k_{11}][1 + (L_1)/k_{12}] \cdots [1 + (L_1)/k_{1n}])_{M^{(0)}} \qquad (15\text{-}78)$$

where k_{1p} denotes the microscopic dissociation constant for L_1 at site p, and the subscript $M^{(j)}$ outside the brackets denotes that the microscopic constants pertain to the species with j bound L_2 molecules. Thus, the microscopic constants for L_1

binding in general can vary according to how many L_2 molecules are bound. Using Equations 15-77 and 15-78, we obtain for Equation 15-76

$$\bar{K}_j = \tilde{K}_{0j} \frac{\prod\limits_{i=1}^{n} \left(1 + (L_1)/k_{1i}\right)_{M^{(0)}}}{\prod\limits_{i=1}^{n} \left(1 + (L_1)/k_{1i}\right)_{M^{(j)}}} \tag{15-79}$$

According to Equation 15-79, if one or more of the microscopic constants for L_1 depends on the degree of saturation with L_2, then $\bar{K}_j$ depends on (L_1).

The linkage between L_1 and L_2 also can be seen by differentiating Equation 15-76 and making use of the results of Equations 15-52 and 15-53 to obtain

$$\frac{d[\ln \bar{K}_j]}{d[\ln(L_1)]} = \frac{d\left\{\ln \sum\limits_{i=0}^{n} [(L_1)^i/\tilde{K}_{i0}]\right\}}{d[\ln(L_1)]} - \frac{d\left\{\ln \sum\limits_{i=0}^{n} [(L_1)^i/\tilde{K}_{ij}]\right\}}{d[\ln(L_1)]} \tag{15-80}$$

or

$$\boxed{\frac{d[\ln \bar{K}_j]}{d[\ln(L_1)]} = -n(\bar{y}_1^{(j)} - \bar{y}_1^{(0)})} \tag{15-81}$$

where $n\bar{y}_1^{(j)}$ is the number of molecules of L_1 bound to molecules of M that have j bound L_2 molecules. Thus the expression on the right-hand side of Equation 15-81 gives the number of L_1 molecules released upon binding of j L_2 molecules to M. However, according to Equation 15-79, if

$$\prod\limits_{i=1}^{n} \left(1 + (L_1)/k_{1i}\right)_{M^{(0)}} = \prod\limits_{i=1}^{n} \left(1 + (L_1)/k_{1i}\right)_{M^{(j)}}$$

then $\bar{K}_j = \tilde{K}_{0j}$, and $d[\ln \bar{K}_j]/d[\ln(L_1)] = 0$. Hence, *linkage between L_1 and L_2 binding occurs if the microscopic constants for binding of one ligand are influenced by the amount of the other ligand that is bound.*

These types of interrelationships between ligand bindings are well illustrated in the case of hemoglobin and the Bohr effect. In this case, there is strong linkage between oxygen binding and the binding of "Bohr" protons (see Chapter 17). Another example is provided by the hydrolysis of ATP, where both reactant and products can complex with Mg^{2+} and H^+. R. A. Alberty (1969) summarizes this interesting system.

15-7 LINKAGE OF LIGAND BINDING FROM AN ENERGETIC VIEWPOINT

Coupling free energy

Gregorio Weber (1975) has used a different viewpoint for examining coupled ligand equilibria, treating them in terms of energetic considerations. This treatment helps us to think concretely in terms of the energies involved in the linked reactions.

For the sake of illustration, consider a simple system in which a macromolecule binds one molecule each of ligands L_1 and L_2. The reactions are

$$M + L_1 \rightleftarrows ML_1 \qquad \Delta G_1^0 \tag{15-82a}$$

$$M + L_2 \rightleftarrows L_2M \qquad \Delta G_2^0 \tag{15-82b}$$

$$L_2M + L_1 \rightleftarrows L_2ML_1 \qquad \Delta G_1^0(2) \tag{15-82c}$$

$$ML_1 + L_2 \rightleftarrows L_2ML_1 \qquad \Delta G_2^0(1) \tag{15-82d}$$

Standard free energies for each of the reactions are indicated on the right-hand side; for example, $\Delta G_1^0(2)$ is the standard free energy change for binding L_1 to the macro-molecule saturated with L_2.

The free energies in Equation 15-82 are not independent, but are tied together because

$$\Delta G_1^0 + \Delta G_2^0(1) = \Delta G_2^0 + \Delta G_1^0(2) = \Delta G^0(1,2) \tag{15-83}$$

where $\Delta G^0(1,2)$ is the standard free energy change for the reaction

$$M + L_1 + L_2 \rightleftarrows L_2ML_1 \tag{15-84}$$

Figure 15-8 is a diagram of these relationships. Note that there is no requirement that $\Delta G_1^0 = \Delta G_1^0(2)$ or that $\Delta G_2^0 = \Delta G_2^0(1)$. From Equation 15-83, we have

$$\Delta G_1^0(2) - \Delta G_1^0 = \Delta G_2^0(1) - \Delta G_2^0 \equiv \Delta G_{12}^0 \tag{15-85}$$

The meaning of this equation is similar to the linkage relationship of Equation 15-65 for the case $m = n = 1$. It says that the effect (in terms of free energy) of L_2 on the binding of L_1 is the same as the effect of L_1 on the binding of L_2. This mutual effect of one ligand on the other can be put in terms of a coupling free energy ΔG_{12}^0 (defined in Eqn. 15-85).

Combining Equations 15-83 and 15-85, we obtain another expression for ΔG_{12}^0:

$$\Delta G_{12}^0 = \Delta G^0(1,2) - \Delta G_1^0 - \Delta G_2^0 \tag{15-86}$$

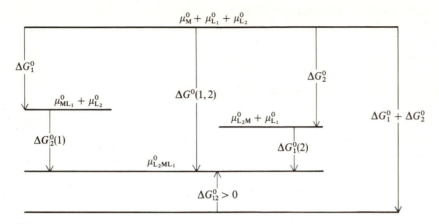

Figure 15-8

Free energy diagram for a system of two ligands, L_1 and L_2, and a macromolecule M. Each ligand has one site on the macromolecule. Standard chemical potentials are designated μ^0 with subscripts referring to particular species. [After G. Weber, *Adv. Protein Chem.* 29:1 (1975).]

Thus, ΔG_{12}^0 is the difference between (a) the standard free energy for the overall reaction $M + L_1 + L_2 \rightleftarrows L_2ML_1$, and (b) the sum of the standard free energies for the reactions $M + L_1 \rightleftarrows ML_1$ and $M + L_2 \rightleftarrows L_2M$. Figure 15-8 shows the definition of ΔG_{12}^0.

Clearly, if $\Delta G_{12}^0 = 0$, there is no interaction between ligands; binding of each proceeds in a truly independent fashion. For other cases, the sign of the coupling free energy determines whether the interaction between ligands is *cooperative* or *antagonistic*. If $\Delta G_{12}^0 < 0$, then binding of either L_1 or L_2 facilitates binding of the other ligand. Conversely, when $\Delta G_{12}^0 > 0$, there is antagonism between the bindings of the ligands.

There is still another way to look at the coupling free energy. To do this, we write out the three relevant equilibria and their associated free energy changes:

$$L_1 + L_2 + M \rightleftarrows L_2ML_1 \qquad \Delta G^0(1,2) \tag{15-87a}$$

$$L_1 + M \rightleftarrows ML_1 \qquad \Delta G_1^0 \tag{15-87b}$$

$$L_2 + M \rightleftarrows L_2M \qquad \Delta G_2^0 \tag{15-87c}$$

Subtracting Equation 15-87b,c from 15-87a, we obtain

$$ML_1 + L_2M \rightleftarrows M + L_2ML_1 \qquad \Delta G^0(1,2) - \Delta G_1^0 - \Delta G_2^0 = \Delta G_{12}^0 \tag{15-88}$$

Thus, ΔG_{12}^0 is the free energy change for a kind of disproportionation reaction. This reaction has an equilibrium constant K_{12} given by

$$K_{12} = e^{-\Delta G_{12}^0/RT} = (L_2ML_1)(M)/(ML_1)(L_2M) \tag{15-89}$$

From this analysis, it is clear that, if $\Delta G^0_{12} < 0$, then $K_{12} > 1$, and the species L_2ML_1 and M are favored over the partially saturated forms ML_1 and L_2M. The reverse holds true when $\Delta G^0_{12} > 0$.

Effect of coupling energy on distribution of bound ligands

It is of interest to inquire what magnitude of ΔG^0_{12} is required to alter significantly the distribution of L_1 and L_2 among the species ML_1, L_2M, and L_2ML_1, as compared with the distribution when there is no coupling ($\Delta G^0_{12} = 0$). For this purpose, we define the fractional saturations $\bar{y}_1$, $\bar{y}_2$, and $\bar{y}_{12}$:

$$\bar{y}_1 = [(L_2ML_1) + (ML_1)]/(M)_{\text{Tot}} \tag{15-90a}$$

$$\bar{y}_2 = [(L_2ML_1) + (L_2M)]/(M)_{\text{Tot}} \tag{15-90b}$$

$$\bar{y}_{12} = (L_2ML_1)/(M)_{\text{Tot}} \tag{15-90c}$$

where $(M)_{\text{Tot}} = (M) + (ML_1) + (L_2M) + (L_2ML_1)$. Clearly, $\bar{y}_1$ and $\bar{y}_2$ are the overall fractional saturations with respect to L_1 and L_2, and $\bar{y}_{12}$ is the degree of double saturation.

Consider a situation where (L_1) and (L_2) are so adjusted that one-half of the L_1 sites and one-half of the L_2 sites are filled. Under these conditions, it is easy to show (see Problem 15-4) that

$$K_{12} = \bar{y}^2_{12}/[(1/2) - \bar{y}_{12}]^2 \tag{15-91}$$

and

$$\bar{y}_{12} = (1/2)K^{1/2}_{12}/(1 + K^{1/2}_{12}) \tag{15-92}$$

Substituting Equation 15-91 into Equation 15-89, we obtain

$$\Delta G^0_{12} = -2RT \ln[2\bar{y}_{12}/(1 - 2\bar{y}_{12})] \tag{15-93}$$

From Equation 15-93 we can obtain a plot of ΔG^0_{12} versus $2\bar{y}_{12}$ (Fig. 15-9). When $2\bar{y}_{12} = 1$, all of the bound ligands are in the form of L_2ML_1. When $2\bar{y}_{12} = 0$, all of the bound ligands are in the form of ML_1 and L_2M. At the point $2\bar{y}_{12} = 0.5$, we see that $\Delta G^0_{12} = 0$ (no coupling); this is the result expected for a simple unbiased statistical distribution of the ligands among ML_1, L_2M, and L_2ML_1, and where each species is present in equal amounts. Note that the plot in Figure 15-9 is symmetric about the point $2\bar{y}_{12} = 0.5$.

When $\Delta G^0_{12} = -2$ kcal mole^{-1}, then $2\bar{y}_{12}$ is greater than 0.8; when $\Delta G^0_{12} = -3$ kcal mole^{-1}, then $2\bar{y}_{12}$ is over 0.9. In the latter case, over 90% of the bound L_1 and L_2 is in the form of the double-saturated species L_2ML_1. In this instance, ligand

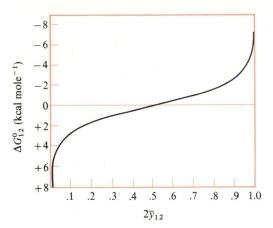

Figure 15-9

Relation between degree of double saturation $(\bar{y}_{12})$ *and free energy coupling* (ΔG^0_{12}) *of the bound ligands, when the degree of saturation of each ligand* $(\bar{y}_1, \bar{y}_2)$ *equals 0.5.* [After G. Weber, *Adv. Protein Chem.* 29:1 (1975).]

binding proceeds largely from the species M to L_2ML_1, with little formation of ML_1 and L_2M. Conversely, when $\Delta G^0_{12} = +2$ or $+3$ kcal mole^{-1}, most of the species at half-saturation are the monoliganded forms ML_1 and L_2M. Thus, a coupling energy of only about ± 2 kcal mole^{-1} is sufficient to cause a substantial skewing of the distribution of liganded forms away from that obtained on a random basis.

Coupling free energies found in biological systems

Table 15-1 gives several examples of values for the coupling free energy between two different ligands that interact with a protein. Both positive and negative energies are found, corresponding to antagonistic and cooperative effects, respectively. The

Table 15-1

Free energy coupling between ligands

Protein	Ligand couple§	ΔG^0_{12} (kcal mole^{-1})
Hemoglobin	Oxygen, 2,3-DPG	+1.3
Hemoglobin	Oxygen, IHP	+2.3
Serum albumin, bovine	ANS, 3,5-dihydroxybenzoate	+1.5
Pyruvate kinase	Phosphoenol pyruvate, K$^+$	−1.2
Pyruvate kinase	K$^+$, Mn^{2+}	−1.4
Pyruvate kinase	Phenylalanine, Mn^{II}	+0.8
Aspartate transcarbamoylase	CTP, succinate	+0.5
Lactate dehydrogenase, chicken heart	NADH, oxalate	−1.5

§ IHP = inositol hexaphosphate; CTP = cytidine triphosphate; 2,3-DPG = 2,3-diphosphoglycerate; ANS = 1-anilinonaphthalene 8-sulfonate.
SOURCE: After G. Weber, *Adv. Protein Chem.* 29:1 (1975).

energies fall in the range of 0 to ± 2.5 kcal mole^{-1}, although it must be recognized that the data are sparse and that further research may turn up a wider range of values. At least in some cases, the coupling energies are large enough to produce quite pronounced effects in the ligand saturation curves, as might be expected from the plot in Figure 15-9 (and as shown in Chapter 17, where we discuss ligand interactions of aspartate transcarbamoylase and hemoglobin).

15-8 INTERACTION OF LARGE LIGANDS WITH LATTICELIKE CHAINS

In considering ligand binding equilibria in biological systems, a case of great interest is the association of large ligands with latticelike chains, such as double-helical nucleic acids. Large ligands include polyamines (such as spermine and spermidine), histones, DNA-unwinding proteins, and large drugs such as actinomycin D. These systems have special statistical features that are quite distinct from the situations described thus far. Moreover, they are amenable to a rather straightforward treatment, which we present along lines developed by J. D. McGhee and P. H. von Hippel (1974).

The homogeneous lattice: statistical features

Consider first a homogeneous lattice constructed of N identical repeating units. For example, in a helical nucleic acid, the repeating units could be phosphate groups or sugar units. Assume that one ligand L occupies l consecutive lattice units; saturation of the lattice with L results in N/l bound ligand molecules per lattice. The ligand is assumed to be able to occupy any l consecutive lattice units. Therefore, in the completely naked lattice, there are $N - l + 1$ potential sites that the first bound ligand can occupy. Thus, at the beginning of a ligand titration there are many more potential sites than the N/l sites that can be occupied at saturation. In this feature, this situation contrasts sharply with our earlier treatment of identical and independent sites (Eqn. 15-12), in which the number of free sites (n) on M_0 corresponds to the number that are filled on M_n. It is this aspect of the lattice that gives rise to a Scatchard plot markedly different from those obtained with the simple system of Equation 15-12.

Figure 15-10 illustrates the statistical complexities of the lattice for the case of $N = 12$ and $l = 3$. Let v represent the moles of L bound per mole of lattice. At saturation $v = N/l = 4$ bound ligand molecules per lattice. At the outset of titration, however, there are $N - l + 1 = 10$ potential sites per lattice. The figure shows that a variety of microspecies are generated after $v = 2$ bound ligands per lattice. Owing to the arrangement of ligands on the lattice, one species can accommodate no additional ligands, whereas others can take on two more. All of these species are in equilibrium with the free-ligand concentration, and they must be accounted for in any attempt to calculate the shape of the binding curve. Of course, as binding proceeds beyond $v = 2$, there must be a continual redistribution of the ligand on the lattice until the final state of $v = 4$ is reached.

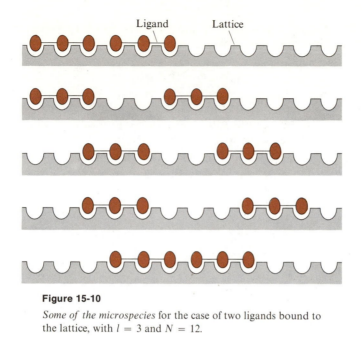

Figure 15-10

Some of the microspecies for the case of two ligands bound to the lattice, with $l = 3$ and $N = 12$.

Calculation of ligand-binding behavior

To calculate the ligand-binding behavior of the lattice, we start with the basic relationship

$$v = \bar{N}_{fl}(L)/k \tag{15-94}$$

where $\bar{N}_{fl}$ is the average number of free ligand sites of length l per lattice, and k is the intrinsic microscopic dissociation constant. Equation 15-94 follows directly from the definition of k. Our main task is to calculate an expression for $\bar{N}_{fl}$. We do this by first calculating the probability p_l that, starting at any position in the lattice, l *consecutive* lattice units are unoccupied with ligand; this probability times N gives the average number of free ligand sites of length l per lattice.

The probability p_l can be written as

$$p_l = p_f p_{ff}^{l-1} \tag{15-95}$$

where p_f is the probability that a lattice unit selected at random is unoccupied, and p_{ff} is the *conditional* probability that a given free unit is followed by another free unit. Because the fraction of occupied lattice units is vl/N, the fraction of unoccupied ones is $1 - vl/N$; therefore, p_b (the probability of occurrence of a bound unit) equals vl/N, and $p_f = 1 - vl/N$.

The conditional probability p_{ff} is also the fraction of the total free lattice units that are preceded by free units. This is [(total number of free lattice units) minus (number of free lattice units that are preceded by an occupied unit)] divided by (total number of free lattice units):

$$p_{ff} = [p_f - (p_b/l)p_{bf}]/p_f \qquad (15\text{-}96)$$

where p_{bf} is the conditional probability that a free lattice unit follows the last (lth) segment of a bound ligand, and p_b/l is the probability that a bound lattice unit is occupied by the last (lth) segment of the ligand. Because there is no interaction between bound ligands, it is clear that $p_{ff} = p_{bf}$; with this relationship, we can rearrange Equation 15-96 to obtain

$$p_{ff} = p_f/(p_f + p_b/l) \qquad (15\text{-}97)$$

or

$$p_{ff} = \frac{1 - lv/N}{1 - (l-1)v/N} \qquad (15\text{-}98)$$

We are now ready to obtain an expression for $\bar{N}_{fl}$:

$$\bar{N}_{fl} = Np_l = N(1 - lv/N)\left(\frac{1 - lv/N}{1 - (l-1)v/N}\right)^{l-1} \qquad (15\text{-}99)$$

Equation 15-99 is obtained by using Equation 15-95 for p_l and the expressions for p_f and p_{ff}. Substituting Equation 15-99 into Equation 15-94, we obtain

$$v/(\text{L}) = \frac{N(1 - lv/N)}{k}\left(\frac{1 - lv/N}{1 - (l-1)v/N}\right)^{l-1} \qquad (15\text{-}100)$$

Equation 15-100 is the desired result. (Note that we ignored end effects in deriving this equation so that, strictly speaking, it is valid only for an infinite lattice. However, as we show in a subsequent section, Eqn. 15-100 affords a sufficiently accurate description for many systems of experimental interest involving finite lattices.)

Nonlinear Scatchard plots resulting from statistical effects

Equation 15-100 may be compared with the corresponding equation (Eqn. 15-29) for binding of L to n identical independent sites in which the "overlap" effect is not operative. When $l = 1$, Equation 15-100 reduces to Equation 15-29, as expected. (Note that, when $l = 1$, the n of Eqn. 15-29 equals the N of Eqn. 15-100; in general,

$n = N/l$.) However, when $l > 1$, then the last factor in Equation 15-100 is always less than unity and varies with v. This gives rise to marked nonlinearity in Scatchard plots of $v/(L)$ versus v; it also means that plots for $l > 1$ always fall below the plot for the case $l = 1$.

Figure 15-11 shows some calculated Scatchard plots for values of l between 1 and 20, with $k = 1$ M. For ease in comparing the plots, the abscissa is given in units

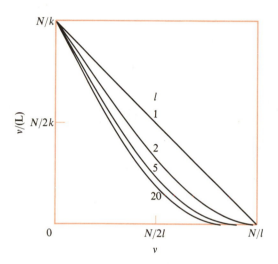

Figure 15-11

Scatchard plots of ligand binding to a lattice for various values of l with $k = 1$ M. [After J. D. McGhee and P. H. von Hippel, J. Mol. Biol. 86:469 (1974).]

of $1/l$. In all cases, complete saturation is achieved when $v = n = N/l$. However, as l increases, the concave-up curvature of the plots becomes increasingly apparent. This happens because the lattice entropically resists being saturated, with the resistance becoming more pronounced as l increases. Thus, the statistically large number of microstates that are generated at a given degree of partial saturation gives a strong entropic contribution to the free energy. This is lost, of course, as binding proceeds to the completely saturated lattice, which is comprised of only one microstate. As a result, for large ligands, saturation is not practically feasible; for example, with $l = 10$ to 20, (L) must change by 10-fold to 100-fold in order to increase the lattice saturation merely from 80% to 90%.

The v intercept of the Scatchard plot is N/l, which corresponds to n in Equation 15-29. However, the $v/(L)$ intercept is N/k, which is not the same as n/k, the intercept given by Equation 15-29. This distinction is important to recognize when obtaining parameters from Scatchard plots involving ligand binding to latticelike chains.

Some results on a real system

Figure 15-12 shows some actual data. This figure gives the Scatchard plot for binding of an ε-dinitrophenyl (ε-DNP) oligomer of lysine, ε-DNP-Lys–(Lys)$_5$, to the

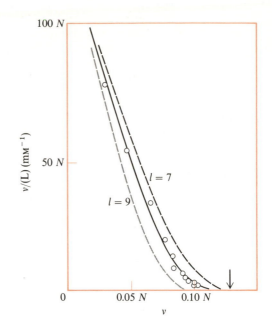

Figure 15-12

Scatchard plot for binding of ε-DNP-Lys–(Lys)$_5$ to the poly (rI):poly (rC) helix. Points are experimental values. The solid curve is calculated from Equation 15-100 with $l = 7.8$ and $k = 7.1$ μM. Dashed curves give calculated results with k held fixed and $l = 7$ or $l = 9$. The arrow denotes the point of lattice saturation. [After J. D. McGhee and P. H. von Hippel, *J. Mol. Biol.* 86:469 (1974).]

poly(rI):poly(rC) helix. The solid curve is calculated from Equation 15-100, with $l = 7.8$, and $k = 7.1$ μM. The arrow denotes the point of complete saturation. An excellent fit to the data is obtained. The dashed curves show the sensitivity of the data to variations in l with k held fixed. It is clear that a change of only about ± 1 from the best-fit value results in a marked departure of the calculated from the observed curve.

If a mixture of ligands bind to the lattice, then Equation 15-100 may be extended in a straightforward manner to handle this situation. Indexing by i the parameters associated with each ligand, we obtain

$$v_i/(L_i) = \frac{N\left(1 - \sum_i l_i v_i \middle/ N\right)}{k_i}\left(\frac{1 - \sum_i l_i v_i \middle/ N}{1 - \sum_i (l_i - 1)v_i \middle/ N}\right)^{l_i - 1} \tag{15-101}$$

This equation can be used when separate means are available to measure the binding of each ligand. For example, if binding to a helix of two different polyamines (such as spermine and spermidine) is studied, the use of a ^{14}C label for one and a ^{3}H label for the other would be appropriate in a dialysis experiment. In other situations, ligands with distinct optical bands might be of use.

Lattices of finite length and end effects

Strictly speaking, Equations 15-100 and 15-101 are valid only in the limit that N goes to infinity. This is true because we ignored end effects in the derivation. We can estimate the correction factor for any finite value of N. For this purpose, let $v/N = v_N$ be the saturation density of a lattice of N units, and v_∞ be that for a lattice with an infinite number of units. Equation 15-100 then can be written as

$$v_\infty/(L) = \frac{1 - lv_\infty}{k}\left(\frac{1 - lv_\infty}{1 - (l-1)v_\infty}\right)^{l-1} \qquad (15\text{-}102)$$

With the infinite lattice, each lattice unit has an average saturation density of v_∞; for the finite lattice, this is true *except* for the $l-1$ units at each end, which can have an average density of less than v_∞. In the limit of $(L) \to \infty$, it is obvious that v_N/v_∞ approaches the ratio (integral part of $N/l)/(N/l)$. For example, if $l = 3$ and $N = 100$, then $v_N/v_\infty = 0.99$. At the other extreme,

$$v_N \to [(N - l + 1)/N]v_\infty \qquad \text{as } (L) \to 0 \qquad (15\text{-}103)$$

where $N - l + 1$ is the number of potential binding sites for a ligand of length l in the naked lattice. Thus, with $l = 3$ and $N = 100$, we have $v_N/v_\infty = 0.98$. From these and similar calculations, we conclude that, for values of $N/l \geqslant 30$, Equation 15-100 should be applicable to finite chains to an accuracy greater than that achievable by experiment.

Ligand–ligand interactions

In some cases, binding of a ligand to a lattice chain can be cooperative. This is true for some of the DNA-binding proteins. These situations too can be handled with the formalism we have developed. For this purpose, we must introduce a parameter that accounts for the ligand–ligand interaction on the lattice. Consider two partially saturated lattice microspecies with identical numbers of bound ligands. Moreover, we assume that the distributions of these ligands are exactly the same, except that a pair of isolated (not contiguous) ligands on lattice A are made contiguous on lattice B, so as to introduce only one new ligand–ligand interaction on lattice B. The lattice equilibrium is A $\rightleftarrows$ B, where

$$\omega = (B)/(A) \qquad (15\text{-}104)$$

Proceeding along lines similar to those used to derive Equation 15-100, we obtain

$$\frac{v}{(L)} = \frac{N(1 - lv/N)}{k}\left(\frac{(2\omega + 1)(1 - lv/N) + v/N - R}{2(\omega - 1)(1 - lv/N)}\right)^{l-1}\left(\frac{1 - (l+1)v/N + R}{2(1 - lv/N)}\right)^2$$

$$(15\text{-}105)$$

where

$$R = \{[1 - (l+1)v/N]^2 + (4\omega v/N)(1 - lv/N)\}^{1/2} \qquad (15\text{-}106)$$

(For a more complete derivation, see McGhee and von Hippel, 1974.)

For the case of $\omega = 1$, Equation 15-105 can be shown to reduce to Equation 15-100. When $\omega < 1$ (anticooperative binding), Scatchard plots according to Equation 15-105 fall below those of the noninteracting case (Eqn. 15-100). When $\omega = 0$ (infinite anticooperativity), Equation 15-105 reduces to Equation 15-100 for non-interacting ligands with a length of $l + 1$ segments. When $\omega > 1$, the Scatchard plots are concave-down, and they fall above the corresponding noninteracting case. Figure 15-13 shows some plots for various values of ω with $l = 1$ and $k = 1$ M. Because these

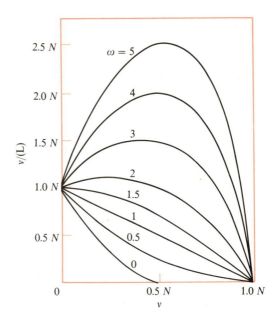

Figure 15-13

Scatchard plots for ligand binding to a homogenous lattice, for $k = 1$ M, and $l = 1$, with various values of the cooperativity parameter ω. [After J. D. McGhee and P. H. von Hippel, *J. Mol. Biol.* 86:469 (1974).]

curves are drawn for $l = 1$, these is no "overlap" effect mixed with the cooperativity. As shown earlier (Fig. 15-11), the effect of increasing l is to introduce an anticooperative feature into the binding. In a case where $\omega > 1$ and $l > 1$, the resulting curve obviously will be a compromise between the two parameters.

The $v/(L)$ intercept of the Scatchard plot of Equation 15-105 is N/k, the same as that for the noninteracting system described by Equation 15-100. Moreover, the v intercept is still N/l. However, when $\omega > 1$, lattice saturation at any given value of n clearly is easier to achieve than in the noninteracting case ($\omega = 1$) where entropic effects inhibit saturation.

This treatment of ligand binding to a homogeneous lattice is applicable to a number of situations that are encountered in practice. Although another approach must be used to handle heterogeneous lattices, which have more than one type of ligand combining site, the above treatment does indicate that nonlinear regions in Scatchard plots can be explained in significant measure by the "overlap" effect. When effects beyond those predicted by Equation 15-100 or 15-105 are observed, lattice heterogeneity must be considered. This requires a further extension of the theory.

In Chapter 23, where ligand binding to nucleic acids is considered further, we develop a different kind of treatment of ligand binding to lattices, based on matrix methods.

Summary

Ligand interactions are widespread in biochemical systems, and a rich and useful formalism has developed for treating the equilibrium properties of the diversity of systems that are encountered. In such treatments it is important to keep firmly in mind the distinction between microscopic and macroscopic constants, and associated statistical features. The simplest systems involve binding of a ligand to a single class of identical, independent sites. These can be analyzed conveniently by a Scatchard plot. An extension of the Scatchard type of analysis can be useful for treating multiple classes of independent sites.

In many biological systems, interactions occur between binding sites for a single kind of ligand. For example, cooperative interactions commonly are found, and they may be treated by established procedures. Another type of interaction is that between binding sites for different kinds of ligands that bind with the same macromolecule. In this case, particularly interesting and useful linkage relationships describe the interaction. The interaction energy that characterizes the linkage commonly is on the order of 0 to ± 2.5 kcal mole^{-1}.

Another case of considerable interest is the interaction of large ligands with latticelike chains, such as helical DNA. In such systems, statistical, entropic effects play a major role in determining the character of the observed binding equilibria.

Problems

15-1. The relationship shown by Equation 15-20 is useful in many calculations. Prove this relationship.

15-2. A ligand binds to four sites on a macromolecule. The apparent macroscopic dissociation constants K_2 and K_3 are identical (within experimental error). Is there an interaction energy between sites 2 and 3? If not, why not? If so, calculate the interaction energy.

15-3. A macromolecule has six sites for the ligand L. An investigator is able to measure three of the macroscopic dissociation constants for ligand binding, K_1, K_4, and K_5. He finds that $K_5 = 15K_1$, and $K_4 = 8K_1$. He then claims that this result clearly demonstrates negative (anticooperative) site–site interactions, and that the Scatchard plot for this system is surely concave-up. A critic disagrees and says that the limited data indicate there are no site–site interactions, and he even ventures to predict relative K_i values for the other constants. Who is right, and why? Is it possible to predict the other K_i values? If so, make the predictions. If not, explain why none can be made.

15-4. Consider a case in which two ligands, L_1 and L_2, bind to a macromolecule. Let half of the L_1 and half of the L_2 sites be filled. Using Equations 15-89 and 15-90, derive Equations 15-91, 15-92, and 15-93.

15-5. The relationship shown by Equation 15-54 is a simple connection between microscopic dissociation constants, k_i, and macroscopic dissociation constants, $\tilde{K}_i$, defined in Equation 15-51. Establish the validity of Equation 15-54 for the case of three microscopic constants, k_1, k_2, and k_3, and of three macroscopic constants, $\tilde{K}_1$, $\tilde{K}_2$, and $\tilde{K}_3$, and reflect on the reason for the particularly simple form of Equation 15-54.

References

GENERAL

Alberty, R. A., and F. Daniels. 1979. *Physical Chemistry*, 5th ed. New York: Wiley. [Chaps. 4 and 6 give a good introduction to chemical equilibrium and biochemical equilibria.]

McGhee, J. D., and P. H. von Hippel. 1974. Theoretical aspects of DNA–protein interactions: cooperative and non-cooperative binding of large ligands to a one-dimensional homogeneous lattice. *J. Mol. Biol.* 86:469. [Includes an extremely clear exposition of ligand interactions with latticelike chains. The statistical complexities of the lattice are well discussed, and they are put in physically understandable terms.]

Wyman, J., Jr. 1964. Linked functions and reciprocal effects in hemoglobin: A second look. *Adv. Protein Chem.* 19:223. [A classical paper on ligand equilibria and linked functions.]

SPECIFIC

Alberty, R. A. 1969. Thermodynamics of the hydrolysis of adenosine triphosphate. *J. Chem. Ed.* 46:713.

Hess, V. L., and A. Szabo. 1979. Ligand binding to macromolecules: Allosteric and sequential models of cooperativity. *J. Chem. Ed.* 56:289. [A nice introduction to the use of generating functions to treat macromolecule–ligand systems.]

Hill, A. V. 1910. The possible effects of the aggregation of the molecules of hemoglobin on its dissociation curves. *J. Physiol.* (*London*) 40:iv.

Scatchard, G. 1949. The attraction of proteins for small molecules and ions. *Ann. N.Y. Acad. Sci.* 51:660.

Schimmel, P. R. 1976. Equilibrium and kinetic studies of the cooperative interaction of cations with transfer RNA. *J. Polymer Sci. Symp. No. 54*, p. 387.

Strandberg, M. W. P. 1979. Linear differential equations in chemical equilibrium calculations. *J. Chem. Phys.* 71:4765. [A differential equations approach is used to show certain general features of macromolecule–ligand systems.]

Tanford, C. 1961. *Physical Chemistry of Macromolecules*. New York: Wiley. [Chap. 8 treats multiple equilibria and complex systems, with many illustrations.]

Weber, G. 1975. Energetics of ligand binding to proteins. *Adv. Protein Chem.* 29:1.

16
Kinetics of ligand interactions

16-1 BIOCHEMICAL KINETIC STUDIES

The study of ligand interactions from an equilibrium standpoint (Chapter 15) gives insight into thermodynamic features and general mechanistic aspects of the ligand reactions. However, equilibrium measurements alone are not sufficient for a detailed mechanistic understanding of ligand reactions. In general, there are many possible reaction schemes that can account for a given set of thermodynamic data. These different schemes often can be sorted out by studies of the dynamics (or kinetics) of ligand reactions. Such investigations not only permit identification of reaction pathways that are followed, but they also can give quantitative insight into the time scale of elementary processes; from this information, it is possible to deduce a clearer picture of events at the molecular level.

All biochemical systems exhibit dynamic behavior. In many cases, the *physiological property* itself is in essence a *dynamic property*, as in enzyme catalysis, active ion transport across membranes, repression and activation of gene expression, and so on. For this reason, studies of the dynamic behavior of biochemical systems have been of great interest for many years.

In this chapter, we discuss some of the essential features of kinetic systems. Many of the ideas and equations developed here have applications in other chapters of the book. For example, in Chapter 21, where mechanisms for the folding of proteins are discussed, kinetic aspects are of great importance. In the following discussion, much of the treatment centers around enzymatic reaction systems, which have been investigated in more depth from a kinetic viewpoint than has any other class of biochemical systems. Also, as a concrete example of how an integration of kinetic and other approaches has given molecular insight into ligand interactions and catalysis, a discussion is given of ribonuclease A, an unusually well-studied protein.

In the case of enzymes, the most common approach is to carry out kinetic studies under conditions where the free enzyme and its intermediate species are in a steady state. This allows an enormous simplification of rate equations, but at the sacrifice of losing much of the detailed information concerning the reaction mechanism and intermediates. For this reason, more sophisticated kinetic studies involve measurements of transient-state features. This approach is technically more difficult and also can require complex and laborious mathematical analysis of kinetic curves. However, a considerable simplification in transient-state kinetic studies is provided by relaxation methods that have gained considerable popularity in recent years.

Before treating some of the important features of steady-state and relaxation kinetics, it is useful to consider first some general features of enzymatic and related kinetic systems. In particular, it is important to appreciate the inherent complexities of even some simple kinetic systems. With this in mind, the simplifications and serious limitations of the steady-state approximation become apparent; similarly, the particular significance and advantages of a special approach, such as relaxation spectrometry, are also grasped more concretely.

16-2   UNIMOLECULAR REACTIONS

One-step reactions

The simplest reversible reactions are unimolecular processes such as protein conformational changes. An example is

$$X_0 \underset{k_{-1}}{\overset{k_1}{\rightleftarrows}} X_1 \tag{16-1}$$

where X_0 and X_1 might be two different conformations of a protein. The rate equation for this mechanism is

$$-d(X_0)/dt = k_1(X_0) - k_{-1}(X_1) \tag{16-2}$$

This equation is easily solved by changing variables; let $\Delta(X_0) = (X_0) - \overline{(X_0)}$, and $\Delta(X_1) = (X_1) - \overline{(X_1)}$, where the overbars denote equilibrium concentrations. Because mass conservation requires that the total X concentration $(X)_0 = (X_0) + (X_1)$ be constant, $\Delta(X_0) = -\Delta(X_1)$. Therefore, Equation 16-2 becomes

$$d\,\Delta(X_0)/dt = -\lambda\,\Delta(X_0) \tag{16-3}$$

or

$$\Delta(X_0) = \Delta(X_0)^0 e^{-\lambda t} \tag{16-4}$$

where $\lambda = k_1 + k_{-1}$, and $\Delta(X_0)^0$ is the deviation at $t = 0$ of (X_0) from its equilibrium value. Equation 16-4 also can be written as

$$(X_0) = \overline{(X_0)} + \Delta(X_0)^0 e^{-\lambda t} \tag{16-5}$$

Thus for the simple, reversible, unimolecular scheme, (X_0) exponentially rises or falls to its equilibrium value, depending on the initial conditions.

Two coupled reactions

The next-most-simple unimolecular scheme involves two coupled reactions:

$$X_0 \underset{k_{-1}}{\overset{k_1}{\rightleftarrows}} X_1 \underset{k_{-2}}{\overset{k_2}{\rightleftarrows}} X_2 \tag{16-6}$$

Using the concentration variables $\Delta(X_0)$, $\Delta(X_1)$, and $\Delta(X_2)$, we can write the rate equations as

$$-d\Delta(X_1)/dt = -k_1 \Delta(X_0) + (k_{-1} + k_2)\Delta(X_1) - k_{-2}\Delta(X_2) \tag{16-7a}$$

$$-d\Delta(X_2)/dt = -k_2 \Delta(X_1) + k_{-2}\Delta(X_2) \tag{16-7b}$$

Making use of the conservation relationship $\Delta(X_0) = -[\Delta(X_1) + \Delta(X_2)]$, we obtain

$$-d\Delta(X_1)/dt = a_{11}\Delta(X_1) + a_{12}\Delta(X_2) \tag{16-8a}$$

$$-d\Delta(X_2)/dt = a_{21}\Delta(X_1) + a_{22}\Delta(X_2) \tag{16-8b}$$

where $a_{11} = k_1 + k_{-1} + k_2$; $a_{12} = k_1 - k_{-2}$; $a_{21} = -k_2$; and $a_{22} = k_{-2}$.

Note that there are only two rate equations that characterize Equation 16-6. Although an expression can be written for $-d\Delta(X_0)/dt$, from mass conservation this is just the sum: $d\Delta(X_1)/dt + d\Delta(X_2)/dt$. In general, the number of rate equations that characterize a mechanism corresponds to the number of *independent* steps.

The expressions of Equation 16-8 are two simultaneous linear differential equations. The standard procedure for solving such equations is to seek solutions of the form

$$\Delta(X_1) = C_1 e^{-\lambda t} \tag{16-9a}$$

$$\Delta(X_2) = C_2 e^{-\lambda t} \tag{16-9b}$$

where C_1, C_2, and λ are constants to be determined. Differentiation of Equation 16-9 shows that

$$-d\Delta(X_1)/dt = \lambda \Delta(X_1) \tag{16-10a}$$

$$-d\Delta(X_2)/dt = \lambda \Delta(X_2) \tag{16-10b}$$

Substitution of Equation 16-10 into Equation 16-8 gives

$$0 = (a_{11} - \lambda)\,\Delta(X_1) + a_{12}\,\Delta(X_2) \qquad (16\text{-}11a)$$

$$0 = a_{21}\,\Delta(X_1) + (a_{22} - \lambda)\,\Delta(X_2) \qquad (16\text{-}11b)$$

The solution to the set of homogeneous simultaneous linear equations (Eqn. 16-11a,b) requires that the determinant of the coefficients vanishes. Thus,

$$\begin{vmatrix} a_{11} - \lambda & a_{12} \\ a_{21} & a_{22} - \lambda \end{vmatrix} = 0 \qquad (16\text{-}12)$$

or

$$\lambda^2 - (a_{11} + a_{22})\lambda + a_{11}a_{22} - a_{12}a_{21} = 0 \qquad (16\text{-}13)$$

The quadratic equation has two roots, given by

$$\lambda_{1,2} = [(a_{11} + a_{22})/2]\{1 \pm [1 - 4(a_{11}a_{22} - a_{12}a_{21})/(a_{11} + a_{22})^2]^{1/2}\} \quad (16\text{-}14)$$

where we assign λ_1 to the positive sign and λ_2 to the negative sign. Thus, the original rate equations are satisfied with $\lambda = \lambda_1$ or with $\lambda = \lambda_2$ in Equation 16-9. The most general solution is the sum of the two particular solutions; it is given by

$$\Delta(X_1) = A_{11}e^{-\lambda_1 t} + A_{12}e^{-\lambda_2 t} \qquad (16\text{-}15a)$$

$$\Delta(X_2) = A_{21}e^{-\lambda_1 t} + A_{22}e^{-\lambda_2 t} \qquad (16\text{-}15b)$$

where the A_{ij} terms are constants determined by initial conditions.

According to Equation 16-15, the rise or decay of $\Delta(X_1)$ and of $\Delta(X_2)$ to the final equilibrium values is a biphasic process, with parameters λ_1 and λ_2 characterizing the two phases. The two phases cannot be assigned to the individual elementary steps in the scheme of Equation 16-6. Inspection of Equation 16-14 shows that each of the two parameters λ_1 and λ_2 involves aggregates of rate constants from both elementary steps. In actuality, λ_1 and λ_2 correspond to the time constants associated with the "normal" concentration variables that are analogous to the "normal" coordinates obtained in a standard analysis of classical mechanics.

A series of sequential reactions

Consider now the more general case of a series of sequential unimolecular changes:

$$X_0 \underset{k_{-1}}{\overset{k_1}{\rightleftarrows}} X_1 \underset{k_{-2}}{\overset{k_2}{\rightleftarrows}} \cdots \underset{k_{-i}}{\overset{k_i}{\rightleftarrows}} X_i \underset{k_{-(i+1)}}{\overset{k_{i+1}}{\rightleftarrows}} \cdots \underset{k_{-n}}{\overset{k_n}{\rightleftarrows}} X_n \qquad (16\text{-}16)$$

The rate equations for this system are

$$-d(X_0)/dt = k_1(X_0) - k_{-1}(X_1)$$
$$\vdots \qquad\qquad \vdots$$
$$-d(X_i)/dt = -k_i(X_{i-1}) + (k_{-i} + k_{i+1})(X_i) - k_{-(i+1)}(X_{i+1}) \qquad (16\text{-}17)$$
$$\vdots \qquad\qquad \vdots$$
$$-d(X_n)/dt = -k_n(X_{n-1}) + k_{-n}(X_n)$$

Although there are $n + 1$ equations involving $n + 1$ concentration variables, one variable may be eliminated by virtue of the mass-conservation relationship:

$$(X)_0 = \sum_{i=0}^{n} (X_i) \qquad (16\text{-}18)$$

Thus, there are only n independent rate equations, because any one of the expressions in Equation 16-17 can be expressed as a linear combination of the others. Choosing to eliminate (X_0), and defining $\Delta(X_i) = (X_i) - (\overline{X_i})$, we can write Equation 16-17 as

$$-d\,\Delta(X_i)/dt = \sum_{j=1}^{n} a_{ij}\,\Delta(X_j) \qquad \text{for } i = 1, 2, \ldots, n \qquad (16\text{-}19)$$

where the a_{ij} terms are parameters composed of rate constants. The solution to a set of linear first-order differential equations is

$$\Delta(X_i) = \sum_{j=1}^{n} A_{ij}e^{-\lambda_j t} \qquad \text{for } i = 1, 2, \ldots, n \qquad (16\text{-}20)$$

where the A_{ij} factors depend on initial conditions, and the λ_j factors are functions of the rate constants (and are given by the solutions to an $n \times n$ secular determinant analogous to that given in Eqn. 16-12). Thus, for an n-step sequential mechanism, the concentration of any given species is a sum of n exponential terms. Moreover, the parameters A_{ij} and λ_j are given by exact relationships. From knowledge of the equilibrium constants for each step in the mechanism, together with an experimental determination of each λ_j, all $2n$ rate constants can be computed.

It is apparent from this discussion that unimolecular reactions can be treated by standard procedures used to solve simultaneous linear differential equations. This is true, regardless of the complexity of the scheme or of the number of kinetic species. In contrast, higher-order reactions introduce complexities that are difficult to surmount, even when the schemes themselves are simple.

16-3   SIMPLE BIMOLECULAR REACTION

Even for the simple bimolecular reaction, the solution to the rate equation is complex. The bimolecular reaction between enzyme E and substrate S is

$$E + S \underset{k_{-1}}{\overset{k_1}{\rightleftharpoons}} X \tag{16-21}$$

The simple rate equation is

$$-d(S)/dt = k_1(E)(S) - k_{-1}(X) \tag{16-22}$$

and the conservation relations are

$$(E)_0 = (E) + (X) \tag{16-23a}$$

$$(S)_0 = (S) + (X) \tag{16-23b}$$

where $(E)_0$ and $(S)_0$ are the concentrations of total enzyme and of total substrate, respectively. Although this is a simple one-step reaction, there is no convenient analytical solution that is valid for general boundary conditions. For the special case that $(E)_0 = (S)_0$, and $(X) = 0$ at $t = 0$, the solution is

$$\ln\left(\frac{2(S) + (k_{-1}/k_1)(1 + \alpha)}{2(S) + (k_{-1}/k_1)(1 - \alpha)}\right) = k_{-1}\alpha t + C \tag{16-24}$$

where $\alpha = [1 + 4(S)_0 k_1/k_{-1}]^{1/2}$, and C is a constant. Thus, even with a restrictive assumption, the solution to the simple bimolecular reaction is complex and unwieldy. It is not surprising, therefore, that severe difficulties are encountered with the simplest enzymatic mechanisms. For this reason, experiments often are designed to utilize conditions where steady-state solutions to rate equations or solutions valid near equilibrium may be used.

16-4   SIMPLE MICHAELIS–MENTEN MECHANISM

Solution to the rate equations for a special case

The simplest enzymatic Michaelis–Menten mechanism is

$$E + S \underset{k_{-1}}{\overset{k_1}{\rightleftharpoons}} X \underset{k_{-2}}{\overset{k_2}{\rightleftharpoons}} E + P \tag{16-25}$$

where P is product. Although there is no convenient, general analytical solution to the rate equations for this mechanism, W. G. Miller and R. A. Alberty (1958) showed that a solution can be obtained when $k_1 = k_{-2}$. In this special case,

$$(X) = \frac{2k_1(E)_0(S)_0(1 - e^{-\lambda_1 t})}{(\lambda_1 - \lambda_2) + (\lambda_1 + \lambda_2)e^{-\lambda_1 t}} \tag{16-26a}$$

$$(P) = \frac{2k_2(S)_0[\lambda_1(1 - e^{-\lambda_3 t}) - \lambda_3(1 - e^{-\lambda_1 t})]}{(k_{-1} + k_2)[(\lambda_1 - \lambda_2) + (\lambda_1 + \lambda_2)e^{-\lambda_1 t}]} \tag{16-26b}$$

where

$$\lambda_1 = (\{k_1[(E)_0 + (S)_0] + k_{-1} + k_2\}^2 - 4k_1^2(E)_0(S)_0)^{1/2} \tag{16-27a}$$

$$\lambda_2 = -\{k_1[(E)_0 + (S)_0] + k_{-1} + k_2\} \tag{16-27b}$$

$$\lambda_3 = k_1(E)_0[2k_1(S)_0 + \lambda_2 - \lambda_1]/(\lambda_2 - \lambda_1) \tag{16-27c}$$

and where the conditions $(S) = (S)_0$ and $(P) = 0$ at $t = 0$ have been assumed. Although this solution is unwieldy for analyzing experimental data, it is useful for gaining insight into the kinetic behavior of the mechanism.

Figure 16-1 plots $(X)/(E)_0$ and $(P)/(\bar{P})$ versus time, for $(E)_0 = (S)_0 = 10^{-6}$ M, and for $(E)_0 = 10^{-6}$ M and $(S)_0 = 10^{-4}$ M, where $(\bar{P})$ is the final equilibrium concentration of P. Common values for the rate parameters were used (see figure legend).

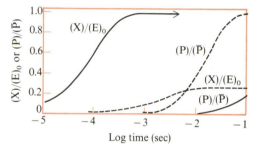

Figure 16-1

Approach to equilibrium of the species in the single-intermediate Michaelis–Menten mechanism. Dashed lines show $(E)_0 = (S)_0 = 1\ \mu\text{M}$; solid lines show $(E)_0 = 1\ \mu\text{M}$ and $(S)_0 = 100\ \mu\text{M}$. The rate parameters are $k_1 = k_{-2} = 10^8\ \text{M}^{-1}\ \text{sec}^{-1}$ and $k_{-1} = k_2 = 10^2\ \text{sec}^{-1}$. [After G. G. Hammes and P. R. Schimmel, in *The Enzymes*, vol. 2, ed. P. D. Boyer (New York: Academic Press, 1970), p. 67.]

It is clear that, when $(E)_0 \ll (S)_0$, the concentration of X rises rapidly to a constant value, reaching that plateau long before a significant fraction of S is transformed to P. In this case, $d(X)/dt \cong 0$ during most of the conversion of S to P. On the other hand, when substrate levels of enzyme are employed, the reaction goes rapidly to completion and, during the early phase, both (X) and (P) are simultaneously rising with time.

Condition for steady-state approximation

When $k_1 \neq k_{-2}$, a convenient analytical solution to the rate equations for Equation 16-25 cannot be obtained. For this reason, it is common to use the steady-state approximation to the rate equations. This approximation assumes that $d(X)/dt \cong 0$ during the course of the reaction. With this assumption, the rate equations are considerably simplified, so that a useful form is achieved. (We show this in a subsequent section, where we discuss steady-state rate equations in some detail.)

Because the steady-state approximation is commonly employed, it is of interest to inquire as to the conditions under which $d(X)/dt \cong 0$. Figure 16-1 suggests that, when $(E)_0 \ll (S)_0$, then (X) is constant during most of the time in which P is produced. Returning to the special case of $k_1 = k_{-2}$, when $(E)_0 \ll (S)_0$, Equation 16-26a becomes

$$(X) = [k_1(E)_0(S)_0/\lambda_1](1 - e^{-\lambda_1 t}) \tag{16-28}$$

where

$$\lambda_1 = k_1(S)_0 + k_{-1} + k_2 \tag{16-29}$$

With $(S)_0 = 10^{-4}$ M, and $k_1 = 10^8$ M^{-1} sec^{-1}, and $k_{-1} = k_2 = 10^2$ sec^{-1}, then we obtain $\lambda_1 = 10^4$ sec^{-1}. Therefore, with $t \gg \lambda_1^{-1}$ (for example, in this case, $t \geqslant 10^{-3}$ sec), it is clear that

$$(X) = k_1(E)_0(S)_0/\lambda_1 = (E)_0/[1 + K_S/(S)_0] \qquad \text{for } t \gg \lambda_1^{-1} \tag{16-30}$$

where K_S is the Michaelis constant for the forward reaction and is defined as

$$\boxed{K_S = (k_{-1} + k_2)/k_1} \tag{16-31}$$

Therefore, in our example, (X) is in a stationary state within 10^{-3} sec. After this transient phase, the steady-state approximation is valid. (As we shall show, Eqn. 16-30 can be derived directly from the steady-state rate equations, without recourse to more complex equations such as Eqns. 16-26 and 16-27.) From these considerations, we conclude that *the steady-state condition is achieved when* $(E)_0 \ll (S)_0$. (Although we derived our example for the case that $k_1 = k_{-2}$, this conclusion is valid for virtually any situation encountered in practice.) Thus, in steady-state kinetic studies, it is common to employ catalytic amounts of the enzyme, so that the steady-state approximation can be used.

16-5 MULTIPLE INTERMEDIATES

It is instructive to consider briefly the behavior of a multiple-intermediate enzyme reaction mechanism such as

$$E + S \underset{k_{-1}}{\overset{k_1}{\rightleftharpoons}} X_1 \underset{k_{-2}}{\overset{k_2}{\rightleftharpoons}} X_2 \underset{k_{-3}}{\overset{k_3}{\rightleftharpoons}} X_3 \underset{k_{-4}}{\overset{k_4}{\rightleftharpoons}} X_4 \underset{k_{-5}}{\overset{k_5}{\rightleftharpoons}} E + P \qquad (16\text{-}32)$$

We can inquire about the behavior of each of the intermediate forms, as well as investigate whether the condition $(E)_0 \ll (S)_0$ will ensure that each of the X_i concentrations is in a stationary state after a short initial phase.

To visualize the behavior of the species in Equation 16-32, we must numerically solve the rate equations with the use of a computer. Figure 16-2 shows an analog computer solution to the rate equations of Equation 16-32. For this solution, $k_{-i} = 0$

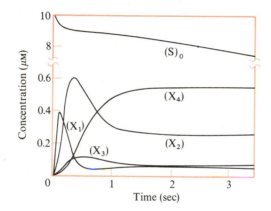

Figure 16-2

Analog-computer solution to four-intermediate Michaelis–Menten mechanism, with $(E)_0 = 1 \ \mu M$; $(S)_0 = 10 \ \mu M$; $k_1 = 0.9 \times 10^6 \ M^{-1} \ sec^{-1}$; $k_2 = 9 \ sec^{-1}$; $k_3 = 1 \ sec^{-1}$; $k_4 = 9 \ sec^{-1}$; $k_5 = 2 \ sec^{-1}$; and $k_{-1} = 0$ for $i = 1, 2, \ldots, 5$. [Courtesy of J. Higgins.]

for all i, so that the reaction is irreversible. Values for the other parameters are given in the figure legend. It is assumed that $(E)_0 = 0.1(S)_0$. For this example, (X_1), (X_2), and (X_3) quickly pass through transient maxima and then fall to virtually constant levels in a little over 1 sec, whereas (X_4) rises steadily to a stationary state that also is reached in just over 1 sec. (Although not shown, at long times each (X_i) decays to zero, because each $k_{-i} = 0$.) During the establishment of the stationary levels of the intermediates, only a minor fraction of S is consumed. Therefore, after a brief period, the steady-state conditions hold for all the intermediates. Hence, after this transient period, it is valid to assume that $d(X_i)/dt = 0$ (for all i) in order to derive the steady-state rate law for this mechanism. And, in general, regardless of the number of intermediates, the steady-state approximation may be applied to each one if $(E)_0 \ll (S)_0$.

16-6 STEADY-STATE KINETICS

In this section, we discuss steady-state enzyme kinetics in some detail. We assume throughout the discussion that $(E)_0 \ll (S)_0$, and that the rate laws are applicable only after passage through the brief transient phase.

Steady-state enzyme kinetics is a vast and well-studied field. In the treatment here, we direct attention only at some of the major features of these special kinetic systems. For further reading, see the references at the end of the chapter.

Single-substrate-to-single-product mechanism with one intermediate

The simplest enzyme reaction mechanism is the case where a single substrate yields a single product P, passing through only a single intermediate:

$$E + S \underset{k_{-1}}{\overset{k_1}{\rightleftharpoons}} X \underset{k_{-2}}{\overset{k_2}{\rightleftharpoons}} E + P \qquad (16\text{-}33)$$

The two rate equations describing this mechanism are

$$-d(S)/dt = k_1(E)(S) - k_{-1}(X) \qquad (16\text{-}34\text{a})$$

$$-d(X)/dt = -k_1(E)(S) + (k_{-1} + k_2)(X) - k_{-2}(E)(P) \qquad (16\text{-}34\text{b})$$

In the steady state, $d(X)/dt \cong 0$. Also, if the reaction initially contains only E and S, then at early times $(P) \cong 0$ and $(S) = (S)_0$. With these assumptions, Equation 16-34b becomes

$$(X) = k_1(E)(S)_0/(k_{-1} + k_2) \qquad (16\text{-}35)$$

Using the mass-conservation relationship for total enzyme and Equation 16-35, we have

$$(E)_0 = (E) + (X)$$
$$= (E)[1 + k_1(S)_0/(k_{-1} + k_2)]$$
$$= (X)[1 + (k_{-1} + k_2)/k_1(S)_0] \qquad (16\text{-}36)$$

Substituting these equations into Equation 16-34a, we obtain the initial reaction velocity $v_i = [-d(S)/dt]_{t=0}$:

$$v_i = V_S/[1 + K_S/(S)_0] \qquad (16\text{-}37)$$

where $V_S = k_2(E)_0$, and $K_S = (k_{-1} + k_2)/k_1$.

The parameter V_S is the maximal velocity; it is easy to see that $v_i \rightarrow V_S$ as $(S)_0 \rightarrow \infty$. Note that, when $(S)_0 = K_S$, then $v_i = V_S/2$; that is, K_S is the substrate concentration at which half-maximal velocity is achieved. From Equation 16-35, it is clear that $(S)_0 = K_S$ also corresponds to "half-saturation"; that is, from Equation 16-36,

$$(X)/(E)_0 = [1 + K_S/(S)_0]^{-1}$$

which equals 1/2 when $(S)_0 = K_S$.

Obtaining steady-state parameters from experimental data

A plot of v_i versus $(S)_0$ is hyperbolic (Fig. 16-3), with $v_i \cong (V_S/K_S)(S)_0$ at low $(S)_0$ (first-order kinetics), and $v_i \cong V_S$ at high $(S)_0$ (zero-order kinetics). The parameters

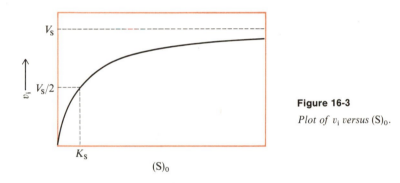

Figure 16-3

Plot of v_i versus $(S)_0$.

K_S and V_S are most conveniently obtained by rearranging Equation 16-37. Two popular forms are the following:

$$1/v_i = (K_S/V_S)[1/(S)_0] + 1/V_S \tag{16-38}$$

$$v_i/(S)_0 = -v_i/K_S + V_S/K_S \tag{16-39}$$

In using Equation 16-38, $1/v_i$ is plotted versus $1/(S)_0$ (Fig. 16-4a). The slope of this Lineweaver–Burk plot gives K_S/V_S, and the intercept gives $1/V_S$. In using Equation 16-39, $v_i/(S)_0$ is plotted versus v_i (Fig. 16-4b). The slope of this Eadie–Hofstee plot gives $-1/K_S$, and the intercept gives V_S/K_S. Because the two methods of graphing treat data points in different ways, it is advisable to use both methods in obtaining K_S and V_S values, and then to compare the results obtained.

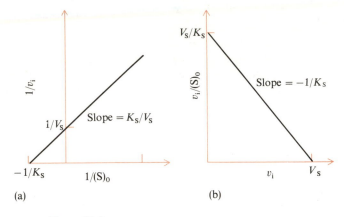

Figure 16-4

(a) Lineweaver–Burk plot. (b) Eadie–Hofstee plot.

A more general solution: the Haldane relationship

If the assumption $(P) \cong 0$ is removed, the steady-state solution to Equation 16-34 is

$$v = \frac{(V_S/K_S)(S) - (V_P/K_P)(P)}{1 + (S)/K_S + (P)/K_P} \qquad (16\text{-}40)$$

where v is the steady-state velocity, $V_P = k_{-1}(E_0)$, and $K_P = (k_{-1} + k_2)/k_{-2}$. Thus V_P is the maximal velocity for the reverse reaction, and K_P is the Michaelis constant for the reverse reaction. Note that, if $(P) = 0$, Equation 16-40 simplifies to Equation 16-37; note also that the equation analogous to Equation 16-37 is obtained for the reverse reaction when $(S) = 0$.

The parameters V_S, K_S, V_P, and K_P generally are obtained by separate measurements on the forward and reverse reactions. The four parameters are interrelated through the equilibrium constant for the overall reaction. At equilibrium, $v = 0$; from Equation 16-40, we have $(V_S/K_S)\,\overline{(S)} - (V_P/K_P)\overline{(P)} = 0$, where the overbars denote equilibrium concentrations. In terms of the equilibrium constant K_{eq}, we obtain the Haldane relationship:

$$K_{eq} = \overline{(P)}/\overline{(S)} = (V_S/K_S)(K_P/V_P) \qquad (16\text{-}41)$$

One of the most valuable features of the Haldane relationship is that, if K_{eq} is known, the equation provides a means of checking the internal consistency of the four experimentally determined steady-state parameters.

Relationship between Michaelis constants and equilibrium constants

The Michaelis constants K_S and K_P sometimes are implicitly assumed to be equivalent to equilibrium constants. However, this is true only in special circumstances. By definition, $K_S = (k_{-1} + k_2)/k_1$. When $k_2 \ll k_{-1}$, then $K_S = k_{-1}/k_1$ and, because $k_1\overline{(E)(S)} = k_{-1}\overline{(X)}$, we have $K_S = \overline{(E)(S)}/\overline{(X)}$. Thus, K_S is an equilibrium dissociation constant when the rate of dissociation of X to E + S is substantially faster than its conversion to E + P. (In this case, the reaction $E + S \rightleftharpoons X$ continuously adjusts itself so as to be always at equilibrium while the overall reaction proceeds and X is converted to P.) However, for the one-intermediate mechanism, K_P cannot also be an equilibrium constant if $k_2 \ll k_{-1}$ (in which case $K_P = k_{-1}/k_{-2}$). On the other hand, as we shall see, in situations where there is more than one intermediate, it is possible for both Michaelis constants to represent equilibrium constants.

Multiple-intermediate form of the steady-state rate equation

The single-intermediate mechanism of Equation 16-33 contains four elementary rate constants. Because four steady-state parameters are obtained from steady-state kinetic measurements, it is possible to solve these parameters for each of the four elementary constants. However, in actuality, a mechanism with only a single intermediate is rarely (perhaps never) encountered.

It is not difficult to show that the form of the steady-state rate equation (Eqn. 16-40) is the same regardless of the number of intermediates. For example, consider the two-intermediate mechanism:

$$E + S \underset{k_{-1}}{\overset{k_1}{\rightleftharpoons}} X_1 \underset{k_{-2}}{\overset{k_2}{\rightleftharpoons}} X_2 \underset{k_{-3}}{\overset{k_3}{\rightleftharpoons}} E + P \qquad (16\text{-}42)$$

To obtain the steady-state rate equation, we set both $d(X_1)/dt$ and $d(X_2)/dt$ equal to zero:

$$-d(X_1)/dt = 0 = -k_1(E)(S) + (k_{-1} + k_2)(X_1) - k_{-2}(X_2) \qquad (16\text{-}43a)$$

$$-d(X_2)/dt = 0 = -k_2(X_1) + (k_{-2} + k_3)(X_2) - k_{-3}(E)(P) \qquad (16\text{-}43b)$$

We also use the mass-conservation relationship:

$$(E)_0 = (E) + (X_1) + (X_2) \qquad (16\text{-}44)$$

The solution is identical in form to Equation 16-40, with

$$K_S = (k_2 k_3 + k_{-1} k_3 + k_{-1} k_{-2})/k_1(k_2 + k_{-2} + k_3) \qquad (16\text{-}45a)$$

$$V_S = k_2 k_3 (E)_0/(k_2 + k_{-2} + k_3) \qquad (16\text{-}45b)$$

$$K_P = (k_2 k_3 + k_{-1} k_3 + k_{-1} k_{-2})/k_{-3}(k_{-1} + k_2 + k_{-2}) \qquad (16\text{-}45c)$$

$$V_P = k_{-1} k_{-2} (E)_0/(k_{-1} + k_2 + k_{-2}) \qquad (16\text{-}45d)$$

In this case, there are six rate constants but only four steady-state parameters. Therefore, it is impossible to obtain unique solutions for the rate constants in terms of the steady-state parameters.

Lower bounds to rate constants

Michaelis constants and maximal velocities can be used to obtain lower bounds to elementary rate constants in multiple-intermediate mechanisms. The general n-intermediate mechanism is

$$E + S \underset{k_{-1}}{\overset{k_1}{\rightleftharpoons}} X_1 \underset{k_{-2}}{\overset{k_2}{\rightleftharpoons}} \cdots \underset{k_{-i}}{\overset{k_i}{\rightleftharpoons}} X_i \underset{k_{-(i+1)}}{\overset{k_{i+1}}{\rightleftharpoons}} \cdots \underset{k_{-n}}{\overset{k_n}{\rightleftharpoons}} X_n \underset{k_{-(n+1)}}{\overset{k_{n+1}}{\rightleftharpoons}} E + P \qquad (16\text{-}46)$$

For this general case, the form of the rate equation is the same as Equation 16-40, but the expressions for the four steady-state parameters in terms of elementary rate constants become increasingly complex with progressively large values of n.

However, it always is possible to calculate lower bounds to all of the $2(n + 1)$ rate constants. The appropriate relationships are the following (Peller and Alberty, 1959):

$$k_1 \geqslant (V_S + V_P)/K_S(E)_0 \qquad (16\text{-}47a)$$

$$k_{-(n+1)} \geqslant (V_S + V_P)/K_P(E)_0 \qquad (16\text{-}47b)$$

$$k_{i+1} \geqslant V_S/(E)_0 \qquad \text{for } i \neq 0 \qquad (16\text{-}47c)$$

$$k_{-i} \geqslant V_P/(E)_0 \qquad \text{for } i \neq n + 1 \qquad (16\text{-}47d)$$

These inequalities are easily verified for the case $n = 2$ by using Equation 16-45. Also, it is simple to show that the equality sign holds when $n = 1$.

Turnover numbers

The parameters $V_S/(E)_0$ and $V_P/(E)_0$ are known as the turnover numbers for the forward and reverse directions, respectively. Table 16-1 lists some typical turnover numbers for enzymatic reactions. These numbers vary from 10^2 sec^{-1} to 10^6 sec^{-1}.

Table 16-1
Approximate turnover numbers for some enzymes

Enzyme	Turnover number (sec^{-1})
Carbonic anhydrase	10^6
Acetylcholinesterase	10^4
Urease	10^4
Fumarase	10^3
Transaminases	10^3
Chymotrypsin	10^2-10^3
Ribonuclease A	10^2-10^4
Carboxypeptidase	10^2

SOURCE: After G. G. Hammes, *Principles of Chemical Kinetics* (New York: Academic Press, 1978).

According to Equation 16-47c,d, these turnover numbers specify lower bounds for the unimolecular rate constants in the forward and reverse directions. This clearly is reasonable, because each elementary step must proceed at least as fast as the maximal rate of the overall reaction. Conversely, the reciprocal turnover numbers specify upper limits to the lifetimes of any intermediate species. For example, if $V_S/(E)_0 = 10^2$ sec^{-1}, then a given intermediate has a mean lifetime of no more than 10^{-2} sec before conversion to another species along the forward reaction path.

Another useful parameter is k_{cat}/K_S (see Box 16-1).

pH dependence of enzyme reactions

The rates of enzymatic reactions commonly show a substantial pH dependence. In many cases, maximal activity is achieved at a single pH value (or within a narrow range of values), and substantially less activity is manifested at other values of pH. These variations in activity are due in turn to a sensitivity to pH on the part of the steady-state parameters.

Figure 16-5 shows a bell-shaped variation in V_S and in K_S. This type of behavior implies that on the enzyme are at least two ionizable groups that have an effect on activity. For example, in the case of V_S, a single protonation leads to an increase in

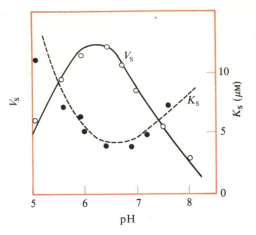

Figure 16-5

The pH dependence of steady-state parameters, illustrated for the fumarase reaction. [After C. Frieden and R. A. Alberty, *J. Biol. Chem.* 212:859 (1955).]

activity, but additional protonation results in a decrease in activity. The simplest alteration of the single-intermediate mechanism that can account for the observed behavior is

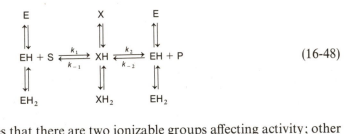

$$(16\text{-}48)$$

This mechanism assumes that there are two ionizable groups affecting activity; other ionizations that do not alter the activity are omitted from Equation 16-48. The mechanism assumes that catalysis proceeds by way of the central pathway involving the singly-protonated species. Clearly, at the extreme of either high or low pH, the activity drops as the amount of the singly-protonated forms decreases.

To derive the steady-state rate expression, we assume that the vertical ionization steps are equilibrated rapidly relative to the rate of the reaction along the horizontal path. This assumption is reasonable because ionization steps commonly are diffusion-controlled, having time constants significantly shorter than those associated with many of the elementary steps in an enzymatic reaction. Therefore, in deriving the steady-state rate expression, we can make use of the dissociation constants that characterize each ionization equilibrium. These constants are

$$K_{aE} = (EH)(H^+)/(EH_2) \qquad (16\text{-}49a)$$

$$K_{bE} = (E)(H^+)/(EH) \qquad (16\text{-}49b)$$

$$K_{aX} = (XH)(H^+)/(XH_2) \qquad (16\text{-}49c)$$

$$K_{bX} = (X)(H^+)/(XH) \qquad (16\text{-}49d)$$

The steady-state rate equation is derived by starting with the initial velocity given by

$$v_i = -d(S)/dt = d(P)/dt = k_2(XH) \tag{16-50}$$

where we assume that $(P) \cong 0$ at early times. The steady-state assumption is

$$d[(X) + (XH) + (XH_2)]/dt = k_1(EH)(S) - (k_{-1} + k_2)(XH) = 0 \tag{16-51}$$

On the left-hand side of Equation 16-51, note that the three intermediate species are treated as one entity; this is a direct consequence of assuming that the three forms equilibrate rapidly relative to the rate of the horizontal reaction pathway. However, on the right-hand side of Equation 16-51, only individual species enter into the rate expression, because there are no rate constants on the horizontal path of Equation 16-48 that can be applied to species other than the singly-protonated forms.

From Equation 16-51, we obtain

$$(XH) = [k_1/(k_{-1} + k_2)](EH)(S)_0 = (EH)(S)_0/K_S \tag{16-52}$$

where, as before, $K_S = (k_{-1} + k_2)/k_1$. The mass-conservation relationship for total enzyme is

$$(E)_0 = (E) + (EH) + (EH_2) + (X) + (XH) + (XH_2)$$

$$= (EH)[1 + (E)/(EH) + (EH_2)/(EH)] + (XH)[1 + (X)/(XH) + (XH_2)/(XH)] \tag{16-53}$$

Box 16-1   CATALYTIC EFFICIENCY

The turnover number sometimes is referred to as k_{cat}. Equation 16-37 can be rewritten in the form

$$v_i = V_S(S)_0/[(S)_0 + K_S]$$

Under conditions where $(S)_0 \ll K_S$, we have

$$v_i = V_S(S)_0/K_S \qquad \text{for } (S)_0 \ll K_S$$

Because $k_{cat} = V_S/(E_0)$, we can write this relation as

$$v_i = (k_{cat}/K_S)(E_0)(S)_0 \qquad \text{for } (S)_0 \ll K_S$$

The ratio k_{cat}/K_S is a useful parameter that has dimensions of a second-order rate constant ($\text{M}^{-1} \text{ sec}^{-1}$). It obviously is a measure of the catalytic efficiency—combining both the maximal velocity and the Michaelis constant into a single rate parameter.

and, from Equations 16-49 and 16-52,

$$(E)_0 = (EH)[1 + K_{bE}/(H^+) + (H^+)/K_{aE}] + (XH)[1 + K_{bX}/(H^+) + (H^+)/K_{aX}]$$
$$= (XH)\{[K_S/(S)_0][1 + K_{bE}/(H^+) + (H^+)/K_{aE}] + 1 + K_{bX}/(H^+) + (H^+)/K_{aX}\}$$

$$(16\text{-}54)$$

Substituting Equation 16-54 into Equation 16-50, we obtain

$$v_i = k_2(E)_0/\{[K_S/(S)_0][1 + K_{bE}/(H^+) + (H^+)/K_{aE}] + 1 + K_{bX}/(H^+) + (H^+)/K_{aX}\}$$

$$(16\text{-}55)$$

By definition, v_i goes to the maximal velocity V'_S when $(S)_0 \to \infty$. Using this definition, we obtain from Equation 16-55

$$V'_S = V_S/[1 + K_{bX}/(H^+) + (H^+)/K_{aX}] \qquad (16\text{-}56)$$

where $V_S = k_2(E)_0$, the maximal velocity obtained earlier for the simple single-intermediate case that lacked ionization equilibria. With this expression for V'_S, we can rewrite Equation 16-55 as

$$v_i = V'_S/[1 + K'_S/(S)_0] \qquad (16\text{-}57)$$

where

$$K'_S = K_S\left(\frac{1 + K_{bE}/(H^+) + (H^+)/K_{aE}}{1 + K_{bX}/(H^+) + (H^+)/K_{aX}}\right) \qquad (16\text{-}58)$$

Understanding the pH dependence

Equation 16-57 is identical in form to Equation 16-37. However, the parameters V'_S and K'_S are pH-dependent. It is easy to see that V'_S has a bell-shaped pH dependence. In fact, the pH dependence of V'_S mimics that of (XH), under conditions where the enzyme is saturated. This is easy to see from Equation 16-54, when $(S) \to \infty$:

$$(XH) = (E)_0/[1 + K_{bX}/(H^+) + (H^+)/K_{aX}] \qquad \text{for } (S) \to \infty \qquad (16\text{-}59)$$

The denominator in Equation 16-59 is identical to that in Equation 16-56.

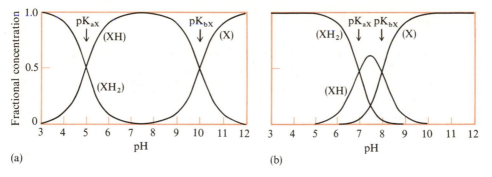

Figure 16-6

The pH dependence of the intermediate species in the single-intermediate Michaelis–Menten mechanism (Eqn. 16-48). **(a)** Curves for $pK_{aX} = 5$ and $pK_{bX} = 10$. **(b)** Curves for $pK_{aX} = 7$ and $pK_{bX} = 8$.

Equations analogous to Equation 16-59 may be derived for (X) and (XH_2). These and Equation 16-59 can be used to sketch the pH dependence of the three intermediate species. Figure 16-6 shows examples for the case $pK_{aX} = 5$ and $pK_{bX} = 10$, and for the case $pK_{aX} = 7$ and $pK_{bX} = 8$. It is clear that the height and breadth of the bell-shaped (XH) curve (or V_S' curve) are sensitive to the differences in the two pK values.

The expression for K_S' in Equation 16-58 contains the ratio of two bell-shaped functions. Therefore, K_S' may not be a simple inverted bell-shaped function of pH (like that sketched in Fig. 16-5). In fact, if $K_{aX} = K_{aE}$ and $K_{bX} = K_{bE}$, it is clear that K_S' will be pH-independent, even though V_S' is pH-dependent. Therefore, to observe a pH dependence in K_S', one or both pK values of the enzyme intermediate must be shifted to values different from those on the free enzyme. The shift, of course, is induced by binding of substrate.

We showed earlier that, for the single-intermediate mechanism with no ionization equilibria (Eqn. 16-33), K_S is a simple equilibrium dissociation constant for the reaction $E + S \rightleftarrows X$ when $k_2 \ll k_{-1}$. If we again assume $k_2 \ll k_{-1}$, then Equation 16-58 becomes

$$K_S' = \frac{k_{-1}}{k_1} \left(\frac{1 + K_{bE}/(H^+) + (H^+)/K_{aE}}{1 + K_{bX}/(H^+) + (H^+)/K_{aX}} \right) \qquad \text{for } k_2 \ll k_{-1}$$

$$= \frac{(EH)(S)}{(XH)} \left(\frac{1 + (E)/(EH) + (EH_2)/(EH)}{1 + (X)/(XH) + (XH_2)/(XH)} \right)$$

$$= [(E) + (EH) + (EH_2)](S)/[(X) + (XH) + (XH_2)] \qquad (16\text{-}60)$$

Therefore, according to Equation 16-60, K_S' (like K_S) can be regarded as an equilibrium dissociation constant when $k_2 \ll k_{-1}$. In the case of K_S', it is a macroscopic dissociation constant that takes into account all ionized forms of the enzyme.

Determination of ionization constants

For the simple single-intermediate case, determination of the four ionization constants is relatively straightforward. The constants K_{aX} and K_{bX} can be determined from V'_S. If a broad maximum in V'_S is observed (pK_{aX} and pK_{bX} are well separated), then the two pK values can be obtained with fair accuracy from the midpoints of the two descending limbs of the curve (pK_{aX} from the midpoint on the acid side, and pK_{bX} from that on the basic side). If the peak in V'_S is sharp, the two pK values can be obtained approximately from the midpoints on each side of the curve; more accurate values of pK_{aX} and pK_{bX} can then be calculated by adjusting each until the best agreement is obtained between the observed and calculated behaviors of V'_S.

The parameter K'_S depends on all four ionization constants. However, it is clear from Equations 16-56 and 16-58 that the ratio K'_S/V'_S depends only on the pK values on the free enzyme:

$$K'_S/V'_S = [K_S/k_2(E)_0][1 + K_{bE}/(H^+) + (H^+)/K_{aE}] \qquad (16\text{-}61)$$

Therefore, K'_S/V'_S is a simple inverted bell-shaped function. The two pK values (pK_{aE} and pK_{bE}) can readily be obtained by procedures analogous to those used to obtain pK_{aX} and pK_{bX}.

Of course, the number of pK values that must be taken into account multiplies as the number of intermediates increases. For the two-intermediate mechanism (Eqn. 16-42), assignment of three protonation states to each enzyme species leads to a total of six pK values. Here V'_S will be a function of the four pK values associated with the intermediate states, whereas K'_S will depend on all six pK values. In this case, attempts to fit the observed pH dependence of K'_S and V'_S to Equations 16-56 and 16-58 for the single-intermediate model may fail. Such failure is evidence that additional states (more intermediates) must be taken into account.

In general, one attempts to find the simplest mechanism (with the fewest ionizations) that will account for all the data. This gives a feeling for the minimal number of ionizable groups that influence the catalytic activity, and it also gives a good estimate of their pK values. From the pK values, some idea of the chemical identities of the ionizable groups also can be obtained. For example, side-chain carboxyl groups tend to ionize in the range of pH 3.5 to 5.0; histidine around pH 5.5 to 7.0; cysteine in the range of pH 7.5 to 9.0; etc. The rather broad range of values observed for each group is due to the variety of chemical microenvironments that can exist on a protein in the area of any particular residue.

Finally, we should mention that, in cases where the substrate also exists in more than one ionized form, these ionizations may need to be considered when attempting to correlate quantitatively the dependence of steady-state parameters on pH. In these instances, the derivation of the appropriate steady-state equations proceeds analogously to that described above where only groups on the enzyme were considered to ionize. A simplifying feature is that pK values for the free substrate may be obtained directly from titrations, rather than by attempting to fit kinetic data. Once these values have been measured, they can be used in the appropriate rate expression.

Limitations of steady-state kinetics

Enzyme mechanisms cannot be adequately explored by the steady-state approach alone. This is chiefly because measurement of the steady-state reaction from substrate to product gives no detailed information on the existence of multiple intermediate states, or on their lifetimes. Furthermore, steady-state conditions typically are achieved by having $(E)_0 \ll (S)_0$; in this situation, the concentrations of the various enzyme species are so small that they escape detection. Although the situation can be remedied by employing substrate levels of enzyme, and by using rapid-mixing devices that permit mixing of enzyme and substrate in $\sim 10^{-3}$ sec, many elementary steps in catalysis are faster than 10^3 sec^{-1}; these quick steps therefore are not observed. Moreover, as we have noted, when using substrate levels of enzyme in non-steady-state conditions, the rate equations can become so complex that special and tedious numerical procedures must be followed to obtain solutions.

16-7 RELAXATION SPECTROMETRY

Relaxation spectrometry is an alternative to steady-state kinetics that has met with considerable success in the study of enzyme mechanisms; this approach was developed by Manfred Eigen (see Eigen and de Maeyer, 1963). The principle of relaxation spectrometry is quite simple. A reaction first is allowed to come to equilibrium. A rapid shift then is made in one of the thermodynamic variables (such as temperature, pressure, or electric field intensity) that affect the equilibrium position of the reaction. This causes the reaction to adjust to its new equilibrium position—that is, the reaction relaxes to its new final equilibrium state. The time course of this relaxation depends upon the rates of the elementary steps in the reaction sequence.

This approach has two principal advantages. First, thermodynamic variables such as temperature can be changed very rapidly (a temperature jump can be made in 10^{-6} sec or less). Any reaction that has rate parameters longer than the time required for the thermodynamic perturbation may be studied; therefore, it is clear that this approach greatly extends the accessible time range well beyond that possible with rapid-mixing procedures.

Second, the thermodynamic perturbation is choosen to be small enough so that the reactions are not perturbed far from their equilibrium positions. This has the considerable advantage that the rate equations describing the relaxation to equilibrium can be linearized—that is, regardless of the complexity of the mechanism, a

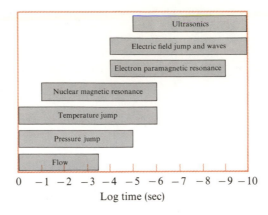

Figure 16-7

Summary of time ranges accessible with various fast-reaction methods. [After G. G. Hammes and P. R. Schimmel, in *The Enzymes*, vol. 2, ed. P. D. Boyer (New York: Academic Press, 1970), p. 67.]

set of first-order rate equations is obtained. This enormous simplification makes it possible to obtain relatively simple solutions for any situation that is encountered.

There are two kinds of relaxation methods generally used: "jump" techniques, whereby a thermodynamic variable is changed abruptly (stepwise) to a new value; and stationary methods, in which a continuous variation (that is, a wave) in a thermodynamic parameter is used. The temperature-jump and pressure-jump methods are examples of the jump approach, and ultrasonics is an example of the stationary approach. Figure 16-7 shows the time ranges that may be explored with the various relaxation methods, and with certain other techniques. (In general, both an upper and a lower limit on time exist. For example, in the case of the temperature-jump method, the lower limit is established by the rate at which a jump can be made, and the upper limit is set by the rate of cooling and of convective disturbances that set in after a pulse of heat has been supplied to the solution.) It is clear that a wide range of times is accessible with these methods.

In the following discussion, we assume a stepwise perturbation such as the temperature jump. This is the relaxation method most commonly used to study biochemical systems. However, the principles discussed are valid in general and are not limited to this single situation.

Thermodynamic principle

In general, the equilibrium concentrations of species that participate in a reaction are dependent on a thermodynamic variable such as temperature. Consider, for example, the reaction

$$E + S \rightleftarrows X \tag{16-62}$$

where the equilibrium constant K is

$$K = \overline{(E)(S)}/\overline{(X)}$$
$$= e^{-\Delta H^0/RT}e^{\Delta S^0/R} \tag{16-63}$$

and ΔH^0 and ΔS^0 are the standard enthalpy and entropy changes for the reaction. We are interested in how the equilibrium concentrations of the reaction species vary with temperature. By differentiation of the expression for $\ln K$, we obtain

$$d(\ln K) = \overline{(E)}^{-1}\,d\overline{(E)} + \overline{(S)}^{-1}\,d\overline{(S)} - \overline{(X)}^{-1}\,d\overline{(X)} \tag{16-64}$$

Because by conservation of mass $d(X) = -d\overline{(E)} = -d\overline{(S)}$, Equation 16-64 becomes

$$d(\ln K) = [\overline{(E)}^{-1} + \overline{(S)}^{-1} + \overline{(X)}^{-1}]\,d\overline{(E)} \tag{16-65}$$

Also, from the van't Hoff relationship, we know that at constant pressure

$$d(\ln K) = (\Delta H^0/RT^2)\,dT \tag{16-66}$$

Substituting Equation 16-66 into Equation 16-65, we obtain

$$d\overline{(E)} = \{1/[\overline{(E)}^{-1} + \overline{(S)}^{-1} + (X)^{-1}]\}(\Delta H^0/RT^2)\,dT$$
$$= \Gamma(\Delta H^0/RT^2)\,dT \tag{16-67}$$

where

$$\Gamma = 1/[\overline{(E)}^{-1} + \overline{(S)}^{-1} + \overline{(X)}^{-1}] \tag{16-68}$$

For a finite change ΔT in T, we have

$$\boxed{\Delta\overline{(E)} = \Gamma(\Delta H^0/RT^2)\Delta T} \tag{16-69}$$

The equilibrium shifts in the other species are obtained from $\Delta\overline{(E)} = \Delta\overline{(S)} = -\Delta\overline{(X)}$.

Equation 16-69 shows that, for a given ΔT, the magnitude of $\Delta\overline{(E)}$ depends critically on Γ and on ΔH^0. Clearly, $\Delta\overline{(E)} \to 0$ when $\Delta H^0 \to 0$ or $\Gamma \to 0$. However, for many biochemical processes, $\Delta H^0 \neq 0$. The parameter $\Gamma \to 0$ when $\overline{(E)} \to 0$, or when $\overline{(S)} \to 0$, or when $\overline{(X)} \to 0$. The condition $\overline{(E)} \to 0$ is obtained when all of the enzyme is bound; similarly, $\overline{(S)} \to 0$ when all of S is bound. In these cases, the equilibrium has been pushed far to the right by having an excess of S or E. Conversely, when the reaction lies far to the left, then $\overline{(X)} \to 0$. Clearly, the optimal value of Γ is achieved when the reaction is at neither extreme. It is not hard to show that, when $(E)_0 = (S)_0$, the maximal value of Γ occurs near the half-saturation point of the reaction.

The principle of the relaxation method is to make a rapid change (ΔT) in T; assuming $\Delta H^{\circ} \neq 0$ and $\Gamma \neq 0$, this rapid change forces $\overline{(E)}$ to adjust to a new value. The time it takes to adjust is determined by rate constants and equilibrium concentrations; it characteristically is known as the "relaxation time."

Figure 16-8 shows the situation schematically. Enzyme is initially at its equilibrium concentration $\overline{(E_1)}$ at $T = T_1$. At time $t = t_0$, a step pulse in temperature is given to the solution, so that T rises to T_2. The shift in temperature means that (E) must change from $\overline{(E_1)}$ to $\overline{(E_2)}$. Assuming that the rate of equilibration of Equation 16-62 is slow compared to the rate of the temperature rise, the rise in (E) from $\overline{(E_1)}$ to $\overline{(E_2)}$ is a smooth exponential that lags behind the temperature rise.

By measuring the time constant associated with the exponential of Figure 16-8 at different values of $\overline{(E)}$ and $\overline{(S)}$, we can obtain accurate values of the rate constants for the reaction, as we shall see.

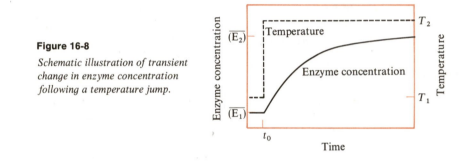

Figure 16-8

Schematic illustration of transient change in enzyme concentration following a temperature jump.

Relaxation kinetics

Near equilibrium, the solution to the rate equation that characterizes the one-step bimolecular reaction is extremely simple. Let k_1 be the bimolecular rate constant for the forward reaction, and k_{-1} be that for the reverse reaction. The rate equation is

$$-d(E)/dt = k_1(E)(S) - k_{-1}(X) \qquad (16\text{-}70)$$

The concentrations may be written as

$$(E) = \overline{(E)} + \Delta(E) \qquad (16\text{-}71a)$$

$$(S) = \overline{(S)} + \Delta(S) \qquad (16\text{-}71b)$$

$$(X) = \overline{(X)} + \Delta(X) \qquad (16\text{-}71c)$$

where $\overline{(E)}$, $\overline{(S)}$, and $\overline{(X)}$ indicate the final equilibrium concentrations—$\overline{(E)}$ is denoted as $\overline{(E_2)}$ in Figure 16-8—and $\Delta(E)$, $\Delta(S)$, and $\Delta(X)$ represent deviations from equilibrium. With these expressions, Equation 16-70 becomes

$$-d\,\Delta(E)/dt = k_1[\overline{(E)} + \Delta(E)][\overline{(S)} + \Delta(S)] - k_{-1}[\overline{(X)} + \Delta(X)] \qquad (16\text{-}72)$$

Mass conservation requires that $\Delta(E) = \Delta(S) = -\Delta(X)$; also, by definition of the equilibrium state, $k_1\overline{(E)(S)} = k_{-1}\overline{(X)}$. Substituting these relationships into Equation 16-72, and ignoring the $\Delta(E)^2$ term relative to the terms linear in $\Delta(E)$, we obtain

$$-d\,\Delta(E)/dt = \Delta(E)/\tau \qquad (16\text{-}73a)$$

$$\Delta(E) = \Delta(E)^0 e^{-t/\tau} \qquad (16\text{-}73b)$$

where

$$1/\tau = k_1[\overline{(E)} + \overline{(S)}] + k_{-1} \qquad (16\text{-}74)$$

and $\Delta(E)^0$ is the deviation in (E) at $t = 0$ from its final equilibrium value; in Figure 16-8, $\Delta(E)^0 = \overline{(E_2)} - \overline{(E_1)}$. The parameter τ is the relaxation time; it is a function of rate constants and equilibrium concentrations. Common order-of-magnitude values for the parameters in Equation 16-74 in an experiment are $k_1 = 10^8$ M^{-1} sec^{-1}; $k_{-1} = 10^4$ sec^{-1}; and $\overline{(E)} = \overline{(S)} = 10^{-4}$ M. With these values for the parameters, $\tau = 33$ μsec.

Equation 16-73 is valid only for small perturbations from equilibrium; the deviation from equilibrium must be small enough to assure that any quadratic terms in $\Delta(E)$ are minor compared to their linear terms.

The rate constants k_1 and k_{-1} may be obtained from a plot of $1/\tau$ versus $[\overline{(E)} + \overline{(S)}]$ (Fig. 16-9). The slope gives k_1, and the intercept yields k_{-1}.

In order to construct Figure 16-9, it is necessary to calculate $\overline{(E)}$ and $\overline{(S)}$. This

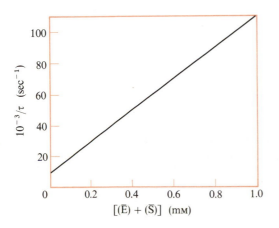

Figure 16-9

Plot of $1/\tau$ *versus* $[\overline{(E)} + \overline{(S)}]$, according to Equation 16-74, with $k_1 = 10^8$ M^{-1} sec^{-1} and $k_{-1} = 10^4$ sec^{-1}. [After G. G. Hammes and P. R. Schimmel, in *The Enzymes*, vol. 2, ed. P. D. Boyer (New York: Academic Press, 1970), p. 67.]

in turn requires a knowledge of the equilibrium constant $K = k_{-1}/k_1$. If K is not known, a trial value may be assumed for the purpose of constructing a plot. The plot yields k_1 and k_{-1}, from which a new value of K is calculated. This is used to construct another plot, and so on, until the trial value used is identical to the one obtained from the graph.

Linearized rate equations for a two-step enzymatic mechanism

In many cases, binding of a small molecule to a protein is followed by a conformational change in the initial complex. This gives a two-step reaction:

$$E + S \underset{k_{-1}}{\overset{k_1}{\rightleftharpoons}} X_1 \underset{k_{-2}}{\overset{k_2}{\rightleftharpoons}} X_2 \qquad (16\text{-}75)$$

This reaction mechanism has two independent steps, and therefore it is described completely by two rate equations. We choose (E) and (X_2) as our variables. The rate equations are

$$-d(E)/dt = k_1(E)(S) - k_{-1}(X_1) \qquad (16\text{-}76a)$$

$$-d(X_2)/dt = -k_2(X_1) + k_{-2}(X_2) \qquad (16\text{-}76b)$$

Proceeding by analogy with our treatment of the one-step mechanism with $(E) = \overline{(E)} + \Delta(E)$ and so on, and making use of the mass-conservation relationship $\Delta(E) = \Delta(S) = -\Delta(X_1) - \Delta(X_2)$, Equation 16-76 becomes

$$-d\,\Delta(E)/dt = \{k_1[\overline{(E)} + \overline{(S)}] + k_{-1}\}\,\Delta(E) + k_{-1}\,\Delta(X_2) \qquad (16\text{-}77a)$$

$$-d\,\Delta(X_2)/dt = k_2\,\Delta(E) + (k_{-2} + k_2)\,\Delta(X_2) \qquad (16\text{-}77b)$$

or

$$-d\,\Delta(E)/dt = a_{11}\,\Delta(E) + a_{12}\,\Delta(X_2) \qquad (16\text{-}78a)$$

$$-d\,\Delta(X_2)/dt = a_{21}\,\Delta(E) + a_{22}\,\Delta(X_2) \qquad (16\text{-}78b)$$

where the a_{ij} terms in Equation 16-78 are defined by comparison with Equation 16-77. Thus, near equilibrium, the rate equations that describe the two-step reaction (Eqn. 16-75) are simple linear first-order differential equations.

The solution to a pair of linear first-order differential equations was taken up in the treatment of the sequential unimolecular reactions of Equation 16-6. Equation 16-78 is exactly analogous to Equation 16-8. Therefore, the solution is given by a sum of two exponentials for each concentration variable $\Delta(E)$ and $\Delta(X_2)$:

$$\Delta(E) = A_{11}e^{-t/\tau_1} + A_{12}e^{-t/\tau_2} \qquad (16\text{-}79a)$$

$$\Delta(X)_2 = A_{21}e^{-t/\tau_1} + A_{22}e^{-t/\tau_2} \qquad (16\text{-}79b)$$

There are two relaxation times: τ_1 and τ_2. The parameters A_{ij} depend on the two ΔH^0 and Γ values that characterize the mechanism, as well as on ΔT (in a temperature-jump experiment, for example). The reciprocal relaxation times are given by the solution to the determinant (compare with Eqn. 16-12)

$$\begin{vmatrix} a_{11} - 1/\tau & a_{12} \\ a_{21} & a_{22} - 1/\tau \end{vmatrix} = 0 \qquad (16\text{-}80)$$

or

$$(1/\tau)^2 - (a_{11} + a_{22})(1/\tau) + (a_{11}a_{22} - a_{12}a_{21}) = 0 \qquad (16\text{-}81)$$

The solutions to this standard quadratic equation are

$$1/\tau_{1,2} = [(a_{11} + a_{22})/2]\{1 \pm [1 - 4(a_{11}a_{22} - a_{12}a_{21})/(a_{11} + a_{22})^2]^{1/2}\}$$

$$(16\text{-}82)$$

where τ_1 is assigned to the positive sign and τ_2 to the negative sign.

The two relaxation times are functions of the equilibrium concentrations $\overline{(E)}$ and $\overline{(S)}$ and of the four rate constants of the two-step mechanism. If each time constant is studied as a function of concentration, all four rate parameters can be obtained.

Simplified relaxation-time expressions for the two-step mechanism

In many cases, the bimolecular step in Equation 16-75 is considerably more rapid than the unimolecular conformational change; that is, $k_1[\overline{(E)} + \overline{(S)}] + k_{-1} \gg k_2 + k_{-2}$. In this case, $a_{11} \gg a_{22}$, and we can simplify the square-root term in Equation 16-82:

$$[1 - 4(a_{11}a_{22} - a_{12}a_{21})/(a_{11} + a_{22})^2]^{1/2} \cong 1 - 2(a_{11}a_{22} - a_{12}a_{21})/a_{11}^2 \qquad (16\text{-}83)$$

[Recall that $(1 - x)^{1/2} \cong 1 - x/2$, when $x \ll 1$.] This simplification, together with the expression for the a_{ij} parameters in terms of rate constants and equilibrium concentrations (see Eqns. 16-77 and 16-78), gives

$$1/\tau_1 = k_1[\overline{(E)} + \overline{(S)}] + k_{-1} \qquad (16\text{-}84a)$$

$$1/\tau_2 = k_2/(1 + k_{-1}/\{k_1[\overline{(E)} + \overline{(S)}]\}) + k_{-2} \qquad (16\text{-}84b)$$

These limiting forms of $1/\tau_1$ and $1/\tau_2$ are very simple. Note that the expression for $1/\tau_1$ is identical to that obtained for the one-step mechanism (Eqn. 16-74). With the second step in the two-step mechanism very slow compared to the first one, the bimolecular process essentially equilibrates independently of the conformational change; that is, the second step is "frozen" during the equilibration time of the first step. For this reason, the relaxation time of the bimolecular process is identical to that for a simple one-step mechanism. In the case of $1/\tau_2$, the relaxation-time expression contains all four rate constants as well as the equilibrium concentrations of E and S. This is because the slow relaxation of the second step is linked to the rapid equilibration of the first one.

The four rate constants are obtained from plots of $1/\tau_1$ and of $1/\tau_2$ versus $[\overline{(E)} + \overline{(S)}]$. The plot of $1/\tau_1$ gives k_1 and k_{-1} (as in Fig. 16-9). Figure 16-10 is a plot of $1/\tau_2$ versus $[\overline{(E)} + \overline{(S)}]$. At low $[\overline{(E)} + (S)]$, the plot is linear, with slope equal to

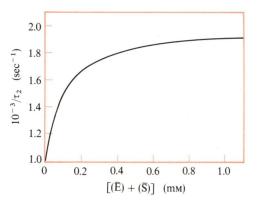

Figure 16-10

Plot of $1/\tau_2$ versus $[\overline{(E)} + \overline{(S)}]$, according to Equation 16-84b, with $k_1 = 10^8 \text{ M}^{-1}\text{ sec}^{-1}$; $k_{-1} = 10^4 \text{ sec}^{-1}$; and $k_2 = k_{-2} = 10^3 \text{ sec}^{-1}$. [After G. G. Hammes and P. R. Schimmel, in *The Enzymes*, vol. 2, ed. P. D. Boyer (New York: Academic Press, 1970), p. 67.]

$k_1 k_2/k_{-1}$. At high $[\overline{(E)} + (S)]$, the value of $1/\tau_2$ asymptotically approaches $k_2 + k_{-2}$, whereas an ordinate intercept of k_{-2} is obtained at $[\overline{(E)} + \overline{(S)}] = 0$. Therefore, we can readily obtain k_2 and k_{-2} from the concentration dependence of $1/\tau_2$.

Another derivation of relaxation-time expressions for the two-step mechanism

An alternative procedure can be used to derive the two relaxation times (Eqn. 16-84) for the condition that the first step equilibrates much more rapidly than the second step. If we regard the second step as frozen during the relaxation of the first, then the expression for τ_1 as the relaxation time of the one-step mechanism follows directly.

In the case of the second step, the rate equation is

$$-d\,\Delta(X_2)/dt = -k_2\,\Delta(X_1) + k_{-2}\,\Delta(X_2) \qquad (16\text{-}85)$$

To solve this equation, we must express $\Delta(X_1)$ in terms of $\Delta(X_2)$. We can accomplish this by using the mass-conservation relations and the fact that the first step is equilibrated at all times during the relaxation of the second one. From mass conservation of total enzyme, we have

$$\Delta(E) + \Delta(X_1) + \Delta(X_2) = 0 \qquad (16\text{-}86a)$$

or

$$\Delta(X_1) = -\Delta(X_2)/[1 + \Delta(E)/\Delta(X_1)] \qquad (16\text{-}86b)$$

The stipulation that the bimolecular step is equilibrated during the relaxation means that the equilibrium-constant relation can be used to solve for $\Delta(E)/\Delta(X_1)$. Therefore, $k_{-1}/k_1 = (E)(S)/(X_1)$, and

$$(k_{-1}/k_1)\,\Delta(X_1) = \overline{(E)}\,\Delta(S) + \overline{(S)}\,\Delta(E) = [\overline{(E)} + \overline{(S)}]\,\Delta(E) \qquad (16\text{-}87)$$

where we have made use of the fact (easily demonstrated from conservation relationships) that $\Delta(E) = \Delta(S)$. Therefore,

$$\Delta(E)/\Delta(X_1) = k_{-1}/\{k_1[\overline{(E)} + \overline{(S)}]\} \qquad (16\text{-}88)$$

Substituting this expression into Equation 16-86, we obtain

$$\Delta(X_1) = -\Delta(X_2)/(1 + k_{-1}/\{k_1[\overline{(E)} + \overline{(S)}]\}) \qquad (16\text{-}89)$$

Using this relationship in Equation 16-85, we get

$$-d\,\Delta(X_2)/dt = \Delta(X_2)/\tau_2 \qquad (16\text{-}90)$$

where $1/\tau_2$ is given by Equation 16-84b.

Spectrum of relaxation times for an *n*-intermediate mechanism

The simple procedures used to treat the two-step mechanism often can be used to treat a variety of other mechanisms in which one step (or more) is very slow compared to other steps in the mechanism. As a general example, consider the *n*-intermediate mechanism

$$E + S \underset{k_{-1}}{\overset{k_1}{\rightleftarrows}} X_1 \underset{k_2}{\overset{k_2}{\rightleftarrows}} \cdots \underset{k_{-n}}{\overset{k_n}{\rightleftarrows}} X_n \underset{k_{-(n+1)}}{\overset{k_{n+1}}{\rightleftarrows}} E + P \qquad (16\text{-}91)$$

This mechanism has $n + 1$ independent steps; therefore, it is characterized by $n + 1$ rate equations and $n + 1$ relaxation times. The linearized rate equations are

$$-d\,\Delta(X_1)/dt = a_{11}\,\Delta(X_1) + a_{12}\,\Delta(X_2) + \cdots + a_{1(n+1)}\,\Delta(E) \qquad (16\text{-}92a)$$

$$-d\,\Delta(X_2)/dt = a_{21}\,\Delta(X_1) + a_{22}\,\Delta(X_2) + \cdots + a_{2(n+1)}\,\Delta(E) \qquad (16\text{-}92b)$$

$$-d\,\Delta(X_n)/dt = a_{n1}\,\Delta(X_1) + a_{n2}\,\Delta(X_2) + \cdots + a_{n(n+1)}\,\Delta(E) \qquad (16\text{-}92c)$$

$$-d\,\Delta(E)/dt = a_{(n+1)1}\,\Delta(X_1) + a_{(n+1)2}\,\Delta(X_2) + \cdots + a_{(n+1)(n+1)}\,\Delta(E) \qquad (16\text{-}92d)$$

where the a_{ij} terms are functions of the $2(n + 1)$ rate constants and of $\overline{(E)}$, $\overline{(S)}$, and $\overline{(P)}$. The solution to these equations is

$$\Delta(X_i) = \sum_{j=1}^{n+1} A_{ij}e^{-t/\tau_j} \qquad \text{for } i = 1, 2, \ldots, n \qquad (16\text{-}93a)$$

$$\Delta(E) = \sum_{j=1}^{n+1} A_{(n+1)j}e^{-t/\tau_j} \qquad (16\text{-}93b)$$

where the A_{ij} terms are constants, and the τ_j parameters are relaxation times. The τ_j values are given by the solution to the secular equation

$$\begin{vmatrix} a_{11} - 1/\tau & a_{12} & \cdots & a_{1(n+1)} \\ a_{21} & a_{22} - 1/\tau & \cdots & a_{2(n+1)} \\ \vdots & \vdots & & \vdots \\ a_{(n+1)1} & a_{(n+1)2} & \cdots & a_{(n+1)(n+1)} - 1/\tau \end{vmatrix} = 0 \qquad (16\text{-}94)$$

In principle, careful measurement of each relaxation time as a function of concentration can yield the various rate constants. In many cases where more than two relaxation times are present, they are separated well enough that simplifying procedures (such as those we used in the two-step mechanism) can be used to obtain relatively simple approximate expressions for the relaxation times.

Matrix methods provide a way of neatly treating relaxation spectra (Box 16-2).

Some conclusions from fast-reaction studies

A large number of systems have been studied by fast-reaction methods. In order to appreciate the significance of some of the rate parameters that have been obtained, it is helpful to consider the theoretical limits placed on the bimolecular rate constant

k_1 (Box 16-3). For this purpose, consider a spherical enzyme and a spherical substrate molecule whose distance of closest approach is r_0 (that is, r_0 is the sum of their radii), and whose diffusion constants are D_E and D_S, respectively. If every E–S collision results in complex formation, then (see Box 16-3) we have

$$k_1 = 4\pi r_0 (D_E + D_S) 10^{-3} N_0 \qquad\qquad (16\text{-}95)$$

where N_0 is Avogadro's number, r_0 is expressed in cm, $(D_E + D_S)$ is expressed in $cm^2\ sec^{-1}$, and k_1 has the typical units of $M^{-1}\ sec^{-1}$. In practice, Equation 16-95 should be modified by a fractional factor to account for the fact that the active site of the enzyme occupies only a small part of the total enzyme surface, and many collisions do not occur at this site. For a small-molecule substrate, $D_S \gg D_E$, and $D_S \cong 10^{-5}\ cm^2\ sec^{-1}$. With a value of r of 2×10^{-7} to 4×10^{-7} cm, and assuming that a small percentage of the collisions are at the active site, we have $k_1 \cong 10^9\ M^{-1}\ sec^{-1}$. This gives a rough idea of the diffusion-controlled limit for k_1; of course, most molecules are not spherical, and an interaction energy between E and S may alter the collision frequency. Nevertheless, Equation 16-95 should give an order-of-magnitude estimate for k_1 (see Box 16-3 and Hammes, 1978, for more details).

Table 16-2 shows rate parameters k_1 and k_{-1} characterizing the initial complex formation between enzyme and substrate (or other small molecule) for a few systems. These data show that, in many cases, the second-order rate constant k_1 for association of the physiological substrate with a given enzyme is in the range of 10^7 to $10^8\ M^{-1}\ sec^{-1}$; this rate is somewhat less than the diffusion-controlled rate. It is also apparent from these data that the "fit" between enzyme and substrate may critically affect the rate constant. For example, for the association of the natural substrate aspartate with aspartate aminotransferase, $k_1 = 10^7$ to $10^8\ M^{-1}\ sec^{-1}$. However, for α-methylaspartate, $k_1 = 10^4\ M^{-1}\ sec^{-1}$. The structures of these two substrates are

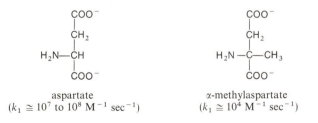

aspartate
($k_1 \cong 10^7$ to $10^8\ M^{-1}\ sec^{-1}$)

α-methylaspartate
($k_1 \cong 10^4\ M^{-1}\ sec^{-1}$)

Thus, mere replacement of the H (van der Waals radius of ~ 1.2 Å) by CH_3 (van der Waals radius of ~ 1.9 Å) affects k_1 by three to four orders of magnitude.

The dissociation rate constants given in Table 16-2 show a range of values. These constants reflect the strengths of the interactions between enzyme and the bound small molecule.

Box 16-2 A MATRIX APPROACH TO LINEARIZED RATE EQUATIONS

A set of linear differential rate equations is treated conveniently by matrix methods. The advantage of this approach is that it allows identification of the *normal* concentration variables associated with any mechanism characterized by linearized rate equations. These variables offer physical insight into the mechanism, and they are also useful for calculating other properties of the system, such as relaxation amplitudes (see Eigen and de Maeyer, 1963).

Consider a general reaction mechanism consisting of n independent chemical species; we shall deal with their concentrations, represented by C_i indexed 1 to n. The linearized rate equations are

$$-d\Delta C_1/dt = a_{11}\Delta C_1 + a_{12}\Delta C_2 + \cdots + a_{1n}\Delta C_n$$
$$-d\Delta C_2/dt = a_{21}\Delta C_1 + a_{22}\Delta C_2 + \cdots + a_{2n}\Delta C_n$$
$$\vdots \qquad\qquad \vdots \qquad\qquad \vdots \qquad\qquad \vdots$$

where the a_{ij} terms are functions of rate constants and equilibrium concentrations (see text for examples). The solution to these equations (cf. Eqn. 16-93) is

$$\Delta C_i = \sum_{j=1}^{n} A_{ij}e^{-t/\tau_j} \qquad \text{for } i = 1, 2, \ldots, n$$

where the A_{ij} terms are constants, and the τ_j parameters are relaxation times obtained by solving the usual secular determinant (cf. Eqn. 16-94).

We seek a set of concentration variables y_i such that their relaxation to equilibrium is given by

$$\Delta y_i = \Delta y_i^0 e^{-t/\tau_i} \qquad \text{for } i = 1, 2, \ldots, n$$

where Δy_i is the deviation of y_i from its final equilibrium value, and Δy_i^0 is the initial deviation at $t = 0$.

The y_i factors are *normal* concentration variables, and each decays to equilibrium with a *single* characteristic time constant τ_i. Thus, each y_i behaves as if it participates in a single, independent unimolecular reaction.

The relationship between ΔC_i and Δy_i can be expressed as

$$\Delta C_i = \sum_{j=1}^{n} T_{ij}\Delta y_j \qquad \text{for } i = 1, 2, \ldots, n$$

where the T_{ij} factors are constants. In matrix notation (see Appendix A), this relationship can be written as

$$\underset{\sim}{\Delta C} = \underset{\sim}{T}\,\underset{\sim}{\Delta y}$$

where $\underset{\sim}{\Delta C}$ and $\underset{\sim}{\Delta y}$ are column vectors of the ΔC_i and Δy_i terms, and $\underset{\sim}{T}$ is a square $n \times n$ matrix of the time-independent T_{ij} terms. Similarly, the linearized rate equations given earlier can be written as

$$-d\underset{\sim}{\Delta C}/dt = \underset{\sim}{a}\,\underset{\sim}{\Delta C}$$

where $\underset{\sim}{a}$ is a square $n \times n$ matrix of the a_{ij} terms. Using the relationship $\Delta C = T\Delta y$, we can write the rate equations as

$$-d(\underset{\sim}{T}\,\underset{\sim}{\Delta y})/dt = \underset{\sim}{a}\underset{\sim}{T}\,\underset{\sim}{\Delta y}$$

or

$$-d\underset{\sim}{\Delta y}/dt = \underset{\sim}{T}^{-1}\underset{\sim}{a}\underset{\sim}{T}\,\underset{\sim}{\Delta y}$$

where $\underset{\sim}{T}^{-1}$ is the inverse of $\underset{\sim}{T}$ (see Appendix A). If $\underset{\sim}{T}$ is constructed from the eigenvectors of $\underset{\sim}{a}$ (see Appendix A), then

$$\underset{\sim}{T}^{-1}\underset{\sim}{a}\underset{\sim}{T} = \underset{\sim}{\Lambda}$$

where

$$\underset{\sim}{\Lambda} = \begin{bmatrix} \lambda_1 & 0 & \cdots & 0 \\ 0 & \lambda_2 & \cdots & 0 \\ \vdots & \vdots & & \vdots \\ 0 & 0 & \cdots & \lambda_n \end{bmatrix}$$

and

$$\lambda_i = 1/\tau_i$$

Thus the matrix $\underset{\sim}{T}$, known as the transformation matrix, diagonalizes the rate equations, so that they become

$$-d\Delta y_i/dt = -\Delta y_i/\tau_i$$

with the simple solution $\Delta y_i = \Delta y_i^0 e^{-t/\tau_i}$. Hence, the exponential decay e^{-t/τ_i} gives the transient behavior of the ith normal concentration variable.

Box 16-3   BIMOLECULAR RATE CONSTANT
FOR A DIFFUSION-CONTROLLED REACTION

The expression for the bimolecular rate constant for a diffusion-controlled reaction is not difficult to calculate. Consider the reaction $A + B \rightleftarrows AB$. Assume for simplicity that A and B are neutral spherical molecules with no potential energy of interaction. (Equations can also be developed for cases of an interaction energy—see Hammes, 1978.) We wish to calculate the rate at which the molecules collide. Imagine that B is stationary, and that A molecules diffuse into and collide with the B molecules. According to Fick's first law the flux $\mathbf{J}_A$ (number of A molecules flowing across 1 cm² per second) is

$$\mathbf{J}_A = -D_A \nabla C_A$$

where D_A is the diffusion coefficient (cm² sec⁻¹) and C_A is the concentration in molecules cm⁻³. (See Chapter 10 for a discussion of Fick's laws and diffusion coefficients.) There is no general solution to Fick's first law, but we can obtain a solution for the steady-state condition whereby $-\partial C_A/\partial t = 0$. First note that, with spherical symmetry, Fick's first law becomes

$$J_A = -D_A \, \partial C_A/\partial r$$

where r is the distance of separation of the molecular centers. The total *inward* flux across a spherical surface of radius r (with area $4\pi r^2$) is

$$\mathscr{J}_A = -4\pi r^2 J_A = 4\pi r^2 D_A \, \partial C_A/\partial r$$

Owing to the steady-state flow, the parameter $\mathscr{J}_A$ is a constant, irrespective of r. Therefore, $\mathscr{J}_A$ gives the number of A molecules that flow into the surface of a B molecule per unit time.

In order to obtain a more convenient form for $\mathscr{J}_A$, we assume that the distance of closest approach of the molecular centers is r_0 (in cm), and that $C_A = 0$ at $r = r_0$, and $C_A = \tilde{C}_A$ (the bulk solution concentration) at $r = \infty$. Rearranging and integrating the expression for $\mathscr{J}_A$, we obtain

$$\mathscr{J}_A \int_{r_0}^{\infty} r^{-2} \, dr = 4\pi D_A \int_0^{\tilde{C}_A} dC_A$$

or

$$\mathscr{J}_A/r_0 = 4\pi D_A \tilde{C}_A$$

or

$$\mathscr{J}_A = 4\pi D_A r_0 \tilde{C}_A$$

This expression gives the number of A molecules that collide with a B molecule per second. To take account of the motion of B molecules, we replace D_A by $D_A + D_B$ and obtain the following expression for the total number of collisions per unit of time with all B molecules (at a bulk concentration $\tilde{C}_B$):

$$\text{Rate} = 4\pi(D_A + D_B)r_0\tilde{C}_A\tilde{C}_B = k'\tilde{C}_A\tilde{C}_B$$

where k' is a second-order rate constant with units of cm^3 molecule^{-1} sec^{-1}. To convert to the standard units of M^{-1} sec^{-1}, we multiply k' by $10^{-3}N_0$ (where N_0 is Avogadro's number), obtaining

$$k = 10^{-3}N_0 k' = 4\pi r_0 (D_A + D_B)10^{-3}N_0$$

For further details and discussion, see the text by G. G. Hammes (1978).

Table 16-2

Association and dissociation rate constants for enzyme–ligand complexes

Enzyme	Ligand	k_1 (M^{-1} sec^{-1})	k_{-1} (sec^{-1})
Aspartate aminotransferase	Glutamate; aspartate	$>10^7$–10^8	$>10^5$–10^6
	Oxalacetate; ketoglutarate	$>10^8$	$>10^4$
	α-Methylaspartate	1.2×10^4	1.3×10^2
	erythro-β-Hydroxyaspartate	3.1×10^6	1.1×10^4
	NH_2OH	3.7×10^6	6.2×10
Chymotrypsin	Proflavin	1.1×10^8	2.2×10^3
	Furylacryloyl-L-tryptophanamide	6.2×10^6	2.7×10^3
Creatine kinase	ADP	2.2×10^7	1.8×10^4
	MgADP	5.3×10^6	5.1×10^3
	CaADP	1.7×10^6	1.2×10^3
Lactate dehydrogenase, rabbit muscle	NADH	$\sim 10^9$	$\sim 10^4$
Lactate dehydrogenase, pig heart	NADH	5.5×10^7	3.9×10
	Oxamate	8.1×10^6	1.7×10
	3-Thio-NAD	5.8×10^6	4.1×10^2
Malate dehydrogenase	NADH	5×10^8	5×10
Old yellow enzyme	FMN	1.5×10^6	$\sim 10^{-4}$
Pyruvate carboxylase–Mn^{2+}	Pyruvate	4.5×10^6	2.1×10^4
Pyruvate kinase–Mn^{2+}	Fluorophosphate	1.3×10^7	3.4×10^4
Ribonuclease A	Cytidine 3'-phosphate	4.6×10^7	4.2×10^3
	Uridine 3'-phosphate	7.8×10^7	1.1×10^4
	Cytidine 2',3'-cyclic phosphate	2–4×10^7	1–2×10^4
	Uridine 2',3'-cyclic phosphate	10^7	2×10^4
	Cytidylyl 3',5'-cytidine	1.4×10^7	7×10^3

SOURCE: After G. G. Hammes and P. R. Schimmel, in *The Enzymes*, vol. 2, ed. P. D. Boyer (New York, Academic Press, 1970), p. 67.

Unimolecular configurational changes of enzyme–small molecule (e.g., substrate) complexes have been commonly observed. Table 16-3 lists some of the time constants that have been measured in various systems. Many of these time constants are in the

Table 16-3
Time constants for unimolecular transitions of enzyme–ligand complexes

Enzyme	Substrate	Approximate time constant (sec)
Alkaline phosphatase	2-Hydroxy-5-nitrobenzyl phosphate	10^{-2}
Aspartate aminotransferase	α-Methylaspartate	10^{-2}
	erythro-β-Hydroxyaspartate	10^{-3} to 10^{-1}
Chymotrypsin	Proflavin	10^{-4}
	Furylacryloyl-L-tryptophanamide	10^{-2}
Creatine kinase	ADP; MgADP; CaADP; MnADP; ATP	10^{-4}
Glyceraldehyde-3-phosphate dehydrogenase	NAD	1
Lactate dehydrogenase, rabbit muscle	NADH	10^{-3}
Alcohol dehydrogenase, liver	NADH-imidazole	10^{-3}
Peroxidase	H_2O_2; methyl and ethyl H_2O_2	10^{-1}
Pyruvate kinase	None; Mg^{2+}; Mn^{2+}	10^{-4}
Ribonuclease A	None; Cytidine 3'-phosphate; Uridine 3'-phosphate; Cytidine 2',3'-cyclic phosphate; Uridine 2',3'-cyclic phosphate; Cytidylyl 3',5'-cytidine	10^{-3} to 10^{-4}

SOURCE: After G. G. Hammes and P. R. Schimmel, in *The Enzymes*, vol. 2, ed. P. D. Boyer (New York: Academic Press, 1970), p. 67.

range of 10^{-2} to 10^{-4} sec. This is sufficiently rapid that the observed conformational change could lie on the main catalytic pathway. If, for example, the reciprocal time constant for a unimolecular step is smaller than the turnover number, then the observed conformational change must not be an obligatory step in the catalytic process. However, a very slow transition might be associated with a regulation mechanism.

It should also be pointed out that, in relaxation kinetic studies, it is not uncommon to observe a number of discrete relaxation processes associated with enzyme–substrate interactions. In fact, in one case (the reaction of *erythro-β*-hydroxyaspartate with aspartate aminotransferase to yield dihydroxyfumarate and the pyridoxamine enzyme), eight relaxation times have been measured. Therefore, multiple intermediate states may be a common feature of the mechanism of enzymatic catalysis.

16-8   RIBONUCLEASE AS AN EXAMPLE

Ribonuclease A is an example of a system that has been studied in some depth by kinetic and other approaches. This enzyme is a single polypeptide chain of 124 amino acids with a molecular weight of 13,600d; it has four disulfide bonds. It is commonly obtained from bovine pancreas.

The enzyme specifically cleaves RNA chains after pyrimidine residues, to give the cyclic 2′,3′-phosphates. Then the cyclic phosphates are specifically hydrolyzed to the 3′-phosphates. The reaction may be schematically written as

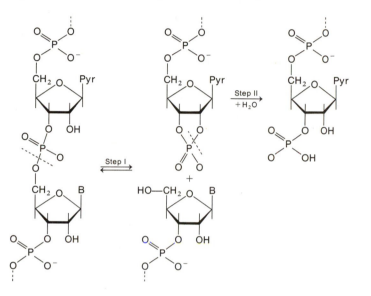

where Pyr is pyrimidine, and B is purine or pyrimidine. The enzyme catalyzes both steps. In addition, it rapidly converts monomeric cyclic UMP and CMP to their corresponding 3′-phosphates, a reaction that at equilibrium lies far to the right. Most studies of the mechanism of the enzyme have focused on the mononucleotide reactions.

Multiple intermediates in kinetic studies

Steady-state kinetic studies of the hydrolysis of cytidine 2′,3′-phosphate to cytidine 3′-phosphate indicate a simple mechanism that accounts for the pH dependence of the kinetic parameters (Herries et al., 1962). The mechanism is given by Equation 16-48 (S = cytidine 2′,3′-phosphate; P = cytidine 3′-phosphate) where the ionization constants (Eqn. 16-49) are $pK_{aE} \cong 5.2$; $pK_{aX} \cong 6.3$; $pK_{bE} \cong 6.8$; and $pK_{bX} \cong 8.1$. The pK values for the free enzyme suggest that histidine residues are involved; alternatively, one pK (pK_{aE}) might correspond to an abnormal carboxyl group.

However, relaxation kinetic studies show that the situation is considerably more complex (see Hammes, 1968). First of all, the enzyme alone has a unimolecular relaxation process that is coupled to a rapid ionization step. This can be written as

$$E'H \underset{k_{-1}}{\overset{k_1}{\rightleftharpoons}} EH \overset{K_a}{\rightleftharpoons} E + H^+ \qquad (16\text{-}96)$$
"slow"   "fast"

The transition $E'H \rightleftharpoons EH$ corresponds to an enzyme conformational change, where $k_1 = 780 \ sec^{-1}$, and $k_{-1} = 2,470 \ sec^{-1}$. The ionization step is associated with a $pK_a \cong 6.1$. This is almost certainly a histidine residue.

When cytidine 3'-phosphate is added to the enzyme, the relaxation effect associated with the conformational change progressively disappears as more of the nucleotide is added. This is due to the preferential binding of the 3'-phosphate to the EH and E forms, so that the reaction (Eqn. 16-96) is pulled to the right. Thus, the E'H form is an inert state of the enzyme.

Because the hydrolysis of the cyclic phosphate to 3'-phosphate lies far to the right, an equilibrium mixture of the nucleotides with enzyme contains largely the free 3'-phosphate and its bound forms. Because the cyclic phosphate is present in such small amounts, relaxation effects associated with its interactions with the enzyme cannot be observed in an equilibrium mixture. This situation was remedied by stopped-flow, temperature-jump studies in which enzyme and cyclic phosphate were rapidly mixed and then subjected to a temperature jump. In this way, the pre-equilibrium reaction was perturbed so that discrete relaxation processes associated with interactions of the cyclic phosphate and the enzyme were detected. In addition, mixtures of the enzyme and cytidine 3'-phosphate were also studied.

The results of these studies are summarized by the following scheme:

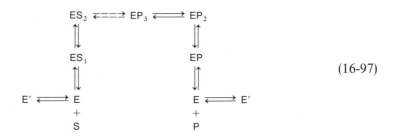

(16-97)

The dashed arrows correspond to the rate-determining step. Including the configuration change of the free enzyme, six discrete steps have been collectively detected on both sides of the rate determining process.

The above scheme omits, for simplicity, the ionization states of the free enzyme and its various complexes. Altogether, three pK values have been implied (one more pK than was obtained from steady-state studies alone); on the free enzyme, these are ~ 5, 6, and 6.7.

These kinetic studies shed considerable light on the details of the elementary steps in catalysis. However, by themselves they leave many critical questions un-answered—such as the location of the active site and the identity of the catalytic groups. For this kind of information, other approaches must be used.

Chemical modification studies

Active sites frequently are identified by means of specific reagents that react with active-site groups and thereby eliminate or sharply reduce the activity. In the case of ribonuclease, it is found that iodoacetate (ICH_2COO^-) reacts under appropriate conditions with a 1:1 stoichiometry (see Crestfield, Stein, and Moore, 1963). Two reaction products are obtained. The major product is 1-carboxylmethyl histidine-119 (1-CM His[119]), and the other is 3-CM His[12]. These adducts are

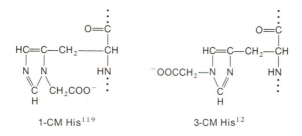

1-CM His[119]   3-CM His[12]

In the case of the 1-CM His[119] adduct, the enzyme is completely inactive, whereas in the other case it retains 15% activity. This observation suggests that the two histidines are at the active site. Confirmation of this idea comes from the observation that the alkylation is inhibited by small molecules that bind to the active site, such as cytidine 3'-phosphate. Also, the observation that a doubly modified enzyme (alkylated at both histidine residues) is not obtained makes it probable that the two histidines—well separated in the sequence—are close spatially in the native structure. Finally, it is likely that one or both of the two histidines modified by iodoacetate are identical to certain of the ionizable groups implicated by the kinetic studies mentioned earlier.

Another interesting modification is produced when fluorodinitrobenzene (FDNB) reacts with the enzyme. Under appropriate conditions, this reagent reacts with the ε-amino group of lysine-41 to give an inactive derivative (Hirs et al., 1965). The reaction is

Thus, this result suggests that lysine-41 may also be at the catalytic center.

Implications of effect of subtilisin on ribonuclease

Further evidence for the importance of His^{12} and His^{119} comes from studies with the subtilisin-modified enzyme. Subtilisin is a protease that cuts ribonuclease at position 20–21 to give two fragments, 1–20 and 21–124 (see Richards and Vithayathil, 1959). The separate fragments are inactive, but when mixed they combine to give active enzyme known as ribonuclease S. These observations are consistent with the idea that both histidines are required for catalytic activity.

Using chemical synthesis, it is possible to make fragments of the enzyme (see Finn and Hoffmann, 1965). It is found that a mixture of 1–11 with 21–124 gives no activity. However, activity is obtained when 1–12 is mixed with 21–124. thus giving strong evidence for the importance of His^{12}.

Active dimers of ribonuclease

A. M. Crestfield, W. H. Stein, and S. Moore (1963) made some interesting observations concerning the dimerization of ribonuclease. When ribonuclease is lyophilized from 50% acetic acid, some dimers of the enzyme are obtained that have twice the activity (per mole of dimer) of monomers. Assuming that His^{12} and His^{119} are both essential for activity, this observation might be explained by a partial unfolding and dimerization of the enzyme according to the scheme in Figure 16-11. This scheme allows the dimer to group together His^{12} and His^{119} in two symmetrically disposed arrays.

A critical test of these ideas is provided by mixing CM His^{12} and CM His^{119} ribonuclease and then carrying out the lyophilization. According to the scheme in Figure 16-12, this should result in the formation of a hybrid dimer containing a modified and an unmodified pair of histidine residues, with the activity corresponding to a monomer (one active site). The experiment indeed gives this result, thus providing rather convincing additional proof for the significance of these residues.

The four histidines resolved and studied by NMR

One of the most informative studies on the histidine residues in ribonuclease resulted from the NMR investigations of Jardetsky and coworkers. The enzyme's four histidine residues can be resolved in a 100 MHz spectrum. (See Chapter 9 for a discussion of NMR.) Each imidazole C^2–H has a distinct chemical shift that depends on pH. By studying the chemical shifts as a function of pH, we can construct a titration curve (Fig. 16-13). The four histidine C^2–H peaks are numbered from 1 to 4. In addition, a peak attributable to a C^4–H of histidine is observed. This has the same pK as that for peak 1 and may, therefore, be assumed to be the same residue as peak 1.

In 3-CM-His^{12} RNase and in 1-CM-His^{119} RNase, peaks 2 and 3 are simultaneously shifted to higher pK values, whereas peaks 1 and 4 are unaffected. That is, carboxymethylation of only one histidine simultaneously affects the environment

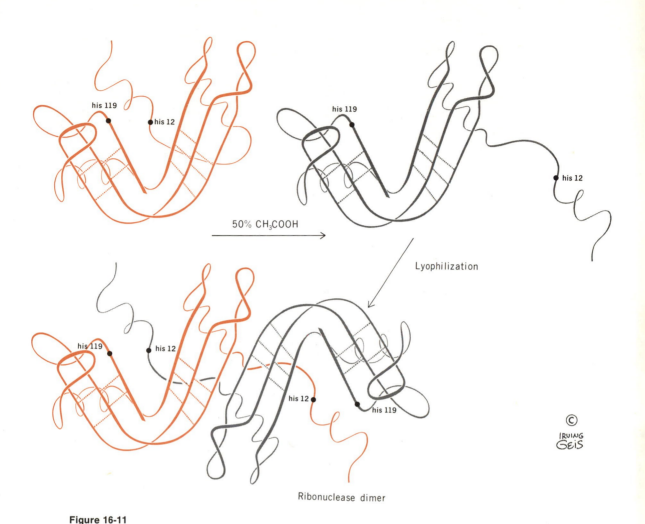

50% CH₃COOH

Lyophilization

Ribonuclease dimer

Figure 16-11

Schematic illustration of the dimerization of ribonuclease. The positions of His[12] and His[119] are indicated. [Drawing by Irving Geis.]

of two histidines. This means that the two histidines must be close together, and suggests that peaks 2 and 3 may be ascribed to His[12] and His[119]. Of course, these data do not establish which peak is His[12] and which is His[119].

The assignment of peaks 2 and 3 was done by J. L. Markley (1975), who studied by NMR the kinetics of the slow exchange of deuterium into the C^2 positions of the four histidines. The kinetics differ for each site, and may be compared with results of a different kind of experiment in which tritium incorporation into the four specific histidines is measured by isolating radioactive peptides from the enzyme (which has been exposed to tritium) and determining directly the amount of exchange at each specific histidine. The two kinds of data on the exchange kinetics may be directly

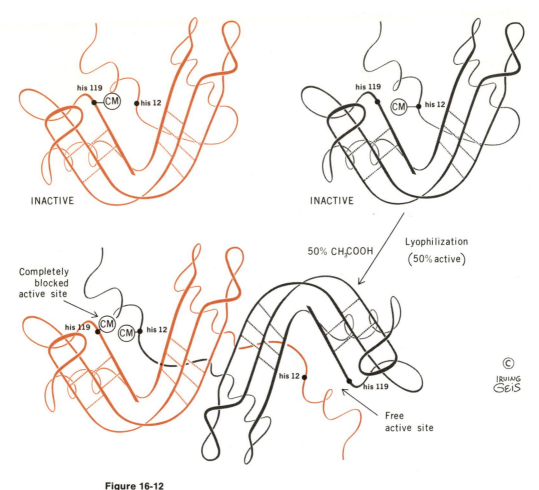

Figure 16-12

Production of active dimers from CM-His[12] RNase and CM-His[119] RNase.
[Drawing by Irving Geis.]

compared so that an NMR-determined rate constant on a specific peak may be matched up with the exchange-rate constant of one of the specific histidine residues, as determined by the other experiment. These results show that His[12] corresponds to peak 3 and His[119] to peak 2. These and other data indicate that the two remaining peaks can be assigned to His[48] (peak 4) and His[105] (peak 1).

Upon binding of cytidine 3'-phosphate the pK values of His[12] and His[119] rise sharply. These sharp increases are consistent with steady-state kinetic results discussed earlier that implicate two pK values on the enzyme that are shifted higher in the enzyme–substrate complex.

The peak corresponding to His[105] is unaffected by binding of cytidine 3'-phosphate, whereas His[48] shows a small shift and line-width change. Also, other data

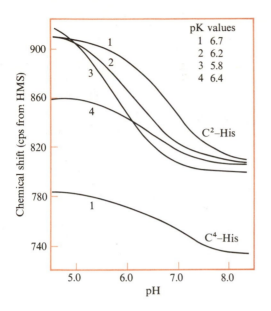

pK values
1 6.7
2 6.2
3 5.8
4 6.4

C²–His

C⁴–His

Figure 16-13

NMR titrations of four histidine C^2–H peaks and a histidine C^4–H peak in RNase A. Data were obtained in 0.2 M deuteroacetate buffer at 32°C. HMS is hexamethyldisiloxane. [After D. H. Meadows, O. Jardetzky, R. M. Epand, H. H. Ruterjans, and H. A. Scheraga, *Proc. Natl. Acad. Sci. USA* 60:766 (1968).]

on His[48] suggest that this residue may flip rapidly between at least two environments, and that this process is affected by cytidine 3′-phosphate binding. Therefore, His[48] may be the residue responsible for the pH-dependent relaxation effect (Eqn. 16-96) studied by Hammes and coworkers.

X-ray structure

The chemical and related studies have provided considerable insight into the nature of the active site of the enzyme. Concrete visualization of the enzyme's three-dimensional structure and its active site has been provided by x-ray crystallographic studies of F. M. Richards, H. W. Wyckoff, G. Kartha, J. Bello, and coworkers. Figure 16-14 shows the structure of ribonuclease S. The enzyme is roughly kidney-shaped, with overall dimensions of $38 \times 28 \times 28$ Å. There is a deep depression in one side, and phosphate crystals show that P_i occurs in this depression. Subsequent studies with substrate analogues have confirmed that the depression corresponds to the active site. His[12] and His[119] are found in this active-site region and also occur close together. In addition, Lys[41] is located in the active-site area. Thus, with respect to identification of active-site residues, the conclusions drawn from the solution studies are correct.

Figure 16-15a is a schematic illustration of the active-site region with a bound dinucleotide substrate B_1pB_2. This figure is based on the molecular model of the active-site area shown in Figure 16-15b. In Figure 16-15a, the ovals labeled B_1, R_1, and p_1 represent the base, ribose, and phosphate moieties of the first unit, and B_2 and R_2 represent base and ribose moieties of the second one. Also, B_2' is an alternate position for the second base, and p_1 is an inferred position for the cyclic phosphate.

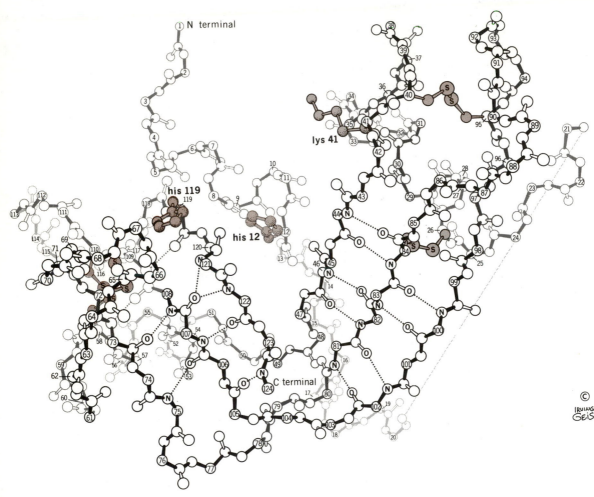

Figure 16-14

Structure of RNase S as determined by x-ray diffraction. The backbone is represented as a continuous tube. Side-chain skeletons are attached to the backbone and are labeled. [Drawing by Irving Geis.]

With the phosphate in position p_1, lysine-41 (to the upper right of the phosphate) cannot make contact with the phosphate group, whereas in position p'_1 it appears that it might make contact. His^{12} is located behind the phosphate, whereas His^{119} can occupy any of four positions that are designated with Roman numerals I to IV. The exact location of His^{119} appears to depend on the molecule that is bound to the enzyme.

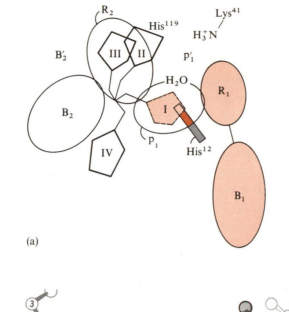

(a)

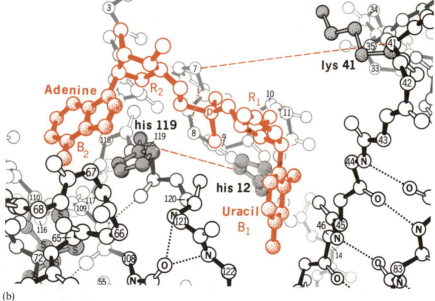

(b)

Figure 16-15

Active-site region of ribonuclease A. (**a**) Schematic illustration of the orientation of the dinucleotide substrate B_1pB_2 in the RNase A active site; see discussion in text. [After F. M. Richards and H. W. Wyckoff, in *Atlas of Molecular Structures in Biology*, vol. 1, ed. D. C. Phillips and F. M. Richards (London: Oxford Univ. Press, 1973).] (**b**) Close-up of the active-site region in the ribonuclease S model. [Drawing by Irving Geis.]

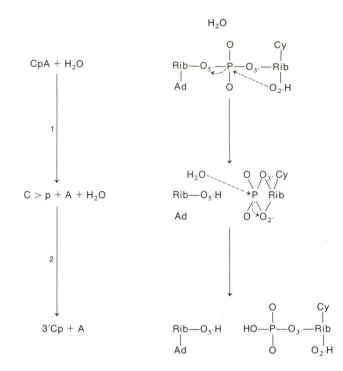

Figure 16-16

Schematic illustration of the mechanism of hydrolysis of the dinucleotide substrate CpA by RNase A; see discussion in text.

Figure 16-16 shows the two steps of the hydrolysis reaction mechanism for the dinucleotide substrate CpA. Surrounding catalytic groups are omitted. In step 1, the cyclic phosphate (C > p) is generated; this can then react in step 2 either with the 5-OH of adenosine to regenerate dinucleotide, or with H_2O to give 3'-CMP.

There are two mechanisms that have been considered for each step of the hydrolytic reaction. One is an "in-line" mechanism. In this mechanism, the attacking nucleophile (either the ribose 2'-OH, in the first step, or water, in the second step) enters on the side opposite to the leaving group. An alternative mechanism is one in which the attacking nucleophile comes in on the same side as the leaving group; this is called an adjacent mechanism. Superficially, the second step of the mechanism (cyclic phosphate ring opening) appears to be the microscopic reverse of the first (ring closure to form cyclic phosphate). On this basis we would assume that the same mechanism (in-line or adjacent) operates in both steps. But the attacking nucleophile in the first step is the ribose 2'-OH, and it is water in the second step. These are not equivalent, so there is no reason for each step to have the identical mechanism. Nevertheless, experiments from the laboratories of F. H. Eckstein and D. A. Usher have established that both steps follow the same mechanism and that this is an in-line mechanism. Figure 16-16 illustrates the in-line case.

It is likely that the His[12] and His[119] act by general acid–base catalysis. For example, in the first step (generation of the cyclic phosphate) His[12] could be a base, assisting the removal of a proton from the ribose 2'-OH that couples to the phosphate to form the cyclic 2',3'-phosphate; His[119] may act as an acid catalyst, protonating the leaving group to give a ribose 5'-OH. In the second step, (opening of the cyclic phosphate ring), these roles would be reversed—His[12] acting as an acid and His[119] as a base. The precise role of Lys[41] is not clear, but its positive charge may stabilize the transient pentacovalent phosphorous that is formed when an OH (from the ribose 2'-OH in the cyclization step or from water in the ring-opening step) attacks the phosphate.

It is clear that our understanding of ribonuclease has advanced tremendously. Our discussion of this enzyme provides only a brief overview of the kinds of multi-pronged approaches that must be used to obtain significant information on an enzyme's mechanism of action; a large number of significant investigations have been omitted from our discussion. It is important to recognize, however, that physical studies (such as kinetics, x-ray structure, and so on), although extremely valuable, are not always the best way (or in fact may be inherently unable) to provide answers to many of the questions that are raised. The greatest understanding clearly is achieved by integrating physicochemical information with the wide variety of results obtained by other approaches, a situation that is well illustrated by the research on ribonuclease.

Summary

The study of the dynamic behavior of biological systems provides critical insight into structure–function–mechanism relationships. A prime objective of such studies is to determine the kinetic rate law for the system under investigation, and thus to ascertain the kinetic mechanism and associated rate parameters. Although rate equations can be solved for sequential unimolecular reactions, simple schemes that include a bimolecular reaction present formidable difficulties. In the case of enzymatic reactions, which are the most widely studied kinetic systems, rate laws usually are obtained in a simplified form by using the steady-state approximation, where it is assumed that each enzyme species is in a steady state. From steady-state studies, considerable insight into enzyme mechanism has been obtained.

Approaches have been developed for studying very fast transient phenomena, although the analysis of such systems sometimes is hindered by difficulties in solving the rate equations. However, even though solutions to rate equations for many kinetic systems cannot be obtained, a solution may always be found near equilibrium by linearizing the rate equations, providing a system of linear differential equations that may be solved by standard procedures. Transient systems near equilibrium may be studied by relaxation spectrometry, where the time-dependent decay to equilibrium of chemical species in the system is followed. This approach has given detailed insight into many processes. In studies of the relaxation spectra of enzyme reactions, it has been found that the binding of substrate (ligand) with enzyme can occur with a frequency within one to two orders of magnitude of the diffusion-controlled limit. Also, protein conformational changes have been observed to occur in the time range of 10^{-4} to 10^{-2} sec.

Ribonuclease is one of the best examples of a system that has been studied by a variety of approaches so that kinetic information can be integrated with other data. With a crystallographic structure of the enzyme also available, it is possible to visualize some events in molecular detail.

Problems

16-1. Derive the expression for $V = d(P)/dt$ for the mechanism

$$E + S \rightleftarrows \begin{array}{c} X_1 \xrightarrow{k_2} \\ \\ X_2 \xrightarrow{k_2'} \end{array} E + P \qquad (A)$$

Express V in terms of $(E)_0$ = total enzyme concentration, and (S) = free substrate concentration. The following reactions may be assumed to be at equilibrium while P is produced:

$$E + S \rightleftarrows X_1 \qquad K_1 = (E)(S)/(X_1)$$

$$E + S \rightleftarrows X_2 \qquad K_2 = (E)(S)/(X_2)$$

so that you need not use the usual steady-state assumption. A bright student claims that the form of the rate law for Equation A is indistinguishable from that for the Michaelis–Menten expression for the following trivial mechanism:

$$E + S \rightleftarrows X_1 \rightarrow E + P \qquad (B)$$

He claims that this is true regardless of the relative values of k_2, k_2', K_1, and K_2 in Equation A. Another student claims that the concentration dependencies for Equations A and B are distinguishable only when $k_2 \gg k_2'$ and $K_1 \ll K_2$. A third student says that the rate laws for the two mechanisms will have different forms under any circumstances. Which student (if any) is correct? Why?

16-2. Enzymatic reactions involving nucleoside phosphates (such as ATP) often are activated by divalent metal ion M. Consider the following steady-state kinetic behavior.
   a. Lineweaver–Burk plots are linear.
   b. The slopes of the Lineweaver–Burk plots are inversely proportional to the concentration of free metal, (M).
   c. The ordinate intercepts of the Lineweaver–Burk plots are independent of metal concentration.
   d. For a given substrate concentration, the rate is much slower (but not zero) in the absence of metal than it is in the presence of metal.

The following mechanism is proposed:

$$S + M$$

$$SM + E \underset{k_{-1}}{\overset{k_1}{\rightleftarrows}} XM \xrightarrow{k_2} E + PM$$

where the reaction $S + M \rightleftarrows SM$ is a rapid equilibrium, with $K_{SM} = (S)(M)/(SM)$. (The dissociation $PM \rightleftarrows P + M$ may be ignored.) Derive the steady-state rate law for this mechanism by expressing the initial velocity [when $(PM) \cong 0$] in terms of $(E)_0 =$ total enzyme concentration, $(S)_0 =$ total substrate concentration $= (S) + (SM)$, and $(M) =$ free metal concentration. Does this rate law satisfactorily account for each of the four kinetic behaviors listed above? If it fails for any or all of the behaviors, suggest an alteration of the mechanism or a new mechanism that might better explain the data.

16-3. For a certain enzyme, the maximal velocity V_S and the Michaelis constant K_S have been studied over a wide range of pH. Between pH 7 and pH 8.5, V_S is at its maximal value and is pH-independent. Below pH 7, V_S decreases, with the midpoint of the decrease occurring at pH 5.0. Above pH 8.5, V_S also decreases, with the midpoint around pH 10.1. Over the entire pH range, K_S is pH-independent, however. From these data alone, calculate the pK value (or values) of the group(s) critical for activity on the enzyme–substrate complex. Also calculate the pK value(s) of the same group(s) on the free (naked) enzyme.

16-4. Consider the reaction to form a Watson–Crick base pair between adenosine (A) and uridine (U). Assume that the reaction can be written as

$$A + U \underset{k_{-1}}{\overset{k_1}{\rightleftarrows}} (A, U) \underset{k_{-2}}{\overset{k_2}{\rightleftarrows}} A === U$$

where (A, U) represents a species in which A and U are close together and have formed one hydrogen bond, and $A===U$ represents the Watson–Crick structure with two hydrogen bonds. We expect (A, U) to be very short-lived and to be present in tiny amounts (why?), so that in a relaxation-kinetics experiment we may assume that $d\Delta((A, U))/dt = 0$, and $\Delta(A) = -\Delta((A, U)) - \Delta(A===U) \cong -\Delta(A===U)$. Show that, near equilibrium under these conditions, the time course of $(A===U)$ is given by $\Delta(A===U) = \Delta(A===U)^0 \exp(-t/\tau)$; that is, a single exponential despite the fact that there now are two reactions. Show also that

$$1/\tau = k_{-1}k_{-2} + [k_1 k_2/(k_{-1} + k_2)][(\bar{A}) + (\bar{U})]$$

A researcher claims that, from a relaxation-kinetics experiment alone, he can distinguish the two-step mechanism given above from a mechanism in which only one step occurs: $A + U \rightleftarrows A===U$. Is he correct? Why, or why not?

16-5. Derive the expression for the relaxation time of the "very slow" step for the following mechanism:

$$E + S \underset{\underset{\text{fast}}{k_{-1}}}{\overset{k_1}{\rightleftarrows}} X_1 \underset{\underset{\text{very slow}}{k_{-2}}}{\overset{k_2}{\rightleftarrows}} X_2 \underset{\underset{\text{fast}}{k_{-3}}}{\overset{k_3}{\rightleftarrows}} X_3$$

References

GENERAL

Bernhard, S. 1968. *The Structure and Function of Enzymes.* New York: Benjamin.

Cleland, W. W. 1970. Steady state kinetics. In *The Enzymes*, vol. 2, ed. P. D. Boyer (New York: Academic Press), p. 1. [A summary of steady-state kinetics, based in part on the author's nomenclature, notation, and work in the field. Much of the article is descriptive and is intended to provide an intuitive understanding of the various features of steady-state kinetics.]

Cornish-Bowden, A. 1976. *Principles of Enzyme Kinetics.* London: Butterworth.

Dixon, M., and E. C. Webb. 1964. *Enzymes.* New York: Academic Press.

Eigen, M., and L. de Maeyer. 1963. Relaxation methods. In *Technique of Organic Chemistry*, vol. 8, part 2, ed. S. L. Friess, E. S. Lewis, and A. Weissberger (New York: Interscience), p. 895. [A classic paper on the theory and methodology of relaxation spectroscopy; a comprehensive and definitive work.]

Fersht, A. 1977. *Enzyme Structure and Mechanism.* San Francisco: W. H. Freeman and Company.

Hammes, G. G., ed. 1974. *Techniques of Chemistry*, vol. 6, part 2. New York: Wiley (Interscience). [A treatment and discussion of elementary reaction steps in solution and very fast reactions.]

———. 1978. *Principles of Chemical Kinetics.* New York: Academic Press. [The chapter on enzyme kinetics and on fast-reaction kinetics—e.g., relaxation spectrometry—is a good introduction to these subjects.]

Hammes, G. G., and P. R. Schimmel. 1970. Rapid reactions and transient states. In *The Enzymes*, vol. 2, ed. P. D. Boyer (New York: Academic Press), p. 67. [A discussion of fast-reaction kinetics with particular emphasis on relaxation spectrometry and enzymatic systems.]

Jencks, W. P. 1969. *Catalysis in Chemistry and Enzymology.* New York: McGraw-Hill. [Kinetics is taken up in significant part from the standpoint of physical organic chemistry.]

Piskiewicz, D. 1977. *Kinetics of Chemical and Enzyme-Catalyzed Reactions.* New York: Oxford Univ. Press.

Walsh, C. 1979. *Enzymatic Reaction Mechanisms.* San Francisco: W. H. Freeman and Company.

SPECIFIC

Crestfield, A. M., W. H. Stein, and S. Moore. 1963. Alkylation and identification of the histidine residues at the active site ribonuclease. *J. Biol. Chem.* 238:2413.

———. 1963. Properties and conformation of the histidine residues at the active site of ribonuclease. *J. Biol. Chem.* 238:2421.

Finn, F. M., and K. Hoffmann. 1965. Studies on polypeptides, XXXIII: Enzymic properties of partially synthetic ribonucleases. *J. Am. Chem. Soc.* 87:645.

Hammes, G. G. 1968. Relaxation spectrometry of biological systems. *Adv. Protein Chem.* 23:1.

Herries, D. G., A. P. Mathias, and B. R. Rabin. 1962. The active site and mechanism of bovine pancreatic ribonuclease, 3: The pH-dependence of the kinetic parameters for the hydrolysis of cytidine 2′,3′-phosphate. *Biochem. J.* 85:127.

Hirs, C. H. W., M. Halmann, and J. H. Kycia. 1965. Dinitrophenylation and inactivation of bovine pancreatic ribonuclease A. *Arch. Biochem. Biophys.* 111:209.

Kartha, G., J. Bello, and D. Harker. 1967. Tertiary structure of ribonuclease. *Nature* 213:862.

Kraut, J. 1977. Serine proteases: Structure and mechanism of catalysis. *Ann. Rev. Biochem.* 46:33. [The serine proteases have been as exhaustively or even more exhaustively studied as ribonuclease.]

Markley, J. L. 1975. Correlation proton magnetic resonance studies at 250 MHz of bovine pancreatic ribonuclease, I: Reinvestigation of the histidine peak assignments. *Biochemistry* 14:3546.

Miller, W. G., and R. A. Alberty. 1958. Kinetics of the reversible Michaelis–Menten mechanism and the applicability of the steady-state approximation. *J. Am. Chem. Soc.* 80:5146.

Peller, L., and R. A. Alberty. 1959. Multiple intermediates in steady state enzyme kinetics, I: The mechanism involving a single substrate and product. *J. Am. Chem. Soc.* 81:5907.

Richards, F. M., and P. J. Vithayathil, 1959. The preparation of subtilisin modified ribonuclease and the separation of the peptide and protein components. *J. Biol. Chem.* 234:1459.

Richards, F. M., and H. W. Wyckoff. 1970. Bovine pancreatic ribonuclease. In *The Enzymes*, vol. 4, ed. P. D. Boyer (New York: Academic Press), p. 647.

Usher, D. A., E. S. Erenrich, and F. Eckstein. 1972. Geometry of the first step in the action of ribonuclease A. *Proc. Natl. Acad. Sci. USA* 69:115.

Usher, D. A., D. I. Richardson, Jr., and F. Eckstein. 1970. Absolute stereochemistry of the second step of ribonuclease action. *Nature* 228:663.

17

Regulation of biological activity

17-1 BIOLOGICAL REGULATION

In Chapters 15 and 16, we discussed equilibrium and kinetic features of ligand interactions with macromolecules. Our emphasis was placed on general features, with the idea of developing insight into basic principles. In this chapter we direct attention to special properties associated with the regulation of biological activity, and to two prominent examples: aspartate transcarbamoylase and hemoglobin. These two well-studied systems beautifully illustrate how multiple-ligand interactions modulate biological activity. In addition, they provide good illustrations and tests of theories that have been developed to explain regulation of the biological activity of a protein.

Feedback inhibition

A good example of biological regulation is the phenomenon of *feedback inhibition*. Consider a pathway in which precursor X_0 is converted by a series of reactions to compound Z, with each step in the pathway catalyzed by a specific enzyme E_i. In many cases it is found that, no matter how much X_0 is supplied, the system does not continuously produce a corresponding amount of Z unless Z is drained off as it is made. This situation arises because Z itself feedback-inhibits the enzyme (E_1) that converts X_0 to the next compound (X_1) on the pathway to Z:

$$\text{Feedback inhibition}$$

$$X_0 \xrightarrow{\ E_1\ } X_1 \xrightarrow{\ E_2\ } X_2 \xrightarrow{\ E_3\ } \cdots \longrightarrow Z \qquad (17\text{-}1)$$

As the concentration of Z rises, it eventually becomes sufficiently high to inhibit E_1. In this way, Z regulates its own synthesis. The inhibition is reversible, so that a decrease in (Z) removes the inhibition on E_1 and allows new synthesis of Z. Note that the cell efficiently avoids the needless expenditure of metabolic energy by having Z act on the first enzyme in the pathway (rather than on a later one).

In most cases, the end product Z is structurally unrelated to X. Therefore, Z is not a competitive inhibitor of E_1; rather E_1 has a special, separate site that is specifically designed to accommodate Z.

An example of the foregoing is provided by the biosynthesis of nucleic acids. In cells that are rapidly dividing and growing, nucleic acids are continuously synthesized through a general pathway of precursors $\rightarrow \cdots \rightarrow$ nucleotides $\rightarrow \cdots \rightarrow$ nucleic acids. However, if for some reason nucleic acid synthesis slows down, nucleotides do not indefinitely accumulate, even though precursors are available. That is, slowing down the conversion of nucleotides to nucleic acids also inhibits the production of nucleotides.

The pyrimidine pathway and aspartate transcarbamoylase

A well-studied example of feedback inhibition is from the pyrimidine pathway, in which the first step is catalyzed by aspartate transcarbamoylase (ATCase). The reaction catalyzed by ATCase is

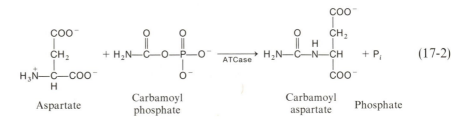

$$ (17\text{-}2) $$

Aspartate   Carbamoyl phosphate   Carbamoyl aspartate   Phosphate

In this reaction, aspartate and carbamoyl phosphate are condensed to give carbamoyl aspartate. After this, there are several steps by which carbamoyl aspartate is converted into the pyrimidine nucleoside triphosphate UTP and, subsequently, to CTP. The end product CTP in turn feedback-inhibits ATCase.

Allosteric proteins

ATCase is known as an allosteric protein ("allo-" means "other"), a name commonly assigned to proteins that have sites (regulatory sites) specific for physiological molecules that regulate the protein's activity; these sites are other than and distinct from the catalytic site. Allosteric proteins have a number of unusual features, which we shall illustrate and subsequently analyze in quantitative terms. Because it is such a well-studied system, we use ATCase as an example. However, the concepts discussed are not limited simply to this particular system.

17-2 SOME FEATURES AND PROPERTIES OF ALLOSTERIC ENZYMES

Sigmoidal curves

A striking feature of allosteric enzymes is that they commonly display nonhyperbolic kinetics. Figure 17-1 compares the profiles of initial reaction velocity, v_i, versus substrate concentration, (S), for a "normal" and an allosteric enzyme. A sigmoidal

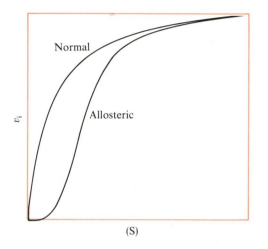

Figure 17-1

Hyperbolic and sigmoidal reaction kinetics. In a plot of initial reaction velocity v_i versus substrate concentration (S), a "normal" enzyme shows a hyperbolic curve, whereas an allosteric enzyme shows a sigmoidal curve.

curve is observed for the allosteric enzyme. At low substrate concentrations, the velocity is weakly responsive to increases in concentration. As substrate concentration is raised, however, a point is reached at which the velocity steeply rises in response to small increases in substrate concentration; that is, cooperative behavior is exhibited. This kind of kinetic behavior is in sharp contrast to the hyperbolic Michaelis–Menten behavior of the normal enzyme.

In the appropriate concentration range, the sigmoidal curve of v_i versus (S) enables the enzyme to produce sharp changes in the rate of product formation in response to small variations in substrate concentration.

Effect of an allosteric inhibitor

Figure 17-2 illustrates the effect of an allosteric inhibitor—CTP in the case of ATCase—on the v_i versus (S) curve. In this example, the inhibitor (CTP) shifts the

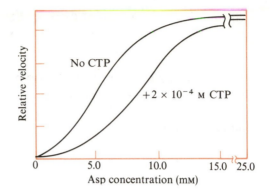

Figure 17-2

Effect of an allosteric inhibitor. Relative reaction velocity is plotted against substrate (aspartate) concentration for ATCase in the presence and absence of the allosteric inhibitor CTP. [After J. C. Gerhart and A. B. Pardee, *Cold Spring Harbor Symp. Quant. Biol.* 28:491 (1963).]

sigmoidal curve to the right, so that higher concentrations of substrate (Asp) are required to achieve the same velocity as in the absence of CTP. At sufficiently high (S), the effect of the inhibitor is overcome. Because high (S) can overcome the inhibition, the situation appears superficially similar to simple competitive inhibition.

However, in the case of ATCase, there is a feature critical to the inhibition that clearly distinguishes it from simple competitive inhibition. We now plot percentage of inhibition against inhibitor concentration (I) at constant (S). For a competitive inhibitor of a normal enzyme, this plot is hyperbolic (Box 17-1) and eventually reaches 100% inhibition (Fig. 17-3). Furthermore, the general shape of the curve remains the same at higher values of (S), although larger amounts of inhibitor are needed to achieve 100% inhibition. The situation is quite different in the case of ATCase. Figure 17-4 shows that the maximal level of inhibition that can be achieved with CTP depends on the aspartate (substrate) concentration; higher (Asp) means lower maximal inhibition levels. This shows that CTP cannot entirely overcome the effects of

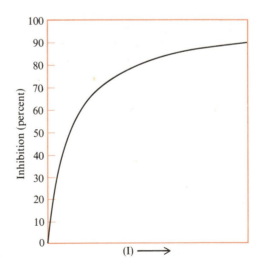

Figure 17-3

Competitive inhibition of a normal enzyme. Percentage of inhibition is plotted against inhibitor concentration.

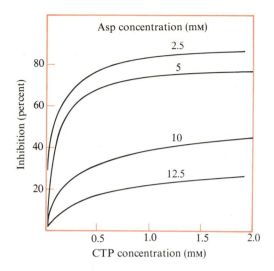

Figure 17-4

Allosteric inhibition. Percentage of inhibition is plotted against inhibitor (CTP) concentration for various concentrations of substrate (aspartate) in the case of ATCase. [After J. C. Gerhart and A. B. Pardee, *Cold Spring Harbor Symp. Quant. Biol.* 28:491 (1963).]

Box 17-1 COMPETITIVE INHIBITION

To the simple mechanism of Equation 16-33, we can add the binding of a competitive inhibitor I:

$$E + I \rightleftharpoons EI \qquad K_I = (E)(I)/(EI)$$

We assume that the inhibitor-binding equilibrium is rapid compared to the overall reaction (Eqn. 16-33), so that the equilibrium constant K_I may be used to calculate concentrations of E and EI. It then is easy to show that, with total inhibitor concentration in great excess over total enzyme concentration, Equation 16-37 is modified to

$$v_1 = V_S/\{1 + [K_S/(S)_0][1 + (I)_0/K_1]\}$$

and the percentage of inhibition ($I_\%$) is given by

$$I_\% = \left(1 - \frac{V_S/\{1 + [K_S/(S)_0][1 + (I)_0/K_1]\}}{V_S/[1 + K_S/(S)_0]}\right) \times 100$$

$$= \left(\frac{[K_S/(S)_0][(I)_0/K_1]}{1 + [K_S/(S)_0][1 + (I)_0/K_1]}\right) \times 100$$

Thus, at constant $(S)_0$, the plot of $I_\%$ versus $(I)_0$ will be a simple hyperbolic curve.

aspartate. On the other hand, because aspartate *can* completely reverse the effects of CTP (Fig. 17-2), the situation is not symmetric with respect to the two ligands. *CTP cannot entirely overcome the effects of aspartate, but aspartate can completely overcome the inhibition by CTP.* This unusual interplay between CTP and aspartate sharply distinguishes this system from common competitive inhibition phenomena.

Effect of a competitive inhibitor

The effects of a true competitive inhibitor also are unusual. Malate and succinate are structural analogs of aspartate that presumably compete with the substrate for sites on the enzyme:

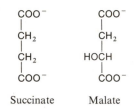

Succinate   Malate

However, the effect of malate, for example, varies with the aspartate concentration (Fig. 17-5). A plot of percentage of activity versus malate concentration (in the range

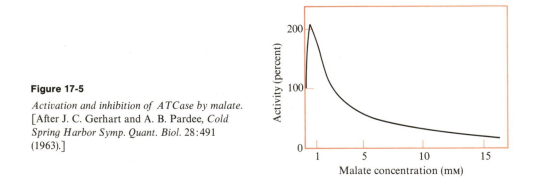

Figure 17-5

Activation and inhibition of ATCase by malate.
[After J. C. Gerhart and A. B. Pardee, *Cold Spring Harbor Symp. Quant. Biol.* 28:491 (1963).]

of low aspartate concentrations) shows high activity at low malate concentration, and progressive inhibition with higher malate concentration until activity is abolished. Thus, depending on its concentration, malate can act either as an activator or as an inhibitor. On the other hand, at much higher levels of aspartate (not shown), addition of malate leads only to inhibition, regardless of the amount of malate already present.

Allosteric proteins invariably are composed of subunits. In ATCase, there are six catalytic (C) chains and six regulatory (R) chains. Each C chain has a molecular weight of about 33,000 d, and each R chain about 17,000 d. In the intact molecule, there are six sites for aspartate and carbamoyl phosphate on the C chains and six sites for CTP on the R chains.

Although a detailed model is not yet available, considerable progress has been made on the structure of ATCase. The work has been directed by William Lipscomb. Figure 2-49 is a schematic illustration of the structure. The enzyme has one 3-fold and three 2-fold axes. The subunits are arranged so that there is one trimer of C chains above and below a belt of three R-chain dimers. With respect to its largest dimensions, the molecule is about $110 \times 110 \times 90$ Å in size.

The reaction properties and structural features clearly show that an allosteric enzyme such as ATCase is quite different from many other enzymes. Consequently, special models have been devised to account for this kind of behavior. These models have been subjected to a number of experimental tests.

17-3   MONOD–WYMAN–CHANGEUX (MWC) MODEL FOR ALLOSTERIC PROTEINS

Models for allosteric proteins invariably assign an important role to the conformational states of the subunits and to subunit–subunit interactions. Because of its simplicity, a popular and frequently cited model is that of J. Monod, J. Wyman, and J.-P. Changeux (1965). Although the MWC model is not applicable to all systems, it contains concepts that are useful for discussing allosteric systems and has an unusually simple algebraic description.

The model distinguishes between protomers and subunits. A protomer is that structural unit that bears one site for each of the various ligands. A subunit is a single polypeptide chain. Therefore, a protomer may be made up of more than one subunit. In the case of ATCase, the protomer is composed of one R and one C chain.

Four main assumptions

The MWC model for an allosteric protein contains four essential assumptions.

1. Identical protomers occupy equivalent positions in the protein. This means the molecule must have at least one symmetry axis.

2. Each protomer contains a unique receptor site for each specific ligand.

3. At least two conformational states are reversibly accessible to the protein. In each of these states, symmetry is conserved. However, the affinity of a given ligand for the protein may be very different in the different conformational states.

4. The binding affinity of a specific ligand depends only on the conformational state of the enzyme and not on the occupancy of neighboring sites.

Homotropic and heterotropic interactions

The model also distinguishes between *homotropic* interactions and *heterotropic* interactions. When the binding of a ligand is influenced by how much of that same ligand is already bound, this is termed a homotropic interaction; the sigmoidal curve of velocity versus substrate concentration in Figure 17-1 is an example. When the binding of one kind of ligand to its stereospecific site affects the binding of a second kind of ligand to its sterospecific site, this is a heterotropic interaction; the effect of CTP on the curve of velocity versus aspartate concentration in Figure 17-2 is an example.

Algebraic treatment of the MWC model

For the purpose of carrying out the algebraic development of the model, we consider a protein composed of four protomers ($n = 4$). Each protomer can exist in either of two reversibly equilibrating conformational states. Following the designations of MWC, these states are denoted R (relaxed) and T (taut) forms. These are symmetrical states that we can schematically illustrate as

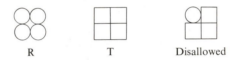

R   T   Disallowed

where each circle or square designates a protomer. Because it lacks symmetry (MWC assumption 3), the structure on the right and other such forms are disallowed as energetically too unfavorable.

According to assumption 4, only two microscopic dissociation constants are required to specify the binding of a ligand F to the enzyme. These microscopic constants are designated k_R and k_T, where k_R may be written, for example, as

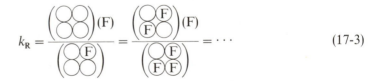

$$(17\text{-}3)$$

Analogous expressions can be written for k_T. Ordinarily, $k_R \neq k_T$; that is, the ligand preferentially binds to one form or the other (if $k_R = k_T$, then the two states are indistinguishable as far as F is concerned).

The binding equilibria of ligand F can be written as

$$R_0 \rightleftarrows T_0$$

$$R_0 + F \rightleftarrows R_1 \qquad\qquad T_1 \rightleftarrows T_0 + F$$

$$R_1 + F \rightleftarrows R_2 \qquad\qquad T_2 \rightleftarrows T_1 + F$$

$$R_2 + F \rightleftarrows R_3 \qquad\qquad T_3 \rightleftarrows T_2 + F \qquad (17\text{-}4)$$

$$R_3 + F \rightleftarrows R_4 \qquad\qquad T_4 \rightleftarrows T_3 + F$$

In this scheme, R_i is defined as the set of all microscopic species of the R conformation that have i bound F molecules. The number of microscopic species that comprise R_i is equal to $\Omega_{n,i}$ with $n = 4$ (see Eqn. 15-14); for example, there are 6 microscopic species for $i = 2$. The same considerations apply to T_i.

In order to calculate the saturation curve of F, it is convenient to work with macroscopic equilibrium constants that are in turn proportional to the appropriate microscopic constants. These macroscopic constants are

$$k_R/4 = (R_0)(F)/(R_1) \qquad (17\text{-}5a)$$

$$2k_R/3 = (R_1)(F)/(R_2) \qquad (17\text{-}5b)$$

$$3k_R/2 = (R_2)(F)/(R_3) \qquad (17\text{-}5c)$$

$$4k_R = (R_3)(F)/(R_4) \qquad (17\text{-}5d)$$

with analogous expressions involving k_T for the T forms of the enzyme. The relationships in Equation 17-5 are easy to derive from the definition of k_R and Equation 15-14 (with $n = 4$). It is convenient to express the concentration as a dimensionless parameter $\alpha = (F)/k_R$, so that the concentration of F is expressed in units of the microscopic dissociation constant k_R. Also, we write $k_T = k_R/c$, where c is a constant. (If $c < 1$, then F binds more strongly to the R state than to the T state.) With these parameters, we obtain the following expressions from Equation 17-5 and analogous expressions for the T species:

$$(R_1) = 4(R_0)\alpha \qquad\qquad (T_1) = 4L(R_0)c\alpha \qquad (17\text{-}6a)$$

$$(R_2) = 6(R_0)\alpha^2 \qquad\qquad (T_2) = 6L(R_0)c^2\alpha^2 \qquad (17\text{-}6b)$$

$$(R_3) = 4(R_0)\alpha^3 \qquad\qquad (T_3) = 4L(R_0)c^3\alpha^3 \qquad (17\text{-}6c)$$

$$(R_4) = (R_0)\alpha^4 \qquad\qquad (T_4) = L(R_0)c^4\alpha^4 \qquad (17\text{-}6d)$$

where L is the conformational equilibrium constant for the $R_0 \rightleftarrows T_0$ interconversion:[§]

$$L = (T_0)/(R_0) \tag{17-7}$$

The equilibrium fractional saturation $\bar{y}_F$ with respect to F is

$$\bar{y}_F = \frac{\sum_i i(R_i) + \sum_i i(T_i)}{4[\sum_i (R_i) + \sum_i (T_i)]} \tag{17-8}$$

where the numerator of the right-hand side is the number of occupied sites, and the denominator is the total number of sites. The various terms in Equation 17-8 are easy to calculate:

$$\sum_{i=0}^{4} (R_i) = (R_0)(1 + 4\alpha + 6\alpha^2 + 4\alpha^3 + \alpha^4) = (R_0)(1 + \alpha)^4 \tag{17-9}$$

$$\sum_{i=0}^{4} (T_i) = L(R_0)(1 + c\alpha)^4 \tag{17-10}$$

$$\sum_{i=0}^{4} i(R_i) = \alpha \frac{d}{d\alpha} \sum_{i=0}^{4} (R_i) = 4(R_0)\alpha(1 + \alpha)^3 \tag{17-11}$$

$$\sum_{i=0}^{4} i(T_i) = 4L(R_0)c\alpha(1 + c\alpha)^3 \tag{17-12}$$

Substituting these relationships into Equation 17-8, we obtain for $\bar{y}_F$ in the case of four protomers

$$\bar{y}_F = \frac{\alpha(1 + \alpha)^3 + Lc\alpha(1 + c\alpha)^3}{(1 + \alpha)^4 + L(1 + c\alpha)^4} \qquad \text{for } n = 4 \tag{17-13}$$

In general, for a protein with n protomers, we have

$$\bar{y}_F = \frac{\alpha(1 + \alpha)^{n-1} + Lc\alpha(1 + c\alpha)^{n-1}}{(1 + \alpha)^n + L(1 + c\alpha)^n} \tag{17-14}$$

[§] In addition to $R_0 \rightleftarrows T_0$, we could have included in Equation 17-4 the conformational equilibria $R_1 \rightleftarrows T_1$, $R_2 \rightleftarrows T_2$, etc. However, these are not thermodynamically independent equilibria; for example, the equilibrium ratio $(R_2)/(T_2)$ is determined by k_R, k_T, and L (see Eqn. 17-6). Therefore, in calculating any *equilibrium* property of the model, we need not consider these other conformational equilibria.

Note that, if $L = 0$, then

$$\bar{y}_F = \alpha/(1 + \alpha) \qquad \text{for } L = 0 \tag{17-15}$$

This is the usual expression that gives a hyperbolic saturation curve for the binding of the ligand F to identical and independent sites on a protein (cf. Eqn. 15-28).

Behavior of $\bar{y}_F$: effects of L and c

It is instructive to consider the behavior of $\bar{y}_F$ under various conditions. For example, consider the case where $n = 4$, $c = 0$, and $L \gg 1$. In this case, the enzyme starts out largely in the T state, but the ligand F binds exclusively to the R state. For this case, Equation 17-13 becomes

$$\bar{y}_F = \frac{\alpha(1 + \alpha)^3}{(1 + \alpha)^4 + L} \qquad \text{for } c = 0 \text{ and } L \gg 1 \tag{17-16}$$

When $\alpha \ll 1$, then $\bar{y}_F \cong \alpha/L$; that is, $\bar{y}_F$ increases linearly with α, but with a shallow slope of L^{-1}. When $1 \cong \alpha \ll L$, then $\bar{y}_F \cong \alpha/L + 3\alpha^2/L + 3\alpha^3/L + \alpha^4/L$; that is, $\bar{y}_F$ begins to change rapidly in response to small changes in L. Finally, when $\alpha \cong L$, then $\bar{y}_F \cong 1$.

Figure 17-6 schematically illustrates this behavior for $n = 4$, $c = 0$, and $L = 10^3$. The sigmoidal features of the curve are quite apparent. It is easy to see that, if a substrate equilibrates rapidly (relative to the rate of catalysis) with its sites on an allosteric enzyme, then the curve of velocity versus substrate concentration will be proportional to $\bar{y}_F$; in such cases, the illustration (Fig. 17-6) is just as applicable to a velocity profile as to a binding curve.

In Equation 17-16, it is clear that both the initial slope, at low α, and the rapidly rising part of the saturation curve are dependent on L. Figure 17-7 gives several

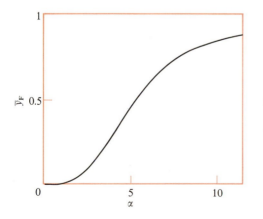

Figure 17-6

Plot of $\bar{y}_F$ versus α, for $n = 4$, $c = 0$, and $L = 10^3$.

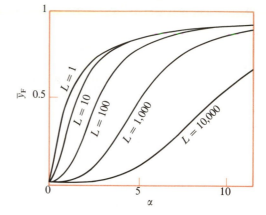

plots of $\bar{y}_F$ versus α for different values of the conformational equilibrium constant L, for the case $c = 0$ and $n = 4$. It is clear that, as L increases, the sigmoidicity associated with the homotropic interactions also increases. Thus, with the ligand binding only to the R state, the greater the initial amount in the T state, the greater is the apparent cooperativity.

Figure 17-8 shows the effect on the saturation curve of varying the parameter c, with $L = 10^3$ and $n = 4$. It is apparent that, as c approaches unity, the sigmoidicity of the saturation curve decreases. As implied earlier, at the limit of $c = 1$, the two conformational forms are indistinguishable as far as the ligand F is concerned.

In general, the greatest cooperativity is achieved when L is large and c is small (or conversely, with F preferentially binding to the T state, the greatest cooperativity occurs with small L and large c). This behavior is a consequence of assumption 3 of the MWC model, which states that symmetry is conserved when the protein switches conformations. This means that all subunits change their conformations in a concerted fashion. Thus, when L is large, most of the protein is initially in the T_0 state.

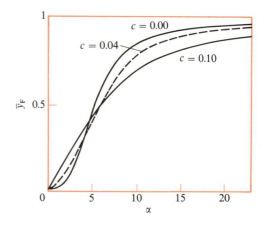

Binding of a ligand F (with high preference for R_0) to give R_1 pulls the overall conformational equilibria of Equation 17-4 slightly to the left, as some T_0 is converted to R_0 to reestablish the $R_0 \rightleftarrows T_0$ equilibrium. However, in the process, a disproportionately large increase in the number of available sites for F has occurred; binding of one F molecule to one site has "pulled" all n protomers (with n sites) into the R state, owing to the concerted nature of the conformational change. In this way, cooperativity results as the number of sites available to F increases many times faster than the number of molecules shifting to the R state. Of course, as more F is added, all the protein eventually shifts to the R form before it is completely saturated; at this point the binding is no longer so sensitive to the F concentration.

These ideas are best visualized by considering the behavior of the parameter $\bar{R}$ (the fraction of molecules in the R state) as a function of (F), and by comparing the behavior of this parameter with that of $\bar{y}_F$.

Comparison of the parameters $\bar{R}$ and $\bar{y}_F$

The parameter $\bar{R}$, the fraction of molecules in the R state, is given by

$$\bar{R} = \sum_{i=0}^{n} (R_i) \bigg/ \sum_{i=0}^{n} [(R_i) + (T_i)] \tag{17-17}$$

$$\bar{R} = (1 + \alpha)^n / [(1 + \alpha)^n + L(1 + c\alpha)^n] \tag{17-18}$$

Equation 17-18 was obtained by substituting into Equation 17-17 the expressions for $\sum_i (R_i)$ and $\sum_i (T_i)$ analogous to those given in Equations 17-9 and 17-10 for the case $n = 4$.

As a concrete example, let $n = 4$, $c = 0$, and $L = 10^3$. The expression for $\bar{y}_F$ is given by Equation 17-16. With $\alpha = 0$, Equation 17-18 gives $\bar{R} \cong L^{-1} = 0.001$; and $\bar{y}_F = 0$. Thus, $\bar{R}$ and $\bar{y}_F$ are essentially equal. However, when $\alpha = 3$, we have $\bar{R} = 0.2$ and $\bar{y}_F = 0.15$; that is, $\bar{R}$ is now significantly larger than $\bar{y}_F$. This means that many molecules are in the R state but are only partially saturated with respect to F; these molecules thus have sites available to F in the conformation to which F exclusively binds (because we assumed $c = 0$). Thus, when $\alpha = 0$, only 0.1% of the molecules (and 0.1% of the total sites) are available for F binding; this results in weak apparent binding when $\alpha = 0$. By the time that $\alpha = 3$, however, about 5% of the total sites are in the R state and are unoccupied ($\bar{R} - \bar{y}_F$ gives the fraction of total sites that are in the R state and unoccupied). Thus, in going from $\alpha = 0$ to $\alpha = 1$, there is an enormous increase in the number of unoccupied sites that are in the proper conformation for binding F. This statistical effect is responsible for the sharp, cooperative rise in the binding. As mentioned, it is a consequence of assuming that all subunits change their conformations in a concerted fashion.

Of course, as α becomes increasingly large, both $\bar{y}_F$ and $\bar{R}$ converge to unity. The completely saturated molecule is entirely in the R state (for the example considered, where $c = 0$).

Effects of allosteric activators and inhibitors

According to the MWC model, allosteric activators and inhibitors bind preferentially to the R (activator) or T (inhibitor) state at sites distinct from the catalytic site so as to give rise to heterotropic interactions.

For simplicity, assume that an activator A binds exclusively to the R form, whereas the inhibitor I binds solely to the T form. Let the microscopic dissociation constants be k_{AR} and k_{IT}, respectively, for activator and inhibitor binding. For generality, assume that both I and A are simultaneously present. We define the parameter L' as

$$L' = \sum_{i=0}^{n} (T_{0,i}) \bigg/ \sum_{i=0}^{n} (R_{0,i}) \tag{17-19}$$

where i in the numerator indexes the number of inhibitor molecules bound to the T form that has no bound F, and i in the denominator indexes the number of activator molecules bound to the R form that has no bound F. Therefore, the expressions in the numerator and denominator of Equation 17-19 are simply the total concentrations of all species that have no bound F; they formerly were designated (T_0) and (R_0), respectively. Hence, L' is exactly analogous to L, and the two parameters are equal when $(I) = (A) = 0$. Proceeding by analogy with Equation 17-9, we obtain

$$\sum_{i=0}^{n} (R_{0,i}) = (R_{0,0})(1 + \gamma)^n \tag{17-20a}$$

$$\sum_{i=0}^{n} (T_{0,i}) = (T_{0,0})(1 + \beta)^n \tag{17-20b}$$

and

$$L' = L(1 + \beta)^n/(1 + \gamma)^n \tag{17-21}$$

where $\beta = (I)/k_{IT}$; $\gamma = (A)/k_{AR}$; and $L = (T_{0,0})/(R_{0,0})$ as before. The expression for $\bar{y}_F$ can now be derived exactly as before (Eqns. 17-8 through 17-14) to give Equation 17-14 with L replaced by L'. Therefore, in the MWC model, the effects of allosteric activators and inhibitors are on the conformational equilibrium constant L.

Figure 17-9 shows the effects of β and γ on the $\bar{y}_F$ versus α plot for the case $n = 4$, $c = 0$, and $L = 10^3$. It is clear that the $\bar{y}_F$ versus (F) profile is very sensitive to (I) and (A).

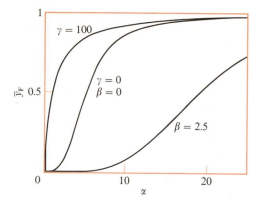

Figure 17-9

Effect of an allosteric activator and an inhibitor on plot of $\bar{y}_F$ versus α, with $c = 0$, $n = 4$, and $L = 10^3$. [After J. Monod, J. Wyman, and J.-P. Changeux, *J. Mol. Biol.* 12:88 (1965).]

17-4 EXPERIMENTAL TESTS OF THE MWC MODEL

Explanation of some data on ligand interactions

The crucial question is whether the MWC model can explain the observed data and can be critically tested by further experimentation. With ATCase, an important observation is that the allosteric inhibitor CTP cannot completely overcome the effects of the substrate aspartate, whereas aspartate can overcome the effects of CTP (Figs. 17-2 and 17-4). This observation can be explained readily by postulating that aspartate binds exclusively to the R state ($c = 0$), and that CTP binds preferentially (but *not* exclusively) to the T state. Therefore, addition of sufficient aspartate will always pull the enzyme entirely into the R state. On the other hand, although CTP will produce inhibition by pulling the enzyme toward the T state, it will not cause a complete conversion to T. Because it is postulated to have some affinity for the R form, the degree to which it pulls the protein into T will be quite sensitive to the concentration of aspartate; at high concentrations of aspartate, the effects of CTP will be reduced because much of it is bound to the R form, whereas at low aspartate concentrations there will be a greater shift toward the T state in the presence of CTP. These effects can be quantitatively calculated from the model by using appropriate values for the various allosteric parameters.

Another important observation is that, at low aspartate concentrations, the competitor malate can be an activator or inhibitor (see Fig. 17-5). The model can account nicely for this behavior. Assuming that L is large, at low aspartate concentrations the enzyme is mainly in the T state. With malate, like aspartate, binding exclusively to the R form, addition of low concentrations of the competitor will promote conversion of some T to R. Moreover, the occupation of only one of n sites ($n = 6$ for ATCase) stabilizes all n sites in the R form. As discussed, at low degrees of saturation the parameter $\bar{R}$ substantially leads the fractional saturation. Therefore, although malate blocks a potential aspartate binding site, it also stabilizes (and shifts the conformational equilibrium toward) R forms that have *unoccupied* aspartate sites. This accounts for the observed activation effect by malate. Of course, as the malate concentration is raised, a greater and greater proportion of the sites are bound to the competitive inhibitor so that inhibition eventually sets in.

Relation between conformational change and fractional saturation

Although the above observations can be explained by the MWC model, a more direct test of some of the features of the model is desirable. Clearly, one of the most important assumptions is that the allosteric phenomena are a consequence of a conformational change. Related to this is the prediction that the conformational change is not colinear with the fractional saturation curve of a cooperatively binding ligand; instead, the conformational change ($\bar{R}$) should lead the fractional saturation. Howard Schachman and colleagues have conducted experiments to explore this question.

An active catalytic trimer (comprised of three C chains) of ATCase can be isolated. When the catalytic trimer is titrated with the substrate analog succinate, a small systematic increase in the sedimentation coefficient s of the trimer is observed. Figure 17-10 plots the percent change in s versus the succinate concentration. It is clear that the sedimentation coefficient shifts between two plateaus as succinate is bound. This indicates that a structural alteration (change in frictional coefficient) in the catalytic trimer occurs upon binding of succinate.

Figure 17-10 also shows a "conformational" titration of the native molecule. In this case, the molecule again undergoes a shift in s between two plateaus. However, a decrease in s, rather than an increase, is observed as succinate is bound; this difference between the intact molecule and the isolated catalytic trimer might be due to

Figure 17-10

Conformational titration of ATCase and the catalytic subunit. The percentage change in sedimentation coefficient is plotted against the log of the succinate concentration. [After M. W. Kirschner and H. K. Schachman, *Biochemistry* 10:1919 (1971).]

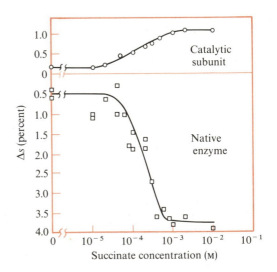

additional changes in the intact structure beyond those that occur in the catalytic trimer. In any event, the data clearly suggest that a conformational change does occur as ligand is bound, and thus they support one feature of the model.

However, the more critical question is that of the relationship between the fractional change in conformation and the fractional saturation with respect to succinate. This question has been approached by introducing an optical probe into the C chains; the probe undergoes spectral changes in response to the binding of succinate. For this purpose, catalytic trimers were treated with tetranitromethane to give approximately one nitrotyrosine residue per C chain. This residue has an absorption maximum at 430 nm, well removed from the major absorption of the protein, which is below 300 nm. To a good approximation, the nitrated trimer catalytically behaves like the unmodified one, thus indicating that nitration does not seriously perturb the molecule.

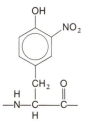

Ordinarily, experiments are done with enough carbamoyl phosphate to saturate the protein. Binding of succinate can be followed spectrophotometrically at 430 nm. Titration of the catalytic trimer to complete saturation with succinate gives an approximately 14% decrease in absorption at 430 nm. Moreover, there is no evidence for cooperativity with the trimer. This means that cooperativity requires contributions from the quaternary structure of the intact enzyme. In addition, the fractional spectral changes due to succinate binding exactly match the fractional sedimentation changes in the catalytic trimer. This indicates that binding of succinate and the conformation change are parallel processes.

When the nitrated catalytic chains are incorporated with the R chains into the native enzyme, the resulting modified enzyme is similar in its catalytic behavior to the unmodified enzyme. Thus, in both the isolated trimer and in the intact enzyme, the introduction of the nitro group into the C chains does not produce a significant perturbation. Moreover, when the enzyme is saturated with succinate, the change in absorption per C chain at 430 nm is almost identical for the intact enzyme and for the isolated catalytic trimer.

However, the profile of the fractional spectral change versus the succinate concentration is quite different for the isolated catalytic trimer than for the intact enzyme. Figure 17-11 compares the hyperbolic binding of succinate to the trimer with the sigmoidal, cooperative binding to the intact molecule. This shows that cooperativity is a feature of the intact molecule, and requires the presence of the R chains.

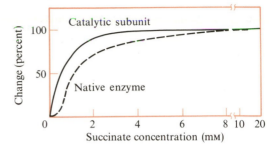

Figure 17-11

Fractional spectral change (at 430 nm) as a function of succinate concentration, for the nitrated catalytic subunit and for nitrated ATCase. [After M. W. Kirschner and H. K. Schachman, *Biochemistry* 12:2997 (1973).]

Figure 17-12 compares, for the intact enzyme, the plot of fractional spectral change versus the succinate concentration with the analogous plot of the fractional sedimentation change. It is apparent that the conformational change, as monitored by sedimentation, leads the spectral change, which monitors binding of succinate. This behavior is exactly that predicted by the MWC model.

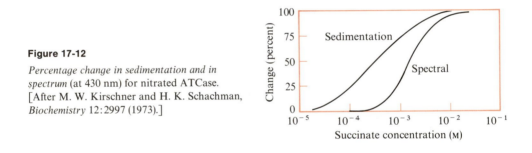

Figure 17-12

Percentage change in sedimentation and in spectrum (at 430 nm) for nitrated ATCase. [After M. W. Kirschner and H. K. Schachman, *Biochemistry* 12:2997 (1973).]

In summary, not only can some kinetic observations be explained by the model, but direct experiments demonstrate the existence of a conformational change and verify a key prediction concerning the relationship between conformational change and fractional saturation.

17-5 ALTERNATIVE MODELS FOR ALLOSTERIC PROTEINS

Even though many observations with ATCase and with other enzymes can be explained by the MWC model, it appears clear that the model cannot account for the behavior of all systems. Moreover, although many of the main features of ATCase are consistent with the MWC proposal, the model does not explain all observations that have been made even on this enzyme.

One of the most obvious limitations of the MWC model is its inability to accommodate *anticooperativity.* In the model, a ligand always pulls the conformational equilibrium toward the form to which it preferentially binds. This acts only to enhance further binding of the same ligand. Under no conditions can a situation be obtained in which binding becomes substantially weaker as saturation proceeds.

Apparent anticooperativity *has* been observed in ligand binding, although there is some question whether true anticooperativity or other effects are responsible for the observed anticooperativity (see Gennis, 1976). Nevertheless, it is clear that, in a truly anticooperative system, an alternative to the MWC model must be used to rationalize the observations.

A sequential model

Because of limitations in the MWC model, some data are interpreted in terms of a *sequential model*. This model has been extensively developed by D. E. Koshland and colleagues (Koshland et al., 1966). The essential idea is that ligand binding produces a sequential set of structural changes in the protein, so that a series of intermediate conformational forms are obtained. The molecule is not restricted to symmetrical conformations, so many possible conformational forms can exist. In principle, any given conformational state can assume its own unique values for the microscopic dissociation constants that characterize ligand-binding sites; as the protein binds ligand and switches from one state to another, the successive dissociation constants can increase or decrease. For this reason virtually any type of binding behavior, including anticooperativity, can be accommodated.

A simple example of a sequential scheme is

$$\square + 4F \rightleftharpoons \square_F + 3F \rightleftharpoons \square_{FF} + 2F \rightleftharpoons \square_{FFF} + F \rightleftharpoons \square_{FFFF} \quad (17\text{-}22)$$

In this example, each subunit is assumed to switch from one conformational state to another as it binds ligand. Thus, the scheme in Equation 17-22 indicates that ligand binding *induces* the conformational switch in the subunit to which it binds. The conformational change may occur in part to achieve a better fit between the binding site and the ligand. When a subunit changes conformation, subunit–subunit interactions may change so that the ligand affinities of unoccupied subunits may be altered. Moreover, in an unsymmetrical intermediate such as

the total subunit–subunit interactions experienced by any given subunit can be unique for each subunit; these interactions in turn can give rise to different ligand dissociation constants for each of the unoccupied sites in an intermediate conformational form. This makes it possible to explain a wide variety of data.

A more general scheme

Figure 17-13 shows a more general scheme for allosteric interactions. This scheme allows the individual subunits to take on freely either of two conformational forms, regardless of the number of ligands that are bound. For a four-subunit protein, this

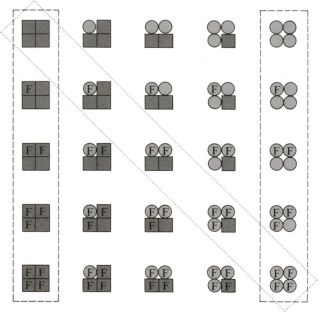

Figure 17-13

A general allosteric scheme. (See text for explanation.)
[After G. G. Hammes and C.-W. Wu, *Science* 172:1205 (1971).]

allows 25 different forms. It is seen that the MWC model is a limiting case of this scheme, involving only the species enclosed by dashed rectangles, whereas the simple sequential scheme involves the forms enclosed by the diagonal dotted rectangle.

More complex schemes (involving more conformational forms of the subunits) than that given in Equation 17-22 or in Figure 17-13 can also be envisioned. In general, one attempts to find the simplest scheme that will accommodate all observations. For this reason, data on allosteric proteins often are interpreted in terms of the MWC model, particularly because it uses only a few parameters. However, when observations cannot be accommodated by the simple MWC framework, schemes that can accommodate more parameters, such as the simple sequential model of Equation 17-22, must be considered.

17-6 HEMOGLOBIN

As another example of a well-studied allosteric protein, we briefly review some of the main features of the hemoglobin system. Some aspects of hemoglobin structure and function are sketched in Chapter 2. Hemoglobin is composed of four polypeptides, two α chains and two β chains, to give an $\alpha_2\beta_2$ structure. The α and β chains are structurally similar; each is about 150 amino acid residues in length and contains a single heme with an iron atom in the Fe^{2+} state. The tetrameric protein combines reversibly with four molecules of oxygen.

Cooperative oxygen binding

Figure 17-14 compares the oxygenation curves of hemoglobin and myoglobin. The latter protein is a single polypeptide, similar to the individual hemoglobin chains and containing only one heme group. It is clear from the curves that, whereas myoglobin

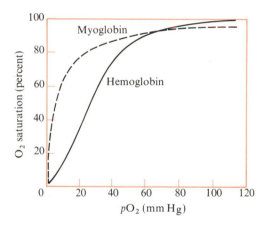

Figure 17-14

Oxygenation curves of hemoglobin and myoglobin at pH 7.4 and 38°C. [After F. Daniels and R. A. Alberty, *Physical Chemistry*, 4th ed. (New York: Wiley, 1975).]

shows the normal behavior for combination of a ligand at a single site, hemoglobin shows cooperative oxygen binding. As oxygen pressure increases, myoglobin combines with oxygen to 30% saturation before any significant association of oxygen with hemoglobin occurs. On the other hand, as a result of the steep rise in the hemoglobin curve, the two curves intersect by 90% saturation.

The steep dependence of the fractional saturation of hemoglobin on oxygen pressure has important physiological consequences. For example, it enables the molecule to respond to small changes in oxygen tension by making sizeable adjustments in the amount of oxygen released or bound.

Treatment of oxygenation equilibrium: the Adair scheme

The oxygenation of hemoglobin can be treated by one or more of the allosteric models, such as the MWC or the sequential model or schemes that incorporate features of both. However, it is instructive to treat the problem according to a simple phenomenological scheme first used by G. S. Adair in 1925. Although the phenomenological description obscures certain of the parameters associated with a more detailed model, it has advantages for giving simple insights into the system.

The binding of oxygen may be described in terms of four macroscopic equilibria:

$$H_0 + O_2 \rightleftarrows H_1 \qquad (17\text{-}23a)$$

$$H_1 + O_2 \rightleftarrows H_2 \qquad (17\text{-}23b)$$

$$H_2 + O_2 \rightleftarrows H_3 \qquad (17\text{-}23c)$$

$$H_3 + O_2 \rightleftarrows H_4 \qquad (17\text{-}23d)$$

where H_i represents the hemoglobin species that have i bound oxygen molecules. For each stage in the reaction, we can assign a macroscopic association constant K_j, where

$$K_j = (H_{j-1})pO_2/(H_j) \qquad (17\text{-}24)$$

and pO_2 is the partial pressure of oxygen. The fractional saturation $\bar{y}$ then is

$$\bar{y} = \sum_{i=0}^{4} i(H_i) \Big/ \left(4 \sum_{i=0}^{4} (H_i) \right) \qquad (17\text{-}25a)$$

$$\bar{y} = \frac{\displaystyle\sum_{i=1}^{4} \left(i(pO_2)^i \prod_{j=1}^{i} (1/K_j) \right)}{4\left[1 + \displaystyle\sum_{i=1}^{4} \left((pO_2)^i \prod_{j=1}^{i} (1/K_j) \right) \right]} \qquad (17\text{-}25b)$$

Equation 17-25b follows from Equation 17-25a through use of the equilibrium-constant relationships in Equation 17-24. The macroscopic constants K_j may be expressed in terms of microscopic constants k_j, where we define k_j as the intrinsic dissociation constant for dissociating one O_2 molecule from any one of the microspecies containing j bound oxygen molecules. Thus, k_j is viewed as identical for all of the microspecies that contain j bound oxygen molecules. The relationship between K_j and k_j is given by an equation analogous to Equation 15-20:

$$K_j = (\Omega_{nj-1}/\Omega_{nj})k_j \qquad (17\text{-}26)$$

with $n = 4$ and Ω_j given by Equation 15-14. For example, with $j = 2$, we obtain $K_j =$

$2k_j/3$. Substituting Equation 17-26 into Equation 17-25b, we obtain

$$\bar{y} = \frac{(pO_2/k_1) + [3(pO_2)^2/k_1k_2] + [3(pO_2)^3/k_1k_2k_3] + [(pO_2)^4/k_1k_2k_3k_4]}{1 + (4pO_2/k_1) + [6(pO_2)^2/k_1k_2] + [4(pO_2)^3/k_1k_2k_3] + [(pO_2)^4/k_1k_2k_3k_4]}$$

(17-27)

Estimates of microscopic constants for oxygen binding

Equation 17-27 may be used in conjunction with accurate measurements of $\bar{y}$ at a large number of pO_2 values in order to obtain estimates of the k_j parameters. Table 17-1 lists some experimental values of k_j obtained at 25°C in solutions buffered to pH 7.4 in the presence and absence of 0.1 M NaCl. In the absence of salt, note the progressive decrease in k_j as oxygenation proceeds; the largest change occurs between k_2 and k_3. It is apparent that the last oxygen binds over 30-fold more strongly than the first one.

Table 17-1

Estimated microscopic dissociation constants for human hemoglobin

	k_1	k_2	k_3	k_4
− NaCl	8.8	6.1	0.85	0.25
+0.1 M NaCl	42	13	12	0.14

NOTE: Values of k_1 are given in units of mm Hg. These values are estimated for a solution with 10 mM Tris buffer (pH 7.4) at 25°C.
SOURCE: Data from I. Tyuma, K. Imai, and K. Shimizu, *Biochemistry* 12:1491 (1973).

If 0.1 M NaCl is added to the solution, each k_j value becomes larger, except for k_4; the oxygenation curve is shifted to the right. As we shall see, this shift can be rationalized in terms of a stabilization of the deoxy and partial-deoxy forms of hemoglobin by anions. Note also that there is a roughly 300-fold difference between k_1 and k_4. This is an enormous increase in affinity for the last oxygen versus the first one.

Concentrations of hemoglobin species during oxygenation

The k_j values are useful for calculating concentrations of the various species H_j. Figure 17-15 plots H_j concentrations versus pO_2 for the case where all k_j values are

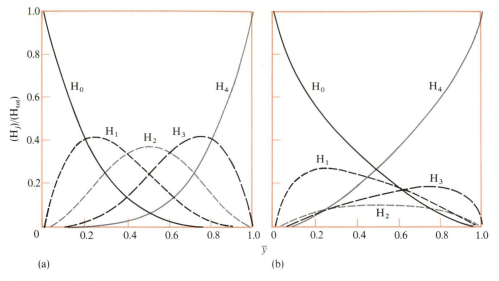

Figure 17-15

Fraction of hemoglobin as H_j *versus* $\bar{y}$. (H_{tot} = total hemoglobin) **(a)** With all k_j values equal. **(b)** For the k_j values ($-$NaCl) given in Table 17-1. [After I. Tyuma, K. Imai, and K. Shimizu, *Biochemistry* 12:1491 (1973).]

equal (Fig. 17-15a) and for the k_j values ($-$NaCl) given in Table 17-1 (Fig. 17-15b). The figure illustrates the physical consequences of the cooperativity. When all k_j values are equal, each of the intermediate species H_1, H_2, and H_3 at some point during the titration rises to about 40% of the total hemoglobin concentration. In contrast, these intermediate forms are suppressed in the real situation where the value of k_j decreases with increasing j. Moreover, (H_4) rises much sooner in Figure 17-15b than it does in Figure 17-15a; for example, when $\bar{y} = 0.5$, the H_4 concentration represents only a few percent of the total in Figure 17-15a, but is about 30% of the total in Figure 17-15b. Thus, to a certain extent, the cooperative hemoglobin oxygenation equilibrium goes directly between the species H_0 and H_4, without building up large amounts of the intermediate forms.

The Hill constant and site–site interaction energy

It is clear from the fit of the oxygenation data to Equation 17-27 that the hemoglobin oxygenation equilibrium does not follow the simple expressions given by Equation 15-43 or 15-45. This means that the apparent Hill constant, as defined by Equation 15-46, will vary the degree of oxygenation. Figure 17-16 is a plot of α versus pO_2 for the case of stripped hemoglobin—that is, hemoglobin treated to remove bound organic phosphates (see below)—in low buffer concentrations. The values of α were calculated from Equation 15-46. The Hill parameter rises and falls in a bell-shaped fashion as oxygenation proceeds. A value of $\alpha = 2.5$ is obtained near the maximum,

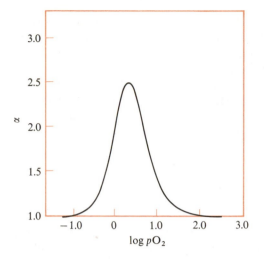

Figure 17-16

Plot of α *versus* $\log pO_2$ for stripped hemoglobin. [After I. Tyuma, K. Imai, and K. Shimizu, *Biochemistry* 12:1491 (1973).]

which occurs at about $\bar{y} = 0.5$. Thus, the greatest degree of apparent cooperativity occurs around the midrange of the saturation curve.

It is of interest to examine the magnitude of the interaction energy between sites i and j that gives rise to the cooperative binding. We did just this in Chapter 15 (following the discussion of Eqn. 15-39) for the case $i = 1$ and $j = 4$. (With these values of i and j, we actually are comparing the affinity of a vacant site when all other sites are occupied versus that when all other sites are vacant.) Using data in Table 17-1, we calculate the value of the interaction energy $\Delta G_{\mathrm{I},ij}$ to be -2.1 kcal mole^{-1} site^{-1} in the absence of NaCl. Thus, for the whole molecule with four sites, the total interaction energy is about -8.4 kcal mole^{-1} (25°C).

Bohr effect

The association of oxygen with hemoglobin is strongly dependent on pH. Figure 17-17 plots the oxygen saturation (in percent) versus pO_2 at three different pH values. As pH

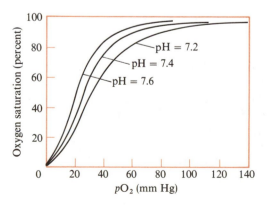

Figure 17-17

The Bohr effect: the effect of pH on the oxygenation curve of hemoglobin. [After R. E. Benesch and R. Benesch, *Adv. Protein Chem.* 28:211 (1974).]

is lowered below 7.6, there is a shift to the right in the oxygenation curves. This shift enables hemoglobin to release more oxygen in the periphery as carbonic acid (from the CO_2 released by respiring tissues) builds up in the red cell.

It is clear that a linkage exists between oxygen binding and the binding of hydrogen ions to certain sites on hemoglobin. This reciprocal relationship between oxygen and H^+ binding is commonly known as the *Bohr effect*. It originates from the variations in certain pK values between deoxyhemoglobin and oxyhemoglobin.

To examine the Bohr effect in more detail, we return to our linkage relationship of Equation 15-70, and let $L_1 = H^+$ and $L_2 = O_2$. We thus obtain

$$\left(\frac{\partial(\ln pO_2)}{\partial[\ln(H^+)]}\right)_{\bar{y}_{O_2}} = -\frac{n}{m}\left(\frac{\partial\bar{y}_{H^+}}{\partial\bar{y}_{O_2}}\right)_{(H^+)} \tag{17-28}$$

where it is assumed that n protons are linked to the binding of m $(=4)$ oxygen molecules. Also, the fractional saturation $\bar{y}_{H^+}$ is only with respect to the n oxygen-linked protons (not with respect to all the protons that can bind to the protein). It is found experimentally that the oxygenation curves are shifted by pH without a significant change in their shapes. However, the magnitudes of the shifts are dependent on pH. This means that the derivative on the left-hand side of Equation 17-28 is virtually independent of $\bar{y}_{O_2}$ and depends only on pH (at any given temperature). It is customary to evaluate the derivative at $\bar{y}_{O_2} = 1/2$. With this in mind, and noting that $\ln(H^+) = -2.3 \times$ pH, Equation 17-28 becomes

$$\left(\frac{\partial[\log(pO_2)_{1/2}]}{\partial(\text{pH})}\right)_{\bar{y}_{O_2}=1/2} = \frac{n}{m}\left(\frac{\bar{y}_{H^+}}{\partial\bar{y}_{O_2}}\right)_{\text{pH}} \tag{17-29}$$

where $(pO_2)_{1/2}$ is the value of pO_2 at half-saturation. The right-hand side of Equation 17-29 is the change in the number of moles of bound H^+ upon binding a mole of O_2. Using notation commonly used in the literature, we rewrite Equation 17-29 as

$$\left(\frac{\partial[\log(pO_2)_{1/2}]}{\partial(\text{pH})}\right)_{y_{O_2}=1/2} = -\Delta\bar{H}^+ \tag{17-30}$$

where $\Delta\bar{H}^+ = \bar{H}^+_{\text{deoxy}} - \bar{H}^+_{\text{oxy}}$ is the difference in bound H^+ per heme between deoxy and oxy forms.

The quantity $\Delta\bar{H}^+$ can be obtained by titrations and compared with that calculated from the observed variation of $\log(pO_2)_{1/2}$ with pH. Two types of titration can be performed. In one, the change in pH accompanying the complete oxygenation of deoxyhemoglobin is measured; this then is used in conjunction with a titration curve of oxyhemoglobin to determine the number of moles of H^+ that were absorbed or released upon oxygenation of deoxyhemoglobin, at the fixed pH of the original deoxyhemoglobin solution. A second procedure is to oxygenate deoxyhemoglobin completely and then titrate back (adding acid or base) to restore the pH to its original value (the value before oxygenation); this procedure gives a direct measure of $\Delta\bar{H}^+$.

Figure 17-18 shows results obtained by titration at 30°C. The figure gives values of $\Delta\bar{H}^+$ over the approximate range of pH 5 to 9. It is apparent that $\Delta\bar{H}^+$ is positive above about pH 6—that is, protons are released upon oxygenation in this pH range. Around pH 7.2, the maximal value is reached—about 0.5 mole H^+ released per mole of oxygen bound.

Equation 17-30 may be integrated to give the relationship between $\log(pO_2)_{1/2}$ and pH, with an arbitrary constant of integration. The integration can be executed numerically with the data from Figure 17-18. The relationship between $\log(pO_2)_{1/2}$ and pH so obtained (which is based on titration data) may then be compared with direct, independent measurements of $\log(pO_2)_{1/2}$ versus pH.

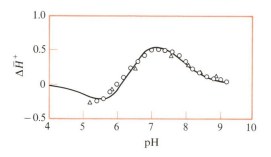

Figure 17-18

The effect of pH on $\Delta\bar{H}^+$ for human hemo-globin at 30°C. The curve is calculated from the constants in Table 17-2. The points correspond to two different types of experimental measurements. [After E. Antonini et al., *J. Biol. Chem.* 240:1096 (1965).]

Figure 17-19 makes such a comparison. The dashed curve shows the behavior predicted from the titration data of Figure 17-18, which were numerically integrated. The points represent direct measurements of $\log(pO_2)_{1/2}$ versus pH. The agreement between the two approaches is excellent, thus confirming the validity of Equation 17-30.

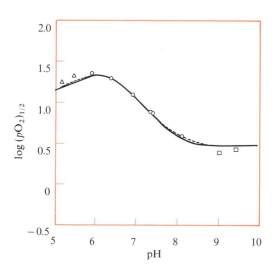

Figure 17-19

The Bohr effect in human hemoglobin at 30°C. The points are experimental values of $\log(pO_2)_{1/2}$ obtained from oxygenation curves measured in three different buffer systems. The solid line is calculated from the constants in Table 17-2. The dashed line comes from a graphical integration of the data in Figure 17-18. [After E. Antonini et al., *J. Biol. Chem.* 240:1096 (1965).]

We should note that, if an additional ion binds preferentially to one form of hemoglobin, the presence of sufficient quantities of that ion could seriously affect the agreement between the two approaches. For example, ions in the buffers used for measuring oxygenation curves (or the organic phosphates such as diphosphoglycerate) could produce such effects. The organic phosphates bind in a pH-dependent fashion, with a strong preference for deoxyhemoglobin. The presence of such a third component clearly invalidates the direct application of Equation 17-30, which is based on the assumption of only two linked components.

Estimating pk values of oxygen-linked ionizations

The data in Figure 17-18 may be used to estimate pk values for the oxygen-linked ionizable groups. The assumption of two ionizable groups per heme generally is sufficient to account for the data. We designate these microscopic dissociation constants as k_1' and k_2' for deoxyhemoglobin, with k_1'' and k_2'' as the constants for the same dissociations from oxyhemoglobin. To calculate the average number of moles of bound Bohr protons per heme in deoxyhemoglobin ($\bar{H}_{deoxy}^+$) and in oxyhemoglobin ($\bar{H}_{oxy}^+$), we use the linkage relationships developed in Chapter 15. In Equations 15-77 and 15-78, we let $M^{(j)}$ (with $j = 4$) and $M^{(0)}$ represent oxyhemoglobin and deoxyhemoglobin, respectively. From Equations 15-52 through 15-54 (see also Eqns. 15-77 through 15-81), we have

$$\bar{H}_{deoxy}^+ = \frac{d\{\ln[1 + (H^+)/k_1'][1 + (H^+)/k_2']\}}{d[\ln(H^+)]} \tag{17-31a}$$

$$= \{(H^+)/[k_1' + (H^+)]\} + (H^+)/[k_2' + (H^+)] \tag{17-31b}$$

$$\bar{H}_{oxy}^+ = \frac{d\{\ln[1 + (H^+)/k_1''][1 + (H^+)/k_2'']\}}{d[\ln(H^+)]} \tag{17-32a}$$

$$= \{(H^+)/[k_1'' + (H^+)]\} + (H^+)/[k_2'' + (H^+)] \tag{17-32b}$$

and
$$\Delta\bar{H}^+ = \bar{H}_{deoxy}^+ - \bar{H}_{oxy}^+ \tag{17-33}$$

Equations 17-31 and 17-32 also can be derived directly by using the expressions for the ionization constants, and without the formalism of Equations 15-52 through 15-54 (or 15-77 through 15-81).

Table 17-2 gives the four pk values for human hemoglobin at 30°C, based on the data of Figure 17-18. Note that, upon oxygenation, pk_1' shifts upward by 0.8 units, whereas pk_2' shifts downward by 1.4 units. The large downward shift in pk_2' is responsible for the liberation of protons when hemoglobin is oxygenated at pH values above about pH 6. Below pH 6, the upward shift in pk_1' upon oxygenation dominates, and there is a net absorption of H^+. The two pk shifts counterbalance each other around pH 6, where $\Delta\bar{H}^+ = 0$.

Table 17-2

pk values for oxygen-linked ionizable
groups in human hemoglobin

pk_1' (deoxy)	pk_1'' (oxy)	pk_2' (deoxy)	pk_2'' (oxy)
5.5	6.3	7.6	6.2

NOTE: Values are for 30°C.
SOURCE: Data from E. Antonini et al., *J. Biol. Chem.*
240:1096 (1965).

Using heats of ionization and other considerations, it has been suggested that
the alkaline group (pk_2') is a histidine, and the acid group (pk_1') is a carboxyl. Crystallographic data suggest that the C-terminal histidine side chain on the β chain and
the N-terminal amino group on the α chain may be responsible for pk_2'; C-terminal
carboxyls (and possibly also internal aspartic acid carboxyl side chains) may account
for pk_1'. The importance of the β-chain histidine and of the C-terminal carboxyls
also has been established by chemical studies.

Influence of organic phosphates on the oxygenation curve

Salts (and particularly organic phosphates) have a profound influence on the oxygenation curve of hemoglobin. This effect was discovered by Reinhold and Ruth Benesch
(1967) and by A. Chanutin and R. R. Curnish (1967). The Beneschs have continued
to study this phenomenon for many years.

Figure 17-20 shows the oxygenation profiles at 30°C of whole blood and of
hemoglobin in 10 mM NaCl, free of all organic phosphates (sometimes called
"stripped" hemoglobin). Although cooperativity is preserved, the entire curve for
whole blood is shifted substantially to the right.

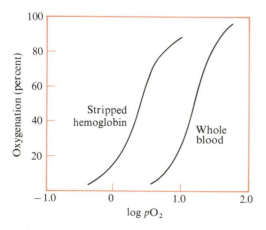

Figure 17-20

*Comparison of the oxygenation curves of
stripped hemoglobin and whole blood.* [After
R. E. Benesch and R. Benesch, *Adv. Protein
Chem.* 28:211 (1974).]

The molecule present in whole blood that most prominently affects hemoglobin's oxygenation equilibrium is 2,3-diphosphoglycerate (DPG):

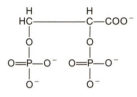

2,3-diphosphoglycerate

Figure 17-21 shows the effect of small concentrations of DPG on the oxygenation curve. It is apparent that DPG concentrations as low as 10^{-4} M profoundly shift the curve, and that increasing amounts of DPG cause successively larger shifts to higher oxygen pressures.

Figure 17-21

Effect of DPG on oxygenation curves of hemoglobin. [After R. E. Benesch et al., *Nature New Biol.* 234:174 (1971).]

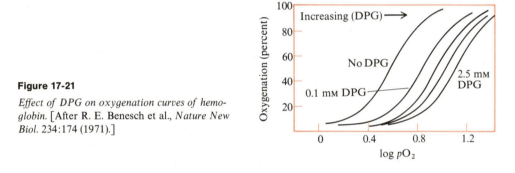

The effect of DPG at low concentrations is due to its strong and preferential binding to deoxyhemoglobin. The dissociation constant for DPG and deoxyhemoglobin is about 15 μM. Because of its marked preference for the deoxy form, DPG exerts a strong influence to shift the hemoglobin oxygenation equilibrium toward the deoxy state. In the red blood cell, the effect is particularly pronounced, because the intracellular concentration of DPG is about 4 mM, which is well above the dissociation constant.

Figure 17-22 shows the oxygenation curve of whole blood and indicates the partial oxygen pressures of venous and arterial blood. At the arterial oxygen pressure, hemoglobin is virtually saturated with oxygen; at the venous pressure, it is about 60% saturated; therefore, approximately 40% of the oxygen is unloaded in the tissues. Because in the absence of DPG the curve is displaced substantially to the left (cf. Figs. 17-20 and 17-21), it is clear that without DPG (or a DPG substitute) little if any oxygen can be released at the venous oxygen pressure. In addition, however, the intracellular concentration of DPG is subject to regulation. This enables the host to adapt to environmental conditions that require a shift in the oxygenation curve.

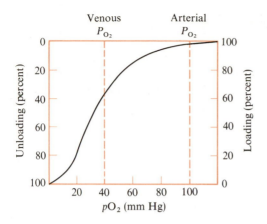

Figure 17-22

Oxygenation curve of whole blood in relation to venous and arterial oxygen pressures. [After R. E. Benesch and R. Benesch, *Adv. Protein Chem.* 28:211 (1974).]

An example of how shifts in DPG levels enable adaptation is provided by the effects on oxygen affinity of exposure to high altitudes. Figure 17-23 plots the partial oxygen pressure at half-saturation, $(pO_2)_{1/2}$, versus time of exposure to high altitude and of exposure to sea-level oxygen pressure. The figure also shows the variation in red blood cell DPG concentration. It is clear that, upon moving from sea level to high altitude, there is a sharp rise in DPG concentration and a concomitant elevation in $(pO_2)_{1/2}$ until a new plateau level in each of these parameters is reached. The shift to the right in the oxygenation curve at high altitude enables hemoglobin to release more oxygen to the tissues than if no shift occurred. Conversely, with a return to sea level, the DPG concentration and $(pO_2)_{1/2}$ fall toward their previous levels.

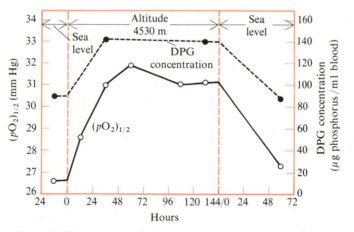

Figure 17-23

Effects of exposure to high altitude on oxygen affinity of blood and on concentration of DPG (2,3-diphosphoglycerate). [After C. Lenfant et al., *J. Clin. Invest.* 47:2652 (1968).]

17-7   INTERACTION OF CARBON DIOXIDE WITH HEMOGLOBIN

Since the early part of this century, it has been known that CO_2 interacts with hemoglobin. In particular, it was found that CO_2 significantly affects the oxygenation curve by shifting it to the right. Although this observation was made many years ago, only recently has it been understood in any depth.

The equilibria that must be considered are

$$CO_2 \text{ (gas)} \rightleftarrows CO_2 \text{ (dissolved)}$$

$$CO_2 + H_2O \rightleftarrows H_2CO_3$$

$$CO_2 + OH^- \rightleftarrows HCO_3^-$$

$$H_2CO_3 \rightleftarrows HCO_3^- + H^+$$

$$HCO_3^- \rightleftarrows CO_3^{2-} + H^+$$

$$HbNH_3^+ \rightleftarrows HbNH_2 + H^+$$

$$CO_2 + HbNH_2 \rightleftarrows HbNHCOOH$$

$$HbNHCOOH \rightleftarrows HbNHCOO^- + H^+$$

In these equations, the first equilibrium concerns the dissolving of CO_2; all subsequent steps with CO_2 are assumed to represent the dissolved species. $HbNH_2$ designates an amino group on hemoglobin (Hb) that combines with CO_2 to give a carbamate.

Work by J. V. Kilmartin and L. Rossi-Bernard (1973) established the identity of the reactive amino groups. They prepared hemoglobin tetramers in which different N-terminal α-amino groups were reacted with cyanate. These tetramers consisted of one species in which all four of the α-amino groups were reacted, one species in which those on the β chains only were reacted, and a third species in which those on the α chains only were blocked. It was found that, with the species in which all four groups were blocked, there is no effect of CO_2 on the oxygenation profile. On the other hand, with the species in which only two of the N-terminal groups are blocked, the effect of CO_2 on the oxygenation curve is about one-half of that observed with the totally unmodified hemoglobin. These results thus establish that the terminal amino groups are the sites for combination with CO_2. They combine more strongly when hemoglobin is in the deoxy form than in the oxy form.

DPG binds to the N-terminal α-amino groups of the β chains (see below). Therefore, a competition between DPG and CO_2 binding is expected. This indeed has been confirmed by experiment. Of course, the α-amino groups on the α chains can still react with CO_2 even in the presence of DPG.

Together, both DPG and CO_2 act to regulate the oxygen affinity of hemoglobin. Figure 17-24 shows the oxygenation curve for whole blood, for stripped hemoglobin, and for stripped hemoglobin plus DPG, plus CO_2, and plus both DPG and CO_2. It is clear that the greatest shift in the curve is achieved with CO_2 and DPG mixed together, and that the resulting oxygenation profile corresponds closely to that of

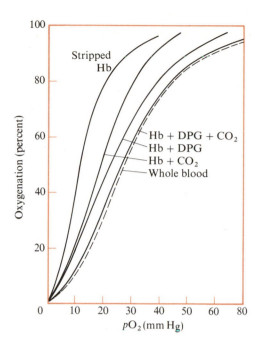

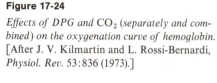

Figure 17-24

Effects of DPG and CO_2 (separately and combined) on the oxygenation curve of hemoglobin. [After J. V. Kilmartin and L. Rossi-Bernardi, *Physiol. Rev.* 53:836 (1973).]

whole blood measured under the same conditions. Thus, it is necessary to consider the effects of both CO_2 and DPG in order to account for the oxygenation curve of whole blood.

Apart from regulating the oxygenation profile of hemoglobin, the reaction with CO_2 provides a mechanism for CO_2 transport by the blood. Thus, after unloading oxygen to the tissues, the deoxyhemoglobin can combine with CO_2 released from tissues and carry it to the lungs where it is eventually expelled. The amount of CO_2 that is transported as a result of its reaction with hemoglobin is estimated as 10% to 15% of the total CO_2 transport.

17-8 HEMOGLOBIN STRUCTURE IN RELATION TO MECHANISM

In recent years considerable progress has been made in elucidating the crystal molecular structure of hemoglobin and in relating the structure to a variety of experimental findings. Max Perutz has directed the crystallographic work on this protein. The structure determinations have been carried out with deoxyhemoglobin and with ferric hemoglobin. The latter is believed to be isomorphous with oxyhemoglobin (which has O_2 bound to Fe^{2+}); however, the oxy form is rapidly oxidized and, therefore, not readily studied.

In both forms the subunits are arranged around a twofold axis of symmetry. Each subunit is similar in structure to myoglobin. There are eight helical segments, which are folded in a specific pattern.

Figure 1-5a gives a schematic view of the horse deoxyhemoglobin molecule, looking down the twofold axis. Figure 1-5b gives the same view for horse oxyhemoglobin. As discussed below, oxygenation produces a significant conformational reorganization that involves the rupture of certain salt bridges. Finally, Figure 17-25 shows another view of deoxyhemoglobin.

A critical difference between deoxyhemoglobin and oxyhemoglobin is that the former has an extensive network of salt bridges that are broken in the latter. In deoxyhemoglobin, each β chain makes an intrachain salt bridge and also makes a bridge to an α chain; this gives a total of four salt bridges involving the two β chains. In addition, there is a total of four salt bridges between the two α subunits. Thus, there are eight salt bridges in deoxyhemoglobin that are absent in oxyhemoglobin. These are all indicated in Figure 17-25. In addition, Table 17-3 lists these eight electrostatic bonds.

Table 17-3

Salt bridges in deoxyhemoglobin

Between Alpha Chains:

Lys^{127} side chain of α_1 with carboxyl group of C-terminal Arg^{141} of α_2
Lys^{127} side chain of α_2 with carboxyl group of C-terminal Arg^{141} of α_1
C-terminal Arg^{141} side chain of α_1 with carboxyl group of Asp^{126} of α_2
C-terminal Arg^{141} side chain of α_2 with carboxyl group of Asp^{126} of α_1

Between Alpha and Beta Chains:

Lys^{40} side chain of α_1 with carboxyl group of C-terminal His^{146} of β_2
Lys^{40} side chain of α_2 with carboxyl group of C-terminal His^{146} of β_1

Intra–Beta Chain:

C-terminal His^{146} side chain with carboxyl group of Asp^{94}

NOTE: Individual α and individual β chains are distinguished by numbers (α_1 and α_2, β_1 and β_2). The salt bridges are shown in Figure 17-25.

In oxyhemoglobin, all four chains are rotated slightly from their positions in deoxyhemoglobin. The major shift is in the β chains, which move toward each other by about 7 Å; this movement is obvious from a comparison of Figure 1-5a with Figure 1-5b.

The cause for the conformational change can be traced to the effect of O_2 as it combines with a heme iron. In deoxyhemoglobin, each heme iron is coordinated with five ligands. Figure 17-26 shows that four of the ligands come from the porphyrin ring, whereas the fifth is from an imidazole nitrogen of a nearby histidine. The iron is in a high-spin form. As a result it is displaced up to 0.3 Å out of the heme plane. Upon oxygenation, an O_2 molecule coordinates with the sixth ligand position to give low-spin iron. Consequently, the iron atom moves toward the heme plane; the porphyrin ring shifts, which in turn precipitates other structural changes in the protein, including the breaking of salt bridges. The transition from the deoxy to the

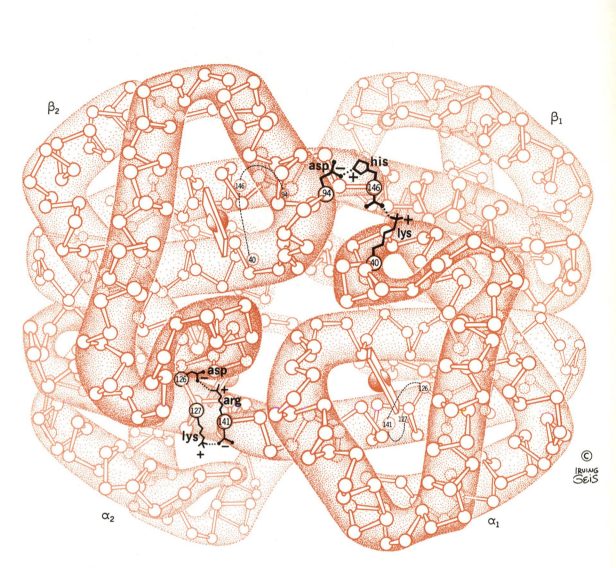

Figure 17-25

A view of deoxyhemoglobin that illustrates important salt bridges; view is 90° with respect to that in Figure 1-5. [Drawing by Irving Geis.]

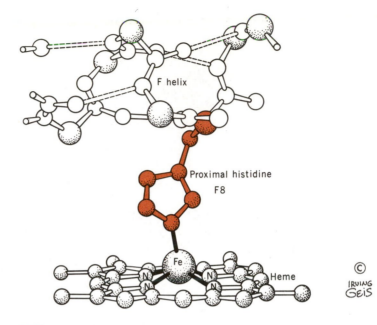

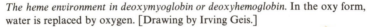

Figure 17-26

The heme environment in deoxymyoglobin or deoxyhemoglobin. In the oxy form, water is replaced by oxygen. [Drawing by Irving Geis.]

oxy structure (in which all salt bridges are broken) may occur progressively, as oxygen is bound and some of the salt bridges are broken. However, the details of intermediate structures are only guesses at this point.

According to these ideas the hemoglobin mechanism is, at least in part, a combination of the sequential, "induced fit," and MWC mechanisms. The deoxy structure is "taut," being highly constrained by the salt bridges; the oxy form is "relaxed," being free of the constraints imposed by the salt bridges. These ideas harmonize nicely with MWC concepts. However, the transition between taut and relaxed forms may be induced by oxygen binding, which causes a movement of heme iron that in turn affects other parts of the protein. This gives an "induced-fit" feature to the mechanism.

The structural information on the two forms also gives plausible possibilities for the origin of the Bohr protons that are released on oxygenation. The β-chain C-terminal histidine side chains and the α-chain N-terminal amino group are possibilities for the alkaline group (pk_2). According to Table 17-2, the alkaline Bohr pk levels shift from about 7.6 to 6.2 upon oxygenation. In the deoxy structure, each C-terminal histidine (His[146]) is bonded to the free carboxyl of Asp[94] on the same subunit (see Figure 17-25 and Table 17-3); as a result, the histidine pK should be greatly elevated. But upon oxygenation, the salt bridge to the carboxyl is broken, and the pK drops, resulting in the release of hydrogen ions (the amount released clearly will depend on the pH). It is thought that 40% of the alkaline Bohr effect is due to the

His^{146}–Asp^{94} interaction. The α-chain amino termini are the source of an additional 30% of the alkaline Bohr protons. There is evidence that these groups may be involved in anion binding, which will affect their pK values. The groups responsible for the remaining 30% have not yet been identified. Finally, it should be mentioned that the molecular origin of the acid Bohr protons is still unclear.

Figure 1-5a illustrates the manner in which DPG is believed to bind. The small organic phosphate fits onto the dyad axis of deoxyhemoglobin, where it binds between the β subunits and neutralizes a specific constellation of positive charges contributed by the β chains. Binding to the dyad axis in this way thus accounts for the stoichiometry of binding one DPG per deoxyhemoglobin tetramer. Upon oxygenation, the configuration of positive charges around the dyad axis changes sharply (see Fig. 1-5b) so that DPG no longer has a strong interaction with this part of the molecule. Hence, DPG can "clamp" only the deoxy structure.

Summary

Multiple-ligand interactions with a macromolecule are a major process of biological regulation. An example is provided by feedback inhibition, whereby a product of a biosynthetic pathways interacts with and alters the activity of the first enzyme in the pathway. A prime example is the pyrimidine biosynthetic pathway, where the first enzyme is ATCase.

ATCase is a multisubunit allosteric protein with six sites for its substrates and six for the allosteric inhibitor CTP. Much effort has been directed at determining the mechanism by which the activity of this and other allosteric proteins is modulated. Experimental data are commonly interpreted within the framework of a scheme involving concerted conformational changes of subunits (Monod–Wyman–Changeaux model) or a scheme involving a sequential series of changes.

Hemoglobin is another well-studied allosteric protein. In this case the oxygen affinity is affected by H^+, organic phosphates, and CO_2. The physiological significance of some of these effects has been explored. Because a high-resolution crystal structure of the protein has been obtained, it is possible to think in molecular terms concerning the relation of structure to mechanism.

Problems

17-1. An allosteric enzyme has been investigated by certain physical methods, and a certain physical property P is measured. This property is sensitive to the conformation or shape of the enzyme molecule. Dr. A studies the binding of a substrate analog S' (a molecule that binds at the substrate sites, but is not converted to product). The binding of S' is normal (hyperbolic). Dr. A measures $\bar{y}_{S'}$ (the fractional saturation of enzyme with S'), and at the same time he measures P at each value of $\bar{y}_{S'}$. Interestingly enough, P changes in a rather straightforward way as S' binds: the fractional change in P exactly follows $\bar{y}_{S'}$. Specifically, the fractional change in P is about 0.30 when $\bar{y}_{S'} = 0.30$; the fractional change in P is about 0.60 when $\bar{y}_{S'} = 0.60$; and so on. Dr. A then studies the binding of an allosteric inhibitor I to the enzyme. The binding of I to enzyme alone is sigmoidal. He finds that, as I binds to the enzyme, there is no change in P whatsoever. Dr. A then gets very excited and exclaims, "This behavior is exactly what I would predict from the Monod–Wyman–Changeaux (MWC) model for allosterism." He writes up his experiments and conclusions, and he sends the manuscript to a scientific journal. The editor writes back to Dr. A: "We can publish your findings, but there is one minor difficulty with your conclusions. Although the behavior of P as the enzyme binds S' is consistent with the MWC model, that model would predict some changes in P when I binds." Who is right, Dr. A or the editor or neither? Explain your answer.

17-2. The binding of a certain allosteric inhibitor I by itself to an allosteric enzyme molecule is only barely sigmoidal—in fact, almost hyperbolic. The same is true for the binding of an allosteric activator A by itself. However, in the presence of A, the binding of I becomes very sigmoidal; conversely, the binding of A is very sigmoidal in the presence of I. Dr. Alpha interprets these results by saying that "the data are consistent with the two-state MWC model if and only if $L \gg 1$ [recall that $L = (T_0)/(R_0)$], I binds only to the T state, and A binds only to the R state." Dr. Omega replies that "the data are not consistent with the two-state MWC model, regardless of what value you assume for L." Who is right, Dr. Alpha or Dr. Omega or neither? Choose and defend one of these three positions, and eliminate the other two with carefully reasoned logic.

17-3. For a particular allosteric enzyme, cooperativity is envisioned as occurring by means of a combination of induced fit and concerted transitions. The scheme is the following, where X denotes the occupancy of a subunit:

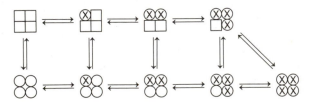

Note that, along the top row, induced fit occurs as X binds; however, constraints operate on the structure to flip all of the subunits into the same conformation, thus giving the bottom row. Let the microscopic dissociation constant for binding X along the bottom row be k_0, and that along the top row be $k_i = k_0/c$. Let L be the ratio of the concentration of the structure at the left end of the top row to that of the structure at the left end of the bottom row. Derive an expression for the fractional saturation ($\bar{y}$) of X as a function of L, c, and $\alpha = (X)/k_0$. Compare this expression with that for the MWC model, and comment. Derive expressions for the equilibrium constants r_1, r_2, and r_3 for each of the vertical steps involving 1, 2, and 3 bound ligands, respectively; express these equilibrium constants in terms of L and c. If $L = 10^3$ and $c = 10^{-2}$, at which point in the binding do the constraints on the structure make the concerted pathway the preferred one—in other words, what is the lowest value of i for which r_i favors the concerted side of the equilibrium? Answer the same question for the case in which $L = 10^4$ and $c = 10^{-2}$. NOTE: In this scheme, for each of the ligand forms there are many different possible microscopic species that must be taken into account in deriving any equations.

17-4. Assume that a protein P interacts with a ligand L according to the following scheme:

$$P + L \rightleftarrows PL_1$$

$$PL_1 + L \rightleftarrows PL_2$$

$$PL_2 + L \rightleftarrows PL_3$$

$$PL_3 + L \rightleftarrows PL_4$$

where the intrinsic dissociation constant k is the same for each site on P. The naked protein P also interacts with a single molecule of a ligand A according to

$$P + A \rightleftarrows PA$$

with a dissociation constant of k_A. The form PA cannot combine with L. Derive an expression for the fractional saturation $\bar{y}_L$ with respect to L in terms of k, k_A, (L), and (A). Make rough sketches of $\bar{y}_L$ versus (L) for different values of (A). Derive expressions for the *apparent* macroscopic dissociation constants K_1, K_2, K_3, and K_4 that are associated with interactions of L with the protein. Note that K_1 is defined as $K_1 = (PL_0)(L)/(PL_1)$, where (PL_0) represents the total concentration of all species with no bound L molecules, and $K_i = (PL_{i-1})(L)/(PL_i)$ for $i = 2, 3, 4$. How do these macroscopic constants depend on (A)? What are their relative values when (A) = 0? when (A) = k_A? when (A) = $9k_A$? Does increasing (A) introduce some apparent cooperativity into the binding of L? Why, or why not?

17-5. The following data have been obtained for an allosteric enzyme with specific and distinct receptor sites for ligands A and B: (1) binding of A alone is sigmoidal; (2) binding of B in the presence of saturating A is sigmoidal; (3) binding of B alone is sigmoidal. It is claimed that these data may be explained by the two-state MWC model with the assumptions that $L = (T_0)/(R_0) \gg 1$, that $c = 0$ for A, and that $c \gg 1$ for B. (Remember that $c = k_R/k_T$.) Do you agree with this claim? Explain.

References

GENERAL

Benesch, R. E., and R. Benesch. 1974. The mechanism of interaction of red cell organic phosphates with hemoglobin. *Adv. Protein Chem.* 28:211. [A clearly written review that gives good coverage of various aspects of the relationship between organic phosphates and hemoglobin.]

Dickerson, R. E., and I. Geis. 1981. *Proteins: Structure, Function, and Evolution.* Menlo Park, Calif.: Benjamin-Cummings. [Chapter 3 gives a vivid account of hemoglobin structure and mechanism.]

Edelstein, S. J. 1975. Cooperative interactions in hemoglobin. *Ann. Rev. Biochem.* 44:209.

Hammes, G. G., and C.-W. Wu. 1971. Regulation of enzyme activity. *Science* 172:1205.

Jacobson, G. R., and G. R. Stark. 1973. Aspartate transcarbamylases. In *The Enzymes*, vol. 9, ed. P. D. Boyer (New York: Adademic Press), p. 225. [A comprehensive review covering cooperative properties, catalytic mechanism, genetics, and more.]

Koshland, D. E., Jr., G. Némethy, and D. Filmer. 1966. Comparison of experimental binding data and theoretical models in proteins containing subunits. *Biochemistry* 5:365. [Includes a detailed treatment of models for allosteric schemes.]

Monod, J., J. Wyman, and J.-P Changeux. 1965. On the nature of allosteric transitions: A plausible model. *J. Mol. Biol.* 12:88. [This well-known paper gives a clear exposition of the MWC model for allosteric proteins.]

Perutz, M. F. 1970. Stereochemistry of cooperative effects of hemoglobin. *Nature* 228:726. [An interesting summary of earlier hemoglobin structural work, and of structure in relation to mechanism.]

———. 1978. Hemoglobin structure and respiratory transport. *Scientific American* 239(6):92. [A narrative discussion of hemoglobin structure and mechanism.]

Schachman, H. K. 1974. Anatomy and physiology of a regulatory enzyme: Aspartate transcarbamylase. *Harvey Lectures* 68:67. [A nice summary, based mainly on the work from Howard Schachman's laboratory.]

SPECIFIC

Adair, G. S. 1925. The hemoglobin system, VI: The oxygen dissociation curve of hemoglobin. *J. Biol. Chem.* 63:529.

Benesch, R., and R. E. Benesch. 1967. The effect of organic phosphates from the human erythrocyte on the allosteric properties of hemoglobin. *Biochem Biophys. Res. Commun.* 26:162.

Chanutin, A., and R. R. Curnish. 1967. Effect of organic and inorganic phosphates on the oxygen equilibrium of human erythrocytes. *Arch. Biochem. Biophys.* 121:96.

Eisenberger, P., R. G. Shulman, B. M. Kincaid, G. S. Brown, and S. Ogawa. 1978. Extended x-ray absorption fine structure determination of iron nitrogen distances in haemoglobin. *Nature* 274:30.

Fermi, G. 1975. Three-dimensional Fourier synthesis of human deoxyhemoglobin at 2.5 Å resolution: Refinement of the atomic model. *J. Mol. Biol.* 97:237.

Gelin, B. R., and M. Karplus. 1977. Mechanism of tertiary structural change in hemoglobin. *Proc. Natl. Acad. Sci. USA* 74:801.

Gennis, L. S. 1976. Negative homotropic cooperativity and affinity heterogeneity: Preparation of yeast glyceraldehyde-3-phosphate dehydrogenase with maximal affinity homogeneity. *Proc. Natl. Acad. Sci. USA* 73:3928.

Kilmartin, J. V., and L. Rossi-Bernardi. 1973. Interaction of hemoglobin with hydrogen ions, carbon dioxide, and organic phosphates. *Physiol. Rev.* 53:836.

Szabo, A., and M. Karplus. 1972. A mathematical model for structure–function relations in hemoglobin. *J. Mol. Biol.* 72:163.

18

Configurational statistics of polymer chains

18-1 CONFORMATION AVERAGING

In Chapters 5 and 6, we gave considerable attention to the factors that play an important role in determining the structures of proteins and nucleic acids. Because of the complexity and enormity of the problem of actually predicting, for example, a protein structure from an amino acid sequence, our treatment was largely descriptive, highlighting the many factors that must be considered. In this chapter, we undertake a more analytical treatment of polymer conformations. In so doing, however, we must omit from consideration some of the complex issues raised in Chapters 5 and 6; for example, there is no good theory of solvent structure and solvent–polymer interactions that allows us rigorously to introduce solvent effects into our mathematical treatment of polypeptide conformations. Instead, our treatment focuses on the configurational statistics of chain molecules, where statistical evaluation of chain properties is made by averaging over all conformations, with weighting factors determined by the chain's bond rotational potentials. The conformational properties so evaluated are a reflection of the chain geometries (bond angles and bond lengths) and of the hindrances to backbone rotations. The procedure of averaging over all conformations means that we are calculating average properties of "statistically coiling" chains, which are to be distinguished from the single, unique conformations that are encountered with many biologically active macromolecules.

In treating chain statistics, attention is directed largely at calculating intrachain distances—that is, the average distance between chain segments, such as between chain termini. This is done by averaging the distance over all conformations in accordance with the principles of statistical mechanics. From the configurational statistics of polypeptides, we gain insight into the chain folding of proteins. In particular, the statistics of homopolymers and copolymers of glycine, alanine, and proline give clues to the ways these residues can influence protein conformation.

An important connection exists between the mean square end-to-end distance and the hydrodynamic and other physical properties of a coil. For example, intrinsic viscosities and sedimentation coefficients commonly are used to evaluate the extensions of chains. In fact, these hydrodynamic properties are prime sources for our understanding not only of flexible coils, but of stiff, wormlike chains such as high-molecular-weight DNA. This chain's conformational properties are particularly amenable to description in terms of a special idealized model for wormlike chains.

Both artificial and real chains are considered in this chapter. The former—such as the freely jointed and freely rotating chains—provide an important reference point for the latter. Attention is directed at the statistical evaluation of chain conformation; the interconnection between conformation and the most commonly measured hydrodynamic properties is taken up in Chapter 19.

18-2   CALCULATION OF CONFORMATION-DEPENDENT PROPERTIES

Figure 18-1 is a schematic illustration of a hypothetical polymer chain, in which the circles denote skeletal atoms or groups. The bonds are serially indexed from 1 to n, with bond i joining skeletal groups $i - 1$ and i. Dihedral rotation angles (see Chapter

Figure 18-1

Schematic illustration of a polymer chain.

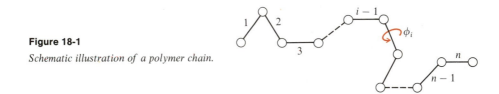

5) are numbered from ϕ_2 to ϕ_{n-1}. As far as backbone conformation is concerned, rotation angles ϕ_1 and ϕ_n obviously are undefined and without meaning or significance, because rotations about these bonds leave the backbone conformation unchanged.

The dihedral rotation angles ϕ are defined and discussed in Chapter 5. In polymer-chain statistics, it is customary to assign the reference state of $\phi = 0°$ to the *trans* conformation, rather than to the *cis* conformation as we did in Chapter 5 for the treatment of conformations of peptides.[§] In order to conform to popular usage in polymer chemistry, in this chapter we adopt the convention that the *trans* conformation is the reference state of $\phi = 0°$.

In treating conformations of real chains, we assume that all bond lengths and bond angles are rigidly fixed. Obviously this is an approximation, but it is not a serious inaccuracy. Deviations in bond angles from their most probable values, for

---

[§] One reason to prefer the *trans* conformation for the reference state is that an all-*trans* chain is easily drawn or constructed from molecular models. Obviously, an all-*cis* chain cannot be represented in any clear fashion.

example, will be small and will tend to occur with opposing signs so that the effects of deviations cancel one another. Thus the conformation of the chain as a whole is determined by the values adopted by the rotation angles ϕ_i. By varying these rotation angles over all possible values (the range of 0 to 2π for each), we can generate all possible conformations of the backbone.

End-to-end distance and radius of gyration

The two conformation-dependent properties most commonly considered are the end-to-end distance and the radius of gyration. Each of these properties is subject to experimental determination. The end-to-end distance vector $\mathbf{r}$ may be calculated as the sum of the bond vectors $\hat{\mathbf{l}}_i$, where each bond vector $\hat{\mathbf{l}}_i$ is colinear with bond i and points from skeletal element $i-1$ to element i:

$$\mathbf{r} = \sum_{i=1}^{n} \hat{\mathbf{l}}_i \tag{18-1}$$

Figure 18-2 illustrates the summation in Equation 18-1. The magnitude r of $\mathbf{r}$ is given

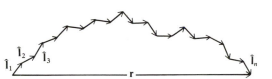

Figure 18-2

The end-to-end distance vector $\mathbf{r}$ *as a summation of bond vectors.*

by $(\mathbf{r} \cdot \mathbf{r})^{1/2}$, where

$$r^2 = \left(\sum_{i=1}^{n} \hat{\mathbf{l}}_i\right)\left(\sum_{j=1}^{n} \hat{\mathbf{l}}_j\right) = \sum_{i=1}^{n}\sum_{j=1}^{n} \hat{\mathbf{l}}_i \cdot \hat{\mathbf{l}}_j \tag{18-2}$$

Factoring the n diagonal terms ($i=j$) out of this sum, and assuming for simplicity that all bonds are of equal length l, we obtain[§]

$$r^2 = nl^2 + 2\sum_{j>i} \hat{\mathbf{l}}_i \cdot \hat{\mathbf{l}}_j \tag{18-3}$$

The mean square end-to-end distance is

$$\langle r^2 \rangle = nl^2 + 2\sum_{j>i} \langle \hat{\mathbf{l}}_i \cdot \hat{\mathbf{l}}_j \rangle \tag{18-4}$$

[§] In Equation 18-3 and other equations in this chapter, the symbol $\sum_{j>i}$ is equivalent to the double summation

$$\sum_{j=i+1}^{n}\sum_{i=1}^{n}$$

where the angle brackets denote average values. The problem of computing the mean square end-to-end distance thus reduces to calculating the average projection of every bond upon every other bond.

This average is calculated by application of the laws of statistical mechanics, which state that the probability $p(\phi_2, \phi_3, \ldots, \phi_{n-1}) d\phi_2 d\phi_3 \cdots d\phi_{n-1}$ of a particular conformation—characterized by rotation angles that lie between ϕ_2 and $\phi_2 + d\phi_2$, between ϕ_3 and $\phi_3 + d\phi_3$, and so on—is determined by a simple Boltzmann distribution in the conformational energy. In particular,

$$p(\phi_2, \phi_3, \ldots, \phi_{n-1}) d\phi_2 d\phi_3 \cdots d\phi_{n-1}$$

$$= \frac{\exp[-E(\phi_2, \phi_3, \ldots)/kT] d\phi_2 d\phi_3 \cdots}{\int_0^{2\pi} \cdots \int_0^{2\pi} \exp[-E(\phi_2, \phi_3, \ldots)/kT] d\phi_2 d\phi_3 \cdots} \qquad (18\text{-}5)$$

where $E(\phi_2, \phi_3, \ldots)$ is the conformational energy of the conformation specified by $\phi_2, \phi_3, \ldots$. For cases discussed here, this energy is calculated from the bond rotational potential functions, but in general it can include other factors such as solvent–polymer interactions and their dependence on conformation.

The evaluation of $\hat{l}_i \cdot \hat{l}_j$ depends upon the value adopted by the rotation angles for the bonds intervening between $\hat{l}_i$ and $\hat{l}_j$—namely, the orientation of $\hat{l}_j$ with respect to $\hat{l}_i$ depends on $\phi_{j-1}, \phi_{j-2}, \ldots, \phi_{i+1}$. Therefore, the sum (over $j > i$) $\sum(\hat{l}_i \cdot \hat{l}_j)$ depends on the entire set of rotation angles, and its average value is

$$\sum_{j>i} \langle \hat{l}_i \cdot \hat{l}_j \rangle = \int_0^{2\pi} \cdots \int_0^{2\pi} \left[\sum_{j>i} (\hat{l}_i \cdot \hat{l}_j) p(\phi_2, \phi_3, \ldots) \right] d\phi_2 d\phi_3 \cdots \qquad (18\text{-}6)$$

Although Equation 18-6 may appear formidable, there are many situations in which its execution is straightforward. Substitution of Equation 18-6 into Equation 18-4 yields $\langle r^2 \rangle$.

The connection between $\langle r^2 \rangle$ and the conformations of the chain is precisely made in Equations 18-5 and 18-6. If $E(\phi_2, \phi_3, \ldots)$ has most of its low-energy minima at large values of r^2, the value of $\langle r^2 \rangle$ will be biased toward large extensions. A sharp and unique minimum in $E(\phi_2, \phi_3, \ldots)$ at $\phi_2', \phi_3', \ldots$ will result in a value of $\langle r^2 \rangle$ that is biased toward that characteristic of the conformation specified by $\phi_2', \phi_3', \ldots$. In the limit that $E(\phi_2, \phi_3, \ldots)$ is constant for all values of $\phi_2, \phi_3, \ldots$ (the freely rotating chain), $\langle r^2 \rangle$ represents the simple arithmetic average of r^2 for all chain conformations. It is clear, therefore, that $\langle r^2 \rangle$ is a reflection of the distribution of conformations and so of the most prevalent conformations adopted by a chain molecule.

The radius of gyration R_G is defined as the root-mean-square distance of an array of atoms or groups from their common center of gravity. With a chain of $n + 1$ identical elements (joined by n bonds—see Fig. 18-1), if R_{Gi} is the distance of skeletal element i from the chain's center of gravity, then for a particular conformation

$$R_G^2 = [1/(n + 1)] \sum_{i=0}^{n} R_{Gi}^2 \qquad (18\text{-}7)$$

It is possible to show, however, (see Box 18-1) that

$$R_G^2 = [1/(n + 1)^2] \sum_{j>i} r_{ij}^2 \qquad (18\text{-}8)$$

where r_{ij} is the distance between elements i and j (see Flory, 1969). Therefore, the calculation of the radius of gyration reduces to the calculation of the intrachain distances, as done in Equation 18-4 for the end-to-end distance.

Average chain dimensions in the unperturbed state

In the following treatment of conformation statistics, we do not take into account the excluded-volume effect. Excluded volume is simply the physical constraint that two segments of a chain cannot simultaneously occupy the same position in space. In treating real chains, conformations involving excluded-volume interactions between neighboring chain units in the sequence are implicitly discarded, by assigning such conformations an inordinately large energy. However, intersections between segments distant in the chain sequence usually are too difficult to account for. Therefore, the conformation statistics given apply to chains *unperturbed* by excluded-volume effects. These effects are treated separately in a later section, where it is indicated that the effects of excluded volume on chain dimensions, and the chain dimensions unperturbed by excluded volume, both can be evaluated experimentally.

   In calculating unperturbed dimensions, we follow a common convention of assigning a subscript 0 to the calculated parameter; for example, $\langle r^2 \rangle_0$ is the unperturbed mean square end-to-end distance.

   Equation 18-4 can be used, of course, to calculate any intrachain distance r_{ij}. In view of Equation 18-8, it is perhaps not surprising, therefore, that $\langle R_G^2 \rangle_0$ and $\langle r^2 \rangle_0$ may, under certain circumstances, be directly related. This occurs in the limit of $n \to \infty$, where

$$\boxed{\langle R_G^2 \rangle_0 = \langle r^2 \rangle_0/6 \qquad \text{for } n \to \infty} \qquad (18\text{-}9)$$

(see Flory, 1969). Thus, *in the limit of infinite chain length, the mean square end-to-end distance and the mean square radius of gyration for the unperturbed chain are directly related.* In the following discussion, we focus attention on calculating $\langle r^2 \rangle_0$ rather than $\langle R_G^2 \rangle_0$. However, methods used to calculate $\langle r^2 \rangle_0$ clearly can be extended to calculate $\langle R_G^2 \rangle_0$.

18-3   THE FREELY JOINTED OR RANDOM-WALK CHAIN

The freely jointed or random-walk chain is an artificial model in which the direction of each bond vector is completely uncorrelated to every other bond vector in the chain. Even the restriction of fixed bond angles therefore is removed, so that the polymer is freely jointed. Thus, the conformations of the chain are described by a random walk in which each step of length l is randomly placed with respect to its predecessor.

An analogy between the freely jointed chain and diffusing gas molecules

Figure 18-3 illustrates the freely jointed chain. The mean square end-to-end distance is easily calculated from Equation 18-4. The quantity $\langle \hat{\mathbf{l}}_i \cdot \hat{\mathbf{l}}_j \rangle$ equals zero for all $i \neq j$,

---

Box 18-1   THE RADIUS OF GYRATION

Consider an object composed of $n + 1$ units. The ith unit is located at the point (x_i, y_i, z_i) and has a mass m_i. The center of mass (or gravity) of the object is a point in space that has the following property: if $\mathbf{s}_i$ is the vector drawn from the center of mass to unit i, then

$$\sum_{i=0}^{n} = m_i \mathbf{s}_i = 0$$

If the object is composed of identical mass units, then one has the even simpler result that

$$\sum_{i=0}^{n} \mathbf{s}_i = 0$$

With the center of mass known, it is simple to calculate the radius of gyration by using Equation 18-7. Frequently, however, it is convenient to have an expression for the radius of gyration that depends only on internal coordinates of a polymer, as given in Equation 18-8 in the text. We shall demonstrate the equivalence of these two equations.

It is easiest to start with the sum in equation 18-8. This can be rewritten as

$$\sum_{j>i} r_{ij}^2 = (1/2) \sum_{j=0}^{n} \sum_{i=0}^{n} \mathbf{r}_{ij} \cdot \mathbf{r}_{ij}$$

where $\mathbf{r}_{ij}$ is the vector between polymer segments i and j, and we have exploited the fact that $\mathbf{r}_{ij} = 0$ if $i = j$. Consider any two polymer segments i and j as shown in the figure. If $\mathbf{s}_i$ and $\mathbf{s}_j$ are vectors to these units from the center of mass, then clearly $\mathbf{r}_{ij} = \mathbf{s}_j - \mathbf{s}_i$. Therefore, the

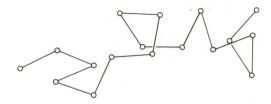

Figure 18-3
The random-walk (freely jointed) chain.
The orientation of each bond with respect
to its predecessor is completely random.

because bond vectors $\hat{\mathbf{l}}_i$ and $\hat{\mathbf{l}}_j$ are randomly disposed with respect to each other. This means that every projection of $\hat{\mathbf{l}}_j$ on $\hat{\mathbf{l}}_i$ has an equal and opposite projection, so that the sum of all such projections is zero. Therefore, from Equation 18-4, we obtain

$$\langle r^2 \rangle_0 = nl^2 \tag{18-10}$$

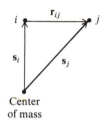

Center
of mass

double sum can be transformed to

$$(1/2) \sum_{j=0}^{n} \sum_{i=0}^{n} (\mathbf{s}_j - \mathbf{s}_i) \cdot (\mathbf{s}_j - \mathbf{s}_i) = (1/2) \sum_{j=0}^{n} \sum_{i=0}^{n} (\mathbf{s}_i^2 + \mathbf{s}_j^2 - 2\mathbf{s}_i \cdot \mathbf{s}_j)$$

Each term can be summed easily. From Equation 18-7, we know that s_i^2 is just $(n+1)R_G^2$. The effect of the sum over j is simply to multiply this term by an additional factor of $n+1$. Similarly, the double sum in s_j^2 is just $(n+1)^2 R_G^2$. The third term can be written as

$$-\sum_{j=0}^{n} \sum_{i=0}^{n} \mathbf{s}_i \cdot \mathbf{s}_j = -\sum_{j=0}^{n} \mathbf{s}_i \cdot \sum_{i=0}^{n} \mathbf{s}_j$$

Each of the two sums can be performed separately. However, from the definition of the center of mass given above, both sums are zero. Thus we have shown that

$$\sum_{j>i} r_{ij}^2 = (n+1)^2 R_G^2$$

Equation 18-10 gives a result that is exactly analogous to the displacement of diffusing gas molecules as a function of time. In the latter case, the root-mean-square displacement is proportional to the square root of the time; in the present case, $\langle r^2 \rangle_0^{1/2}$ is proportional to $n^{1/2}$, where n is analogous to time in the diffusing-gas case.

The characteristic ratio

According to Equation 18-10, $\langle r^2 \rangle_0$ is a linear function of n for all values of n. In the case of real chains, a linear relationship between $\langle r^2 \rangle_0$ and n exists only in the limit of sufficiently large n. Even in this limit, however, $\langle r^2 \rangle_0 / nl^2 \neq 1$ generally for real chains. The ratio $\langle r^2 \rangle_0 / nl^2$ is known as the characteristic ratio. In most cases, the asymptotic value $(n \to \infty)$ of the characteristic ratio is greater than unity. Its precise value is a reflection of the preferred conformations and of the stiffness of the chain under consideration.

Calculation of the distribution of end-to-end distances

It is of considerable interest to calculate the distribution of end-to-end distances—that is, the relative number of conformations that have a particular end-to-end distance. To accomplish this, we define a Cartesian coordinate system and first compute the average projection of a bond vector on any given Cartesian axis. We then assume that the average projection of each bond occurs with a positive or negative sign, as dictated by random-walk statistics. This enables us then to calculate the statistical distribution of conformations, by a rather straightforward procedure.

Consider the average projection length on the x axis of a bond of length l (Fig. 18-4a). The mean square projection length $\overline{l_x^2}$ is $l^2 \overline{\cos^2 \theta}$, where θ is the angle between the bond vector and the x axis, and an overbar denotes "average." The average value of $\cos^2 \theta$ is easily calculated. The distribution of values of $\cos \theta$ is determined by the fact that all bond directions are equally probable. Therefore, the number of vectors between θ and $\theta + d\theta$ is determined by the area of the annular ring obtained by rotating the bond vector about the x axis, with the fixed angle θ between the vector and the x axis (Fig. 18-4b). This area is $2\pi l^2 \sin \theta \, d\theta$ and, obviously, the number of projections between θ and $\theta + d\theta$ will be proportional to $2\pi l^2 \sin \theta$. Therefore, the mean square projection length is

$$\overline{l_x^2} = \frac{\int_0^\pi l^2 \cos^2 \theta (2\pi l^2 \sin \theta) \, d\theta}{\int_0^\pi 2\pi l^2 \sin \theta \, d\theta} = l^2/3 \tag{18-11}$$

The average root-mean-square projection along a given axis clearly is $l/\sqrt{3}$.

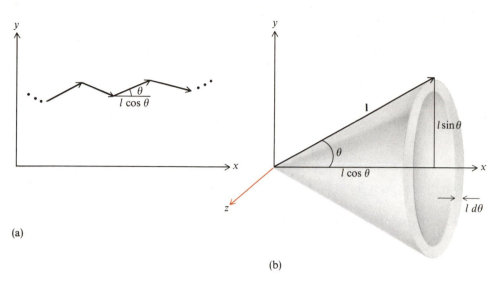

(a)

(b)

Figure 18-4

Components of bond vectors along the x axis. **(a)** The x components of bond vectors are given by $l \cos \theta$. **(b)** Calculation of distribution of values of $\cos \theta$ for randomly oriented bonds.

We now assume that the average steps $l/\sqrt{3}$ along the chosen axis occur with a random frequency in the negative and positive directions. If there are n steps (corresponding to n bonds), the distance traversed along the x axis is

$$x = (n_+ - n_-)l/\sqrt{3} \qquad (18\text{-}12)$$

where n_+ and n_- are the number of steps in the positive and negative directions, respectively, and $n = n_+ + n_-$. The probability $W(n_+, n_-, n)$ of n_+ and n_- positive and negative steps out of a total of n steps is

$$W(n_+, n_-, n) = (1/2)^n n!/n_+! n_-! \qquad (18\text{-}13)$$

The factor $(1/2)^n$ is the probability of generating a particular sequence of $+$ and $-$ steps (for example, $+ - + -$ as one sequence for $n = 4$; $n_+ = 2$; $n_- = 2$), and the factor $n!/n_+! n_-!$ is the number of particular sequences of steps that have the same parameters n_+ and n_- (such as the six sequences possible for $n_+ = n_- = 2$: $+ + - -$; $+ - + -$; $+ - - +$; $- + + -$; $- + - +$; and $- - + +$). Equation 18-13 can be cast as a function of a continuous variable by introducing Stirling's approximation:

$$\ln N! = N \ln N - N + (1/2) \ln 2\pi + (1/2) \ln N \qquad (18\text{-}14)$$

Equation 18-14 is valid for large N; when we apply it to Equation 18-13, we assume that n, n_+, and n_- are large. Substituting $m = n_+ - n_-$ and the conservation equation $n = n_+ + n_-$ into Equation 18-13 and introducing Stirling's approximation, we obtain

$$\ln W(m, n) = n \ln(1/2) + n \ln n - n + (1/2) \ln 2\pi + (1/2) \ln n$$
$$- \{[(n + m)/2] \ln[(n + m)/2] - (n + m)/2 + (1/2) \ln 2\pi$$
$$+ (1/2) \ln[(n + m)/2] + [(n - m)/2] \ln[(n - m)/2]$$
$$- (n - m)/2 + (1/2) \ln 2\pi + (1/2) \ln[(n - m)/2]\} \qquad (18\text{-}15)$$

After some simplification, Equation 18-15 becomes

$$\ln W(m, n) = (1/2) \ln n - \ln(n/2) - (1/2) \ln 2\pi$$
$$- \{(n/2) \ln[(1 + m/n)(1 - m/n)] + (m/2) \ln[(1 + m/n)(1 - m/n)]$$
$$+ (1/2) \ln[(1 + m/n)(1 - m/n)]\} \qquad (18\text{-}16)$$

The first three factors in the right-hand side of Equation 18-16 simplify to $\ln(n\pi/2)^{-1/2}$. The remaining factors may be simplified for the case $|m/n| \ll 1$ (that is, where the excesses of n_+ over n_-, or vice versa, are small compared to n). Physically, this corresponds to relatively small chain extensions, which are the most abundant configurations in any event. Under these circumstances, the following relationships are useful: $\ln(1 + x) \cong x$; and $\ln[(1 + x)/(1 - x)] \cong 2x$ for $|x| \ll 1$. Using these relationships, we can simplify the latter part of Equation 18-16 to

$$m^2/2n - m^2/n + m^2/2n^2 \cong -m^2/2n \qquad (18\text{-}17)$$

Thus we obtain

$$\ln W(m, n) = \ln(2/n\pi)^{1/2}(-m^2/2n) \qquad (18\text{-}18)$$

or

$$W(m, n) = \sqrt{2/n\pi}\, e^{-m^2/2n} \qquad (18\text{-}19)$$

This is the familiar Gaussian distribution for the excesses of positive steps over negative ones. or vice versa.

To transform $W(m, n)$ into $W(x, n)$, we first note that $m^2 = 3x^2/l^2$. Furthermore, the exact meaning of $W(x, n)$ and $W(m, n)$ must be kept firmly in mind. $W(x, n)$ is the probability *per unit interval of distance* that x takes on a value between x and $x + dx$, but $W(m, n)$ is the probability that $n_+ - n_- = m$, an integer. Because, for fixed n, the value of m changes in units of two,[§] the relation $W(x, n) = W(m, n)/(\text{interval})$

[§] Because $m = n_+ - n_-$, it is obvious from Equation 18-12 that, when m changes by ± 2, then x changes by $\pm 2l/\sqrt{3}$.

requires an interval equivalent to two units of m—that is, $2l/\sqrt{3}$. Therefore,

$$W(x, n) = (\sqrt{3}/2l)W(m, n)$$
$$= \sqrt{3/2\pi}\,(1/n^{1/2}l)\exp(-3x^2/2nl^2) \qquad (18\text{-}20)$$

Equation 18-20 is the desired result.

Similar expressions may be derived for $W(y, n)$ and $W(z, n)$. For small extensions, it is legitimate to assume that $W(x, y, z)\,dx\,dy\,dz$—the probability that x lies between x and $x + dx$, that y lies between y and $y + dy$, and that z lies between z and $z + dz$— may be taken as the product of the independent probabilities $W(x, n)\,dx$, $W(y, n)\,dy$, and $W(z, n)\,dz$; that is,

$$W(x, y, z)\,dx\,dy\,dz = (\beta/\sqrt{\pi})^3 e^{-\beta^2 r^2}\,dx\,dy\,dz \qquad (18\text{-}21)$$

where $\beta = (3/2nl^2)^{1/2}$, and $r^2 = x^2 + y^2 + z^2$. Note that $W(x, y, z)$ depends on n as well as on x, y, and z.[§]

Two kinds of distribution function

There are two kinds of distribution functions that may be based on Equation 18-21. If we let one end of our polymer chain be the origin of the coordinate system, then Equation 18-21 gives the probability that the end-to-end distance vector **r** will fall in the volume element $dx\,dy\,dz$ (see Fig. 18-5a). This distribution function is plotted in Figure 18-5b. It is clear that the highest density of conformations occurs near the origin, and that higher extensions have a steadily decreasing likelihood of occurrence. In fact, when $r = \beta^{-1} = \sqrt{(2/3)nl^2}$, which is much less than half of maximal extension, then $W(x, y\,z)$ has decayed to e^{-1} of its maximal value.

A second kind of distribution function is the *radial* distribution function $W(r)\,dr$, the probability of occurrence of an end-to-end distance r and $r + dr$, *irrespective* of direction. It includes all volume elements $dx\,dy\,dz$ in Equation 18-21 (see also Fig. 18-5a) that lie at a distance r from the origin. These volume elements all fall in the volume $4\pi r^2\,dr$ of a spherical shell of radius r and thickness dr (see Fig. 18-6a). Therefore, the radial distribution function is

---

[§] It is important to review the assumptions made in deriving Equation 18-21, and to examine their consequences. First, we have assumed that bond projections along a given axis can assume only values of $\pm l/\sqrt{3}$. However, it is clear that all values between 0 and $\pm l$ can occur. Second, the distance travelled along one axis cannot be truly independent of that along the other axes as we have assumed; for example, consider the restrictions placed on the allowable projections on the other axes when the chain's projection along one axis approaches the maximal extension, nl. These two assumptions, however, do not introduce serious error for large n and for $r \ll nl$, which are the very conditions assumed in using Equation 18-14 and in deriving Equation 18-19 from Equation 18-16.

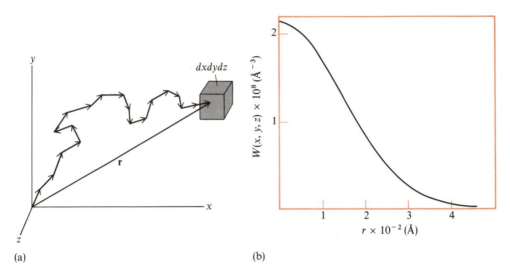

(a)

(b)

Figure 18-5

The Gaussian distribution function. **(a)** The end-to-end distance vector **r** and the volume element $dxdydz$. **(b)** The Gaussian density distribution $W(x, y, z)$ versus end-to-end distance r for a freely jointed chain composed of 10^4 bonds of length 2.5 Å. [After P. J. Flory, *Principles of Polymer Chemistry* (Ithaca, N.Y.: Cornell Unv. Press, 1953), p. 406.]

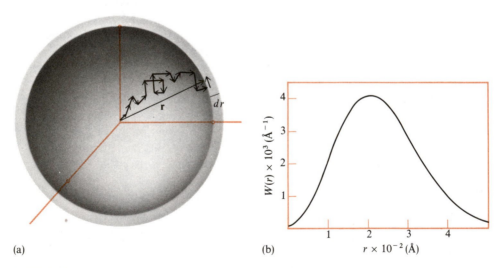

(a)

(b)

Figure 18-6

The radial distribution function. **(a)** The end-to-end distance vector **r** and the spherical shell of volume $4\pi r^2 dr$. **(b)** The distribution function $W(r)$ versus r for the same chain as that in Figure 18-5b. [After P. J. Flory, *Principles of Polymer Chemistry* (Ithaca, N.Y.: Cornell Univ. Press, 1953), p. 406.]

$$W(r)dr = (\beta/\sqrt{\pi})^3 e^{-\beta^2 r^2} 4\pi r^2 \, dr \tag{18-22}$$

This distribution function is plotted in Figure 18-6b. In this case, the distribution function does not peak at the origin, but peaks just below the root-mean-square end-to-end distance $\langle r^2 \rangle^{1/2}$. The reason that $W(r)$ does not peak at the origin, as does $W(x, y, z)$, is that the spherical shell of radius r (Fig. 18-6b) encompasses an increasingly larger volume as r increases, according to $4\pi r^2 \, dr$. This tends to offset the exponentially decreasing function $e^{-\beta^2 r^2}$.

The distribution function of Equation 18-22 may be used to calculate $\langle r^2 \rangle_0$, in accordance with

$$\langle r^2 \rangle_0 = \int_0^\infty r^2 W(r) \, dr \bigg/ \int_0^\infty W(r) \, dr \tag{18-23}$$

Note that the normalization integral in the denominator is equal to unity; that is, the distribution function is normalized. From Equation 18-23, we obtain (Problem 18-1)

$$\langle r^2 \rangle_0 = nl^2 \tag{18-24}$$

Equation 18-24 agrees with our prior calculation of $\langle r^2 \rangle_0$ (Eqn. 18-10), which was done by a different method.

The Gaussian distribution functions shown in Figure 18-5b and 18-6b are of considerable interest. Note that the distribution is rather broad, with most conformations being encompassed in a tenfold range of r. These conformations tend to fall in a range of r that is far below that of maximal extension of the freely joined chain, $r = nl$. Of course, the distribution functions were derived under the assumption that $r \ll nl$, but it is clear from Figures 18-5b and 18-6b that $W(x, y, z)$ and $W(r)$ have largely died out long before full extensions are realized. A more exact treatment, valid for higher extensions, shows that the distribution functions of Equations 18-21 and 18-22 are a fair approximation up to $r = nl/2$ (see Flory, 1969).

Significance of the Gaussian distribution function for real chains

Perhaps the greatest significance of the Gaussian distribution function lies in the fact that all *real* chains—with fixed bond geometries, rotational hindrance, and unperturbed by excluded-volume interactions—may be described by a Gaussian distribution function when $n \to \infty$. The reasons for this are discussed in a later section. For shorter lengths, departures from Gaussian behavior for unperturbed real chains may become severe. The distribution function for a stiff polymer of finite length, for example, will be much narrower and more sharply peaked.

18-4 THE FREELY ROTATING CHAIN

Although it provides useful insights and serves as a convenient reference point, the freely jointed chain clearly is imbued with artificiality. Some of this artificiality can be removed by imposing the constraint of fixed bond angles. The result is the freely rotating chain, in which rotations about the skeletal bonds are completely unhindered.

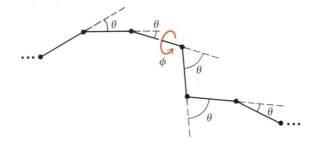

Figure 18-7

The freely rotating chain. Bond-angle supplements θ are fixed, but the rotation angle ϕ is free to vary.

Figure 18-7 shows an example where bond angles (and bond-angle supplements θ) are fixed, but the dihedral angle ϕ is free to vary. (The θ's are not all the same in Fig. 18-7, but will be considered identical in the treatment below.)

Calculation of $\langle r^2 \rangle_0$

The chain is composed of n bonds. The calculation of $\langle r^2 \rangle_0$ requires an evaluation of $\langle \sum_{j>i} (\hat{\mathbf{l}}_i \cdot \hat{\mathbf{l}}_j) \rangle$ (see Eqn. 18-4), with averaging over all conformations accomplished by Equation 18-6. Clearly, however, all conformations are equally probable, so that $p(\phi_2, \phi_3, \ldots)$ is a constant. Consequently, the evaluation is easily accomplished by the following procedure. It is obvious that $\langle \hat{\mathbf{l}}_i \cdot \hat{\mathbf{l}}_{i+1} \rangle = l^2 \cos \theta$. Consider now the quantity $\langle \hat{\mathbf{l}}_i \cdot \hat{\mathbf{l}}_{i+2} \rangle$, the average projection on $\hat{\mathbf{l}}_i$ of a "second-neighbor" bond. As the rotation angle ϕ about $\hat{\mathbf{l}}_{i+1}$ is varied, all possible orientations of $\hat{\mathbf{l}}_i$ and $\hat{\mathbf{l}}_{i+2}$ are generated (Fig. 18-8). It should be clear that all components of $\hat{\mathbf{l}}_{i+2}$ transverse to $\hat{\mathbf{l}}_{i+1}$

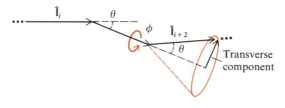

Figure 18-8

Section of a freely rotating chain. As ϕ is freely varied, the transverse components of $\hat{\mathbf{l}}_{i+2}$ cancel.

cancel, as ϕ is varied. The component of $\hat{\mathbf{l}}_{i+2}$ parallel to $\hat{\mathbf{l}}_{i+1}$ is $l \cos \theta$, and this does not cancel. The projection of this component onto $\hat{\mathbf{l}}_i$ is $l(\cos \theta)(\cos \theta) = l \cos^2 \theta$, and $\langle \hat{\mathbf{l}}_i \cdot \hat{\mathbf{l}}_{i+2} \rangle = l^2 \cos^2 \theta$. By induction, we conclude that $\langle \hat{\mathbf{l}}_i \cdot \hat{\mathbf{l}}_j \rangle = l^2 \cos^{j-i} \theta$. Therefore, we can write Equation 18-4 as

$$\langle r^2 \rangle_0 = nl^2 + 2l^2 \sum_{j>i} \alpha^{j-i} \tag{18-25}$$

where $\alpha = \cos \theta$.

There are $n - k$ combinations of bond vectors in Equation 18-25, which are separated by k bond angles. This enables us to write Equation 18-25 in terms of a single summation variable:

$$\langle r^2 \rangle_0 = nl^2 + 2l^2 \sum_{k=1}^{n-1} (n - k)\alpha^k \tag{18-26}$$

where $k = j - i$.

A familiar procedure is used for solving Equation 18-26. We define $S_m = \sum \alpha^k$, where the sum is taken from $k = 1$ to $k = m$, and we note that $S_{m+1} = \alpha(1 + S_m)$, and $S_{m+1} - S_m = \alpha^{m+1}$. These equations, when solved for S_m, give

$$S_m = (\alpha - \alpha^{m+1})/(1 - \alpha) \tag{18-27a}$$

Furthermore, it is clear that

$$\sum_{k=1}^{m} k\alpha^k = \alpha \, dS_m/d\alpha$$

$$= \alpha[1 - (m + 1)\alpha^m]/(1 - \alpha) + \alpha(\alpha - \alpha^{m+1})/(1 - \alpha)^2 \tag{18-27b}$$

Using the relationships of Equations 18-27a,b in Equation 18-26 and simplifying the result, we obtain

$$\langle r^2 \rangle_0 = nl^2 \left(\frac{1 + \alpha}{1 - \alpha} - \frac{2\alpha(1 - \alpha^n)}{n(1 - \alpha)^2} \right) \tag{18-28}$$

Behavior of the characteristic ratio

Two important facts emerge from Equation 18-28. First, the ratio $\langle r^2 \rangle_0/nl^2$ is dependent on n, which was not true for the random-walk chain (Eqn. 18-10); this dependence exists for all statistically coiling, real chains. Second, the ratio $\langle r^2 \rangle_0/nl^2$ at large values of n reaches an asymptotic value independent of n; however, unlike the random-walk

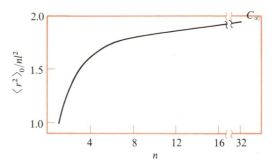

Figure 18-9

The characteristic ratio. Plot of $\langle r^2 \rangle_0/nl^2$ versus n for a tetrahedrally bonded, freely rotating chain.

chain, the asymptotic limit for the freely rotating chain is not simply equal to nl^2.

Figure 18-9 is a plot of the dimensionless characteristic ratio $\langle r^2 \rangle_0/nl^2 = C_n$ versus n for the case $\alpha = 1/3$; this case corresponds to a tetrahedrally bonded chain such as polymethylene. It is clear that $\langle r^2 \rangle_0/nl^2$ comes close to its asymptotic value by $n = 30$. This asymptotic limit is given by

$$(\langle r^2 \rangle_0/nl^2)_{n \to \infty} = (1 + \alpha)/(1 - \alpha) \tag{18-29}$$

For the case illustrated, the asymptotic limit of the characteristic ratio is 2. From Equation 18-29, it is evident that $(\langle r^2 \rangle_0/nl^2)_{n \to \infty} > 1$ for most freely rotating chains with bond angles corresponding to those in real chains. This is due to the fact that bond angles tend to be greater than $90°$ (and bond-angle supplements less than $90°$), which means that $\alpha > 0$.

The physical significance of the plot in Figure 18-9 bears some comment. For a rodlike chain that is completely inflexible, $\langle r^2 \rangle_0 \propto n^2 l^2$, so that $C_n \propto n$; thus the characteristic ratio rises without bound as n increases. The fact that C_n for the freely rotating chain at first increases with n but subsequently becomes independent of n suggests that over the short range the chain has some stiffness (as C_n increases with n), but that over the long range it behaves as a random-walk chain (as C_n becomes independent of n). By contrast, the random-walk chain's behavior is such that C_n is independent of n for all n. These facts collectively imply that the rate of convergence of C_n to its asymptotic value C_∞ is a measure of the stiffness of a chain. This relationship becomes more clear in Sections 18-9 and 18-10, where we discuss the concepts of statistical segment and persistence length.

18-5   REAL CHAINS AND THE ROTATIONAL ISOMERIC-STATE MODEL

Real chains differ sharply from the two models just considered, in that the real chains have complicated bond-rotational potential functions. This aspect of real chains makes the calculation of $p(\phi_2, \phi_3, \ldots)$ and $\langle r^2 \rangle_0$ much more difficult and complex. It is possible, however, to circumvent some of these difficulties by using the rotational isomeric-state model.

Consider a simple chain molecule of three bonds (Fig. 18-10a). Let us assume that the energy of rotation about the central bond is represented by the plot of energy versus ϕ shown in Figure 18-10b. We envision three well-spaced relative minima

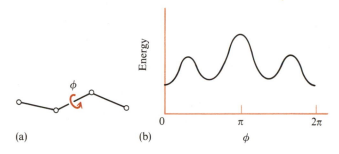

(a)   (b)

Figure 18-10

Dihedral angle and energy. (**a**) A hypothetical chain. The dihedral angle ϕ describes rotations about the central bond. In this chapter, $\phi = 0°$ corresponds to the *trans* conformation. (See Chapter 5 for further discussion of dihedral angles.) (**b**) Plot of energy versus ϕ for the hypothetical chain.

separated by significant barrier heights. In accordance with simple Boltzmann statistics, the probability of ϕ lying between ϕ and $\phi + d\phi$ is

$$p(\phi)\,d\phi = e^{-E(\phi)/kT}\,d\phi \bigg/ \int_0^{2\pi} e^{-E(\phi)/kT}\,d\phi \qquad (18\text{-}30)$$

This probability function will have its largest values when ϕ adopts positions in the minima of Figure 18-10b; for values of $E(\phi)$ greater than a few times kT, the value of $p(\phi)$ becomes vanishingly small. Thus, the chain spends the vast majority of its time in or near one of the three minima. Of course, if we were to calculate the average end-to-end distance of our simple three-bond chain, we would find that only the conformations in the vicinity of the minima in Figure 18-10b would make significant contributions to the average. The same is true for the calculation of the average of any property that depends on conformation (such as ϕ). Therefore, we can approximate the states of the chain by considering only three states: those that are represented by each minimum, designated ϕ_1, ϕ_2, and ϕ_3. The three states so chosen are the rotational isomeric states.

The probability of occurrence of any given state ϕ_i in this model is

$$p(\phi_i) = e^{-E(\phi_i)/kT} \bigg/ \sum_{i=1}^{3} e^{-E(\phi_i)/kT} \qquad (18\text{-}31)$$

Equation 18-31 is considerably simpler than Equation 18-30. Conformation-dependent average properties now are calculated by using this simple probability function.

In most situations, the rotational potential functions associated with each skeletal bond in a long polymer are complicated by the fact that the rotations are interdependent. That is, the hindrances to rotations ϕ_i depend upon the values adopted by ϕ_{i-1} and ϕ_{i+1}, the adjacent neighboring states. The reason for this is that interactions (such as steric contacts) can take place between atoms separated by more than one rotation angle (Fig. 18-11; see also Chapter 5 for energy-contour diagram of polypeptides). Therefore, the rotational isomeric states that are chosen usually

Figure 18-11

Steric contacts between groups whose distance of separation depends on two rotation angles.

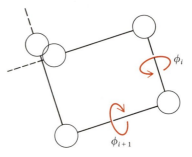

represent the low-energy states of pairs of neighboring bonds, rather than individual bonds. Average values of conformational properties generally are calculated by mathematical methods involving matrices, which are ideally suited for this type of analysis. Energy functions required for the calculations are constructed along lines discussed in Chapter 5, in connection with polypeptides. Calculations based on the rotational isomeric scheme are in many cases remarkably capable of accounting for experimental data.

18-6 CALCULATION OF INTRACHAIN DISTANCES IN POLYPEPTIDES

Here we consider the calculation of intrachain distances in polypeptides. This treatment illustrates the methods and considerations that enter into the statistical evaluation of configurations of real chains. It should be appreciated, however, that the polypeptide case is one in which the full power and utility of the rotational isomeric-state scheme does not emerge.

Figure 18-12 is a schematic representation of the polypeptide chain in its planar zigzag conformation, where all dihedral rotation angles ϕ and ψ are set equal to zero. The amino acid repeating unit is serially indexed from 0 to n. Rotations about the bonds adjoining the ith alpha-carbon atom C_i^α are designed ϕ_i and ψ_i. (See Chapter 5 for a discussion of ϕ_i and ψ_i. Remember that in the present chapter we use the convention that $\phi_i = 0°$ and $\psi_i = 0$ correspond to the *trans* configurations about the $N\!-\!C_i^\alpha$

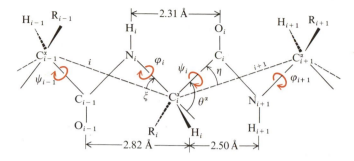

Figure 18-12

Schematic illustration of the polypeptide backbone in its planar zigzag form. The virtual bonds i and $i + 1$ are shown as dashed lines. Fixed geometric parameters and the variable rotation angles ϕ and ψ are indicated. θ^{α} is the supplement of the N_i–C_i^{α}–C_i bond angle, η is the angle between virtual bond $i + 1$ and the C_i^{α}–C_i bond, and ξ is the angle between virtual bond i and the N_i–C_i^{α} bond. [After P. J. Flory, *Statistical Mechanics of Chain Molecules* (New York; Wiley, 1969), p. 251.]

and C_i^{α}—C_i' bonds, respectively. See Section 5-2 for further discussion. Table 5-1 shows values for the various skeletal bond lengths and bond angles.)

We point out in Chapter 5 that the distance between successive α-carbon atoms is fixed at 3.8 Å because of the intervening planar *trans* amide group. Thus, for the purposes of calculating intrachain distances, the polypeptide may be represented as a sequence of virtual bonds, where virtual bond i joins C_{i-1}^{α} and C_i^{α}. These virtual bonds are shown as dashed lines in Figures 18-12. A chain of $n + 1$ residues has n such bonds, which run from C_0^{α} to C_n^{α}. The chain termini, in this scheme, are C_0^{α} and C_n^{α} rather than an amino and a carboxyl group.

● Calculation of $\langle r^2 \rangle_0$ with virtual bonds

The mean square unperturbed end-to-end distance $\langle r^2 \rangle_0$ is calculated according to Equation 18-4, where $\hat{\mathbf{l}}_i$ now refers to the *i*th *virtual* bond vector. To calculate $\langle r^2 \rangle_0$, all configurations of the chain must be generated by varying each ϕ_i and ψ_i. When these rotational angles are varied, the orientations of the virtual bonds are changed. In particular, the mutual orientation of $\hat{\mathbf{l}}_{i+1}$ and $\hat{\mathbf{l}}_i$ is determined by ϕ_i and ψ_i.

The calculation of $\langle \hat{\mathbf{l}}_i \cdot \hat{\mathbf{l}}_j \rangle$ is done by using the conformational energy functions developed in Chapter 5. There we show that rotations ϕ_i, ψ_i within the *i*th unit are essentially independent of those in adjacent units, because of the intervening *trans* amide group. Hindrances to rotations ϕ_i, ψ_i are represented by energy-contour diagrams such as those in Figures 5-7 and 5-8. (Because we are here using a different set of conventions for the reference state of $\phi = 0°$ and $\psi = 0°$, the steric and conformational energy diagrams of Figs. 5-3, 5-4, 5-7, 5-8, 5-10, 5-11, and 5-12 should be renumbered on both the ordinates and abscissas by adding 180° to the values given there.) The conformational energy of the chain as a whole, $E(\phi_1, \psi_1, \phi_2, \psi_2, \ldots)$, thus may be written as a sum of the energies E_i associated with rotations within each residue:

$$E(\phi_1, \psi_1, \phi_2, \psi_2, \ldots) = E_1(\phi_1, \psi_1) + E_2(\phi_2, \psi_2) + \cdots + E_{n-1}(\phi_{n-1}, \psi_{n-1}) \quad (18\text{-}32)$$

[Energies $E_0(\phi_0, \psi_0)$ and $E_n(\phi_n, \psi_n)$ are of no interest, because rotations ϕ_0, ψ_0 and ϕ_n, ψ_n do not affect the distance between terminal α-carbons.] The essentially independent rotations within individual residues greatly simplify the calculation of $\sum_{j>i}\langle \hat{\mathbf{l}}_i \cdot \hat{\mathbf{l}}_j \rangle$, according to Equations 18-5 and 18-6. However, the quantity $\hat{\mathbf{l}}_i \cdot \hat{\mathbf{l}}_j$ is not as easily calculated for polypeptides as for freely jointed and freely rotating chains. The considerably greater complexities of the polypeptide case are self-evident. The required quantity can be computed by the procedure we describe next.

● A coordinate system for each bond

A coordinate system $x_i y_i z_i$ is defined for each bond i; we shall sometimes refer to this as coordinate system i. Each bond i is then given a vectorial representation $\hat{\mathbf{l}}_i$ in coordinate system i. However, the scalar $\hat{\mathbf{l}}_i \cdot \hat{\mathbf{l}}_j$ can be calculated only if both $\hat{\mathbf{l}}_i$ and $\hat{\mathbf{l}}_j$ are expressed in the same coordinate system. This is done by transformation of the coordinate system j into coordinate system i. These transformations are done with matrices, and we assume some familiarity with matrix methods in the following discussion. (See Appendix A for a brief review of matrix algebra.)

Figure 18-13 illustrates the assignment of a Cartesian coordinate system $x_i y_i z_i$ to each bond (of length l), whether real or virtual. The axis x_i is taken to lie along bond i. Axis y_i is located in the plane of bonds $i - 1$ and i; it points in the direction that renders positive its projection on bond $i - 1$. Axis z_i is chosen to complete a right-handed coordinate system. Note that, with these conventions, successive z axes point alternately into and out of the plane of the page, when the chain is in the planar zigzag conformation. It is also important to note that choice of x_i along $\hat{\mathbf{l}}_i$ means that the column vector representation of each $\hat{\mathbf{l}}_i$ in coordinate system i is

Figure 18-13

*Assignment of a Cartesian coordinate system
to bond i in a hypothetical chain.*

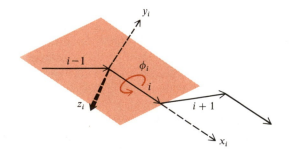

$$\hat{\mathbf{l}}_i = \begin{pmatrix} l \\ 0 \\ 0 \end{pmatrix} = l \begin{pmatrix} 1 \\ 0 \\ 0 \end{pmatrix} \qquad \text{for } i = 1, 2, \dots, n \qquad (18\text{-}33)$$

where l, 0, and 0 are the x_i, y_i, and z_i components, respectively, of $\hat{\mathbf{l}}_i$.

When $\hat{\mathbf{l}}_i$ and $\hat{\mathbf{l}}_j$ are expressed in the same coordinate system, their scalar product is $\hat{\mathbf{l}}_i^T \cdot \hat{\mathbf{l}}_j$, where $\hat{\mathbf{l}}_i^T$ is the transpose or row form of $\hat{\mathbf{l}}_i$. This product is

$$(l_{ix}, l_{iy}, l_{iz}) \begin{pmatrix} l_{jx} \\ l_{jy} \\ l_{jz} \end{pmatrix} = l_{ix}l_{jx} + l_{iy}l_{jy} + l_{iz}l_{jz} \qquad (18\text{-}34)$$

where l_{ix}, l_{iy}, and l_{iz} are the x, y, and z components of $\hat{\mathbf{l}}_i$ in the common coordinate system, and l_{jx}, l_{jy}, and l_{jz} are corresponding components of $\hat{\mathbf{l}}_j$. The general procedure (for $j > i$) is to choose the $x_i y_i z_i$ system (coordinate system i) as the common coordinate system and to transform $\hat{\mathbf{l}}_j$ into this system.

● **Transformation matrices**

Consider now the transformation of a vector from its representation in one coordinate system into its representation in another coordinate system. The exact relationship between systems $x_{i+1} y_{i+1} z_{i+1}$ and $x_i y_i z_i$, for example, depends upon the fixed parameters, ξ, η, and θ^α, and upon the variables ϕ_i and ψ_i (Fig. 18-12). As ϕ_i and ψ_i are varied, the relationship changes between the axes of the two coordinate systems. Let $\mathbf{T}_i(\phi_i, \psi_i)$ be the matrix that transforms a vector in $x_{i+1} y_{i+1} z_{i+1}$ into its representation in $x_i y_i z_i$; clearly, this matrix depends on the variable parameters ϕ_i and ψ_i. A detailed derivation of $\mathbf{T}_i$ is given in Section 18-7. For the moment, we note simply that it is a 3×3 matrix, and we proceed with our calculation of $\langle r^2 \rangle_0$.

The expression of $\hat{\mathbf{l}}_{i+1}$ in system $x_i y_i z_i$ is $\mathbf{T}_i \hat{\mathbf{l}}_{i+1}$. The representation of $\hat{\mathbf{l}}_{i+1}$ in $x_{i-1} y_{i-1} z_{i-1}$, therefore, is $\mathbf{T}_{i-1}(\mathbf{T}_i \hat{\mathbf{l}}_{i+1}) = \mathbf{T}_{i-1} \mathbf{T}_i \hat{\mathbf{l}}_{i+1}$. In general, the representation of $\hat{\mathbf{l}}_j$ in system $x_i y_i z_i$ is $\mathbf{T}_i \mathbf{T}_{i+1} \cdots \mathbf{T}_{j-1} \hat{\mathbf{l}}_j$. Therefore, by multiplication of the matrices $\mathbf{T}_i$, the desired transformations are effected, and $\hat{\mathbf{l}}_i \cdot \hat{\mathbf{l}}_j$ may be calculated as

$$\hat{\mathbf{l}}_i^T \mathbf{T}_i \mathbf{T}_{i+1} \cdots \mathbf{T}_{j-1} \hat{\mathbf{l}}_j = l^2 (1, 0, 0) \mathbf{T}_i \mathbf{T}_{i+1} \cdots \mathbf{T}_{j-1} \begin{pmatrix} 1 \\ 0 \\ 0 \end{pmatrix} \qquad (18\text{-}35)$$

where l has been factored out of $\hat{\mathbf{l}}_i$ and $\hat{\mathbf{l}}_j$. Equation 18-35 is identical to Equation 18-34. In Equation 18-34, $\hat{\mathbf{l}}_j$ is expressed in a common coordinate system and has components $\hat{\mathbf{l}}_{jx}, \hat{\mathbf{l}}_{jy}, \hat{\mathbf{l}}_{jz}$; in Equation 18-35, $\hat{\mathbf{l}}_j$ is written in its own coordinate system (coordinate system j) with components $l, 0, 0$ and is transformed by $\mathbf{T}_i \mathbf{T}_{i+1} \cdots \mathbf{T}_{j-1}$

into coordinate system i; thus, coordinate system i is the common coordinate system, in this example.

● Products of averaged transformation matrices

Evaluation of $\langle \hat{l}_i \cdot \hat{l}_j \rangle$ now is fairly straightforward. In accordance with Equation 18-35, we have

$$\langle \hat{l}_i \cdot \hat{l}_j \rangle = l^2 \mathbf{e}^{\mathrm{T}} \langle \underline{T}_i \underline{T}_{i+1} \cdots \underline{T}_{j-1} \rangle \mathbf{e} \tag{18-36}$$

where $\mathbf{e}^{\mathrm{T}} = (1, 0, 0)$. The average of the product of the $\underline{T}$ matrices is

$$\langle \underline{T}_i \underline{T}_{i+1} \cdots \underline{T}_{j-1} \rangle$$

$$= \frac{\int_0^{2\pi} \cdots \int_0^{2\pi} \underline{T}_i \cdots \underline{T}_{j-1} \exp[-(E_i + \cdots + E_{j-1})/kT] \, d\phi_i \, d\psi_i \cdots d\phi_{j-1} \, d\psi_{j-1}}{\int_0^{2\pi} \cdots \int_0^{2\pi} \exp[-(E_i + \cdots + E_{j-1})/kT] \, d\phi_i \, d\psi_i \cdots d\phi_{j-1} \, d\psi_{j-1}} \tag{18-37}$$

where $E_i = E_i(\phi_i, \psi_i)$, and $E_{j-1} = E_{j-1}(\phi_{j-1}, \psi_{j-1})$, etc.

The significance is now clear of the fact that rotations ϕ_k, ψ_k within each residue are independent of those in neighboring residues. Each of the double integrations over $d\phi_k \, d\psi_k$ may be carried out independently, because the total energy is a simple sum of terms $E_k(\phi_k, \psi_k)$, and the matrix $\underline{T}_k$ is a function only of ϕ_k and ψ_k. Furthermore, because $\exp(A + B + C + \cdots) = e^A e^B e^C \cdots$, Equation 18-37 may be simplified to

$$\langle \underline{T}_i \underline{T}_{i+1} \cdots \underline{T}_{j-1} \rangle$$

$$= \frac{\left(\int_0^{2\pi} \int_0^{2\pi} \underline{T}_i \exp\left(\frac{-E_i}{kT}\right) d\phi_i \, d\psi_i \right) \cdots \left(\int_0^{2\pi} \int_0^{2\pi} \underline{T}_{j-1} \exp\left(\frac{-E_{j-1}}{kT}\right) d\phi_{j-1} \, d\psi_{j-1} \right)}{\left(\int_0^{2\pi} \int_0^{2\pi} \exp\left(\frac{-E_i}{kT}\right) d\phi_i \, d\psi_i \right) \cdots \left(\int_0^{2\pi} \int_0^{2\pi} \exp\left(\frac{-E_{j-1}}{kT}\right) d\phi_{j-1} \, d\psi_{j-1} \right)} \tag{18-38}$$

$$= \langle \underline{T}_i \rangle \langle \underline{T}_{i+1} \rangle \cdots \langle \underline{T}_{j-1} \rangle \tag{18-39}$$

where again $E_k = E_k(\phi_k, \psi_k)$, and

$$\langle \underline{T}_k \rangle = \frac{\int_0^{2\pi} \int_0^{2\pi} \underline{T}_k \exp(-E_k/kT) \, d\phi_k \, d\psi_k}{\int_0^{2\pi} \int_0^{2\pi} \exp(-E_k/kT) \, d\phi_k \, d\psi_k} \tag{18-40}$$

The fact that the average of a product of matrices (Eqn. 18-37) is equal to the product of the averaged matrices (Eqn. 18-39) is a great simplification. It is important to appreciate also that $\langle \underset{\sim}{T}_k \rangle$ depends only on the nature of amino acid k, because E_k is a function only of rotations ϕ_k, ψ_k within residue k. Compare, for example, the different energy functions that describe rotations within a glycyl residue and within an alanyl residue (Figs. 5-7 and 5-8). Therefore, in calculating the mean square end-to-end distance, each amino acid residue takes on a matrix representation, which is an averaged transformation matrix. The value of $\langle \underset{\sim}{T}_k \rangle$ may be evaluated by the rotational isomeric-state approximation, whereby only discrete low-energy states of ϕ_k, ψ_k are considered (see, for example, the local minima in the energy-contour diagrams of Figs. 5-7 and 5-8), or it may be evaluated by a more extensive numerical integration.

● **Summing matrix products**

To complete our calculation of $\langle r^2 \rangle_0$ according to Equation 18-4, consider (for simplicity) the case of homopolymers, where each amino acid unit is identical. In this case,

$$\langle \underset{\sim}{T}_i \rangle \langle \underset{\sim}{T}_{i+1} \rangle \cdots \langle \underset{\sim}{T}_{j-1} \rangle = \langle \underset{\sim}{T} \rangle^{j-i} \tag{18-41}$$

where $\langle \underset{\sim}{T} \rangle$ is the averaged transformation matrix for the amino acid in question. Therefore, we obtain

$$\langle r^2 \rangle_0 = nl^2 + 2l^2 \sum_{j>i} e^T \langle \underset{\sim}{T} \rangle^{j-i} e \tag{18-42}$$

There are $n - k$ terms in Equation 18-42 for which $j - i = k$. Therefore, Equation 18-42 can be simplified by grouping terms together to give

$$\langle r^2 \rangle_0 = nl^2 + 2l^2 \sum_{k=1}^{n-1} (n-k) e^T \langle \underset{\sim}{T} \rangle^k e \tag{18-43}$$

Equation 18-43 is completely analogous to Equation 18-26. The scalar α^k in Equation 18-26 is analogous to $\langle \underset{\sim}{T} \rangle^k$ in Equation 18-43, for example. Matrices can be summed in exactly the same way as scalar quantities. Referring to Equation 18-28 and replacing α by $\langle \underset{\sim}{T} \rangle$, we obtain the result of executing the summation in Equation 18-43:

$$\langle r^2 \rangle_0 = nl^2 e^T \underset{\sim}{H} e \tag{18-44}$$

$$= nl^2 H_{11} \tag{18-45}$$

where

$$\underset{\sim}{H} = (\underset{\sim}{I} + \langle \underset{\sim}{T} \rangle)(\underset{\sim}{I} - \langle \underset{\sim}{T} \rangle)^{-1} - (2\langle \underset{\sim}{T} \rangle / n)[(\underset{\sim}{I} - \langle \underset{\sim}{T} \rangle^{n})(\underset{\sim}{I} - \langle \underset{\sim}{T} \rangle)^{-2}]$$ (18-46)

In Equation 18-46, superscript -1 denotes the inverse of a matrix, superscript -2 indicates the square of an inverse matrix, and $\underset{\sim}{I}$ is the identity matrix:

$$\underset{\sim}{I} = \begin{pmatrix} 1 & 0 & 0 \\ 0 & 1 & 0 \\ 0 & 0 & 1 \end{pmatrix}$$

The parameter H_{11} is the element in the first row and first column of $\underset{\sim}{H}$. This is the only element that enters into the calculation of $\langle r^2 \rangle_0$, because $e^T \underset{\sim}{H} e = H_{11}$.

The preceding treatment, culminating in Equations 18-44 through 18-46, is an impressive illustration of the power of matrix methods in executing complex conformational calculations for real chains. Far more complex situations also can be treated by these methods. In the next section, we derive the expression for the transformation matrix $\underset{\sim}{T}$. Following this derivation, we discuss the results of conformational calculations for polypeptides, and their implications for protein structure.

● 18-7 CALCULATION OF THE COORDINATE TRANSFORMATION MATRIX

Calculation of coordinate transformation matrices is invariably necessary in any rigorous treatment of real chains. The objective is to express a vector $\mathbf{v}$ in terms of its components in coordinate system i, where we are given its components in coordinate system $i + 1$ and the relationship between systems $i + 1$ and i. This is accomplished as follows. Let the vector $\mathbf{v}$ in coordinate system $i + 1$ be given by

$$\mathbf{v} = \hat{\mathbf{i}}_{i+1} v_x + \hat{\mathbf{j}}_{i+1} v_y + \hat{\mathbf{k}}_{i+1} v_z$$ (18-47)

where subscript $i + 1$ denotes the coordinate system $i + 1$. In coordinate system i, the vector $\mathbf{v}$ is

$$\mathbf{v} = \hat{\mathbf{i}}_i v_x' + \hat{\mathbf{j}}_i v_y' + \hat{\mathbf{k}}_i v_z'$$ (18-48)

where the primed quantities are to be determined. In column-vector form, in coordinate system $i + 1$, $\mathbf{v}$ is

$$\mathbf{v} = \begin{pmatrix} v_x \\ v_y \\ v_z \end{pmatrix}$$ (18-49)

Let the $\hat{\mathbf{i}}_i$, $\hat{\mathbf{j}}_i$, and $\hat{\mathbf{k}}_i$ components of the unit vector $\hat{\mathbf{i}}_{i+1}$ be represented[§] by the elements R_{11}, R_{21}, and R_{31} in the column vector

$$\begin{pmatrix} R_{11} \\ R_{21} \\ R_{31} \end{pmatrix}$$

Similarly, the $\hat{\mathbf{i}}_i$, $\hat{\mathbf{j}}_i$, and $\hat{\mathbf{k}}_i$ components of unit vector $\hat{\mathbf{j}}_{i+1}$ are

$$\begin{pmatrix} R_{12} \\ R_{22} \\ R_{32} \end{pmatrix}$$

and the $\hat{\mathbf{i}}_i$, $\hat{\mathbf{j}}_i$, and $\hat{\mathbf{k}}_i$ components of $\hat{\mathbf{k}}_{i+1}$ are

$$\begin{pmatrix} R_{13} \\ R_{23} \\ R_{33} \end{pmatrix}$$

It follows, then, that the matrix $\underset{\sim}{\mathbf{R}}$ transforms the components of $\mathbf{v}$ from system $i+1$ to system i, where

$$\underset{\sim}{\mathbf{R}} = \begin{pmatrix} R_{11} & R_{12} & R_{13} \\ R_{21} & R_{22} & R_{23} \\ R_{31} & R_{32} & R_{33} \end{pmatrix} \tag{18-50}$$

and the transformation is expressed as

$$\begin{pmatrix} v'_x \\ v'_y \\ v'_z \end{pmatrix} = \begin{pmatrix} R_{11} & R_{12} & R_{13} \\ R_{21} & R_{22} & R_{23} \\ R_{31} & R_{32} & R_{33} \end{pmatrix} \begin{pmatrix} v_x \\ v_y \\ v_z \end{pmatrix} \tag{18-51}$$

$$= \begin{pmatrix} R_{11}v_x + R_{12}v_y + R_{13}v_z \\ R_{21}v_x + R_{22}v_y + R_{23}v_z \\ R_{31}v_x + R_{32}v_y + R_{33}v_z \end{pmatrix} \tag{18-52}$$

According to Equation 18-52, $v'_x = R_{11}v_x + R_{12}v_y + R_{13}v_z$, and so also for v'_y and v'_z. Our task now is to compute the transformation matrix $\underset{\sim}{\mathbf{R}}$.

[§] It should be kept firmly in mind that $\hat{\mathbf{i}}_{i+1}$ means a unit vector in the x direction of coordinate system $i+1$.

Figure 18-14

One possible relative orientation of coordinate systems i and $i + 1$. The axes x_i, x_{i+1}, and y_{i+1} are in the plane of the page, but y_i makes an acute angle with the page (projecting out from the page).

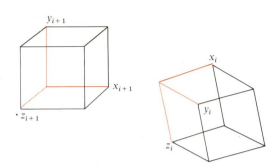

Figure 18-14 illustrates a relationship between two right-hand coordinate systems. We assume that the relationship between the two systems is such that rotation of system $i + 1$ about the z_{i+1} axis through the angle α brings x_{i+1} and x_i into coincidence. Subsequently, rotation of $i + 1$ about the common x axis through the angle β brings the y axes and the z axes into coincidence (Fig. 18-15, which should be studied carefully). Coordinate systems related as are the ones in Figure 18-15 are the ones of interest in chain conformation calculations.

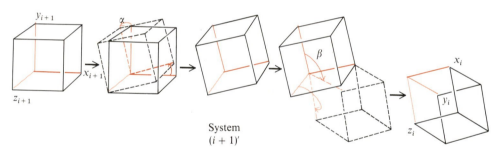

System
$(i + 1)'$

Figure 18-15

Transformation of coordinate systems. Two rotations transform coordinate system $i + 1$ into system i.

To accomplish our transformation from coordinate system $i + 1$ to system i, we first transform from coordinate system $i + 1$ to system $(i + 1)'$, the system obtained by rotating system $i + 1$ through the angle α about the z_{i+1} axis. This transformation matrix is designated as $\mathbf{R}_1$. We now have a matrix that transforms unit vectors along $\hat{\mathbf{i}}_{i+1}$, $\hat{\mathbf{j}}_{i+1}$, and $\hat{\mathbf{k}}_{i+1}$ into their components in system $(i + 1)'$. The vector $\mathbf{R}_1\mathbf{v}$, for example, is now the vector $\mathbf{v}$ expressed in coordinate system $(i + 1)'$. If $\mathbf{R}_2$ transforms coordinate system $(i + 1)'$ to system i, then $\mathbf{R}_2\mathbf{R}_1\mathbf{v}$ is the vector $\mathbf{v}$ in coordinate system i. Therefore, $\mathbf{R}_2\mathbf{R}_1 = \mathbf{R}$. The reader can easily verify that

$$\mathbf{R}_1 = \begin{pmatrix} \cos \alpha & \sin \alpha & 0 \\ -\sin \alpha & \cos \alpha & 0 \\ 0 & 0 & 1 \end{pmatrix} \tag{18-53}$$

and

$$\underset{\sim}{\mathbf{R}}_2 = \begin{pmatrix} 1 & 0 & 0 \\ 0 & \cos \beta & \sin \beta \\ 0 & -\sin \beta & \cos \beta \end{pmatrix} \tag{18-54}$$

Remember that the three elements of the first column of $\underset{\sim}{\mathbf{R}}_1$ are, from top to bottom, the $\hat{\mathbf{i}}_{(i+1)'}$, $\hat{\mathbf{j}}_{(i+1)'}$, and $\hat{\mathbf{k}}_{(i+1)'}$ components of a unit vector along the x_{i+1} axis. Similarly, the $\hat{\mathbf{i}}_{(i+1)'}$, $\hat{\mathbf{j}}_{(i+1)'}$, and $\hat{\mathbf{k}}_{(i+1)'}$ components of a unit vector along the y_{i+1} axis are given by the three elements in the second column of $\underset{\sim}{\mathbf{R}}_1$. For example, $\hat{\mathbf{i}}_{i+1}$ has a component $\cos \alpha$ along the $x_{(i+1)'}$ axis and a component $-\sin \alpha$ along the $y_{(i+1)'}$ axis.

The matrix $\underset{\sim}{\mathbf{R}} = \underset{\sim}{\mathbf{R}}_2\underset{\sim}{\mathbf{R}}_1$ is

$$\underset{\sim}{\mathbf{R}}(\alpha, \beta) = \begin{pmatrix} \cos \alpha & \sin \alpha & 0 \\ -\cos \beta \sin \alpha & \cos \beta \cos \alpha & \sin \beta \\ \sin \beta \sin \alpha & -\sin \beta \cos \alpha & \cos \beta \end{pmatrix} \tag{18-55}$$

Equation 18-55 gives the desired transformation matrix $\underset{\sim}{\mathbf{R}}$, which transforms coordinate system $i + 1$ into system i.

● Transformation between the coordinate systems of virtual bonds

For our polypeptide case, we wish to transform the coordinate system associated with virtual bond $i + 1$ to that of virtual bond i (see Fig. 18-12). Coordinate systems for virtual bonds are defined in the same way as those for real bonds (see Section 18-6). This transformation is done in three steps: transforming virtual bond system $i + 1$ to that of the C_i^α–C_i bond; then transforming to the system of the N_i–C_i^α bond; and finally transforming to the coordinate system of virtual bond i. The coordinate system of virtual bond $i + 1$ is brought into coincidence with that of the C_i^α–C_i bond by rotations $-\eta$ and $-\psi$.[§] Therefore, $\underset{\sim}{\mathbf{R}}(-\eta, -\psi_i)$ effects this transformation. Subsequent transformation to the system of the N_i–C_i^α bond is accomplished by $\underset{\sim}{\mathbf{R}}[\theta^\alpha, (\pi - \phi_i)]$, and the final transformation to the coordinate system of virtual bond i is done by $\underset{\sim}{\mathbf{R}}(\xi, 0)$. Therefore, the matrix $\underset{\sim}{\mathbf{T}}_i(\phi_i, \psi_i)$ that transforms virtual bond coordinate system $i + 1$ to system i is

$$\underset{\sim}{\mathbf{T}}_i(\phi_i, \psi_i) = \underset{\sim}{\mathbf{R}}(\xi, 0)\underset{\sim}{\mathbf{R}}[\theta^\alpha, (\pi - \phi_i)]\underset{\sim}{\mathbf{R}}(-\eta, -\psi_i) \tag{18-56}$$

This is the matrix that is used in the preceding treatment of polypeptide statistics (Eqn. 18-35 and following).

[§] Referring to Equation 18-55, this means that $\alpha = -\eta$, and $\beta = -\psi$. For example, recall that α is the angle through which the x axis of virtual bond $i + 1$ must be rotated in order to bring it into coincidence with the x axis of the coordinate system of the C_i^α–C_i bond. This rotation is a right-hand rotation about the z axis of the coordinate system of virtual bond $i + 1$. Inspection of Figure 18-12 shows that a rotation of $-\eta$ is required to bring into coincidence the x axis of virtual bond $i + 1$ and the x axis of the C_i^α–C_i bond.

● **Averaged transformation matrices for glycine and L-alanine**

It is interesting to note, for illustrative purposes, the averaged transformation matrices for a glycyl and an L-alanyl residue situated within a polypeptide chain. These are computed according to Equation 18-40 by using the residue energy functions in Figures 5-7 and 5-8. The result is

$$\langle \mathbf{T}_{Gly} \rangle = \begin{pmatrix} 0.36 & -0.077 & 0 \\ -0.092 & -0.037 & 0 \\ 0 & 0 & -0.12 \end{pmatrix} \tag{18-57}$$

$$\langle \mathbf{T}_{Ala} \rangle = \begin{pmatrix} 0.51 & 0.20 & 0.59 \\ -0.046 & -0.61 & 0.21 \\ 0.65 & -0.23 & 0.30 \end{pmatrix} \tag{18-58}$$

The symmetry of the glycyl residue gives rise to the four zero elements in $\langle \mathbf{T}_{Gly} \rangle$. These matrices are used in the configurational calculations of polyglycine and poly-L-alanine homopolymers according to Equations 18-44 through 18-46. They are employed also in the calculations of polypeptide copolymers.

18-8   RESULTS OF CONFORMATIONAL CALCULATIONS OF POLYPEPTIDES

Many interesting calculations on polypeptides have been carried out. These have given significant insights into the random-coil (or statistical-coil) conformation of these important polymers, and have shed light on protein folding as well.

Table 18-1 summarizes values of the limiting characteristic ratio $C_\infty = (\langle r^2 \rangle_0 / nl^2)_{n \to \infty}$, for polyglycine, poly-L-alanine, poly-L-proline (where the rotation angle ϕ is fixed by the pyrrolidine ring), and poly-(L-lactic acid). The latter is a

Table 18-1
Limiting characteristic ratios C_∞

Polymer	Free rotation	Dipolar interaction omitted	Dipolar interaction included	Experimental
Polyglycine	1.93	1.79	2.16	——
Poly-L-alanine	1.93	2.97	9.27	9.0
Poly-L-proline II	1.86	116	——	——
Poly(L-lactic acid)	1.92	1.24	2.13	2.1

SOURCE: Data from P. J. Flory, *Statistical Mechanics of Chain Molecules* (New York: Wiley, 1969), p. 277.

polypeptidelike chain in which the amide link is replaced by an ester link. Calculations in Table 18-1 were carried out as we have described (see Eqns. 18-44 through 18-46). Note that, in the limit $n \to \infty$, Equation 18-46 simplifies considerably. The table gives calculations for cases where free rotation about every ϕ_i and ψ_i (just ψ_i for poly-L-proline) is assumed to occur, and for the cases where E_d, the dipole–dipole interaction energy between successive amide (or ester) dipoles (see Chapter 5), either is omitted or is included in the energy function $E_i(\phi_i, \psi_i)$ used to calculate $\langle \underset{\sim}{\mathbf{T}} \rangle$ (Eqn. 18-40). Two experimental values also are included in Table 18-1.

The amide dipolar interaction

It is apparent from Table 18-1 that agreement between calculated and experimental values is satisfactory only when E_d is included in the calculation. This observation underscores the significance of the dipolar interaction in biasing polypeptide conformation. In the case of poly-L-alanine, the calculated mean square dimensions are more than threefold larger when E_d is included. This result can be rationalized by referring to the energy-contour diagram for an L-alanyl residue (Fig. 5-8). In the absence of dipolar effects, the preferred conformations lie in both regions I and III. The dipolar interaction gives a definite preference to region III, however, so that successive residues tend to adopt ϕ, ψ values in this area. The consequence of this is that the chain's trajectory has a greater tendency to follow along one given direction, and has fewer bends or turns than would occur if other distinct regions of the ϕ, ψ plane (such as region I) were more competitive.

Differences between polymers of glycine, L-alanine, and L-proline

In the case of polyglycine, the chain dimensions are much smaller than for poly-L-alanine, thus indicating the prevalence of many compact conformations. In fact, the dimensions are similar to those of the freely rotating polyglycine chain, even though the allowed conformations of the real chain by no means include all values of ϕ and ψ (see Fig. 5-7). The effects of E_d, in this instance, are not pronounced. These results are a consequence of the fact that *two* symmetrically disposed minima in $E(\phi, \psi)$ must occur, owing to the symmetry of the glycyl residue (see Fig. 5-7). This means that successive residues in the chain can adopt alternate conformations, thus giving rise to highly tortuous configurations.

Poly-L-proline II[§] is a special case in which residue rotations ψ are severely hindered. The result is a very stiff chain of large extensions. (Effects of E_d on chain dimensions have not been estimated quantitatively, but they are believed to be small.) In fact, for short chains, the dimensions of the coil are similar to those of the poly-L-proline II helix, which has been extensively studied.

[§] Form II has all imide links in the *trans* configuration. They are all *cis* in Form I. (See Chapter 20 for illustrations.)

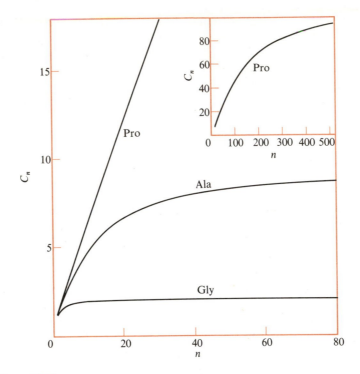

Figure 18-16

Characteristic ratio for various polypeptide homopolymers. The characteristic ratio C_n is plotted versus n for polyglycine, poly-L-alanine, and poly-L-proline. [After P. R. Schimmel and P. J. Flory, *Proc. Natl. Acad. Sci. USA* 58:52 (1967).]

Figure 18-16 illustrates further differences in the conformational characteristics of polyglycine, poly-L-alanine, and poly-L-proline. This figure plots C_n versus n. It is clear that the rate of approach of the dimensionless ratio to its asymptotic value differs significantly for the three polymers. Polyglycine most rapidly reaches its limiting value, whereas poly-L-proline only slowly approaches its asymptote. These characteristics are a reflection of the relative stiffnesses of the chains. Recall that C_n is independent of n for all n, in the case of the random-walk chain (cf. Eqn. 18-10). Therefore, when C_n becomes approximately independent of n, a real chain is behaving (in this limited sense) as a random-walk polymer. (Note that C_n is not simply equal to unity in the case of the real chain, as it is in the random-walk case. See Section 18-9 for further clarification.)

Comparisons between calculated and experimental values

Table 18-2 shows a number of comparisons between calculated and experimental values of $\langle r^2 \rangle_0^{1/2}$. A large range of molecular weights is covered with several different

Table 18-2

Experimental and calculated values of intrachain distances

Polypeptide	Number of amino acid residues	RMS distance between chain termini (Å)		Experimental method
		Experimental	Calculated	
Poly-L-lysine	550[§]	260[a]	271.4[b]	Viscosity
Poly-β-benzyl-L-aspartate	912[§]	360[a]	349.4[b]	Viscosity
Poly-γ-benzyl-L-glutamate	1,534[§]	440[c]	453.2[b]	Viscosity, light scattering
Polyglycine	3	7.8[d]	7.9[e]	Dielectric increment–dipole moment
Polyglycine	7	12.1[d]	12.0[e]	Dielectric increment–dipole moment
Poly-L-proline	7	20[f]	18.5[g]	Spectroscopic–energy transfer
Poly-L-proline	12	34[f]	33.4[g]	Spectroscopic–energy transfer

NOTE: In the case of high-molecular-weight species, experimental values have been corrected for excluded-volume effects.
[§] Average number of residues.
SOURCE: [a] D. A. Brant and P. J. Flory, *J. Am. Chem. Soc.* 87:2788 (1965). [b] D. A. Brant and P. J. Flory, *J. Am. Chem. Soc.* 87:2791 (1965).
[c] P. Doty, J. H. Bradbury, and A. M. Holtzer, *J. Am. Chem. Soc.* 78:947 (1956). [d] J. T. Edsall and J. Wyman, *Biophysical Chemistry*, vol. 1 (New York: Academic Press, 1958), p. 372. [e] P. J. Flory and P. R. Schimmel, *J. Am. Chem. Soc.* 89:6807 (1967). [f] L. Stryer and R. P. Haugland, *Proc. Natl. Acad. Sci. USA* 58:719 (1967). [g] P. R. Schimmel and P. J. Flory, *Proc. Natl. Acad. Sci. USA* 58:52 (1967). [The experimental value in these cases is actually $\langle r^6 \rangle^{1/6}$, which for short, stiff proline oligomers should be close to $\langle r^2 \rangle^{1/2}$.

polypeptides. In all cases, agreement is excellent between calculated and observed values of $\langle r^2 \rangle_0^{1/2}$, from $n = 3$ to $n \cong 1,500$. Note the large difference in $\langle r^2 \rangle_0^{1/2}$ for polyglycine and poly-L-proline when $n = 7$. Calculations on poly-L-lysine, poly-β-benzyl-L-aspartate, and poly-γ-benzyl-L-glutamate were obtained by using the energy functions for an L-alanyl residue. This is legitimate, as the data in the table indicate, because hindrances to rotations ϕ, ψ are essentially the same in alanine as in residues bearing side chains that are unbranched, or that are branched only beyond the β carbon atom (see Chapter 5).

Polypeptide copolymers

The results of the calculations just described illustrate the significant influence of the nature of the side chain on the conformation of polypeptide homopolymers. This influence suggests that a broad range of average chain dimensions may be achieved simply by varying the amino acid composition of a copolymer, for example. Therefore, an interesting question to explore is the exact relationship between the dimensions of a polypeptide copolymer and its amino acid composition.

This question has been approached for several copolymers. Figure 18-17 shows a plot of C_∞ versus the mole percentage of glycine or L-alanine, for glycine-L-alanine random copolymers. The figure shows that introduction of a small mole percentage of glycyl residues into a poly-L-alanine chain produces a disproportionately large drop in the chain dimensions. Experimental results are consistent with this prediction. The reason for this behavior is that the greater diversity of conformations

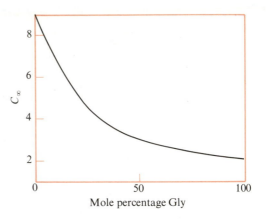

Figure 18-17

Limiting values for the characteristic ratio. The limiting characteristic ratio C_∞ is plotted versus the mole percentage of glycine for random copolymers of L-alanine and glycine. [After W. G. Miller, D. A. Brant, and P. J. Flory, *J. Mol Biol.* 23:67 (1967).]

afforded by a glycyl residue gives considerable flexibility to the chain at points where the glycyl residues are introduced into a stiff poly-L-alanine chain. This results in the generation of more compact chain conformations. Therefore, the occurrence of glycine residues in a small mole percentage in a protein, for example, may be very important for the generation of a compact, folded form. Interesting conclusions have also been obtained when other copolymers are examined.

The polypeptide calculations illustrate the power of the matrix method, and the ability of such calculations to account for experimental results. These studies give considerable insight into factors important in polypeptide and protein conformations. However, they do not resolve the problem of predicting the unique, native conformation of a protein, given the amino acid sequence. A discussion of the further complexities of this problem is given in Chapter 5.

18-9 STATISTICAL SEGMENT

For all statistically coiling real chains, the ratio $\langle r^2 \rangle_0 / nl^2$ approaches a constant, C_∞, as $n \to \infty$. This is clear in the plots of C_n versus n for the three polypeptide chains in Figure 18-16. As mentioned, when C_n is essentially independent of n, the polymer resembles the random-walk chain for which C_n is independent of n, for all n (Eqn. 18-10). This relationship between $\langle r^2 \rangle_0$ and n for the random-walk model is due to the random orientation of each bond with respect to every other bond—that is, $\langle \hat{\mathbf{l}}_i \cdot \hat{\mathbf{l}}_j \rangle = 0$, for all $i \neq j$. In the case of real chains, $\langle \hat{\mathbf{l}}_i \cdot \hat{\mathbf{l}}_j \rangle \to 0$ for sufficiently large $|j - i|$; that is, bonds sufficiently far apart are randomly oriented with respect to each other. This gives rise to the concept of a "statistical segment" for a real chain. The idea is that the chain may be approximated as a polymer of freely jointed statistical segments, each of which is randomly oriented with respect to all other segments. The chain segment is chosen to be long enough to assure its random orientation. In this approximation, the treatment of the distribution of conformations of the random-walk chain is now carried over to the real chain, which is viewed as a polymer of statistical segments.

Relationship between statistical segments and real bonds

The relationship between the length l_e and number N_e of statistical segments and the length l and number n of actual bonds may be computed as follows. When the value of $\langle r^2 \rangle_0 / nl^2$ has reached the limiting characteristic ratio C_∞, then

$$\langle r^2 \rangle_0 = C_\infty nl^2 \qquad (18\text{-}59)$$

and, in terms of the equivalent statistical segments (cf. Eqn. 18-10),

$$\langle r^2 \rangle_0 = N_e l_e^2 \qquad (18\text{-}60)$$

A second relationship is required, and this may be obtained from the constraint that the maximal extension r_m is

$$r_m = N_e l_e \qquad (18\text{-}61)$$

and

$$r_m = fnl \qquad (18\text{-}62)$$

The factor f depends on bond angles in the real chain because, when fully extended (in the all-*trans* zigzag form, for example), the bond vectors do not simply add in a straight-line fashion. [In all situations, of course, $f \leqslant 1$. If all bond-angle supplements θ are identical, $f = \cos(\theta/2)$, for even-numbered n.[§]] These relationships (Eqns.18-59 through 18-62) may be solved easily to give

$$N_e = (f^2/C_\infty)n \qquad (18\text{-}63)$$

$$l_e = (C_\infty/f)l \qquad (18\text{-}64)$$

Note that l_e is proportional to C_∞, whereas N_e is inversely proportional to C_∞. Because the magnitude of C_∞ can reflect the stiffness of the chain, we expect it to be related to l_e and N_e in this way.

[§] This is easy to show. Draw a chain (even n) in its fully extended zigzag form, and draw the straight line that connects one end with the other. Note that the projection of each bond vector on the straight line is $\cos(\theta/2)$. Can you see why f must be slightly modified if n is odd?

Sizes of statistical segments for real chains

In the case of polymethylene, $f = 0.83$, and $C_\infty = 6.7$. Therefore, $N_e \cong 0.1\, n$, or $n/N_e \cong 10$; that is, the "statistical segment" spans approximately 10 real bonds. For a polypeptide of n virtual bonds, $f = 0.95$, and the values of C_∞ depend upon the side chain (see Table 18-1). Values of n/N_e of 2.4, 10.3, and 129, respectively, are obtained for polyglycine, poly-L-alanine, and poly-L-proline. The corresponding values of l_e are about 8.6 Å, 37 Å, and 460 Å.

A word of caution. The fact that a real chain can be cast into an equivalent random-walk chain implies that a Gaussian function will describe the distribution of end-to-end distances of the real chain, with N_e and l_e replacing n and l, respectively, in the parameter β of Equations 18-21 and 18-22. It is important to recognize, however, that the Gaussian behavior of the equivalent chain of statistical segments holds only for very large n. Significant departures from Gaussian behavior may occur at smaller n, even though the chain is large enough to encompass several statistical segments.

18-10   PERSISTENCE LENGTH

Another useful configurational parameter is the persistence length a. This can be defined as the average projection of the end-to-end distance vector $\mathbf{r}$ on the first bond of the chain $\hat{\mathbf{l}}_1$, in the limit of infinite chain length (Fig. 18-18). Thus,

$$a = \left\langle (\hat{\mathbf{l}}_1/l_1) \cdot \sum_{i=1}^{n} \hat{\mathbf{l}}_i \right\rangle \qquad \text{as } n \to \infty \qquad (18\text{-}65)$$

where $\hat{\mathbf{l}}_1/l_1$ is a unit vector in the direction of $\hat{\mathbf{l}}_1$. The persistence length clearly is a measure of the length over which the chain "persists" in the same direction as the first bond. For an infinitely long chain, of course, this persistence also may be measured with respect to any bond $\hat{\mathbf{l}}_i$ remote from the last bond, so that a is simply

Figure 18-18

Persistence length. The average projection of $\mathbf{r}$ onto $\hat{\mathbf{l}}_1$, in the limit $n \to \infty$, gives the persistence length. The chain is shown in a particular conformation; the value of the persistence length is obtained by averaging over all conformations (Eqn. 18-65).

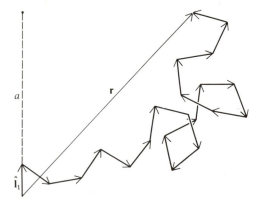

the sum of the projections of all bonds $j \geq i$ on $\hat{\mathbf{l}}_i / l_i$:

$$a = \left\langle (\hat{\mathbf{l}}_i / l_i) \cdot \sum_{j=i}^{n} \hat{\mathbf{l}}_j \right\rangle \qquad \text{as } n \to \infty \qquad (18\text{-}66)$$

For chains composed of identical bonds of length l, it is easy to show that the persistence length is closely related to the limiting characteristic ratio C_∞. This may be accomplished by first rewriting Equation 18-4 as

$$\langle r^2 \rangle_0 = nl^2 + 2 \left\langle \sum_{j=i+1}^{n} \sum_{i=1}^{n} (\hat{\mathbf{l}}_i \cdot \hat{\mathbf{l}}_j) \right\rangle \qquad (18\text{-}67)$$

$$= nl^2 + 2 \left\langle \sum_{j=2}^{n} (\hat{\mathbf{l}}_1 \cdot \hat{\mathbf{l}}_j) \right\rangle + 2 \left\langle \sum_{j=i+1}^{n} \sum_{i=2}^{n} (\hat{\mathbf{l}}_i \cdot \hat{\mathbf{l}}_j) \right\rangle \qquad (18\text{-}68)$$

In the limit $n \to \infty$, the first summation on the right-hand side of Equation 18-68 is identical to the one in Equation 18-65, except that the latter runs from 1 to n, whereas the one in Equation 18-68 runs from 2 to n. Therefore, when $n \to \infty$, the first summation is

$$\left\langle \sum_{j=2}^{n} (\hat{\mathbf{l}}_1 \cdot \hat{\mathbf{l}}_j) \right\rangle = al - \hat{\mathbf{l}}_1 \cdot \hat{\mathbf{l}}_1 = al - l^2 \qquad \text{as } n \to \infty \qquad (18\text{-}69)$$

Similarly, from Equation 18-66, it is clear that, for each value of i in the second summation of Equation 18-68, the summation over j gives $al - l^2$. Therefore, we have

$$\langle r^2 \rangle_0 = nl^2 + 2n(al - l^2) \qquad \text{as } n \to \infty \qquad (18\text{-}70)$$

$$= 2nla - nl^2 \qquad \text{as } n \to \infty \qquad (18\text{-}71)$$

Finally, we obtain

$$(\langle r^2 \rangle_0 / nl^2)_{n \to \infty} = C_\infty = 2a/l - 1 \qquad (18\text{-}72)$$

and

$$\boxed{a = (C_\infty + 1)l/2} \qquad (18\text{-}73)$$

The simple relationship between a and C_∞ gives us a physical feeling for the limiting characteristic ratio. For the polypeptide chains listed in Table 18-1, a may be calculated from the C_∞ values tabulated. Values of $a \cong 6$ Å, 20 Å and 220 Å, respectively, are obtained for polyglycine, poly-L-alanine, and poly-L-proline. This illustrates the close connection between the persistence length and the "stiffness" of a chain, and it gives us another quantitative measure of the differences in the flexibilities of the polypeptide chains.

18-11   EXCLUDED-VOLUME EFFECT AND THE UNPERTURBED STATE

The preceding treatment of polymer statistics did not take into account the well-known excluded-volume effect. In calculations of real chains, backbone geometries and rotational hindrances were considered as the factors that determine the distribution of conformations—that is, end-to-end distances. The rotational hindrances arise from short-range interactions between neighboring groups in the chain. Long-range interactions, between chain elements far apart in the sequence, were not considered. Likewise, interactions between different polymer molecules were not taken into account.

The effects of intermolecular interactions may be virtually eliminated by using sufficiently dilute solutions. Intramolecular effects cannot be eliminated by dilution, however. Chief among such effects is that of excluded volume, the fact that two segments of the chain cannot occupy the same volume of space. In generating all conformations of the chain for the purpose of computing $\langle r^2 \rangle$, we will find many conformations in which the chain crosses itself as $n \to \infty$. Clearly, such conformations are physically impossible and should be excluded from consideration in computing $\langle r^2 \rangle$.

There are certain circumstances in which excluded-volume effects are not serious and may be ignored. Consider, for example, a rather extended, stiffly coiling chain, such as poly-L-proline. Excluded-volume effects probably may be ignored for chain lengths of up to 100 or more units. More flexible chains, such as polyglycine, will experience the effects of excluded volume at much shorter chain lengths. In data cited in Table 18-2 for short glycine oligomers, however, excluded volume is of little consequence in affecting the chain dimensions.

It is not yet practical (for large n) to compute those conformations of the chain that must be discarded because of the interpenetration of two chain segments. An alternate procedure must be sought, therefore, to compare the calculated dimensions (which are not corrected for excluded volume) with experimental results. The procedure used is rather straightforward and emerges from a careful consideration of the nature of the problem.

Qualitative discussion of effects of excluded volume on chain dimensions

The excluded-volume interaction perturbs the chain dimensions $\langle r^2 \rangle$ to a value different from that calculated by the preceding analyses, which we designate $\langle r^2 \rangle_0$, the dimensions of the unperturbed chain. This is a hypothetical chain in the sense that calculations of $\langle r^2 \rangle_0$ encompass configurations forbidden by excluded-volume effects. On the other hand, there are experimental conditions (see below) under which $\langle r^2 \rangle = \langle r^2 \rangle_0$; that is, the effects of excluded volume are exactly counterbalanced by other factors, so that the unperturbed-chain dimensions are achieved. In this sense, therefore, the unperturbed chain is a physical reality.

We may designate the relationship between the perturbed-chain and the unperturbed-chain dimensions (see Flory, 1953) as

$$\langle r^2 \rangle = \alpha^2 \langle r^2 \rangle_0 \tag{18-74}$$

Because self-intersections of the chain are more likely in the compact conformations, it is expected that, most usually, $\alpha > 1$. The exact value of α will depend on molecular weight and solvent, however, The effects of molecular weight are obvious when it is considered that longer chains have a greater opportunity for self-intersections than do shorter ones. Therefore, α should increase with molecular weight.

Solvents affect the dimensions in an entirely different way. In a good solvent, extended or more open conformations are given some bias, because these can be most highly solvated. Therefore, the *effective* covolume (or excluded volume) of a pair of chain segments is increased beyond that of the actual covolume of the segments themselves. In these cases, α may exceed unity considerably. In an indifferent solvent, the segment–segment interactions are as favorable as segment–solvent interactions. In this situation, the effective covolume of the segment is unaffected by the solvent, and excluded-volume effects are less severe than in a good solvent. That is, α exceeds unity, but less so than in the case of the good solvent. On the other hand, in a poor solvent, the tendency for interactions between close segments is increased, and compact configurations are favored over the more open, extended ones. The result is that the effects of excluded volume are compensated for by the increased relative preference for compact structures (in which no segment intersections occur) over more open conformations. When these opposing effects exactly counterbalance, $\langle r^2 \rangle = \langle r^2 \rangle_0$, and $\alpha = 1$.

Theta solvents and theta-point temperature

Solvents in which $\alpha = 1$ are called *theta* solvents. In general, α depends upon temperature, so that a "theta-point" temperature is identified with that temperature at which $\alpha = 1$ in the theta solvent. At this temperature, the second virial coefficient of the osmotic pressure (which depends on intermolecular interactions) vanishes, because effects due to intermolecular excluded volume are eliminated under the same conditions that nullify intramolecular excluded volume. Under these circumstances, the osmotic pressure follows the van't Hoff relation over a significant range of polymer concentration (see Chapter 25). Thus, the theta point for polymers is analogous to the Boyle point for real gases. At the Boyle point, the attractive potential between gas molecules exactly counterbalances their mutual exclusion, and the gas will follow the ideal gas law over a large range of pressures.

In actual practice, it may be difficult to find a solvent and temperature in which $\alpha = 1$. Therefore, the parameter α must be evaluated. This may be calculated from polymer-solution theory, which relates α to the molecular weight, intrinsic viscosity, and second osmotic virial coefficient (see Flory, 1953).

Experimental values for $\langle r^2 \rangle_0$ in Tables 18-1 and 18-2 have been corrected for excluded-volume effects, in the case of the high-molecular-weight species. For the short chains listed in Table 18-2, the excluded-volume effect is of little significance.

Summary

In the quantitative treatment of polymer chain conformations, methods are developed for calculating conformation-dependent properties such as the unperturbed mean square end-to-end distance $\langle r^2 \rangle_0$. An analysis of artificial model chains, such as the random-walk and freely rotating chains, provides a useful reference frame for discussing the properties of real chains.

The complexities associated with real chains are greatly simplified by means of the rotational isomeric-state approximation. With this approximation, the calculation of intrachain distances is efficiently accomplished through matrix methods. In the case of polypeptide homopolymers, the calculated chain extensions are sensitive to amino acid type; for example, the extensions of polyglycine $<$ poly-L-alanine $<$ poly-L-proline. This result can be rationalized according to the conformational energy diagrams of the residues. Also, calculated dimensions of copolymers are sensitive to amino acid composition.

Experimental measurements of $\langle r^2 \rangle_0$ have been made for several different kinds of polypeptides, and these measurements agree well with the computed values.

Other conformation-dependent chain parameters also may be calculated. For this purpose, the concepts of the statistical segment and of the persistence length are sometimes useful. Computed values of statistical segment sizes and of persistence lengths give further insights into the conformational features of different polypeptide homopolymers. For example, relatively large statistical segment sizes or persistence lengths typically are associated with stiff chains.

Excluded-volume interactions generally perturb chain dimensions from the calculated unperturbed dimensions. However, the effective covolume (or excluded volume) of a pair of chain segments is dependent on the solvation characteristics of the solvent. It is sometimes possible to find a solvent (a theta solvent, which typically is a poor solvent for the polymer) such that at a particular temperature (theta point) the effects of excluded volume are exactly counterbalanced by the solvent-induced preference of the chain for relatively compact conformations. In this situation, the effects of excluded volume are nullified, and the unperturbed chain dimensions are adopted. In cases where "theta conditions" cannot be found, polymer-solution theory affords a means of correcting for effects of excluded volume. Therefore, it usually is possible to compare experimentally determined chain dimensions with calculated ones.

Problems

18-1. Prove the relationship of Equation 18-24 by performing the integration of Equation 18-23 with the distribution function of Equation 18-22.

18-2. Assume you have a group of chains of varying molecular weight. The configuration may be described as N_e equivalent statistical segments of length l_e, where l_e is the same for each chain, and N_e is directly proportional to the molecular weight M (that is, $M = $ const N_e). In all cases, $N_e > 300$. You are interested in the probability $p(N_e)$ that, for a chain of N_e segments, the two ends are within one statistical segment length l_e (or less) of each other. That is, you want the probability that the end-to-end distance is $0 \leqslant r \leqslant l_e$. Show that

$$p(N_e) = \text{const } M^{-3/2}$$

That is, that the probability is inversely proportional to the 3/2 power of the molecular weight.

18-3. A polypeptide consists of $n + 1$ amino acid residues, numbered from 0 to n. You wish to calculate the average mean square distance $\langle r_{kp}^2 \rangle$ between residues k and p in the sequence, where k and p are any two arbitrary residues, and $p > k$. A professor claims that, to accomplish this, you must formally evaluate the expression

$$\langle r_{kp}^2 \rangle = (p - k)l^2 + 2\left\langle \sum_{j=i+1}^{p} \sum_{i=k}^{p} (\hat{\mathbf{l}}_i \cdot \hat{\mathbf{l}}_j) \right\rangle$$

where l is the virtual bond length, and $\hat{\mathbf{l}}_i$ and $\hat{\mathbf{l}}_j$ refer to the ith and jth virtual bonds, respectively. Is the professor correct? Why, or why not? If not, give the correct expression.

18-4. Is it possible for a statistically coiling polymer with restricted rotations about its constituent bonds to have a mean square end-to-end distance *smaller* than what it would have if there were free rotation about each of the bonds? (Assume that there is a large number of bonds.) Why, or why not?

18-5. Short polypeptide chains, in their zwitterionic form, have been studied in aqueous solution; end-to-end distances have been obtained, and these have been compared with calculated values. Calculations are done in the usual fashion, but ignoring any possible electrostatic attraction between the charged chain ends. It is found that calculated and observed values of $\langle r^2 \rangle$ are in good agreement for short oligomers where $n = 2$ or 3, but the agreement becomes much worse at larger values of n. Offer an explanation for such a finding. (Note: $n = $ number of residues.)

References

GENERAL

Birshtein, T. M., and O. B. Ptitsyn. 1966. *Conformations of Macromolecules* (English translation). New York: Wiley-Interscience.

Flory, P. J. 1953. *Principles of Polymer Chemistry*. Ithaca, N.Y.: Cornell Univ. Press. [A classic text that covers diverse topics and gives a historical perspective.]

————. 1969. *Statistical Mechanics of Chain Molecules*. New York: Wiley. [An advanced monograph that gives thorough coverage of and sharp insight into calculations of conformation-dependent properties, especially using matrix methods.]

Tanford, C. 1961. *Physical Chemistry of Macromolecules*. New York: Wiley. [Chapter 3 gives an introduction to the calculation of dimensions of model chains, and to related material.]

Volkenstein, M. V. 1963. *Configurational Statistics of Polymeric Chains* (English translation). New York: Wiley.

————. 1977. *Molecular Biophysics* (English translation). New York: Academic Press. [Chapter 3 treats some aspects of polymer-chain statistics.]

Yamakawa, H. 1971. *Modern Theory of Polymer Solutions*. New York: Harper and Row. [Chapters 2 and 3 treat some aspects of polymer-chain statistics.]

SPECIFIC

Brant, D. A. 1976. Conformational theory applied to polysaccharide structure. *Quart. Rev. Biophys.* 9:527.

Brant, D. A., and P. J. Flory. 1965. The configuration of random polypeptide chains, II: Theory. *J. Am. Chem. Soc.* 87:2791.

Miller, W. G., D. A. Brant, and P. J. Flory. 1967. Random coil configurations of polypeptide copolymers. *J. Mol. Biol.* 23:67.

Miller, W. G., and C. V. Goebel. 1968. Dimensions of protein random coils. *Biochemistry* 7:3925.

Ullman, R. 1978. An unsophisticated calculation of the excluded volume effect. *Macromolecules* 11:1292. [A short note giving an approximate calculation of excluded volume and references to earlier literature dealing with this problem.]

Yevich, R., and W. K. Olson. 1979. The spatial distributions of randomly coiling polynucleotides. *Biopolymers* 18:113.

Elementary polymer-chain hydrodynamics and chain dimensions

19-1 POLYMER-CHAIN HYDRODYNAMICS

Polymer-chain conformations are closely related to hydrodynamic properties such as the intrinsic viscosity and frictional coefficient, and measurements of such properties commonly are the basis for obtaining a picture of chain conformations in solution. For example, in Chapter 18 we compared calculated chain dimensions of high-molecular-weight polypeptides with those deduced from intrinsic viscosity measurements (see Table 18-2). As shown in this chapter, hydrodynamic measurements also have provided us with information on chain dimensions of single-stranded and double-stranded nucleic acids in solution. In particular, the analysis of hydrodynamic properties of high-molecular-weight double-stranded DNA has provided strong evidence that it is a stiff, wormlike coil.

Polymer-chain hydrodynamics is a broad and sophisticated subject. In this chapter, we direct attention toward some of the elementary features, with the aim of conveying insight into the relationships between chain dimensions and hydrodynamic properties. For this purpose, we utilize the basic hydrodynamic concepts discussed in Chapters 10 to 12. For more advanced treatments of polymer-chain hydrodynamics, the reader is referred to the references cited at the end of this chapter.

19-2 THE FREE-DRAINING COIL

To illustrate the physics of the intrinsic viscosity of random coils, we consider the "free-draining" model. The polymer is viewed as a string of beads, through which the solvent streams during viscous flow. In the free-draining limit, the perturbation of the flow of solvent by one bead in the chain has no effect on the flow around any other bead. That is, the flow of solvent around bead j is the same as if there were no other beads present. Although real chains do not approach this limit (as we shall see), the treatment of the free-draining model offers considerable insight into the underlying physical features of polymer viscosities.

Rotatory motion of the coil in viscous flow

Consider a random coil, modeled as a string of beads, moving in viscous flow. A solvent velocity gradient is created by moving a plate in the x direction, at a fixed velocity, over a stationary plate directly below. A velocity gradient, $\gamma = dv_x/dz$, is created along the z axis (Fig. 19-1). The random coil finds itself crossing flow lines

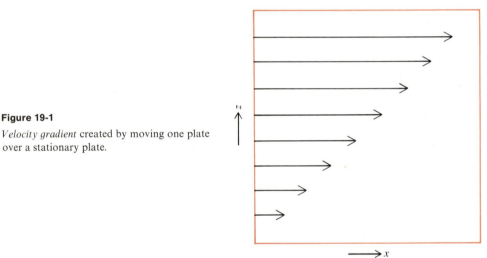

Figure 19-1

Velocity gradient created by moving one plate over a stationary plate.

of different v_x. Obviously, it cannot adjust simultaneously to the velocities of all the flow lines unless it rotates as it translates through the solvent. Therefore, we envision the center of mass of the polymer moving with the solvent flow line at that point, and the rest of the molecule tumbling about the center of mass. We now take the center of mass of the polymer as the origin of our coordinate system (Fig. 19-2).

The polymer itself is viewed as a string of n hydrodynamically linked beads. Each bead has a frictional coefficient ζ. As a result of the beads' frictional interactions with the gradient, the molecule experiences a torque that sets it into rotatory motion about the y axis. As it rotates about this axis, however, opposing torques are created as the molecule cuts *across* the velocity gradient. Eventually, dynamic equilibrium is established, so that the molecule experiences a net torque of zero and rotates at a constant angular velocity ω. From this constant rate of angular rotation and the

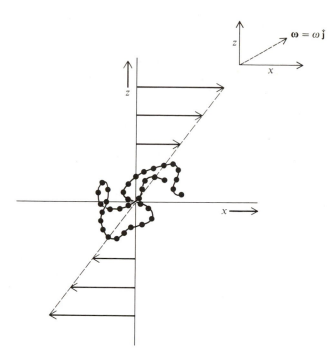

Figure 19-2

Rotation of the polymer. Velocity of solvent relative to the center of mass
of the polymer produces a torque and, subsequently, rotation of the
polymer. The inset shows the angular velocity vector $\boldsymbol{\omega}$.

frictional coefficients of the beads, we can calculate the polymer's contributions to
the viscosity.

Angular velocity of rotation and shear gradient

The velocity $\mathbf{v}_j$ of the solvent at bead j is

$$\mathbf{v}_j = \hat{\mathbf{i}}\gamma z_j \qquad (19\text{-}1)$$

where $\hat{\mathbf{i}}$ is the unit vector along the x axis, and z_j is the z-axis displacement of bead
j from the center of mass (see Fig. 19-2). The angular velocity ω can be given a vector
representation $\boldsymbol{\omega} = \omega\hat{\mathbf{j}}$, where $\hat{\mathbf{j}}$ is the unit vector along the y axis; in Figure 19-2,
this vector is perpendicular to the x–z plane and points into the page. If $\mathbf{s}_j(x_j, y_j, z_j) = \hat{\mathbf{i}}x_j + \hat{\mathbf{j}}y_j + \hat{\mathbf{k}}z_j$ is the distance from the center of mass to bead j, then the velocity
$\mathbf{u}_j$ of bead j is tangential to $\mathbf{s}_j$ and is

$$\mathbf{u}_j = \boldsymbol{\omega} \times \mathbf{s}_j = \hat{\mathbf{i}}\omega z_j - \hat{\mathbf{k}}\omega x_j \qquad (19\text{-}2)$$

The velocity $\mathbf{u}_{rel}$ of the solvent relative to the velocity of bead j is

$$\mathbf{u}_{rel} = \mathbf{v}_j - \mathbf{u}_j \tag{19-3}$$

and the frictional force $\mathbf{F}_j$ acting on bead j is

$$\mathbf{F}_j = \zeta(\mathbf{v}_j - \mathbf{u}_j) = \zeta[\hat{\mathbf{i}}(\gamma - \omega)z_j + \hat{\mathbf{k}}\omega x_j] \tag{19-4}$$

where ζ is the frictional coefficient of the bead.[§] The torque τ_j on bead j is

$$\tau_j = \mathbf{s}_j \times \mathbf{F}_j \tag{19-5}$$

$$= \hat{\mathbf{i}}\zeta\omega x_j y_j + \hat{\mathbf{j}}\zeta[(\gamma - \omega)z_j^2 - \omega x_j^2] - \hat{\mathbf{k}}\zeta(\gamma - \omega)y_j z_j \tag{19-6}$$

where the cross product in Equation 19-5 has been expanded in Equation 19-6. The requirement for dynamic stability is that $\langle \sum_j \tau_j \rangle = 0$, where the sum is taken over all beads, and the angle brackets denote "average value." Because torques around the x and z axes must equal zero (because the molecule is rotating about y only), we conclude that $\sum_j x_j y_j = 0$, and $\sum_j y_j z_j = 0$. Therefore,

$$\left\langle \sum_j \tau_j \right\rangle = \hat{\mathbf{j}}\zeta\left[(\gamma - \omega)\left\langle \sum_j z_j^2 \right\rangle - \omega\left\langle \sum_j x_j^2 \right\rangle \right] = 0 \tag{19-7}$$

Assuming the shear forces are not sufficient to perturb the average distribution of chain segments, the coil must be spherically symmetric about the center of mass, and $\langle \sum z_j^2 \rangle = \langle \sum x_j^2 \rangle = \langle \sum y_j^2 \rangle = (1/3)\langle \sum s_j^2 \rangle$; therefore, we obtain

$$\left\langle \sum_j \tau_j \right\rangle = \hat{\mathbf{j}}\zeta(\gamma - 2\omega)(1/3)\left\langle \sum_j s_j^2 \right\rangle = 0 \tag{19-8}$$

Hence, to satisfy Equation 19-8,

$$\omega = \gamma/2 \tag{19-9}$$

Thus, the polymer rotates with an angular velocity equal to half the shear gradient.

The average energy dissipation per unit of time due to the action of bead j is simply the product of the frictional force and the relative solvent velocity in the area of bead j. From Equations 19-1, 19-2, 19-3, 19-4, and 19-9, we obtain

$$\mathbf{F}_j \cdot \mathbf{u}_{rel} = \zeta(\gamma^2/4)(x_j^2 + z_j^2) = \zeta(\gamma^2/4)(2/3)\langle s_j^2 \rangle = \zeta(\gamma^2/6)\langle s_j^2 \rangle \tag{19-10}$$

[§] Equations 10-29 and 19-4 are identical, except for opposite signs. In Equation 10-29, $\mathbf{F}_j$ is the force exerted by bead (or segment) j on the fluid, whereas, in Equation 19-4, $\mathbf{F}_j$ is the force acting on bead j.

Intrinsic viscosity of the free-draining coil

From Equation 19-10, it is a simple matter to calculate the intrinsic viscosity of the polymer, by first returning to the definition of the viscosity coefficient η. If the force F is exerted on the top plate of our viscosity apparatus of Figure 19-1, then the viscosity coefficient η is defined by

$$F = \eta \gamma A \tag{19-11}$$

where A is the area of the plate. The total energy dissipated per unit of time is

$$Fu = \eta \gamma^2 A h \tag{19-12}$$

where h is the distance between the plates, and $u = \gamma h$ is the speed of the moving plate. From Equation 19-12, we note that η is the energy loss per unit of volume per unit of time divided by the square of the shear rate:

$$\eta = Fu/\gamma^2 A h \tag{19-13}$$

Returning to Equation 19-10, we see that the total contribution to the energy dissipation by one polymer molecule is $(\zeta/6)\gamma^2 \sum_j \langle s_j^2 \rangle$, where the sum is taken over all beads. If η_0 is the viscosity of the solvent, we obtain the following expression for the contribution to the viscosity from all of the polymer molecules in one cubic centimeter:

$$\eta - \eta_0 = (\zeta/6)\left(\sum_j \langle s_j^2 \rangle \right) c N_0/M \tag{19-14}$$

where $c N_0/M$ is the number of polymer molecules per cm^3, because c is the polymer concentration in g cm^{-3}; N_0 is Avogadro's number; and M is the molecular weight. Equation 19-14 may easily be converted into a simple form for the intrinsic viscosity $[\eta]$ by noting that $\sum_j \langle s_j^2 \rangle = n \langle R_G^2 \rangle$, where $\langle R_G^2 \rangle$ is the mean square radius of gyration (see Chapter 18), and also that $M/n = M_0$, the molecular weight of a monomer (bead) unit. Therefore,[§]

$$\lim_{c \to 0} \left[(\eta - \eta_0)/\eta_0 c \right] = [\eta] = (\zeta/6\eta_0)(N_0/M_0)\langle R_G^2 \rangle \tag{19-15}$$

Because $\langle R_G^2 \rangle_0 \propto \langle r^2 \rangle_0$ for polymers with large n (Eqn. 18-9), and $\langle r^2 \rangle_0 \propto M$ (Chapter 18), we conclude that $[\eta] \propto M$ for an unperturbed free-draining polymer.

The physical meaning of these results bears some comment. Note that the con-

---

[§] The extrapolation of the quantity on the left-hand side of Equation 19-15 to $c \to 0$ is necessary in experimental practice, in order to eliminate polymer-polymer interactions. In our treatment of the free-draining coil, we have implicitly ignored these interactions.

tribution to the viscosity by a single polymer molecule is $(\zeta/6)\sum_j\langle s_j^2\rangle = (\zeta/6)n\langle R_G^2\rangle$. The simple dependence on $n\zeta$ is easy to explain; it merely reflects the fact that, the longer the polymer, the more beads are present to interact with the solvent and dissipate energy. However, $\langle R_G^2\rangle$ itself is dependent on n, and this n-dependent contribution to the energy dissipation arises because, as n increases, a greater number of beads are located farther and farther from the center of mass. These outer beads dissipate more energy than those closer to the center of mass, because of their higher velocities ($\mathbf{u}_j$ for bead j). Hence, as n increases, the average energy dissipated *per bead* also increases.

Limitations of the free-draining approximation

The preceding treatment applies to the so-called free-draining polymer, in which there is no hydrodynamic interaction between beads—that is, the flow at bead j is totally undisturbed by the presence of other beads. Clearly, this is an unreasonable assumption, because the velocity of solvent in the interior of the polymer must be greatly modified by the beads, and in such a way that the flows around each bead mutually interact and perturb each other (see Chapter 10). In fact, the untenable nature of the free-draining approximation is seen in the observation that, for random coils, $[\eta]$ generally is proportional to a power of M considerably smaller than one; typically, $[\eta]$ is proportional to $M^{0.5}$ to $M^{0.7}$.

19-3   INTRINSIC VISCOSITY OF THE NON-FREE-DRAINING COIL

In the non-free-draining case, solvent near the center of the macromolecule moves almost in concert with the polymer. Thus, these interior solvent molecules behave as if they were trapped by the high density of polymer segments. The result is that frictional energy losses in this central region are smaller than those for the free-draining case. The rigorous derivation of the proper relationship between $[\eta]$ and M for the non-free-draining case is considerably more complex than that for the free-draining polymer; we include here only a simple treatment.

For solid particles, the intrinsic viscosity $[\eta]$ is

$$[\eta] = vV_m \tag{19-16}$$

where v is a shape factor, and V_m is the volume per unit of mass (see Chapter 12). For a solid, rigid sphere, $v = 2.5$, and V_m is a constant, independent of molecular weight.[§] Therefore, $[\eta]$ is independent of the size (or molecular weight) of the spherical particle. In the case of a random coil, we may model the chain as a spherical cloud of segments. This cloud is not uniform, of course, as the radial distribution function

---

[§] Equation 19-16 is identical to Equation 12-21, with $V_m = \bar{V}_2 + \delta_1 V_1$.

for the random-walk chain clearly indicates (see Fig. 18-6b). We then assume that this cloud of segments can be replaced by an impenetrable (to solvent) hydrodynamic sphere whose intrinsic viscosity is identical to that of the random coil. The volume of this sphere is taken to be proportional to $\langle r^2 \rangle^{3/2}$, and the volume per unit mass is proportional to $\langle r^2 \rangle^{3/2}/M$, where M is the molecular weight. We thus conclude that

$$[\eta] \propto \langle r^2 \rangle^{3/2}/M \qquad (19\text{-}17)$$

More rigorous considerations (see Flory, 1953) show that

$$[\eta] = \Phi_c \langle r^2 \rangle^{3/2}/M \qquad (19\text{-}18)$$

where $\Phi_c \cong 2.1 \times 10^{23}$ if $[\eta]$ is in units of cm^3 g^{-1} and $\langle r^2 \rangle^{3/2}$ is in units of cm^3. The parameter Φ_c is a universal constant, presumably the same for all random coils. Thus, from knowledge of M and a determination of $[\eta]$, the dimensions of the coil can be calculated directly. Equation 19-18 is very commonly employed to calculate chain dimensions from viscosity measurements.

In contrast to that of the solid sphere, the intrinsic viscosity of a random coil does depend on the size (or molecular weight) of the coil. Consider, for example, the unperturbed chain; at sufficiently large molecular weight, $\langle r^2 \rangle_0^{3/2} = (C_\infty n l^2)^{3/2} \propto M^{3/2}$ (see Chapter 18). Therefore, in accordance with Equation 19-18, $[\eta] \propto M^{1/2}$ for unperturbed chains of sufficiently high molecular weight. The proportionality of $[\eta]$ to $M^{1/2}$ is a characteristic feature of the viscosity behavior of unperturbed chains.

For chains perturbed by excluded-volume effects, $\langle r^2 \rangle^{3/2} = \alpha^3 \langle r^2 \rangle_0^{3/2}$, where the molecular expansion factor α^3 (Chapter 18) is proportional to a power of M between 0.0 and 0.3. Thus, for sufficiently long chains, $\langle r^2 \rangle^{1/2} \propto M^a$, where $0.5 \leqslant a \leqslant 0.6$. In accordance with Equation 19-18, we obtain for the intrinsic viscosity

$$[\eta] = \Phi_c \langle r^2 \rangle^{3/2}/M = \Phi_c \alpha^3 \langle r^2 \rangle_0^{3/2}/M \propto M^x \qquad \text{where } x = 3a - 1 \quad (19\text{-}19)$$

Therefore, for random coils in general, $[\eta] \propto M^x$, where x is a constant under a particular set of conditions; generally, $0.5 \leqslant x \leqslant 0.7$, but less commonly x may be as large as 0.8.

Table 19-1 illustrates the rather characteristic dependence of $[\eta]$ on M for random coils. The table lists the proportionality between $[\eta]$ and M for a solid sphere, a long rod, a wide thin disk, a random coil, and a hypothetical free-draining polymer. The relation between $[\eta]$ and M for the long rod and the wide disk may be calculated from the Simha factors given in Chapter 12. Note that the dependence of $[\eta]$ on M is distinctly different for most cases. Adherence to the relationship $[\eta] \propto M^x$, where $0.5 \leqslant x \leqslant 0.7$, over a wide range of molecular weight is indicative of random-coil behavior.

Table 19-1
Relationship of intrinsic viscosity to molecular weight for different macromolecular shapes

Macromolecule model	Relation between $[\eta]$ and M
Solid sphere	$[\eta]$ independent of M
Very long rod (length/diameter $\geqslant$ 100)	$[\eta] \propto M^{1.8}$
Wide, thin disk (width/thickness $\geqslant$ 100)	$[\eta] \propto M^{1.0}$
Long random coil	$[\eta] \propto M^x$, where $0.5 \leqslant x \leqslant 0.8$
Long random coil (unperturbed chain)	$[\eta] \propto M^{0.5}$
Hypothetical free-draining polymer (unperturbed chain)	$[\eta] \propto M$

19-4   FRICTIONAL AND SEDIMENTATION COEFFICIENTS

The frictional coefficient is another very useful hydrodynamic parameter. It generally is evaluated from diffusion or sedimentation-coefficient measurements. For a solid, rigid sphere of radius r, the frictional coefficient f is

$$f = 6\pi\eta_0 r \tag{19-20}$$

(see Chapter 10). Because $r \propto (\text{volume})^{1/3}$, the frictional coefficient of a solid sphere varies as $M^{1/3}$. In the case of a random-coil polymer, we again envision a spherical cloud of segments, which can be replaced by an equivalent solid hydrodynamic sphere of radius r_e. The frictional coefficient then is $6\pi\eta_0 r_e$. We expect the radius of this equivalent sphere to be proportional to $\langle r^2 \rangle^{1/2}$, but it is not possible to obtain the proportionality constant without an elaborate hydrodynamic treatment. A rigorous treatment yields

$$f = P_c \eta_0 \langle r^2 \rangle^{1/2} \tag{19-21}$$

where $P_c = 5.1$ (see Flory, 1953; Kirkwood and Riseman, 1948). Because $\langle r^2 \rangle^{1/2} \propto M^a$, we conclude that $f \propto M^a$, where $0.5 \leqslant a \leqslant 0.6$.

The dependence of the sedimentation coefficient on dimensions and molecular weight may be obtained by substituting Equation 19-21 into the equation for the sedimentation coefficient s^0 at infinite dilution (see Chapter 11):

$$s^0 = M(1 - \bar{V}_2\rho)/N_0 f \propto M^{1-a} \tag{19-22}$$

where $\bar{V}_2$ is the partial specific volume, and ρ is the solvent density. As a basis for comparison to the random chain, Table 19-2 tabulates the dependence of f and s^0

Table 19-2

Relationship of frictional and sedimentation coefficients to molecular weight for different macromolecular shapes

Macromolecule model	Relation between f and M	Relation between s^0 and M
Solid sphere	$f \propto M^{1/3}$	$s^0 \propto M^{2/3}$
Very long rod (length/diameter $\geqslant$ 100)	$f \propto M^{0.8}$	$s^0 \propto M^{0.2}$
Wide, thin disk (width/thickness $\geqslant$ 100)	$f \propto M^{2/3}$	$s^0 \propto M^{1/3}$
Long random coil	$f \propto M^a$	$s^0 \propto M^{1-a}$
	where $0.5 \leqslant a \leqslant 0.6$	
Long random coil (unperturbed chain)	$f \propto M^{0.5}$	$s^0 \propto M^{0.5}$
Hypothetical free-draining polymer (unperturbed chain)	$f \propto M$	s^0 independent of M

on M for several different structures. The results for the rod and disk are obtained from Perrin factors given in Chapter 10. Note that, as in the case of $[\eta]$, both f and s^0 for random coils have dependencies on M that are distinct from those of other kinds of particles.

Frictional coefficient and intrinsic viscosity

It is of interest to examine the relationship between the frictional coefficient and the intrinsic viscosity. By combining Equations 19-18 and 19-21, we obtain

$$f = \eta_0 P_c \Phi_c^{-1/3} (M[\eta])^{1/3} \tag{19-23}$$

Recall that $[\eta]$ can be regarded as proportional to the volume per unit of mass (Eqn. 19-16); therefore, $M[\eta]$ is proportional to the effective hydrodynamic volume of the coil. On the other hand, the frictional coefficient is proportional to the effective hydrodynamic radius of the chain (Eqn. 19-20 and following discussion). Keeping these facts in mind, we see that Equation 19-23 merely says that the effective hydrodynamic radius in translation (frictional coefficient) is proportional to the cube root of the effective volume in shear (intrinsic viscosity). The following question arises: how close is the numerical value of the proportionality constant $(P_c\Phi_c^{-1/3})$ between f and $(M[n])^{1/3}\eta_0$ in Equation 19-23 to the value obtained by *assuming* that the equivalent hydrodynamic spheres for translation and for shear have the same radius. To answer this, we start with Equation 19-16, and we obtain

$$[\eta] = v V_e N_0 / M \tag{19-24}$$

$$= 2.5 N_0 (4/3) \pi r_e^3 / M \tag{19-25}$$

where the intrinsic viscosity $[\eta]$ is in units of $cm^3\ g^{-1}$, and V_e and r_e are the volume and radius, respectively, of the equivalent sphere (*in shear*). The frictional coefficient of the equivalent sphere is $f = 6\pi\eta_0 r_e$, where r_e is the radius of the equivalent sphere *in translation*. Solving Equation 19-25 for r_e (in shear) and using this value of r_e in the expression $f = 6\pi\eta_0 r_e$ gives

$$f = 6\pi(2.5 N_0 4\pi/3)^{-1/3}\eta_0([\eta]M)^{1/3} \tag{19-26}$$

$$= (1.02 \times 10^{-7})\eta_0([\eta]M)^{1/3} \tag{19-27}$$

With $P_c = 5.1$ and $\Phi_c = 2.1 \times 10^{23}$, the value of $P_c\Phi_c^{-1/3}$ (Equation 19-23) is about 0.86×10^{-7}. The fact that the proportionality constant in Equation 19-27 is as close as it is to $P_c\Phi_c^{-1/3}$ is interesting, because the equivalent spheres for translation and for shear are not required, in our treatment, to have similar radii.

The Mandelkern–Flory–Scheraga constant

There is another important aspect of Equation 19-23. Using Equation 19-22, it may be rewritten as

$$\Phi_c^{1/3}P_c^{-1} = \beta_s = \eta_0(M[\eta])^{1/3}/f \tag{19-28}$$

$$= \eta_0 N_0 s^0 M^{-2/3}[\eta]^{1/3}/(1 - \bar{V}_2\rho) \tag{19-29}$$

(see Mandelkern and Flory, 1952; Scheraga and Mandelkern, 1953). The significance of Equation 19-29 is this: it says that a particular combination of η_0, s^0, M, $[\eta]$, $\bar{V}_2$, and ρ is equal to a constant β_s for all random-coil polymers, regardless of their chemical constitution. The constant β_s is sometimes called the Mandelkern–Flory–Scheraga constant.[§] Table 19-3 shows some experimental values for β_s for several different polymers and solvent systems, including polysacrosine (a polypeptide; sacrosine = N-methylglycine). These polymer–solvent systems are widely different, and yet β_s does assume a constant value, within experimental error. The average value of β_s is 1.2×10^7, in agreement with $P_c = 5.1$ and $\Phi_c = 2.1 \times 10^{23}$. In view of the constancy of β_s, it is clear that Equation 19-29 can usefully be employed to calculate the molecular weight from measurements of $[\eta]$ and s^0.

[§] Equations 19-28 and 19-29 were first derived by L. Mandelkern and P. J. Flory (1952) and applied to random-coil polymers. H. A. Scheraga and Mandelkern (1953) subsequently showed that Equation 19-29 can be applied to globular, ellipsoidal proteins; in this case, the value of β_s depends on the asymmetry of the ellipsoid (see Chapter 12).

Table 19-3

Values of β_s for various polymer–solvent systems

Polymer	Solvent	$\beta_s \times 10^{-7}$
Polystyrene	Methyl ethyl ketone	1.2
Polystyrene	Toluene	1.1
Cellulose acetate	Acetone	1.3
Polysarcosine	Water	1.1
Polyisobutylene	Cyclohexane	1.2
Poly-(methyl methacrylate)	Acetone	1.2
	Average	1.2

SOURCE: Data from P. J. Flory, *Principles of Polymer Chemistry* (Ithaca, N.Y.: Cornell Univ. Press, 1953), p. 628.

19-5 CONFORMATIONAL ANALYSIS OF SINGLE-STRANDED POLYNUCLEOTIDES

Single-stranded polynucleotides provide a good illustration of the relationship between hydrodynamic properties and chain conformations. The statistical treatment of polynucleotide chain configurations is considerably more complex than that for the polypeptide case. There are six backbone bonds in each repeat unit, and the analysis of conformational energies as a function of rotation angles is complicated (see Chapter 6). Nevertheless, the calculation of $\langle r^2 \rangle_0$ proceeds along the same general lines as we have indicated for the polypeptide case (Chapter 18), with certain differences due to the additional complexities of the polynucleotides. We do not carry out these calculations here, but we consider the major conclusions obtained and we compare them with experimental findings.

The dimensions of several single-stranded polynucleotides have been investigated by light-scattering, intrinsic-viscosity, and sedimentation-coefficient measurements. Figure 19-3 shows plots of log $[\eta]$ versus log n_w and of log s^0 versus log n_w for polyadenylic acid under different solution conditions; n_w is the weight-average chain length. Figure 19-4 shows similar plots for polyuridylic acid. All of these data indicate an excellent linear relationship between log $[\eta]$ or log s^0 and log n_w. Therefore, $[\eta] = K_\eta n_w^{a_\eta}$, and $s^0 = K_s n_w^{a_s}$, where the proportionality constants K_η and K_s and the exponents a_η and a_s depend upon the solvent conditions. From our hydrodynamic analysis, it is clear that x in Equation 19-19 may be identified with a_η, and that $1 - a$ in Equation 19-22 is equivalent to a_s. Figure 19-5 shows plots of a_η and a_s versus temperature for polyadenylic acid. Each parameter varies with temperature in a smooth, continuous fashion. The two curves intersect at $a_\eta = a_s \cong 0.5$, at about 27°C. This temperature is very close to the theta-point temperature for this system, as determined by other measurements.

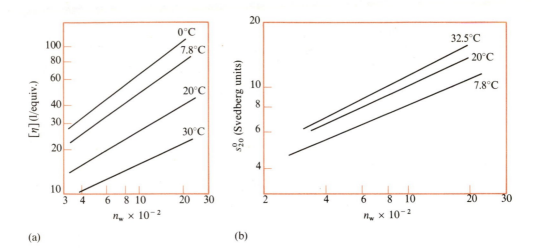

(a)

(b)

Figure 19-3

Dependence of intrinsic viscosity and sedimentation coefficient on poly A chain length. (a) Double log plots of $[\eta]$ versus n_w for poly A in 1 M NaCl, 0.01 M sodium phosphate (pH 7.5). (b) Double log plots of s_{20}^0 versus n_w for poly A in 1 M NaCl, 0.01 M sodium phosphate (pH 7.5). [After H. Eisenberg and G. Felsenfeld, *J. Mol. Biol.* 30:17 (1967).]

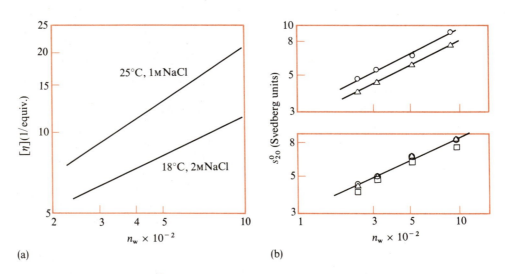

(a)

(b)

Figure 19-4

Dependence of intrinsic viscosity and sedimentation coefficient on poly U chain length. (a) Double log plots of $[\eta]$ versus n_w for poly U under two sets of conditions with 0.01 M sodium phosphate (pH 7.1). (b) Double log plots of s_{20}^0 versus n_w for poly U under various conditions; each type of symbol refers to a particular set of conditions. [After L. D. Inners and G. Felsenfeld, *J. Mol. Biol.* 50:373 (1970).]

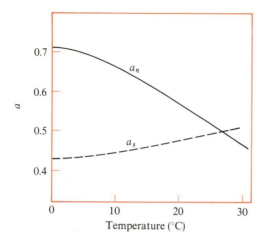

Figure 19-5

Dependence of exponents a_n and a_s for poly-adenylic acid on temperature. [After H. Eisenberg and G. Felsenfeld, *J. Mol. Biol.* 30:17 (1967).]

An analysis of these and other data leads to the conclusion that, in the absence of base stacking, the limiting value of $\langle r^2 \rangle_0 / n\overline{l^2}$ is 10 to 15 for polyadenylic acid, polyuridylic acid, and apurinic acid (a polydeoxyribonucleotide from which nearly all of the purine residues have been removed). The parameter $\overline{l^2}$ is the mean square bond length (~ 2.3 Å^2). For comparison, the limiting characteristic ratio $\langle r^2 \rangle_0 / n\overline{l^2}$ for poly-L-alanine based on $\overline{l^2}$ is 20, as compared to 9 when the characteristic ratio is based on the virtual bond length (Table 18-1).

Theoretical calculations that assign equal weights to all sterically allowed conformations give characteristic ratios in the range of 2.0 to 4.5, in strong disagreement with the experimental values (Olson and Flory, 1972). This observation indicates that energy differences between sterically allowed conformations must bias the chain toward higher extensions. When conformational-energy estimates are carried out and used in the calculation of chain dimensions, considerably larger values are found for the limiting characteristic ratio. For chains with all ribose units $C^{3'}$-*endo*, one obtains $(\langle r^2 \rangle_0 / n\overline{l^2})_{n \to \infty} \cong 9$. For chains with all $C^{2'}$-*endo*, $(\langle r^2 \rangle_0 / n\overline{l^2})_{n \to \infty} \cong 25$. Agreement with experimental results is achieved simply by treating the chain as a random copolymer of $C^{2'}$-*endo* and $C^{3'}$-*endo* units (Olson and Flory, 1972). Unfortunately, in the $C^{2'}$-*endo* chain, the characteristic ratio is very sensitive to rotations about the $C^{3'}$–$O^{3'}$ bond, thus making it difficult to interpret in detail the experimentally observed dimensions.

19-6 CONFORMATION OF DNA

As numerous studies have shown, double-helical DNA is not a perfect rodlike molecule at large molecular weights. Instead, the molecule behaves as a very stiff, wormlike chain. This means that over short lengths it is rodlike, but over long contours the chain gradually curves and bends in arbitrary directions.

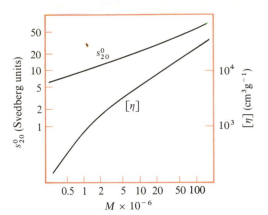

Figure 19-6

Sedimentation and viscosity data for native DNA as a function of molecular weight. [After J. Eigner and P. Doty, *J. Mol. Biol.* 12:549 (1965).]

Several excellent theoretical and experimental studies of the hydrodynamic behavior of DNA have been carried out, and these have provided us with much of our information on DNA conformation in solution. Figure 19-6 shows plots of log s^0 and log $[\eta]$ versus log M over the range of 300,000 d to 10^8 d for double-stranded DNA from various sources. Because the molecular weight of a base pair is about 600, this range of molecular weight spans chain lengths varying from about 500 to about 165,000 base pairs. Both plots show curvature over this wide range of molecular weight. However, the exact dependence of s^0 and $[\eta]$ on M in different ranges of molecular weight gives some useful insights. For $M = 3 \times 10^5$ to $M = 2 \times 10^6$, the viscosity data are described by $[\eta] = K_\eta M^{1.32}$. The exponent 1.32 approaches, but falls short of, the value of 1.8 for a perfect rod (Table 19-1). For the range of molecular weights above $M = 2 \times 10^6$, we find $[\eta] = K'_\eta M^{0.70}$, where $K'_\eta \neq K_\eta$. At these larger molecular weights, the dependence of $[\eta]$ on M shows the result expected for a coil (Table 19-1).

These viscosity data thus suggest that, up to $M \cong 10^6$ d, the DNA chains are quasi rodlike, but that their behavior at larger molecular weights is more like that of a coil. The sedimentation data also support this interpretation. Over the range of $M = 3 \times 10^5$ to $M = 4 \times 10^6$, we find $s^0 = K_s M^{0.325}$, whereas above $M = 4 \times 10^6$, we find $s^0 = K'_s M^{0.405}$. Thus, the sedimentation coefficient also changes from quasi rod behavior below about $M = 10^6$ d to quasi coil behavior at larger molecular weights (see Table 19-1). These sedimentation and viscosity results beautifully illustrate the concept of the stiff coil.

These sedimentation and viscosity data can be used to calculate the parameter β_s (Eqn. 19-29). This parameter is close to 1.2×10^7 over the entire range of molecular weight shown in Figure 19-6. Some small variations in β_s actually are expected at the lower molecular weights ($M < 10^7$ d), because the molecule is not a true coil in this region. However, the variations in β_s are small, and the close identity between the β_s value for the DNA molecules and that obtained for the chemically different polymers in Table 19-3 underscores the validity of the hydrodynamic treatment of s^0 and $[\eta]$ in Section 19-4.

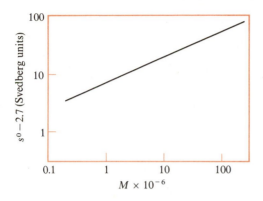

Figure 19-7

A single relationship between sedimentation coefficient and molecular weight. Log $(s^0 - 2.7)$ is plotted against log M. [After D. M. Crothers and B. H. Zimm, *J. Mol. Biol.* 12:525 (1965).]

According to a theoretical treatment of the sedimentation coefficient of an idealized wormlike chain, ignoring excluded-volume effects, the relationship between s^0 and M is $s^0 = b + K_s M^{1/2}$, where b is a constant (Hearst and Stockmayer, 1962). Thus, the theory predicts that addition of a constant term to the general expression $s^0 = K_s M^{1-a}$ (see Eqn. 19-22) permits representation of all the sedimentation data by a *single* equation in which $a = 1/2$. Experimental data are in general agreement with this prediction, over the entire range of molecular weights from 2×10^5 to 1.3×10^8 d. Figure 19-7 shows a plot of $\log(s^0 - b)$ versus $\log M$, covering this entire range of M. The chosen value of b is 2.7, which agrees well with the theoretical calculation of this parameter. The plot is quite linear, but its slope is 0.445 rather than 0.500. However, this discrepancy can be explained by excluded-volume effects (Gray et al., 1967). The theoretical treatment of the sedimentation data also leads to an estimate of ~450 Å for the persistence length (see Chapter 18) of DNA at an ionic strength of ~0.2 M (Hearst and Stockmayer, 1962; Gray et al., 1967).

19-7 MODEL FOR STIFF COILS: THE POROD–KRATKY CHAIN

Stiff chains sometimes are described in terms of an artificial model known as the Porod–Kratky (P–K) chain (see Kratky and Porod, 1949). The conformation of DNA has been represented by this model in the treatment of the chain's hydrodynamic behavior. The P–K model envisions the coil as a continuously curving chain, as opposed to the jagged, uneven contours of highly flexible chains (such as polyglycine). The direction of curvature at any point is assumed to be random.

Treatment of the Porod–Kratky chain

Figure 19-8 is a schematic illustration of the P–K chain. It is described by a series of x hypothetical bonds of length l_b, each of which makes an angle Υ (the "bond-angle" supplement) with its neighbors. Random changes in trajectory along the chain are

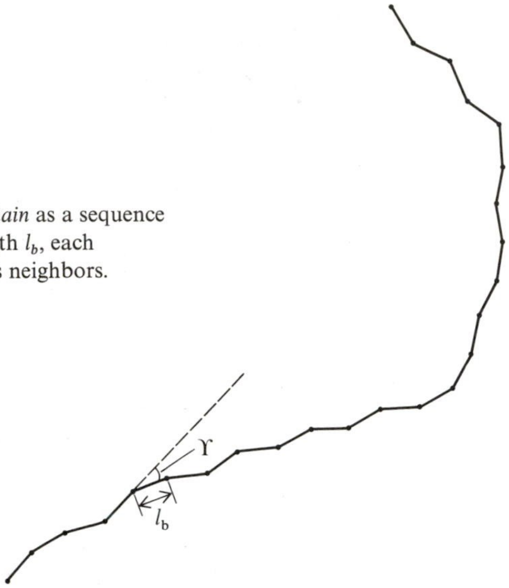

Figure 19-8

Representation of wormlike chain as a sequence of hypothetical bonds of length l_b, each making an angle of Υ with its neighbors.

allowed by assuming that the hypothetical bonds are freely rotating. Therefore, the average projection ρ of the end-to-end distance vector $\mathbf{r}$ on the first bond (see Chapter 18) is[§]

$$\rho = l_b \sum_{k=0}^{x-1} (\cos \Upsilon)^k \qquad (19\text{-}30)$$

For an infinitely long chain, $x \to \infty$, and $\rho \to a$ (the persistence length; see Chapter 18), and the summation converges to $(1 - \cos \Upsilon)^{-1}$, so that

$$a = l_b/(1 - \cos \Upsilon) \qquad (19\text{-}31)$$

Returning to Equation 19-30, we now allow l_b to become smaller and smaller, in such a way that a and the contour length L_c are held constant at their respective predetermined values. Therefore, as l_b decreases, the change in the chain's trajectory at each bond becomes smaller; this means that Υ decreases, $1 - \cos \Upsilon$ must approach

---

[§] The end-to-end distance vector $\mathbf{r} = \sum_i \mathbf{l}_i'$, where $\mathbf{l}_i'$ is the vector representation of hypothetical bond i (Eqn. 18-1), primes are used to distinguish hypothetical bonds from real ones, and the sum is taken from $i = 1$ to $i = x$. Each bond is of equal length l_b. Using the results of our analysis of the freely rotating chain (Section 18-4),

$$\langle (\mathbf{l}_1'/l_b) \cdot \mathbf{r} \rangle = \langle (\mathbf{l}_1'/l_b) \cdot \mathbf{l}_1' + (\mathbf{l}_1'/l_b) \cdot \mathbf{l}_2' + \cdots + (\mathbf{l}_1'/l_b) \cdot \mathbf{l}_x' \rangle$$
$$= l_b[1 + \cos \Upsilon + (\cos \Upsilon)^2 + \cdots + (\cos \Upsilon)^{x-1}]$$

zero and, of course, x approaches infinity. In accordance with Equation 19-31, we obtain

$$\lim_{l_b \to 0} [l_b/(1 - \cos \Upsilon)] = l_b/[-\ln(\cos \Upsilon)] = a \tag{19-32}$$

where we have used the result $\ln x \cong x - 1$, for $x \cong 1$. This procedure thus converts our chain into one that is *continuously* curving, with the direction of curvature being random at any point. Equation 19-32 defines l_b and $\cos \Upsilon$ in such a way that $\cos \Upsilon = \exp(-l_b/a)$. Thus, according to Equation 19-30,

$$\rho = l_b \sum_{k=0}^{x-1} \exp(-kl_b/a) \tag{19-33}$$

Because l_b becomes very small, the sum may be replaced by an integral in which, for large x, we first replace $x - 1$ by x, and then define $L = l_b k$ to be a continuous variable:

$$\rho = l_b \int_0^x \exp(-kl_b/a)\, dk \tag{19-34}$$

$$= \int_0^{L_c} e^{-L/a}\, dL \tag{19-35}$$

$$= a(1 - e^{-L_c/a}) \tag{19-36}$$

where the relation $dL = l_b\, dk$ has been used in converting Equation 19-34 to Equation 19-35.

The relationship between $\langle r^2 \rangle_0$ and L_c is obtained by noting that $\mathbf{r} \cdot d\mathbf{r} = \rho\, dL_c$. This is apparent from Figure 19-9, where the segment $d\mathbf{r}$ is added as an extension of

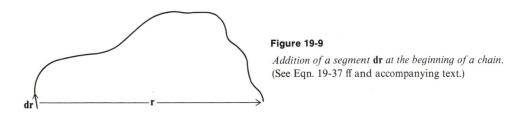

Figure 19-9

*Addition of a segment **dr** at the beginning of a chain.* (See Eqn. 19-37 ff and accompanying text.)

the beginning of the chain (same direction as the first bond). Therefore,

$$d\langle \mathbf{r} \cdot \mathbf{r} \rangle = 2\langle \mathbf{r} \cdot d\mathbf{r} \rangle \tag{19-37}$$

$$= 2\rho\, dL_c \tag{19-38}$$

$$= 2a(1 - e^{-L_c/a})\, dL_c \tag{19-39}$$

and

$$\langle r^2 \rangle_0 = 2a \int_0^{L_c} (1 - e^{-L_c/a}) \, dL_c \tag{19-40}$$

$$= 2aL_c[1 - a/L_c(1 - e^{-L_c/a})] \tag{19-41}$$

Equation 19-41 is the desired result. In the limit of large L_c,

$$(\langle r^2 \rangle_0 / L_c)_{L_c \to \infty} = 2a \tag{19-42}$$

(Note that *formally* Eqn. 19-42 can be derived from Eqn. 18-71 by letting nl in Eqn. 18-71 equal L_c, and by replacing l in Eqn. 18-71 by l_b, where l_b is vanishingly small.) For small values of L_c, Equation 19-41 may be expanded for $L_c/a < 1$ to yield

$$\langle r^2 \rangle_0 = L_c^2[1 - 1/3(L_c/a) + 1/12(L_c/a)^2 - \cdots] \tag{19-43}$$

Behavior of chain dimensions

For $L_c \ll a$, Equation 19-43 states that $\langle r^2 \rangle_0 \cong L_c^2$; that is, the unperturbed end-to-end distance is equal to the contour length. This is exactly the behavior of a rodlike molecule; thus, at short chain lengths, the stiff coil behaves as a rod. As L_c increases, however, $\langle r^2 \rangle_0 < L_c^2$ (see Eqn. 19-43), as the effect of curvature becomes manifest. At very high values of L_c, the ratio $\langle r^2 \rangle_0 / L_c^2 = 2a/L_c$ (Eqn. 19-42); that is, the deviation of $\langle r^2 \rangle_0$ from the value expected for the rod is determined by the ratio of the persistence length to the contour length.

Application of the Porod–Kratky model to DNA

For the Watson–Crick double helix, a distance of 34 Å is traversed for each turn of the helix. Because there are 10 base pairs per turn, the axial translation per base pair is 3.4 Å. These are the data that are used to compute L_c as a function of the number of base pairs. Thus, for a chain of 1,000 base pairs ($M \cong 600,000$ d), the contour length is estimated as 3,400 Å, or 0.34 μm. Under certain conditions, the persistence length has been estimated as roughly 450 Å, as mentioned earlier. Therefore, we calculate from Equation 19-43 that the contour distance below which DNA is rodlike in solution ($\langle r^2 \rangle_0 \geqslant 0.9L_c^2$ or $\langle r^2 \rangle_0^{1/2} \geqslant 0.95L_c$) corresponds to a value of $(1/3)(L_c/a) = 0.1$; from this we estimate that $L_c = 0.3a = 135$ Å, which is equivalent to about 40 base pairs. At large molecular weights, the discrepancy between $\langle r^2 \rangle_0^{1/2}$ and L_c becomes very large. Consider, for example, the DNA of T2 bacteriophage, which has a contour length of 60 μm. With a persistence length of 450 Å, the value of $\langle r^2 \rangle_0^{1/2}$ is 2.3 μm, as calculated by Equation 19-41 or 19-42.

Summary

Hydrodynamic properties of chain molecules are sensitive to chain conformation. In particular, the intrinsic viscosity of random coils is related to the mean square end-to-end distance, so that viscosity measurements provide a convenient way to evaluate chain dimensions of coils. Frictional and sedimentation coefficients also are sensitive to chain dimensions. Moreover, the dependence on molecular weight of the intrinsic viscosity (or frictional or sedimentation coefficient) can be used to determine whether a chain behaves as a random coil or as another shape and character. Finally, some of the hydrodynamic parameters are interrelated through the Mandelkern–Flory–Scheraga constant in a way that is particularly useful for calculating molecular weights from a combination of viscosity and sedimentation measurements.

Both single-stranded and double-helical nucleic acids have been investigated by hydrodynamic methods. Configurational-statistics calculations on single-stranded polynucleotides account fairly well for the experimentally observed chain dimensions. In the case of double-stranded DNA, the molecular-weight dependence of the intrinsic viscosity and of sedimentation coefficient has been studied. These studies show that, up to molecular weights of roughly 10^6 d, the molecule exhibits quasi rod behavior and that, above this molecular weight, it is more like a coil. This suggests that the chain is a stiff, wormlike coil. Such coils are conveniently treated in terms of an artificial model known as the Porod–Kratky chain. This model gives a useful description of DNA conformation and dimensions.

Problems

19-1. Given that $\langle r^2 \rangle$ for a random coil in water is 10^4 Å^2, calculate the radius r_e of an equivalent hydrodynamic sphere that has the same sedimentation coefficient as the random coil. (Assume that the partial specific volume of the equivalent sphere is the same as that of the coil.)

19-2. An investigator studies the sedimentation and viscosity behavior of a polymer as a function of molecular weight. He finds a curious result: the quantity $s^0[\eta]^{1/3}$ is proportional to $M^{2/3}$. "This must mean my polymer is essentially a solid sphere," he exclaims. "Not necessarily," says Joe Smart; "There is another simple explanation for your result." Then Sally Sharp states, "It definitely is not a solid sphere." Who is right (if anyone), and why?

19-3. Consider a chain molecule containing a large number of bonds. The conformational energy of the chain is given by $U(r) = \alpha r^2$, where r is the end-to-end distance, and α is a constant with units of energy per unit of length squared. The number of configurations that give a value of r between r and $r + dr$ is proportional to $4\pi r^2$.   (a) Derive an expression for $\langle r^2 \rangle$ as a function of α and T (temperature), using the approximation that r can vary from 0 to ∞.   (b) The intrinsic viscosity of this chain is measured as 90 cm^3 g^{-1} at 300 K. Calculate $\langle r^2 \rangle^{1/2}$ in angstroms and the value of α from these data. Assume that the chain has a molecular weight of 10^5 d.

19-4. In the text, a calculation of $\langle r^2 \rangle_0^{1/2}$ for DNA from phage T2 uses a persistence length of 450 Å and a contour length of 60 μm in conjunction with the Porod–Kratky model. Execute this calculation, and compare your answer with that given in the text.

19-5. A random-coil polymer has a molecular weight of 10^7 d and a contour length of 30,000 Å. The intrinsic viscosity is measured, under certain conditions, to be 540 cm^3 g^{-1}. From this information, obtain an estimate of the persistence length.

References

GENERAL

Flory, P. J. 1953. *Principles of Polymer Chemistry*. Ithaca, N.Y.: Cornell Univ. Press. [Chap. 14 gives a good account of hydrodynamic properties of polymers in dilute solution.]

Tanford, C. 1961. *Physical Chemistry of Macromolecules*. New York: Wiley. [Chap. 6 treats viscosity, sedimentation, and diffusion of both globular particles (proteins) and flexible polymers.]

Yamakawa, H. 1971. *Modern Theory of Polymer Solutions*. New York: Harper and Row. [An advanced treatment.]

SPECIFIC

Crothers, D. M., and B. H. Zimm. 1965. Viscosity and sedimentation of the DNA from bacteriophages T2 and T7 and the relation to molecular weight. *J. Mol. Biol.* 12:525.

Debye, P. 1946. Viscosity of polymer solutions. *J. Chem. Phys.* 14:636.

Debye, P., and A. M. Bueche. 1948. Intrinsic viscosity, diffusion, and sedimentation rate of polymers in solution. *J. Chem. Phys.* 16:573.

Eigner, J., and P. Doty, 1965. The native, denatured and renatured states of deoxyribonucleic acid. *J. Mol. Biol.* 12:549.

Gray, H. B., Jr., V. A. Bloomfield, and J. E. Hearst. 1967. Sedimentation coefficients of linear and cyclic wormlike coils with excluded-volume effects. *J. Chem. Phys.* 46:1493.

Gray, H. B., Jr., and J. E. Hearst. 1968. Flexibility of native DNA from the sedimentation behavior as a function of molecular weight and temperature. *J. Mol. Biol.* 35:111.

Harrington, R. 1978. Opticohydrodynamic properties of high-molecular-weight DNA. III. The effects of NaCl concentration. *Biopolymers* 17:919. [Analysis of excluded volume and persistence length in T2 DNA over a range of salt concentrations.]

Hearst, J. E., and W. H. Stockmayer. 1962. Sedimentation constants of broken chains and wormlike coils. *J. Chem. Phys.* 37:1425.

Kirkwood, J. G., and J. Riseman. 1948. The intrinsic viscosities and diffusion constants of flexible macromolecules in solutions. *J. Chem. Phys.* 16:565.

Kratky, O., and G. Porod. 1949. Röntgenuntersuchung gelöster fadenmolekule. *Rec. Trav. Chim.* 68:1106.

Mandelkern, L., and P. J. Flory. 1952. The frictional coefficient for flexible chain molecules in dilute solution. *J. Chem. Phys.* 20:212.

Olson, W. K., and P. J. Flory. 1972. Spatial configurations of polynucleotide chains, I: Steric interactions in polyribonucleotides: A virtual bond model. *Biopolymers* 11:1.

———. 1972. Spatial configuration of polynucleotide chains, II: Conformational energies and the average dimensions of polyribonucleotides. *Biopolymers* 11:25.

Roberts, T. M., G. D. Lauer, and L. C. Klotz. 1975. Physical studies on DNA from "primitive eucaryotes." *CRC Critical Reviews in Biochemistry* 3:349.

Scheraga, H. A., and L. Mandelkern, 1953. Consideration of the the hydrodynamic properties of proteins. *J. Am. Chem. Soc.* 75:179.

Weissman, M., H. Schindler, and G. Feher. 1976. Determination of molecular weights by fluctuation spectroscopy: Application to DNA. *Proc. Natl. Acad. Sci. USA* 73:2776.

Zimm, B. H. 1956. Dynamics of polymer molecules in dilute solution: Viscoelasticity, flow birefringence and dielectric loss. *J. Chem. Phys.* 24:269.

20

Conformational equilibria of polypeptides and proteins: the helix–coil transition

20-1 CONFORMATIONAL STABILITY AND CHANGE

Despite the large configuration space available, proteins and polypeptides show a remarkable propensity to adopt unique, well-defined conformations. The diversity among these ordered conformations is very great, ranging from compact globular particles to fibrous, rodlike molecular structures. The many factors involved in stabilizing ordered structures are discussed in some detail in Chapter 5.

Although stable ordered forms are readily adopted in solution, it is also true that, under some conditions, polypeptides can interconvert among these various ordered states, as well as between ordered and random forms. Such conformational changes generally occur in response to variations in an external parameter such as temperature or solvent. In many cases, only small alterations in the external variable are required to bring about very dramatic structural rearrangements. These rearrangements are readily monitored through measurements of a suitable physical parameter such as ORD, CD, or a hydrodynamic property.

The rich diversity of structural changes that can occur is apparent from a few examples. Figure 20-1 shows the ultraviolet ORD spectra of random copolymers of hydroxypropyl-L-glutamine and L-alanine at 5°C (curves A, B, C) and at 76°C (curves A', B', C'). The three curves at each temperature refer to copolymers of differing chain length and composition. The striking feature of these curves is the remarkable alteration brought about by the temperature change. The spectra taken at 5°C are characteristic of α-helical structures, with the degree of helicity being

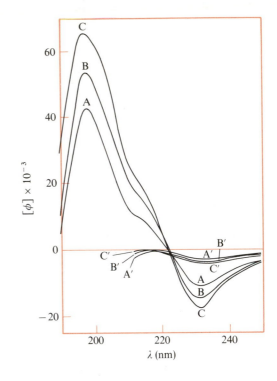

Figure 20-1

Conformational changes in polypeptides.
Ultraviolet ORD spectra of random copolymers of hydroxypropyl-L-glutamine and
L-alanine at 5° C (curves A, B, C) and at 76° C
(curves A′, B′, C′). The three curves at each
temperature refer to copolymers of different
chain length (degree of polymerization 422,
536, and 1,102, respectively) and composition
(14%, 30%, and 49% alanine, respectively).
[After K. E. B. Platzer et al., *Macromolecules*
5:177 (1972).]

greatest for curve C. At 76°C, on the other hand, the chains are largely in disorganized
random forms, thus accounting for the rather weak rotatory disperson. Between the
two temperature extremes, these copolymers must undergo a helix–coil transition.
Details of the temperature-induced helix–coil transition of polypeptides are examined
in subsequent sections of this chapter.

Figure 20-2a shows the ultraviolet and visible ORD of conformational forms I
and II of poly-L-proline. These are fairly rigid helical conformations; in form I the
imide bonds are *cis*, whereas in form II they are *trans* (Fig. 20-2b). The rigid helical
structures are stabilized by strict stereochemical constraints and solvent–polymer
interactions. Intramolecular hydrogen bonds cannot occur in either form, owing
to the absence of an amide N–H. Interconversion between forms I and II is easily
accomplished by varying the solvent composition. Figure 20-2c plots $[\phi]_{546}$ versus
solvent composition. It is clear that the optical rotation undergoes a very sharp,
cooperative transition as the polymer, in response to a small change in solvent composition, flips between forms I and II. It should be noted that the structure and conformational changes of poly-L-proline have been of much interest because of their
connection with structure-function relationships of collagen.

Another very interesting and dramatic structural change can be observed in
poly-L-lysine at alkaline pH, where the ε-amino groups are not protonated. Under
such conditions, ordered intramolecular and intermolecular structures can form.

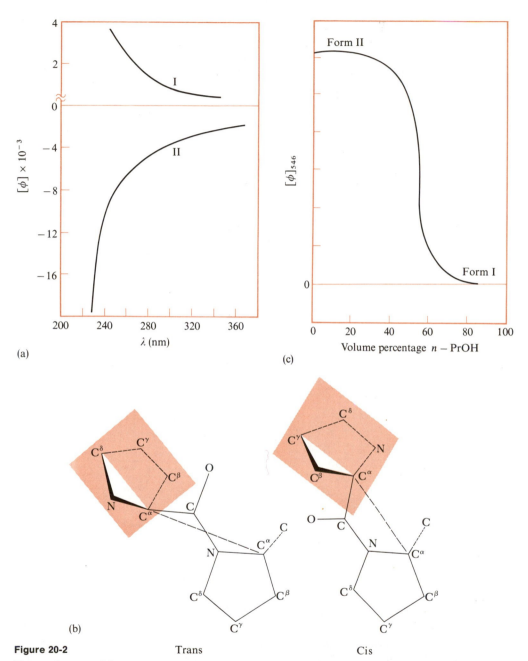

Figure 20-2   Trans   Cis

Two conformational forms of poly-L-proline. (**a**) ORD spectra of the two conformations. [After G. D. Fasman and E. R. Blout, *Biopolymers* 1:3 (1963).] (**b**) The *cis* and *trans* forms of the imide linkage. (**c**) Optical rotation at 546 nm versus solvent composition in acetic acid–1-propanol mixtures containing poly-L-proline. [After F. Gornick et al., *J. Am. Chem. Soc.* 86:2549 (1964).]

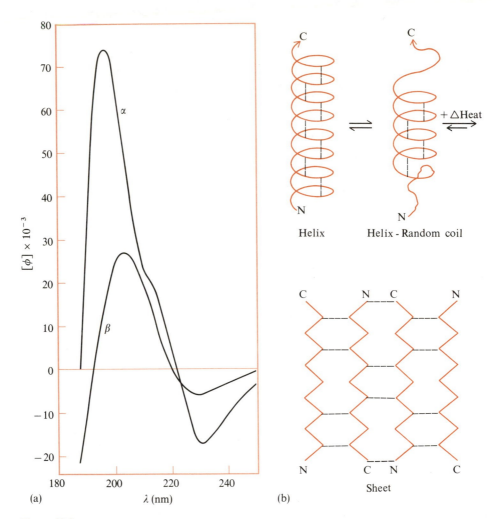

Figure 20-3

Ordered intramolecular and intermolecular structures in poly-L-lysine. (**a**) ORD spectra in the α-helical conformation and in the β-sheet structure. (**b**) Schematic illustration of the heat-induced transition from the α-helical to the β-sheet structure. [After B. Davidson and G. D. Fasman, *Biochemistry* 6:1616 (1967).]

Figure 20-3a shows the ORD spectrum at 22.5°C. It is characteristic of a right-handed α-helical structure (see Chapter 8). Upon heating the solution to 51°C for a period of time, slow changes in ORD are observed as a new structure is formed. The structure may then be stabilized and studied by cooling the solution back to 22.5°C. The resulting ORD curve, also shown in Figure 20-3a, is characteristic of a β-sheet structure. Thus, the polypeptide switches between two ordered forms (Fig. 20-3b) by passing

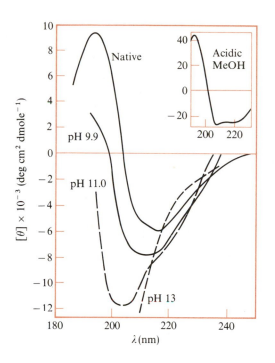

Figure 20-4

Conformational changes in a protein. The far-UV CD spectra of β-lactoglobulin A in the native state and in other states. [After S. N. Timasheff et al., in *Conformation of Biopolymers*, vol. 1, ed. G. N. Ramachandran (New York: Academic Press, 1967), p. 173.]

from an intramolecularly stabilized helix to an ordered aggregate stabilized by intermolecular interactions.

In view of the remarkable structural changes undergone by polypeptides in solution, it is not surprising that rather dramatic changes also occur in the conformations of proteins. In such cases, and even where structural changes are small, there generally is an important change in biological activity. Figure 20-4 displays the far-ultraviolet CD spectrum of β-lactoglobulin in its native state and in various states of denaturation. The inset shows the spectrum in acidic methanol; this spectrum is characteristic of proteins containing considerable amounts of α-helix. The spectral changes produced are quite dramatic, just as for polypeptides. However, exact interpretation of the spectra in terms of specific structures is much more difficult in the case of proteins.

In the present chapter, an example of a conformational change is discussed in depth. This is the well-studied helix–coil transition of polypeptides, which is the most thoroughly examined structural transition of polypeptides, from both the experimental and the theoretical viewpoints. A great deal of insight into the characteristics of polypeptides and of cooperative phenomena has been obtained from study of this transition.

The reversible folding and unfolding of proteins also has been subjected to many careful investigations, but understanding is still fairly primitive. This subject is taken up in Chapter 21. Protein folding–unfolding reactions are interesting, not only because they give important insights into the factors stabilizing (and destabilizing) the native form, but also because they shed light on the question of the mechanism of the assembly of specific, organized protein conformations from a newly synthesized polypeptide chain. Although the protein folding–unfolding reactions are more complicated than the helix–coil transitions of polypeptides, the treatment of the latter in the present chapter serves as a useful background and reference point for understanding the former.

A prime purpose in considering in depth these two conformational transitions is to illustrate the kinds of questions that arise, and the varied approaches that may be used, in pursuing these problems. Of course, many of these questions and approaches are rather general and may, therefore, be carried over to the study of other problems.

20-2   HELIX–COIL TRANSITION: EARLY OBSERVATIONS

It is well known (from x-ray crystallography, infrared spectra, optical rotation, viscosity, and other physical data) that polypeptides can adopt the α-helical conformation in the crystalline state as well as in solution. Figure 2-23 illustrates the structure of the right-handed α helix as determined by x-ray crystallography. The structure involves a remarkable interlacing system of hydrogen bonds, in which the carbonyl oxygen of residue i is bonded to the amide hydrogen of residue $i + 4$. Using the conventions of Chapter 5, we fix each ϕ, ψ rotation angle in the helix at about $-50°$ (right-handed helix) or about $+50°$ (left-handed helix) (see Table 5-2).

As mentioned, the helical structure also has been characterized in solution. However, a variation of solution parameters (temperature, solvent composition, pH) often leads to remarkable changes in the physical characteristics of polypeptides. Consider, for example, some early data obtained for poly-L-glutamate. This polymer contains ionizable carboxyl groups, so that its total charge is a function of pH. A variation in charge might reasonably be expected to produce a change in conformation, because configurations in which ionized carboxyl groups are in close proximity will be suppressed.

Figure 20-5 plots the specific rotation at 589 nm, the intrinsic viscosity, and the degree of ionization of poly-L-glutamate versus pH, in a water–dioxane mixed solvent. (Dioxane is added to maintain the solubility of the uncharged polypeptide.) The optical rotation and intrinsic viscosity undergo abrupt and dramatic changes as the pH is raised. In both cases, much smaller values of these physical parameters are found at the higher pH. These changes are entirely consistent with a transition from an ordered helical rod at low pH to a random coil at high pH. Recall, for example, the high intrinsic viscosity that is characteristic of a high-molecular-weight rod, as opposed to the lower viscosity for a coil of the same high molecular weight (see Chapter 19). (The small rise in intrinsic viscosity after the precipitous drop is due to the effects of increased ionization; the higher charge density encourages greater chain extensions of the coil and also a greater frictional interaction between polymer and solvent.) The degree of ionization steadily increases as the pH is raised. However,

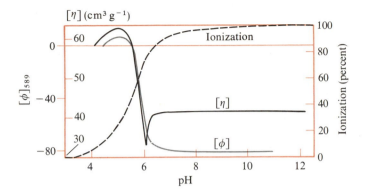

Figure 20-5

The helix–coil transition in a polypeptide. The specific rotation, intrinsic viscosity, and degree of ionization are plotted against pH for poly-L-glutamic acid in 0.2 M NaCl-dioxane (2:1) at 25°C. [After P. Doty et al., *J. Polymer Sci.* 23:851 (1957).]

the big changes in conformation do not occur until approximately 40% of the carboxyl groups have ionized. The midpoint of the ionization occurs at about pH 5.8, which is higher than a typical carboxyl pK in water, owing in part to the presence of dioxane, to electrostatic effects that make it more difficult to remove a proton from a chain bearing negative charges, and to the preference in the helix for un-ionized residues.

The changes in optical rotation and intrinsic viscosity are so abrupt that the transition is termed a "cooperative" process—that is, a kind of "all-or-none" phenomenon whereby the molecule prefers all-helical or all-coil states, but avoids intermediate states. The term "phase transition" is sometimes also applied, by rough analogy to the abrupt changes in properties that take place when a substance undergoes a transformation of phase, such as melting of a solid to a liquid. The latter are truly discontinous phenomena, however, whereas the helix–coil transition retains continuity.

Helix–coil transitions can be induced also by small variations in temperature. In addition, the formation of helices shows a significant dependence on chain length. Figure 20-6 shows the molar ellipticity of the circular dichroism band at 222 nm (see Chapter 8) as a function of chain length n for N-carbobenzoxy-γ-ethyl-L-glutamate oligomers in trifluoroethanol. Strong negative ellipticity of this band is characteristic of α helices. Note that the band is slightly positive and is independent of chain length up to $n = 5$. At $n = 6$, the molar ellipticity begins to drop steeply, and by $n = 10$ it has changed manyfold, indicating substantial α-helix structure in the decamer. As shown later in the chapter, the sharpness (apparent cooperativity) of a helix–coil transition and the thermal midpoint (temperature at which the fractional helicity is 1/2) are also dependent on chain length. The reasons for these features become apparent from a consideration of the molecular mechanism of the transition, and a quantitative treatment of its equilibrium features. From these considerations, the reader should be able to rationalize easily the data in Figure 20-6.

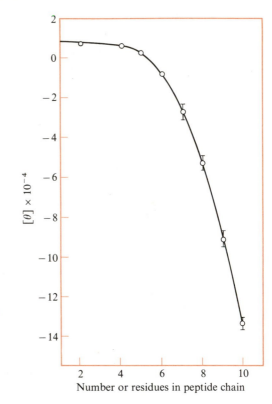

Figure 20-6

Helix formation and chain length. Molar ellipticity at 222 nm is plotted against chain length (number of residues) for oligomers of *N*-carbobenzoxy-γ-ethyl-L-glutamate. [After M. Goodman et al., *Proc. Natl. Acad. Sci. USA* 64:444 (1969).]

20-3 MOLECULAR MECHANISM OF THE TRANSITION

A question naturally arises as to the physical reason for the cooperativity of the transition. Put another way, we want to explain why it is difficult to initiate the transition to a helix, but the process proceeds readily once it is started. To answer this question, we must closely examine the structure of the helix and the underlying forces that stabilize it.

Figure 20-7 is a schematic illustration of the interlacing hydrogen-bond network of an α helix. Because a given hydrogen bond goes from the ith to the $(i + 4)$th residue, six dihedral rotation angles are spanned by a single hydrogen bond. Consider, however, the difficulty this presents when an α helix is to be started from a random coil. The chain initially is in a statistical distribution of configurations where the individual residues adopt a wide variety of ϕ, ψ angles (see Chapter 5). If incipient hydrogen-bond formation is encouraged by lowering the temperature, for example, difficulty is encountered in making the first hydrogen bond, because six rotation angles must be immobilized or frozen at particular values. This represents a sizeable loss in conformational entropy, with the only compensation coming from whatever

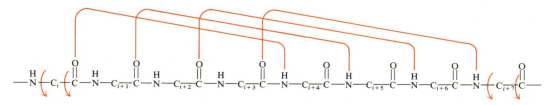

Figure 20-7

The interlacing hydrogen-bonded network of the α helix. Dihedral rotation angles
about bonds attached to α-carbons C_i and C_{i+7} are indicated in this schematic
drawing; side chains have been omitted for simplicity.

favorable interactions arise as a result of the single hydrogen bond. However, the
second and additional hydrogen bonds need not overcome such an unfavorable
entropic barrier if they successively form *adjacent* to the first one. This is because,
once the initial bond is made, adjacent ones form with the attendant immobilization
of only two rotation angles (one ϕ, ψ pair) per hydrogen bond (Fig. 20-8). Obviously,

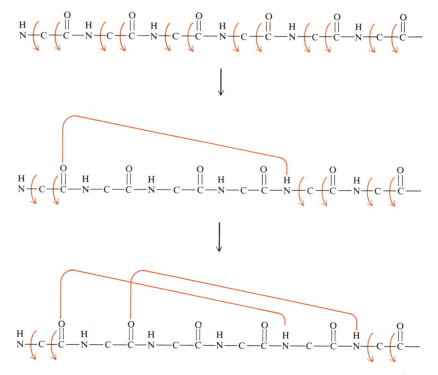

Figure 20-8

Helix initiation. This schematic illustration shows that six rotation angles are
immobilized in making the first hydrogen bond, but only two angles are immobilized in making a bond adjacent to an existing one.

the tendency is to form long stretches of neighboring hydrogen bonds, so that the unfavorable effects of the first, initiating bond are overcome. Likewise, there will be a tendency to confine helical units to one long sequence, rather than to scatter short helical sequences throughout the chain. These features have the effect, of course, of making the overall transition acquire an "all-or-none" character.

There is another very basic reason for the cooperativity, however. The squat, compact structure of the helix gives rise to significant dipole–dipole interactions between amide groups separated by several residues (see Chapter 5 for a discussion of the dipolar interaction). The total dipolar energy of a helix can be substantial. Moreover, the dependence on helix length of this contribution to the total stabilization has a significant effect on the cooperativity.

In an α helix, the amide dipoles align themselves approximately parallel to the axis of the helix (Fig. 20-9). Therefore, the interaction between successive amide

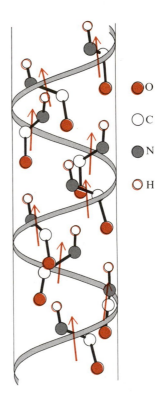

Figure 20-9

Orientation of the amide dipoles (arrows) *in the right-handed α helix.* [After A. Wada, in *Polyamino Acids, Polypeptides, and Proteins,* ed. M. A. Stahman (Madison: Univ. of Wisconsin Press, 1962), p. 131.]

O O
O C
O N
O H

dipoles is repulsive, because they are roughly side by side. However, the interaction of a given dipole with successively more distant neighbors becomes favorable as the dipole pairs adopt a head-to-tail alignment. Consequently, third-neighbor interactions, for example, are very favorable. In a long α helix, the total dipolar interaction of a given residue with all succeeding residues is negative, and is estimated to be -1 to -2 kcal per mole of residues (Brant, 1968). Therefore, the total contribution of the dipolar interaction in a long helix is very large. David Brant (1968) has estimated that a given residue will achieve maximal dipolar stabilization if it is followed by roughly ten more helical units. Viewed in this way, residues near the end of a helical sequence are deficient in energy. Hence, the effect of the dipolar interaction is to discourage the accumulation of many separated helical sequences (where the number of "ends" is large) and to encourage the build-up of long helical runs (where dipolar interactions are optimized). This gives another important contribution to the co-operativity of helix formation.

20-4 SIMPLE THERMODYNAMIC TREATMENT OF THE TRANSITION

A simple thermodynamic treatment gives further insight into the helix–coil transition. Although the following treatment is overly simple, some of its qualitative features are instructive. This treatment is a variation of that given by John Schellman (1955).

Each chain is assumed to be either all-helix or all-coil, with no structures in which the two kinds of regions coexist within the same chain. This means that, when the fractional helicity θ is $1/2$, half of the molecules are all-helix and half are all-coil. (This is a rather stringent assumption.) The free-energy change ΔG_r for converting a coil residue to a hydrogen-bonded helical residue is

$$\Delta G_r = \Delta H_r - T\,\Delta S_r \qquad (20\text{-}1)$$

where ΔH_r and ΔS_r are the per-residue changes in entropy and enthalpy, respectively. The enthalpy change is assumed to be associated with hydrogen-bond formation, and the entropy change is assumed to occur as a result of the greater conformational freedom of a coil unit (as compared to an immobilized helical unit). In practice, solvent effects also may contribute to these parameters, but do not consider these effects explicitly in this simple treatment. The total free-energy change for the transition of a chain of n residues is not simply $n\,\Delta G_r$, however. Because of end effects, there is a deficit of four hydrogen bonds in a helix of n residues (see Fig. 20-7). At the amino terminus are four N—H donors without partners; likewise, at the carboxyl terminus are four unpaired C=O acceptors. Also, the conformational freedom (entropy) of each end residue is independent of the helical or coil state of the chain. Based on these simple considerations, the total free-energy change ΔG_{Tot} is

$$\Delta G_{\text{Tot}} = (n-4)\,\Delta H_r - (n-2)T\,\Delta S_r \qquad (20\text{-}2)$$

The equilibrium constant K for the overall reaction is

$$K = ck^n = (\text{helix})/(\text{coil}) \qquad (20\text{-}3)$$

where

$$\ln k = -(\Delta H_r - T\,\Delta S_r)/RT \tag{20-4}$$

$$\ln c = -(-4\,\Delta H_r + 2T\,\Delta S_r)/RT \tag{20-5}$$

If n is large, it is easy to see from the center term of Equation 20-3 that small changes in k (induced by a change in T, for example) may cause very sharp changes in the equilibrium constant K. The parameter c takes into account the deficit of hydrogen bonds at both ends of the chain and the conformational freedom of the terminal residues. (Note that the signs multiplying ΔH_r and ΔS_r in Eqn. 20-4 are opposite to those in Eqn. 20-5. Therefore a change in k, such as one caused by varying T, will be partly offset by the change in c.)

We can easily find the dependence of the "midpoint" temperature T_m on chain length n. At $T = T_m$, we have $\Delta G_{Tot} = 0$, and $K = 1$. Therefore,

$$\Delta G_{Tot} = 0 = (n-4)\,\Delta H_r - T_m(n-2)\,\Delta S_r \tag{20-6}$$

and

$$T_m = [(n-4)/(n-2)](\Delta H_r/\Delta S_r) \qquad \text{for } n \geqslant 5 \tag{20-7}$$

The stipulation that $n \geqslant 5$ is necessary because it takes a minimum of five residues to make one hydrogen bond. From Equation 20-7, it is apparent that T_m is dependent on chain length. When $n = 5$, we have $T_m = (1/3)\Delta H_r/\Delta S_r$; when n is large, T_m asymptotically approaches $T_m = \Delta H_r/\Delta S_r$. It is clear that the predicted variation of T_m with n is not significant above $n \cong 30$. The prediction of the increase in T_m with increasing n is in qualitative accord with experimental results, but the quantitative agreement is very poor. Generally, T_m changes over a much wider range of n than is indicated by this simple theory.

We also may evaluate the dependence of the "sharpness" of the transition on chain length. The sharpness is simply defined as the slope of a θ versus T plot at $\theta = 1/2$, or $T = T_m$. Therefore, we must obtain an expression for $(d\theta/dT)_{T=T_m}$. We accomplish this by differentiating the expression for the equilibrium constant, where θ is the fraction of residues that are helical and $1 - \theta$ is the fraction that are coil:

$$\ln K = \ln[\theta/(1-\theta)] = \ln\theta - \ln(1-\theta) \tag{20-8}$$

$$= -[(n-4)\,\Delta H_r - (n-2)T\,\Delta S_r]/RT \tag{20-9}$$

and

$$\frac{1}{\theta}\left(\frac{d\theta}{dT}\right)_{T=T_m} - \frac{1}{1-\theta}\left(-\frac{d\theta}{dT}\right)_{T=T_m} = \left(\frac{d(\ln K)}{dT}\right)_{T=T_m} \tag{20-10}$$

Because $\theta = 1/2$ at $T = T_{\mathrm{m}}$ and, from the van't Hoff relation, $[d(\ln K)/dT]_{T=T_{\mathrm{m}}} = \Delta H_{\mathrm{Tot}}/RT_{\mathrm{m}}^2 = (n-4)\Delta H_{\mathrm{r}}/RT_{\mathrm{m}}^2$, we have

$$4(d\theta/dT)_{T=T_{\mathrm{m}}} = (n-4)\Delta H_{\mathrm{r}}/RT_{\mathrm{m}}^2 \tag{20-11}$$

or

$$(d\theta/dT)_{T=T_{\mathrm{m}}} = (1/4)(n-2)^2(\Delta S_{\mathrm{r}})^2/(n-4)\Delta H_{\mathrm{r}} \tag{20-12}$$

where Equation 20-12 was obtained from Equation 20-11 by substituting Equation 20-7 for T_{m}. According to Equation 20-12, the sharpness of the transition should increase with chain length; this prediction is in qualitative agreement with observations. This increase in sharpness is due simply to the increase in the overall enthalpy change as the number of residues increases. However, Equation 20-12 also predicts that the sharpness increases without limit as n increases; this prediction does not agree with experimental results.

The failure of this simple thermodynamic treatment to account quantitatively for experimental results is a consequence of the overly restrictive assumption that each chain exists in either an all-helix or an all-coil form. Despite the thermodynamically unfavorable aspects of helix–coil interfaces (Section 20-3), there is a finite probability that a molecule will contain helical and coil regions simultaneously. The longer a chain, the greater the probability of occurrence of helical regions separated by coil sections. In order to treat accurately the realistic case, a more elaborate statistical thermodynamic approach must be used. Two more accurate theories can be employed. In the first, helical and coil units are considered to coexist in the same chain, but only one helical region is allowed in a given chain. Thus, a helical region may start in any part of the chain, but new helical segments are added only to the ends of the established one. This "zipper" model can give a good account of the transition for smaller polypeptide chains. In the most exact treatment, any number of helical and coil regions are permitted to coexist in the same chain.

These more accurate models require a statistical thermodynamical treatment. Before considering these treatments, we take up the method of constructing partition functions for these situations.

20-5 PARTITION FUNCTIONS FOR CONFORMATIONAL EQUILIBRIA OF SIMPLE LINEAR CHAINS

The more exact treatment of helix–coil equilibria requires the construction of partition functions by the method of statistical weights. This procedure is useful for treating conformational equilibria in chain molecules. In effect, the partition function is written in terms of equilibrium constants for elementary reactions.

A partition function q of a molecule can be defined as

$$q = \sum_{i} g_i e^{-\varepsilon_i/kT} \tag{20-13}$$

where ε_i is the energy of the ith energy level, and g_i is the number of states that have the energy ε_i. The parameter g_i also is called the *degeneracy* of the ith level. In this formulation, we envision the various possible energy levels to be enumerated as ε_0, $\varepsilon_1, \varepsilon_2, \ldots$. In some circumstances, there is more than one way to fill a particular energy level, and the degeneracy factor takes this multiplicity into account. For example, different combinations of vibrational, rotational and electronic energy levels may add up to the same total energy ε_i, and the total number of such combinations would be incorporated into the degeneracy factor g_i. Thus, the partition function is the sum over all microstates of the molecule, with a Boltzmann factor assigned to each state.

Consider now a hypothetical chain molecule consisting of four elements A, each of which can exist in one of two conformational states, A_+ or A_-. (In their classic treatment, B. H. Zimm and J. K. Bragg used 1 and 0 to designate the two distinct states; see Zimm and Bragg, 1959.) The possible conformational states of the chain include $A_+A_-A_-A_-$, $A_-A_+A_+A_-$, $A_-A_-A_-A_+$, and so on. To avoid the need for symmetry numbers, we assume that the chain has directional character, so that configuration $A_+A_-A_-A_-$ is distinct from $A_-A_-A_-A_+$. Each element in state A_- or A_+ has its own set of energy levels and accompanying degeneracies, but we assume that the details of the vibrational, electronic, and other states of the molecule are unknown. Nevertheless, we can express the partition function of Equation 20-13 in a useful form even without knowing these details.

We may divide up the sum in Equation 20-13 as follows. We assume that all terms in the sum involving the $A_-A_-A_-A_-$ state collectively contribute q_0 to the total partition function q. Likewise, we assume that all terms involving the $A_+A_-A_-A_-$ state collectively contribute q_1 to q. Furthermore, for the sake of our simple illustration, we assume that all conformations involving i A_+ elements contribute the same quantity q_i to the total partition function. Therefore, because there are four one-A_+ states for the chain, the total contribution of these states is $4q_1$. There are six distinct ways of putting two A_+ elements in the chain of four elements. Each distinct arrangement contributes q_2 to the partition function, so the six two-A_+ arrangements contribute $6q_2$. Similarly, the four three-A_+ states contribute $4q_3$, and the single four-A_+ state contributes q_4. Therefore, the partition function is

$$q = q_0 + 4q_1 + 6q_2 + 4q_3 + q_4 \qquad (20\text{-}14)$$

$$= q_0(1 + 4k_1 + 6k_2 + 4k_3 + k_4) \qquad (20\text{-}15)$$

where $k_i = q_i/q_0$. The k_i parameters are microscopic equilibrium constants for the change of a molecule from the all-A_- state to any particular state. For example,

$$k_1 = (A_+A_-A_-A_-)/(A_-A_-A_-A_-) = (A_-A_+A_-A_-)/(A_-A_-A_-A_-) = \cdots \qquad (20\text{-}16)$$

$$k_2 = (A_+A_+A_-A_-)/(A_-A_-A_-A_-) = (A_+A_-A_+A_-)/(A_-A_-A_-A_-) = \cdots \qquad (20\text{-}17)$$

These relationships follow directly from the fact that the fraction of molecules in a particular arrangement of i A_+ units is q_i/q, and of those in the $A_-A_-A_-A_-$ state is q_0/q; the ratio $(q_i/q)/(q_0/q) = q_i/q_0$ thus gives the equilibrium constant k_i.

We have thus succeeded in writing the partition function as a sum of equilibrium constants (Eqn. 20-15). The parameter q_0 may be ignored for most practical applications because it drops out of the calculations. For example, a quantity of typical interest is the fraction of molecules with a particular number of A_+ elements. Consider, for example, the calculation of the fraction θ_{2+} of molecules containing two A_+ elements:

$$\theta_{2+} = 6q_2/q \tag{20-18}$$

$$= q_0(6k_2)/q_0(1 + 4k_1 + 6k_2 + 4k_3 + k_4) \tag{20-19}$$

$$= 6k_2/(1 + 4k_1 + 6k_2 + 4k_3 + k_4) \tag{20-20}$$

Thus, the parameter q_0 cancels out of the calculation. If we regard the all-A_- state as the reference state, then q_0 is that part of the partition function associated with the reference state. It is customary to assign unity as the contribution of the reference state to the partition function; that is, $q_0 = 1$. This assumption in no way affects the values of the k_i parameters, because they all are scaled relative to the reference state. Thus, $\Delta G_i = -RT \ln k_i$ is the free-energy change for converting a molecule from the $A_-A_-A_-A_-$ reference state to a particular state containing i A_+ units.

Statistical weights

As a further simplification of the partition function, assume that the free-energy change ΔG can be assigned for the conversion of an A_+ to an A_- element, regardless of the element's location in the sequence. Then

$$k_1 = \exp(-\Delta G/RT) = s \tag{20-21}$$

$$k_2 = \exp(-2\Delta G/RT) = s^2 \tag{20-22}$$

and so on, where s is called the *statistical weight* of a unit in the A_+ state. With $q_0 = 1$, Equation 20-15 for the partition function now becomes

$$q = 1 + 4s + 6s^2 + 4s^3 + s^4$$

$$= \sum_{i=0}^{4} \Omega_i s^i$$

$$= (1 + s)^4 \tag{20-23}$$

where Ω_i is the number of ways of putting i A_+ elements in a chain of four units. It is easy to see that $\Omega_i = 4!/(4 - i)!i!$; this follows directly[§] from the binomial expansion of $(1 + s)^4$. Thus, the partition function is now expressed as a simple sum of terms involving powers of a single statistical-weight parameter s, which is assigned to every A_+ element, with a factor of unity assigned to each unit in the A_- state.

Rules for constructing the partition function

Although the preceding illustration is extremely simple, this general picture of the construction of a partition function as a sum of equilibrium constants or statistical weights is valid for the treatment of many chain-conformation problems. More complex problems may involve a greater number of statistical-weight parameters. These parameters in turn (as we shall see) can be obtained from experimental data. The following is the general procedure for constructing a partition function by these means.

1. Consider all possible conformations of the chain.

2. Establish a reference state for a chain element, and assign this state a statistical weight of unity.

3. For each conformation of the chain, assign an appropriate statistical weight for each element that is not in the reference state.

4. Construct the partition function by summing over all conformations, with appropriate products of statistical-weighting factors assigned to each conformation.

20-6   THE ZIPPER MODEL FOR THE HELIX–COIL TRANSITION

The zipper model assumes that each residue exists in one of two possible states—helix (h) or coil (c)—and that all of the helical units occur contiguously in a single region. Initiation of a helical sequence can occur at any point, and the hydrogen-bonded network grows from there. Thus, the conformations $\cdots$ ccchhhhhhccc $\cdots$ and $\cdots$ hhhhhcccccccc $\cdots$ are allowed, but $\cdots$ hhhhhhhccccchhhhhhhh $\cdots$ and $\cdots$ ccchhhhcccchhhhhccchhhhhhhccc $\cdots$ are forbidden conformations in this model.

In an actual α helix of m residues, there are a maximum of $m - 2$ residues that adopt the unique ϕ, ψ coordinates of the helix (because the two end residues are unconstrained), and $m - 4$ hydrogen bonds that can form (see Section 20-4). For the sake of our treatment, however, we simply regard the chain as composed of n units, each of which can exist in either the helical or the coil state. This simplification does not materially affect the results obtained for the cases of interest here.

[§] See also Equation 15-14 and the accompanying discussion.

Assignment of statistical weights

The partition function is constructed simply by applying the rules given in Section 20-5. We follow here an adaptation of the procedure set forth by J. A. Schellman (1958). We designate the coil (c) configuration of a unit as the reference state and therefore assign it a statistical weight of unity. Let s be the equilibrium constant (statistical weight) for making a new helical unit at the *end* of a helical sequence:

$$s = \frac{(\cdots \text{cccchhhhhhh}\underline{h}\text{cccc} \cdots)}{(\cdots \text{cccchhhhhh}\underline{c}\text{cccc} \cdots)} \qquad (20\text{-}24)$$

where the underscores mark the unit in question. This is the *propagation* step. The initiation of a helical section within an all-coil chain is much more difficult than propagation, for reasons discussed earlier. Let σs be the equilibrium constant for the initiation or *nucleation* step, where the nucleation constant $\sigma \ll 1$:

$$\sigma s = \frac{(\cdots \text{cccc}\underline{h}\text{cccc} \cdots)}{(\cdots \text{cccc}\underline{c}\text{cccc} \cdots)} \qquad (20\text{-}25)$$

In actuality, three residues must be ordered in making the first hydrogen bond (Section 20-3), but we ignore this fact in the present treatment. The important feature is that the nucleation step is much less favorable than the propagation step.

Construction and evaluation of the partition function

The partition function now is easily constructed by considering all possible chain configurations and assigning to each the appropriate statistical weight. For example, the configuration $\cdots$ ccchhhhhccc $\cdots$ contributes σs^5 to the partition function. The configuration hhhccccc $\cdots$ contributes σs^3. For a chain of n units,

$$q = 1 + \sum_{k=1}^{n} \Omega_k \sigma s^k \qquad (20\text{-}26)$$

where Ω_k is the number of ways of putting k helical units together in one sequence in a chain of n residues. The parameter Ω_k is found to be

$$\Omega_k = (n - k + 1) \qquad (20\text{-}27)$$

(this is easy to show by induction on k). By substituting Equation 20-27 into Equation 20-26, we can evaluate the partition function directly. The terms that must be summed are of the forms $\sum_k s^k$ and $\sum_k k s^k$, where the sums are taken from $k = 1$ to $k = n$. The first summation is a geometric series that yields

$$\sum_{k=1}^{n} s^k = (s^{n+1} - s)/(s - 1) \tag{20-28}$$

Differentiation of Equation 20-28 yields the other required series:

$$s\frac{d}{ds}\sum_{k=1}^{n} s^k = \sum_{k=1}^{n} ks^k \tag{20-29}$$

$$= [s/(s - 1)^2][ns^{n+1} - (n + 1)s^n + 1] \tag{20-30}$$

Substitution of these results (Eqns. 20-28 and 20-30) into Equation 20-26 yields

$$q = 1 + [\sigma s^2/(s - 1)^2][s^n + ns^{-1} - (n + 1)] \tag{20-31}$$

Fractional helicity and probabilities

The fractional helicity θ now is easily obtained from the partition function. The probability $p(k)$ that the chain has k helical units is

$$p(k) = (n - k + 1)\sigma s^k/q \tag{20-32}$$

The numerator in Equation 20-32 simply corresponds to the term in the partition function associated with all species having k hydrogen bonds. Using this equation, we easily derive the expression for θ:

$$\theta = \langle k \rangle/n = \left(\sum_{k=1}^{n} kp(k)\right)\Big/n \tag{20-33}$$

$$= \left(\sum_{k=1}^{n} k(n - k + 1)\sigma s^k\right)\Big/nq \tag{20-34}$$

$$= s(\partial q/\partial s)/nq \tag{20-35}$$

$$= (1/n)[\partial(\ln q)/\partial(\ln s)] \tag{20-36}$$

Equation 20-36 demonstrates the elegance of the partition-function method; it shows that an average conformational property, θ, can be obtained as a simple derivative of the partition function.

In general, an advantage of the partition-function approach is that certain equilibrium properties can be calculated as appropriate derivatives of the partition function (see Davidson, 1962). Execution of the differentiation yields

$$\theta = \frac{\sigma s}{(s-1)^3}\left(\frac{ns^{n+2} - (n+2)s^{n+1} + (n+2)s - n}{n\{1 + [\sigma s/(s-1)^2][s^{n+1} + n - (n+1)s]\}}\right) \quad (20\text{-}37)$$

It is possible to show that $\theta = 0$ when $s = 0$, and that $\theta \to 1$ as s becomes large. The transition from helix to coil occurs at intermediate values of s. For long chains, it can be shown that the midpoint of the transition ($\theta = 0.5$) occurs at $s = 1$, according to this model. Because s is an equilibrium constant, its value can easily be varied—for example, by temperature changes. The relationship between the equilibrium constant s and observed experimental parameters is discussed in Section 20-7.

Equation 20-32 may be used to calculate values of $p(k)$, the distribution of helical lengths, as a molecule progresses through the helix–coil transition. Figure 20-10 shows examples of such calculations for a chain of 16 units. Bar graphs of the

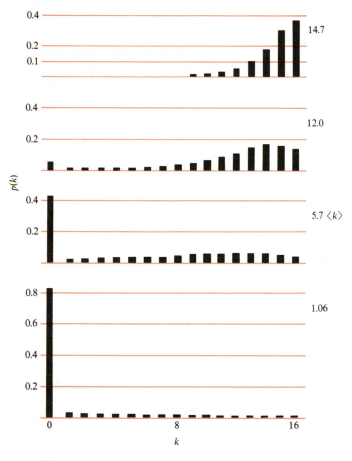

Figure 20-10

Distributions of helical lengths at various values of the average helical length.
[After J. A. Schellman, *J. Phys. Chem.* 62:1485 (1958).]

distribution of $p(k)$ values are shown for cases where $\langle k \rangle = 14.7, 12.0, 5.7$, and 1.1 (or $\theta = 0.92, 0.75, 0.36$, and 0.07, respectively). A value of $\sigma = 1.5 \times 10^{-3}$ was used in these calculations. It is clear that the tendency is to form fairly long helical sequences, or none at all. Thus, even when $\theta = 0.36$, the largest values of $p(k)$—other than the big population at $k = 0$—occur at values of k in the vicinity of 12. The form of these distributions is regulated by the parameter σ: smaller values of σ (implying a greater barrier to initiation) will tend to enlarge further the inhomogeneity of the distribution—that is, even longer helical sequences and the all-coil state will tend to be favored over the shorter helical sequences. This is because, the smaller the value of σ, the longer the helical sequence required to overcome the unfavorable aspects of initiation. Thus (for $s > 1$), the smaller the value of σ, the greater the value of k required in the product σs^k to obtain a significant value of $p(k)$ (see Eqn. 20-32). In this way, the parameter σ regulates the cooperativity of the transition (this topic is discussed further in subsequent sections).

This simple statistical thermodynamic treatment of the helix–coil transition offers a fair amount of useful insight into the underlying features of the transition. It is a reasonably accurate description for short chains, where the tendency is to form only one helical sequence in a partially helical molecule. Longer chains can form two or more helical sequences separated by coil sections, so that the zipper model is then invalid. A more general analysis then is needed for longer chains; such an analysis is conveniently accomplished by the matrix methods we outline next.

20-7 MATRIX METHOD FOR THE PARTITION FUNCTION

We now consider the calculation of the partition function for the general case where helical and coil units are free to occur anywhere in the chain. We no longer restrict the helical elements to a single sequence as we did in the zipper model. Again, we picture a chain composed of n units, each of which can take on the h or c configuration, where statistical weights of 1, σs, and s are used as before (cf. Eqn. 20-26). If h and c units are allowed to occur in any combination whatsoever, then the partition function q is

$$q = \sum_{j,k} \Omega_{j,k} \sigma^j s^k \qquad (20\text{-}38)$$

where $\Omega_{j,k}$ is the number of ways of putting k helical units into j distinct helical sequences, where each helical sequence is separated by one or more coil units. Thus, if $j = 3$ and $k = 50$, the factor s^{k-3} is assigned to the 47 h units that follow another h unit, and $\sigma^3 s^3$ is assigned to the 3 h units located at the very beginnings of the three helical sequences. The contribution of one of these ($j = 3$; $k = 50$) states thus is $\sigma^3 s^{50}$. However, each of the three helical sequences can take on any length between 1 and 48 h units, subject to the restriction that the total number of h units equals 50. The parameter $\Omega_{3,50}$ thus gives the number of ways of putting 50 helical residues into three distinct sequences.

Statistical-weight matrix

This type of problem has a long history in physical chemistry. It can be identified with the one-dimensional Ising lattice, whose partition function is particularly amenable to calculation by matrix methods, where each residue of the chain is given a matrix representation. (See Appendix A for a brief discussion of matrix algebra.) In essence, the four possible configurations of a given residue relative to its predecessor are assigned statistical weights, which are tabulated in a table called a statistical-weight matrix. These four states are c following c, c following h, h following c, and h following h; the corresponding statistical weights are 1, 1, σs and s, respectively (see Eqns. 20-24 and 20-25 and the discussion preceding those equations). The statistical-weight matrix $\underset{\sim}{\mathbf{M}}$ is

$$\underset{\sim}{\mathbf{M}} = \begin{matrix} & \begin{matrix} c & \quad h \end{matrix} \\ \begin{matrix} c \\ h \end{matrix} & \begin{pmatrix} 1 & \sigma s \\ 1 & s \end{pmatrix} \end{matrix} \tag{20-39}$$

This matrix may be interpreted as follows. A particular element in the matrix is given the designation m_{ij}, where i denotes the row and j the column of the element. The columns also index the state (c or h) of the residue of interest, whereas the rows index the state of the predecessor. Thus, the element m_{12} is equal to σs, because it is the statistical weight for a helical unit following a coil unit.

Calculation of the partition function
from the statistical-weight matrix

The partition function for a chain of n units may be generated by raising $\underset{\sim}{\mathbf{M}}$ to the nth power. This process generates all possible chain configurations with the appropriate statistical weights assigned to each. The validity of this procedure is completely a consequence of the rules for matrix multiplication, which combine the elements of the $\underset{\sim}{\mathbf{M}}$ matrices in exactly the required way.

Consider, for example, a two-element chain. The four possible configurations are cc, ch, hc, and hh. We assume that the "predecessor" of the first unit is analogous to a coil state. Therefore, according to the rules for constructing a partition function (Section 20-5), the four statistical weights are 1×1, $1 \times \sigma s$, $\sigma s \times 1$, and $\sigma s \times s$, respectively, and the partition function is $q = 1 + 2\sigma s + \sigma s^2$. This partition function also may be calculated by the matrix product:

$$q = (1, 0)\underset{\sim}{\mathbf{M}}\underset{\sim}{\mathbf{M}}\begin{pmatrix} 1 \\ 1 \end{pmatrix} \tag{20-40}$$

$$= (1, \sigma s)\underset{\sim}{\mathbf{M}}\begin{pmatrix} 1 \\ 1 \end{pmatrix} \tag{20-41}$$

$$= (1 + \sigma s, \ \sigma s + \sigma s^2)\binom{1}{1} \tag{20-42}$$

$$= 1 + 2\sigma s + \sigma s^2 \tag{20-43}$$

In Equation 20-40, the row vector $(1, 0)$ premultiplies the first $\underset{\sim}{M}$ matrix to give $(1, \sigma s)$ in Equation 20-41. This row vector $(1, \sigma s)$ contains the two possible statistical weights for the first element in the chain. Because this element can only be c "following c" or h "following c," the statistical weights are 1 and σs. This is why the row vector that premultiplies $\underset{\sim}{M}\underset{\sim}{M}$ (Eqn. 20-40) is $(1, 0)$ rather than $(1, 1)$. The row vector obtained in Equation 20-42 contains the statistical weights for all four possible configurations, and the column vector merely extracts the statistical weights and constructs the proper sum. Thus, the partition function for this simple case is easily calculated by squaring the matrix $\underset{\sim}{M}$ and then using the appropriate row and column vectors to obtain the scalar sum.

By generalization, we can easily see that, for a chain of n units,[§] the partition function is

$$q = (1, 0)\underset{\sim}{M}^n\binom{1}{1} \tag{20-44}$$

For large n, the process of matrix multiplication becomes very laborious, so that numerical calculations are more conveniently done on a computer. However, there is an easier way to solve the problem. The matrix $\underset{\sim}{M}$ may be diagonalized by a similarity transformation. Because the nth power of a diagonal matrix is simply the diagonal elements raised to the nth power, a great simplification is achieved in calculating q.

Diagonalization of the statistical-weight matrix

Let $\underset{\sim}{T}$ be the transformation matrix that diagonalizes $\underset{\sim}{M}$ by a similarity transformation:

$$\underset{\sim}{T}^{-1}\underset{\sim}{M}\underset{\sim}{T} = \underset{\sim}{\Lambda} = \begin{pmatrix} \lambda_1 & 0 \\ 0 & \lambda_2 \end{pmatrix} \tag{20-45}$$

where $\underset{\sim}{T}^{-1}$ is the inverse of $\underset{\sim}{T}$ (so that $\underset{\sim}{T}\underset{\sim}{T}^{-1} = \underset{\sim}{I}$, the unit matrix), and λ_1 and λ_2 are the eigenvalues of $\underset{\sim}{M}$. The matrix $\underset{\sim}{T}$ is constructed from the eigenvectors of $\underset{\sim}{M}$ (see Appendix A). Returning to Equation 20-44, we premultiply and postmultiply by $\underset{\sim}{T}\underset{\sim}{T}^{-1}$ every matrix factor $\underset{\sim}{M}$ in the product $\underset{\sim}{M}^n$. The partition function is not altered

---

[§] Recall that, for our simplified treatment, we allow each of the n units to adopt either the h or the c state. This accounts for the factor $\mathbf{M}^n$ in Equation 20-44. N. Davidson (1962) uses a factor of $\mathbf{M}^{n-1}$ because he considers the first element to be always in the c state. This difference is immaterial for our purposes here.

by this procedure, because $\underset{\sim}{\mathbf{T}}\underset{\sim}{\mathbf{T}}^{-1} = \mathbf{I}$, and multiplication of any matrix by the unit matrix leaves the original matrix unchanged. Thus,

$$q = (1,0)(\underset{\sim}{\mathbf{T}}\underset{\sim}{\mathbf{T}}^{-1}\underset{\sim}{\mathbf{M}}\underset{\sim}{\mathbf{T}}\underset{\sim}{\mathbf{T}}^{-1})^{n}\begin{pmatrix}1\\1\end{pmatrix} \tag{20-46}$$

$$= (1,0)\underset{\sim}{\mathbf{T}}\underset{\sim}{\Lambda}^{n}\underset{\sim}{\mathbf{T}}^{-1}\begin{pmatrix}1\\1\end{pmatrix} \tag{20-47}$$

The matrix $\underset{\sim}{\Lambda}^{n}$ is simply

$$\underset{\sim}{\Lambda}^{n} = \begin{pmatrix}\lambda_1^{n} & 0\\0 & \lambda_2^{n}\end{pmatrix} \tag{20-48}$$

Hence, the matrix-diagonalization procedure greatly facilitates the calculation of the partition function. The only operations that must be carried out are those of computing the eigenvalues of $\underset{\sim}{\mathbf{M}}$ and constructing the matrices $\underset{\sim}{\mathbf{T}}$ and $\underset{\sim}{\mathbf{T}}^{-1}$ from the associated eigenvectors.

The eigenvalues of $\underset{\sim}{\mathbf{M}}$ are obtained from the determinant

$$0 = \begin{vmatrix}1 - \lambda & \sigma s\\1 & s - \lambda\end{vmatrix} \tag{20-49}$$

$$= (1 - \lambda)(s - \lambda) - \sigma s \tag{20-50}$$

$$= \lambda^2 - (1 + s)\lambda + s(1 - \sigma) \tag{20-51}$$

The two roots of Equation 20-51 are

$$\lambda_{1,2} = \{(1 + s) \pm [(1 - s)^2 + 4\sigma s]^{1/2}\}/2 \tag{20-52}$$

where λ_1 is given by the positive sign and λ_2 by the negative sign. The transformation matrix $\underset{\sim}{\mathbf{T}}$ is

$$\underset{\sim}{\mathbf{T}} = \begin{pmatrix}1 - \lambda_2 & 1 - \lambda_1\\1 & 1\end{pmatrix} \tag{20-53}$$

and its inverse is

$$\underset{\sim}{\mathbf{T}}^{-1} = \frac{1}{\lambda_1 - \lambda_2}\begin{pmatrix}1 & \lambda_1 - 1\\-1 & 1 - \lambda_2\end{pmatrix} \tag{20-54}$$

(Calculation of $\underset{\sim}{\mathbf{T}}$ and $\underset{\sim}{\mathbf{T}}^{-1}$ is demonstrated in Appendix A.) Substituting these $\underset{\sim}{\mathbf{T}}$ and

$\mathbf{\underset{\sim}{T}}^{-1}$ matrices into Equation 20-47, we obtain

$$q = [\lambda_1^{n+1}(1 - \lambda_2) - \lambda_2^{n+1}(1 - \lambda_1)]/(\lambda_1 - \lambda_2) \tag{20-55}$$

Because $\lambda_1 > \lambda_2$, we have $\lambda_1^{n+1} \gg \lambda_2^{n+1}$ for large n. Under these conditions,

$$q \cong \lambda_1^{n+1}(1 - \lambda_2)/(\lambda_1 - \lambda_2) \qquad \text{for large } n \tag{20-56}$$

and

$$\ln q \cong n \ln \lambda_1 \qquad \text{for large } n \tag{20-57}$$

where the error in the approximation of Equation 20-57 becomes vanishingly small as n increases.

Calculating parameters from the partition function

The average number $\langle k \rangle$ of helical units and the fractional helicity θ are easily calculated by differentiating $\ln q$ in Equation 20-57 (cf. Eqn. 20-36);

$$\langle k \rangle = \partial(\ln q)/\partial(\ln s) = n\,\partial(\ln \lambda_1)/\partial(\ln s) \tag{20-58}$$

$$\theta = \langle k \rangle/n$$

$$= (s/\lambda_1)(1/2)\{1 + [(s - 1) + 2\sigma]/[(1 - s)^2 + 4\sigma s]^{1/2}\} \tag{20-59}$$

where Equation 20-52 for λ_1 has been substituted into Equation 20-58 in order to obtain Equation 20-59. The characteristics of the transition can easily be deduced from Equation 20-59 and the expression for λ_1 (Eqn. 20-52). We assume $\sigma \ll 1$. When s is near (but slightly less than) unity, such that $(1 - s)^2 > 4\sigma s$, then the value of the expression in braces in Equation 20-59 approaches zero, and $\theta \cong 0$. [Note that, if s is near unity, then $4\sigma s \cong 4\sigma$, and the condition $(1 - s)^2 > 4\sigma$ means that $(1 - s) > 2\sigma^{1/2} \gg 2\sigma$ (for $\sigma \ll 1$). Therefore, $|s - 1| \gg 2\sigma$; the quotient within the braces in Eqn. 20-59 is approximately equal to -1; and the entire expression within the braces is, approximately, 0.] These conditions require that $s < (1 - 2\sigma^{1/2})$. When s is near (but greater than) unity, and $(1 - s)^2 > 4\sigma s$, then $\lambda_1 \cong 1$ (from Eqn. 20-52) and $\theta \cong 1$ (from Eqn. 20-59). [Note that, with s greater than unity, the requirement that $(1 - s)^2 > 4\sigma$ means that $(s - 1) > 2\sigma^{1/2} \gg 2\sigma$; therefore, with $(s - 1) \gg 2\sigma$, the quotient within the braces of Eqn. 20-59 becomes, approximately, equal to 1, and the entire expression in braces is, approximately, 2.] These conditions require that $s > (1 + 2\sigma^{1/2})$. Therefore, the transition occurs over the narrow range from approximately $s = 1 - 2\sigma^{1/2}$ to $s = 1 + 2\sigma^{1/2}$; this is the transition zone. Its width is determined by $\sigma^{1/2}$. Figure 20-11 clearly illustrates the dependence of the sharpness on σ; the figure shows plots of θ versus $\ln s$ for $\sigma = 10^{-2}, 10^{-3}$, and 10^{-4}.

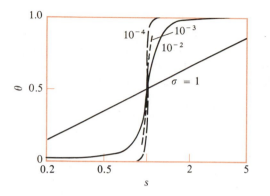

Figure 20-11

Sharpness of transition. The calculated fractional helicity θ is plotted against s (on a natural logarithm scale) for various values of σ. [After B. H. Zimm and J. K. Bragg, *J. Chem. Phys.* 31:526 (1959).]

It is also of interest to note that, when $\sigma = 1$, Equation 20-59 simplifies to

$$\theta = s/(1 + s) \qquad \text{for } \sigma = 1$$

This is the noncooperative case, in which each unit behaves independently. Figure 20-11 illustrates this case also. When $\sigma > 1$, the chain behaves in an anticooperative fashion. In this case, the plot of θ versus $\ln s$ becomes progressively less steep as $\ln s$ increases. Although helix-coil transitions do not show anticooperative behavior, anticooperativity might be observed in other kinds of situations (see Chapter 17).

Another parameter of interest is the average number $\langle j \rangle$ of helical regions. From our general expression for the partition function (Eqn. 20-38), it is straightforward to show that the average value of j may be calculated as a simple derivative. We start with the definition of $\langle j \rangle$:

$$\langle j \rangle = \left(\sum_{k=1}^{n} \sum_{j=1}^{k} j\Omega_{j,k}\sigma^j s^k \right) \bigg/ q \tag{20-60}$$

$$= (\sigma/q)(\partial q/\partial \sigma) \tag{20-61}$$

$$= \partial(\ln q)/\partial(\ln \sigma) \tag{20-62}$$

$$= n\,\partial(\ln \lambda_1)/\partial(\ln \sigma) \tag{20-63}$$

where we have used Equation 20-57 for $\ln q$. Substituting Equation 20-52 into Equation 20-63 and performing the differentiation, we obtain

$$\langle j \rangle = n\sigma s/\lambda_1[(1 - s)^2 + 4\sigma s]^{1/2} \tag{20-64}$$

The value of $\langle j \rangle$ has its maximum at the midpoint of the transition ($s = 1$), where

$$\langle j \rangle = (n/2)\sigma^{1/2} \tag{20-65}$$

For a typical realistic case, $\sigma = 10^{-4}$, so that at the midpoint $\langle j \rangle = n/200$, and a separate helical region occurs every 200 residues, on the average. Because $\langle k \rangle = n/2$ at the midpoint, the ratio $\langle k \rangle / \langle j \rangle = \sigma^{-1/2}$ (at the midpoint), which is a kind of average length of a helical sequence. For $\sigma = 10^{-4}$, the average length of a helical sequence is 100 residues. Thus, at the midpoint of the transition with $\sigma = 10^{-4}$, a molecule has (on the average) helical sequences 100 units in length interspersed with coil sequences of the same length. Note that the average length $(\sigma^{-1/2})$ is independent of n for large n (the conditions under which all equations following Eqn. 20-55 have been derived). This independence simply means that helical regions do not grow indefinitely, but reach a finite limit. The reason for this limit is that in long chains the entropic advantages to having more than one helical sequence counterbalance the barrier to helix initiation. It has long been recognized that, in a one-dimensional system such as the linear polypeptide, a truly discontinuous phase transition does not occur.

20-8 A DESCRIPTION OF THE TRANSITION

An interesting description of the transition at various chain lengths and s values is provided by the n–s plane, which is sketched in Figure 20-12. This sketch is based on an analysis similar to that in Section 20-7, but somewhat more exact. The transition occurs largely in the shaded area, where θ values lie between 0.1 and 0.9. On either side of the shaded area are the states that are largely all-helix or all-coil. The line running diagonally through the helical region indicates conditions under which half the chains contain only one helical section. The most obvious point to note is that the transition tends to occur at higher s values as n becomes smaller. This is because, when n is small, the helical sequences (unlike those of long chains) are too short to counterbalance the adverse effects of helix initiation unless s is sufficiently large.

The diagram also shows clearly that the range of s over which the transition occurs is much broader at small n than at large n; that is, the transition is sharper in the longer chains. This is a reflection of the fact that the change in the overall equilibrium constant—for example, σs^n between all-helix and all-coil—becomes increasingly sensitive to small changes in s as n increases (compare our treatment of the simple two-state model in Section 20-4). Similarly, when n is large, there are many terms such as σs^k in the partition function where k also is large ($k < n$); these terms also are very sensitive to small changes in s near $s = 1$. The net result is that the partition function as a whole is more sensitive to s when n is large.

Consider now the course of the transition at small n, along the line A → B → C in Figure 20-12. In moving from A to B, random chains are converted to a mixture of random chains and chains with single helical regions. Near C, the chains are composed almost entirely of a single helical region, with occasional disorder at the ends. In passing through the helical region from C to D (with increasing n), the helices begin to incorporate coil sections at scattered intervals between the helical regions that pre-

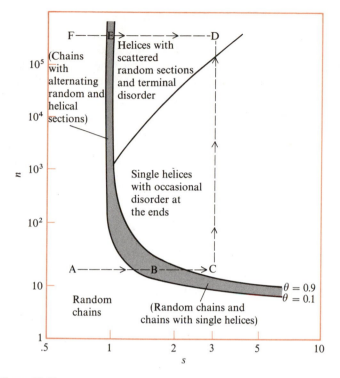

Figure 20-12

The n–s plane for σ = 10⁻⁴. The contours of $\theta = 0.9$ and $\theta = 0.1$ have been chosen arbitrarily as the boundaries of the transition. Along the solid diagonal line, half of the chains contain a single unbroken helical section. [After B. H. Zimm and J. K. Bragg, *J. Chem. Phys.* 31:526 (1959).]

dominate under these conditions. If we now pass from the helical to the coil state at large n (along D → E → F), we move through a transition region in which the chains have alternating random and helical sections. Thus, the microscopic species present in the transition vary considerably with n.

The treatment in Section 20-7 is inaccurate on one point. In actuality, a helix sequence starts by ordering three residues simultaneously, so that the first hydrogen bond can be formed. We ignored this fact, but it can be taken into account by increasing the dimensions of the matrix $\underset{\sim}{\mathbf{M}}$. However, the approximate model used here is quite accurate for large n. Furthermore, the accuracy improves around $\theta = 1/2$. Therefore, in many practical situations, there is no need to consider higher-order matrices.

20-9 COMPARISON WITH EXPERIMENTAL RESULTS

How well does the theory outlined here describe actual experimental data? To compare theory and experiment, we must relate the parameters σ and s to actual experimental variables. To do this, we assume that the equilibrium constant s can be expressed in terms of a van't Hoff relation:

$$d(\ln s)/dT = \Delta H_{\mathrm{g}}/RT^2 \qquad (20\text{-}66)$$

where ΔH_{g} is the enthalpy change for adding one more helical unit onto a preexisting helical section. Therefore, by integration of Equation 20-66 between the limits of T_m and T:

$$\ln s = (\Delta H_{\mathrm{g}}/R)[(T - T_{\mathrm{m}})/T T_{\mathrm{m}}] \qquad (20\text{-}67)$$

where T_{m} is the temperature at the midpoint of the transition. Thus, if ΔH_{g} is known, the connection between s and temperature is established. The parameter σ is taken to be constant. This is not an unreasonable approximation, because σ may be expected to reflect in large part the conformational factors associated with the difficulties of helix initiation. The value of σ may be adjusted to give the best fit between theory and experiment. In practice, it also is necessary to adjust ΔH_{g}, because generally it is not known for the case of interest.

Figure 20-13 compares experimental points and theoretical transition curves for poly-γ-benzyl-L-glutamate chains with average lengths of 1,500, 46, and 26 residues.

Figure 20-13

Theoretical curves and experimental results.
The curves show optical rotation calculated
from predicted fractional helicity as a function
of $T - T'_{\mathrm{m}}$ for poly-γ-benzyl-L-glutamate chains
of three different lengths. The points represent
experimental optical-rotation data. [After
B. H. Zimm et al., *Proc. Natl. Acad. Sci. USA*
45:1601 (1959).]

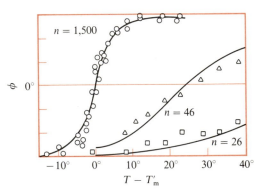

The curves correspond to calculated fractional helicities, whereas the points were obtained from optical-rotation measurements. The temperature axis corresponds to $T - T'_{\mathrm{m}}$, where T'_{m} is the midpoint temperature of the transition curve of the sample of highest molecular weight. Values of $\sigma = 2 \times 10^{-4}$ and $\Delta H_{\mathrm{g}} = 890$ cal mole^{-1} have been used in calculating the curves. (Note that in the solvent used—80% dichloroacetic acid, 20% ethylene dichloride—helix formation occurs as temperature is *increased*. This is rationalized as being due to the ability of dichloroacetic acid to form hydrogen bonds with the amide groups in the coil. Thus, helix formation is endo-thermic, because the hydrogen bonds with solvent molecules are broken, and helix formation has a positive entropy change arising from the liberation of previously bound solvent molecules.) The agreement between calculated and observed behavior is quite good. In particular, the predicted dependence on n of the midpoint tempera-ture T_m and of the sharpness is well confirmed by the experiments. Thus, the theory provides a very adequate description of the transition.

20-10 HELIX PARAMETERS FOR AMINO ACIDS

A question of considerable interest is whether or not helices can form more readily with certain amino acids than with others. The notion is that, if certain amino acids have a greater tendency to adopt the α-helical state, then those residues might be the ones that tend to occur in helical sections of globular proteins. Verification of this idea would represent a major step forward in understanding how a particular amino acid sequence gives rise to a specific native protein structure.

H. A. Scheraga and colleagues have done a considerable amount of work in this area. They suggest that short-range interactions are key factors in determining whether or not a residue has a helix-forming tendency. Short-range interactions are taken to mean those of a given residue with the chain backbone in the immediate vicinity of that residue. These interactions are different for the different amino acids, according to calculations. For example, a glycine residue in a polypeptide chain has considerable conformational freedom (see the ϕ, ψ map for glycine in Fig. 5-7). Because there are not sufficiently favorable energetic interactions in an α-helical conformation (for example, $\phi \cong -50°$ and $\psi \cong -50°$), entropic factors make a coil conformation more favorable for glycine. However, if a CH_3 side chain is attached to the α-carbon (to make alanine), the nonbonded interactions tend to enhance the preference for the α-helical state. On the other hand, with asparagine, the interaction of the polar side chain with the amide dipole tends to destabilize the α-helical confor-mation for this amino acid. In an attempt to assess the relative tendencies of the various amino acids to adopt a helical conformation in a polypeptide chain, calcula-tions have actually been carried out with all 20 naturally occurring amino acids.

The crucial question, of course, is whether these predictions can be tested experi-mentally. In this regard, it is not possible to examine helix–coil transitions for many polypeptide homopolymers in aqueous solution because they are insoluble. However, the helix-forming tendency of a residue can be studied by the "host–guest" method. This involves investigating the effect of introducing the amino acid in question (the guest) into a water-soluble host homopolymer that has a well-characterized helix–coil transition. The effect of the guest residues on the transition can be used to obtain a quantitative measure of the parameters σ and s associated with the guest amino acid (in accordance with appropriate theoretical analysis).

Figure 20-14

Host–guest experimental data. Fractional helicity is plotted against temperature for various random copolymers of polyhydroxypropylglutamine (host) and L-alanine (guest). The curves are calculated from theory; the points are experimental. The calculated curves are subject to some variation because of experimental uncertainties in determining the degree of polymerization and the chain composition. The curves represent the following chain lengths (as degree of polymerization) and compositions (as percentage of alanine): curve 1, 422, 14%; curve 2, 880, 21%; curve 3, 1,010, 32%; curve 4, 1,413, 38%; curve 5, 1,102, 49%. [After K. E. B. Platzer et al., *Macromolecules* 5:177 (1972).]

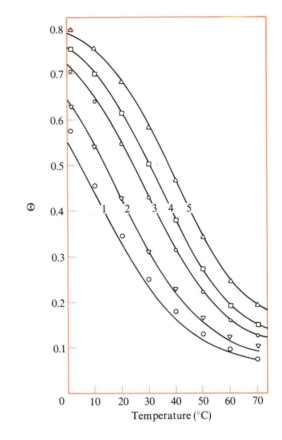

Figure 20-14 shows an example of this approach, where the fractional helicity is plotted against temperature for various random copolymers of polyhydroxypropylglutamine (the water-soluble host) and L-alanine (the guest). The prediction is that introduction of the L-alanyl residue into the host should lead to an elevation of T_m. The experimental data in Figure 20-14 bear out this prediction. The curves in the figure are based on theoretical calculations and are seen to agree quite well with the experimental points. Interestingly enough, the parameters σ and s that are obtained for the L-alanyl residue from these data agree well with those deduced for the same residue in sandwich-type block copolymers of L-alanine and D,L-lysine in aqueous solution.

Table 20-1 summarizes the experimental values of σ and s at 30°C obtained for glycyl, L-alanyl, L-seryl, and L-leucyl residues. These data were obtained by the host–guest technique. The data indicate that glycine has the least tendency toward initiation of helix formation. In agreement with predictions, L-leucine has the greatest helix-forming tendency. The relative preference for the α-helical state has also been evaluated or estimated for other residues, as mentioned earlier. These results then have been used to study protein structure. For example, examination of the sequences

Table 20-1

Host–guest experimental values of σ and s at 30°C

	σ	s
Glycine	1×10^{-5}	0.62
L-Alanine	8×10^{-4}	1.06
L-Serine	7.5×10^{-5}	0.79
L-Leucine	33×10^{-4}	1.14

SOURCE: Data from H. A. Scheraga, *Pure and Applied Chemistry* 36:1 (1973).

of a large number of cytochrome c proteins shows that amino acid substitutions tend to occur so that the helix-making or helix-breaking character of a given position in the sequence is preserved. Moreover, helical regions in the crystal structures of two cytochrome c proteins tend to involve those residues with the greatest probability of helix formation, as adduced from host–guest studies or other estimations. If these observations prove to be generally valid, it may indeed be possible to predict a fair amount about the structure of a protein, based on knowledge of its sequence and short-range interactions. (See Chapter 5 for a discussion of approaches to predicting protein structure.)

Summary

Conformational transitions commonly are observed in proteins and polypeptides; these transitions in many cases have important biological consequences. The α-helix–coil transition of polypeptides is the best-characterized conformational change, and analysis of it provides a useful reference frame for understanding conformational equilibria in other systems. The helix–coil change can be induced by alterations in temperature or solvent composition; for high-molecular-weight polypeptides, it occurs over a narrow range of variation of those parameters. Thus, the transition is cooperative. The cooperativity can be qualitatively rationalized by steric, energetic, and statistical constraints in the polypeptide α helix and coil that encourage growth (or propagation) of helical units next to preexisting ones, in preference to nucleation of helical sites at separated, scattered sites throughout the structure.

Several theoretical treatments can be used to put the thermodynamic features of the transition on a quantitative foundation. The zipper model assumes that, in partially helical chains, all helical units are contiguous (like the interlocked links in a partially closed zipper). The zipper model is particularly useful and easy to analyze in mathematical terms; its predictions are fairly accurate for short chains. A more accurate treatment, which accounts for all possible statistical arrays of helical and coil units within a chain, is accomplished by matrix methods. This treatment offers a concise formulation of the problem and clear insight into its most essential features. Predictions made from the matrix analysis are in good accord with experimental results.

In the matrix treatment, the nucleation constant σ and the propagation parameter s determine the characteristics of the transition. In particular, σ regulates the sharpness of the transition and the average length of helical sections in partially helical structures, whereas s plays a role analogous to that of temperature. On the hypothesis that the tendency to form helices may vary from one amino acid to another, these parameters have been estimated for various amino acids situated within polypeptides. A variation is found in helix-forming tendency, and there is some suggestion that residues with the greatest helix-making ability generate helical sections in proteins. These correlations, if they can be firmly established, will further encourage attempts to predict helical regions in a protein structure from knowledge of amino acid sequence.

Problems

20-1. Using the zipper model, calculate the macroscopic equilibrium constants for the following reactions for a chain of 100 units (each of which can be helix or coil):

$$50\% \text{ helix} \rightleftarrows 20\% \text{ helix} \qquad K_1 = (20\% \text{ helix})/(50\% \text{ helix})$$

$$75\% \text{ helix} \rightleftarrows 25\% \text{ helix} \qquad K_2 = (25\% \text{ helix})/(75\% \text{ helix})$$

In your calculation, assume $s = 1$ and $\sigma = 10^{-4}$.

20-2. Assume that a chain molecule can exist in three conformations: helix 1, helix 2, and coil. The individual transitions helix $1 \rightleftarrows$ coil and helix $2 \rightleftarrows$ coil are highly cooperative and can be described in terms of the zipper model. Let σ and s be statistical weights applicable to the helix $1 \rightleftarrows$ coil transition, and σ' and s' those applicable to the helix $2 \rightleftarrows$ coil transition. Calculate the value of σ', given that the difference in free energy between helix 1 and helix 2 is zero, that $s'/s = 10^{-0.02}$, and that $\sigma = 10^{-4}$. Assume that the chain has 100 units, each of which is capable of being in a helical or coil state.

20-3. Given $\underline{\mathbf{T}}$ and $\underline{\mathbf{T}}^{-1}$ by Equations 20-53 and 20-54, use Equation 20-47 to verify the expression for q given in Equation 20-55.

20-4. By executing the appropriate differentiations, verify Equations 20-59 and 20-64.

20-5. Assume you have a linear polypeptide copolymer made up of two types of residues: A and B. It is a strictly alternating copolymer of structure ABABAB $\cdots$ AB. You wish to write the partition function for the chain using the following rules. For A residues: (a) a coil unit has a statistical weight of 1; (b) a helical unit preceded by a nonhelical unit has a statistical weight of $\sigma_A s_A$; (c) a helical unit preceded by a helical unit has a statistical weight of s_A. For B residues, the same rules apply, but with σ_B replacing σ_A, and s_B replacing s_A. The chain is composed of $n/2$ A residues and $n/2$ B residues for a total of n residues. Can you write the partition function q as

$$q = (1,0)\underset{\sim}{\mathbf{Q}}{}^x \begin{pmatrix} 1 \\ 1 \end{pmatrix}$$

where $\mathbf{Q}$ is a 2×2 matrix, and x is an integer? If so, give all four elements of $\mathbf{Q}$, and also give x. If not, explain carefully why the partition function cannot be written in this form.

References

GENERAL

Bovey, F. A. 1969. *Polymer Conformation and Configuration.* New York: Academic Press. [Chap. 4 deals with NMR and optical studies of the α helix and the helix–coil transition.]

Davidson, N. 1962. *Statistical Mechanics.* New York: McGraw-Hill. [A good introductory text; Chap. 16 covers order–disorder transitions and the matrix method.]

Flory, P. J., and W. G. Miller. 1966. A general treatment of helix–coil equilibria in macromolecular systems. *J. Mol. Biol.* 15:284.

Poland, D., and H. A. Scheraga. 1970. *Theory of Helix–Coil Transitions in Biopolymers.* New York: Academic Press. [A broad discussion of the topic; a good reference work.]

Schellman, J. A. 1958. The factors affecting the stability of hydrogen-bonded polypeptide structures in solution. *J. Phys. Chem.* 62:1485. [A treatment of the zipper model.]

Volkenstein, M. V. 1977. *Molecular Biophysics* (English translation). New York: Academic Press. [Chapter 4 treats the helix–coil transition.]

Zimm, B. H., and J. K. Bragg. 1959. Theory of the phase transition between helix and random coil in polypeptide chains. *J. Chem. Phys.* 31:526. [A classic paper on the matrix method.]

SPECIFIC

Brant, D. A. 1968. Conformational energy estimates for helical polypeptide molecules. *Macromolecules* 1:291.

Finkelstein, A. V., O. B. Ptitsyn, and S. A. Kozitsyn. 1977. Theory of protein molecule self-organization. II. A comparison of calculated thermodynamic parameters of local secondary structures with experiments. *Biopolymers* 16:497. [Evaluates constants for the helix–coil transition for the natural amino acids.]

Gibbs, J. H., and E. A. DiMarzio. 1959. Statistical mechanics of helix–coil transitions in biological macromolecules. *J. Chem. Phys.* 30:271.

Hill, T. L. 1959. Generalization of the one-dimensional Ising model applicable to helix transitions in nucleic acids and proteins. *J. Chem. Phys.* 30:383.

Lifson, S., and A. Roig. 1961. On the theory of the helix–coil transition in polypeptides. *J. Chem. Phys.* 34:1963.

Peller, L. 1959. On a model for the helix–coil transition in polypeptides. *J. Phys. Chem.* 63:1194.

Schellman, J. A. 1955. The stability of hydrogen bonded peptide structures in aqueous solution. *Compt. Rend. Trav. Lab. Carlsberg. Ser. Chim.* 29:230.

Scheraga, H. A. 1978. Use of random copolymers to determine the helix–coil stability constants of the naturally occurring amino acids. *Pure and Applied Chemistry* 50:315.

Conformational equilibria of polypeptides and proteins: reversible folding of proteins

21-1 PROTEIN CONFORMATIONAL TRANSITIONS

The cooperative order–disorder transition, discussed for single-stranded polypeptides in Chapter 20, is also characteristic of proteins, which can convert reversibly from native to denatured states. In the case of most proteins that have been studied, however, the transition between ordered and disordered forms is not primarily a helix–coil transition, because many native protein molecules are not helical rods but are closely packed globular structures (though they may contain some helical sections). Thus, although cooperative order–disorder transitions are features common to polypeptides and proteins, it must be kept firmly in mind that the molecular picture is quite different in the two cases and, therefore, that one expects different mechanisms to account for the cooperativity in each instance.

A major goal of studying the folding and unfolding of proteins is to understand in detail the principles that dictate formation of the native structure. Therefore, many of the issues discussed in Chapter 5 (concerning the forces that stabilize native structures) are closely related to the present discussion. However, in this chapter, we focus attention chiefly on the physical chemical characterization of the folding–unfolding reaction. Another of the major goals is to determine the *mechanism* by which folding occurs. This is particularly interesting because, as discussed in Chapter 5, a remarkable feature of proteins is their tendency to adopt a single specific structure—the native form—even though an astronomically large number of conformations are possible in principle for the average protein. By taking a close look at the thermodynamic and kinetic features of the folding and unfolding reactions, we can hope to determine whether partially folded intermediates occur and if so whether their structural features can be characterized.

One possibility, for example, is that α helices or β structures form easily in certain regions of an unfolded chain, and that these structures are nucleation centers upon which the rest of the structure can build. One also can imagine that small amounts of other kinds of order form with relative ease in certain segments of a chain (such as a specific clustering of hydrophobic side chains), and that these also are nucleation centers. If such nucleated structures do form in proteins, and if it can be demonstrated that these are associated with particular kinds of sequences, then the problem of predicting three-dimensional structure from sequence might be significantly simplified. For example, one could identify in a given sequence those areas that have a high probability of forming specific local structures; from this point, an attempt could be made to build three-dimensional conformations that use the previously identified local structures as a structural foundation.

Many of these considerations boil down to whether or not intermediates can be detected in the pathway between folded and unfolded forms. In this chapter, attention is first directed to the general features and thermodynamic aspects of the folding–unfolding reaction. A rather close analysis then is made of equilibrium measurements, with an eye toward detecting the existence (or lack) of intermediates in the pathway between native and denatured states. Kinetic approaches, reporter groups, and NMR also are shown to be powerful methods of obtaining highly specific information. From all of these considerations, a fairly good (albeit incomplete) picture emerges of the characteristics and mechanism of the cooperative folding process in proteins.

21-2   NARROW TEMPERATURE RANGE OF FOLDING–UNFOLDING EQUILIBRIA

Figure 21-1 shows a fairly typical change in a physical property that occurs as a protein is thermally unfolded; the change in the extinction coefficient at 287 nm of ribonuclease A is plotted versus temperature. This is a low-molecular-weight protein

Figure 21-1

Change in a physcial property as a protein is thermally unfolded. The change in extinction coefficient at 287 nm is plotted against temperature for ribonuclease at five pH values. [After J. F. Brandts and L. Hunt, *J. Am. Chem. Soc.* 89:4826 (1967).]

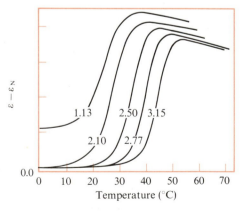

($M \cong 14{,}000$ d) consisting of a single polypeptide chain and four intrachain disulfide linkages (see discussion in Chapter 16). Curves are shown for five pH values between 1.1 and 3.2 The extinction coefficient at this wavelength is sensitive to changes in the environment of the six tyrosine residues of this enzyme (there are no tryptophans) and, in each case, it undergoes an abrupt change over a fairly narrow range of temperature. Above and below the transition zone, the extinction coefficient changes only gradually with temperature.[§]

The abrupt changes in extinction coefficient are associated with a major change in the conformation as the temperature is raised. This may be proven by measurements of, for example, the intrinsic viscosity below and above the transition region. Such measurements demonstrate that the molecule has undergone a major change in shape—from a compact to an expanded, unfolded form. The unfolded form still retains its four disulfide linkages, however.

The sharp change in extinction coefficient is due to an abrupt alteration in the environment of the tyrosine residues. The gradual variations in absorption outside the transition region are primarily due to the small intrinsic dependence of the extinction coefficient on temperature.

21-3   OBTAINING THERMODYNAMIC PARAMETERS: THE SIMPLE TWO-STATE MODEL

Assuming a simple two-state model for denaturation, we can analyze the data of Figure 21-1 to obtain the equilibrium constant K in accordance with the relationships

$$K = (D)/(N) \tag{21-1}$$

$$= [\varepsilon(T) - \varepsilon_N(T)]/[\varepsilon_D(T) - \varepsilon(T)] \tag{21-2}$$

where D and N refer to the denatured and native species, $\varepsilon(T)$ is the observed extinction coefficient at temperature T, and $\varepsilon_N(T)$ and $\varepsilon_D(T)$ are extinction coefficients of the native and denatured forms, respectively, at temperature T. Thermodynamic parameters ΔG^0, ΔH^0, and ΔS^0 may be obtained from the usual relationships:

$$\Delta G^0 = -RT \ln K \tag{21-3}$$

$$\Delta H^0 = -R[\partial(\ln K)/\partial(1/T)]_p \tag{21-4}$$

$$\Delta S^0 = (\Delta H^0 - \Delta G^0)/T \tag{21-5}$$

[§] The problem of the gradual changes in extinction coefficient above and below the transition can be circumvented by using thermal-difference spectra as described by Privalov and associates (see Khechinashvili et al., *FEBS Lett.* 30:57, 1973).

and the change in the constant-pressure heat capacity ΔC_p^0 is

$$\Delta C_p^0 = (\partial\,\Delta H^0/\partial T)_p = T(\partial\,\Delta S^0/\partial T)_p \qquad (21\text{-}6)$$

Application of the two-state model to ribonuclease

Table 21-1 gives values for the various thermodynamic functions associated with the ribonuclease unfolding transition, for three low pH values at 30°C. (At low pH values, the relative stability of the native form is sufficiently reduced that the unfolding re-

Table 21-1
Thermodynamic parameters for ribonuclease unfolding at 30°C

pH	ΔG^0 (kcal mole^{-1})	ΔH^0 (kcal mole^{-1})	ΔS^0 (cal deg^{-1} mole^{-1})	ΔC_p^0 (cal deg^{-1} mole^{-1})
1.13	$-1{,}085$	60.3	202	2,072
2.50	912	57.2	186	1,985
3.15	3,094	53.0	165	1,987

NOTE: Data are calculated on the assumption of a two-state reaction.
SOURCE: J. F. Brandts and L. Hunt, *J. Am. Chem. Soc.* 89:4826 (1967).

action can be studied in an experimentally convenient temperature range.) Under these conditions, both ΔH^0 and ΔS^0 are quite large and positive, whereas ΔG^0 is relatively small. The large increase in enthalpy upon unfolding shows that, at 30°C, the native form is much lower in energy than the denatured form. The increase in entropy upon denaturation is consistent with an increase in disorder for the unfolding process. Very striking in Table 21-1 is the large increase in heat capacity, about 2,000 cal deg^{-1} mole^{-1}, that occurs upon unfolding. This large change is brought about by the exposure to solvent of aliphatic and aromatic groups upon denaturation. Aqueous solutions of these kinds of nonpolar molecules are known to have large heat capacities. Generally, these groups are partly or completely buried in the native structure.

The large value of ΔC_p^0 reflects a substantial temperature dependence of ΔH^0 and ΔS^0, in accordance with Equation 21-6. In fact, ΔC_p^0 itself is temperature dependent. Figure 21-2 shows these variations with temperature; values (at pH 2.5) of ΔH^0, ΔS^0, and ΔC_p^0 are plotted over a wide range of temperature. It is clear that the temperature dependence of these parameters is very great. These effects can be ascribed in part to changes in the aqueous solvent and in the mode of solvation. For example, at low temperatures, water is organized into structures of solvation around nonpolar groups in the denatured state. At higher temperatures, such structures are less stable; a more random, disorganized solvation interaction occurs. As a result, solvation effects tend to give a more positive contribution to the entropy of unfolding as the temperature is raised.

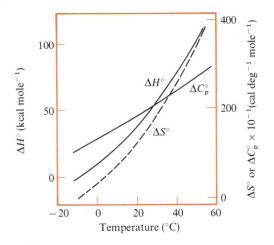

Figure 21-2

Temperature dependence of thermodynamic parameters for the ribonuclease transition at pH 2.5. [After J. F. Brandts and L. Hunt, *J. Am. Chem. Soc.* 89:4826 (1967).]

A prediction of "cold" denaturation

The fact that the thermodynamic parameters vary sharply with temperature leads to the interesting prediction that the protein should show "cold" denaturation as well as heat denaturation, because ΔG^0 itself has a maximum. Figure 21-3 illustrates this point for ribonuclease at the pH values used in Figure 21-1. [To emphasize that the maximum in ΔG^0 corresponds to a minimum in the free energy of the native form (relative to the denatured state), the vertical axis in Fig. 21-3 is constructed to show

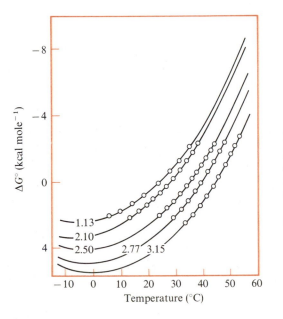

Figure 21-3

Minima in the temperature-dependence curves of ΔG^0 for the ribonuclease transition at five pH values. Points are based on experimental results; the curves are calculated from theory. [After J. F. Brandts and L. Hunt, *J. Am. Chem. Soc.* 89:4826 (1967).]

progressively larger (more positive) values of ΔG^0 as the origin is approached.] The plots show ΔG^0 versus temperature; the points are derived from the experimental data of Figure 21-1, and the curves have been calculated over a range wider than that attained in the experimental measurements. Note that, at these pH values, the free energy of stabilization of the native form is only 0 to 5 kcal mole^{-1} around room temperature. Note also that, in each case, ΔG^0 reaches a maximum around 0°C to -10°C; from this maximum, ΔG^0 decreases with decreasing temperature. Unfortunately, the temperature of maximal stability is just outside the experimentally accessible region, and it is not possible to demonstrate directly the cold denaturation for ribonuclease under these conditions.

A direct demonstration of a temperature of maximal stability has been obtained in the case of β-lactoglobulin in aqueous urea solution. Figure 21-4 plots the optical rotation at 365 nm versus temperature for β-lactoglobulin under three different conditions. Each curve shows a minimum from which $[\phi]$ increases with either decreasing or increasing temperature. Assuming a two-state transition, we can calculate values of the equilibrium constant K from these data. Figure 21-5 plots these computed values of log K versus $1/T$ for the same solvent conditions shown in Figure 21-4. Each curve has a minimum, reflecting the fact that ΔH^0 actually undergoes a change in sign. In this instance, a value of $\Delta C_p^0 \cong 2{,}100$ cal deg^{-1} mole^{-1} has been estimated.

Other proteins also have been examined with regard to these considerations. Table 21-2 lists thermodynamic parameters near room temperature and lists the temperature of maximal stability for ribonuclease, chymotrypsin, myoglobin, and β-lactoglobulin under specified conditions. Note that, in each case, a temperature of maximal stability does exist; such generality suggests that this property may be rather characteristic of proteins. However, in each instance in Table 21-2, behavior producing a temperature of maximal stability is observed only under special pH conditions or in the presence of a denaturant.

Table 21-2

Thermodynamic parameters for protein denaturation

Protein (conditions)	ΔG^0 (kcal mole^{-1})	ΔH^0 (kcal mole^{-1})	ΔS^0 (cal deg^{-1} mole^{-1})	ΔC_p^0 (cal deg^{-1} mole^{-1})	Temperature of maximal stability (°C)
Ribonuclease (pH 2.5, 30°C)	+0.9	+57	+185	+2,000	−9
Chymotrypsinogen (pH 3, 25°C)	+7.3	+39	+105	+2,600	10
Myoglobin (pH 9, 25°C)	+13.6	+42	+95	+1,400	<0
β-Lactoglobulin (5 M urea, pH 3, 25°C)	+0.6	−21	−72	+2,150	35

SOURCE: Data from C. Tanford, *Adv. Protein Chem.* 23:121 (1968).

Figure 21-4

Temperature of maximal stability for β-lactoglobulin in aqueous urea solution. Optical rotation at 365 nm is plotted against temperature in solutions of three different urea concentrations. The top and bottom curves give the rotations for completely unfolded and completely native protein. [After N. C. Pace and C. Tanford, *Biochemistry* 7:198 (1968).]

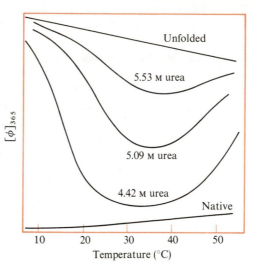

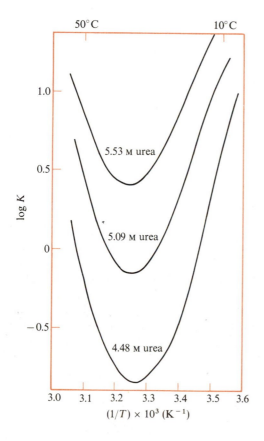

Figure 21-5

Typical van't Hoff plots. The data are derived from the heating curves of Figure 21-4. [After N. C. Pace and C. Tanford, *Biochemistry* 7:198 (1968).]

21-4 SEARCHING FOR INTERMEDIATES: EQUILIBRIUM STUDIES

Fractional denaturation and changes in physical properties

In Section 21-3, we assumed that the denaturation process is a two-state transition. The data given conform well to this assumption. Nevertheless, there is no reason to assume a priori that the transition should conform to a simple two-state process or that, just because one set of equilibrium measurements conforms well to this assumption, the reaction is truly two-state.

What is the actual mechanism of folding and unfolding? Do intermediates exist in the transition? These are interesting questions. Figure 21-6 shows three possible schemes for the mechanism. In mechanism I, the randomly organized chain simply

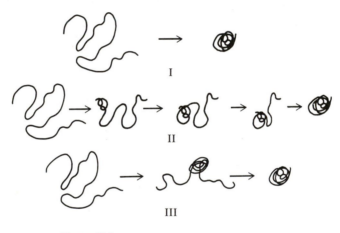

Figure 21-6

Three hypothetical schemes for the folding of a protein, starting from the random-coil state.

undergoes a concerted reaction to the native state without passing through any intermediates; this is a true two-state process. In mechanism II, the protein is viewed as gradually "winding up" from one end (say, the N-terminus) and passing through a continuous series of intermediate states. In mechanism III, a "core" or nucleation unit of organized structure is built within the chain, and the protein then folds the rest of its structure around this core. The nucleated structure itself might be regarded as an intermediate in this mechanism. Obviously, other mechanisms in addition to these three can be postulated.

Assume that the general reaction may be written schematically as

$$N \rightleftarrows X_1 \rightleftarrows X_2 \rightleftarrows \cdots \rightleftarrows D \qquad (21\text{-}7)$$

where each X_i is a stable intermediate.[§] If a physical property y is measured as a function of temperature, it will take on various values that are averages of the characteristic properties of the various species in Equation 21-7, weighted by the respective frequency of occurrence of each species. We want to know what is computed when the "fraction denatured" is calculated from experimental data in such a case, and whether it is possible to deduce the presence of intermediates in such situations from the measurements of an equilibrium physical property.

Assume that the observed value of the physical property y is given by

$$y = f_N y_N + f_D y_D + \sum_i f_i y_i \tag{21-8}$$

where the subscripts i, D, and N refer to the ith intermediate, the denatured form, and the native form, respectively; and the f terms are the fractions of each species. Substituting the conservation relationship

$$f_N = 1 - f_D - \sum_i f_i \tag{21-9}$$

into Equation 21-8 and rearranging, we obtain the following expression for the apparent fractional denaturation, f_{app}:

$$f_{app} = (y - y_N)/(y_D - y_N) = f_D + \sum_i f_i d_i \tag{21-10}$$

where

$$d_i = (y_i - y_N)/(y_D - y_N) \quad \text{and} \quad 0 \leqslant d_i \leqslant 1 \tag{21-11}$$

By stipulating that the parameter d_i lies between zero and unity, we have implicitly assumed that the value of the physical property of the ith intermediate falls between that of the native and that of the denatured species. (Clearly, this is a restrictive assumption.) A value of d_i near unity implies that the physical property of the ith intermediate is close to the denatured (D) state; conversely, a value of d_i near zero implies that it is near the native (N) state.

The apparent equilibrium constant $K_{app} = (\text{"D"})/(\text{"N"})$ is

$$K_{app} \quad f_{app}/(1 - f_{app}) \tag{21-12}$$

[§] The treatment that follows is based in part on that given by Charles Tanford (1968). It should be recognized that this treatment assumes that the denatured (or unfolded) state behaves as a single species. As pointed out in Section 21-7, this may not be true, at least for ribonuclease and possibly for other systems as well. However, the scheme of Equation 21-7 is a good starting point for the analysis of protein folding–unfolding reactions, and further elaborations can be introduced by modifying this simple scheme.

Note that $1 - f_{app}$ is the *apparent* fraction of native material. Now let us introduce the true equilibrium constants, $K_i = f_i/f_N$ and $K_D = f_D/f_N$. Using Equations 21-9, 21-10, and 21-12, we obtain

$$f_{app}/f_N = K_D\left(1 + \sum_i d_i K_i/K_D\right) \qquad (21\text{-}13)$$

$$(1 - f_{app})/f_N = 1 + \sum_i (1 - d_i)K_i \qquad (21\text{-}14)$$

$$K_{app} = K_D\left(1 + \sum_i d_i K_i/K_D\right)\Bigg/\left(1 + \sum_i (1 - d_i)K_i\right) \qquad (21\text{-}15)$$

It must be kept firmly in mind that only the apparent equilibrium constant K_{app}, and not K_D, typically is measured. According to Equation 21-15, $K_{app} = K_D$ when $K_i = 0$ for all i—that is, when there are no detectable intermediates. If $K_i \neq 0$, then $K_{app} \neq K_D$, even if all of the intermediates are nativelike in the value of their physical property ($d_i = 0$) or are identical to the completely denatured species ($d_i = 1$). Furthermore, it is clear that *in general* the numerical value of K_{app} (or f_{app}) depends upon physical-property parameters (d_i) and not just upon thermodynamic equilibrium constants. Only when no intermediates exist will K_{app} be independent of the actual physical parameter being measured. Therefore, a test for intermediates should be easy to construct, based on these conclusions. When f_{app} versus temperature is measured by several different physical properties, the values of f_{app} obtained should vary with the parameter being measured if intermediates do exist; this is because a given d_i is not expected to be the same for each distinct physical property measured.

Figure 21-7 is a plot of f_{app} versus temperature for ribonuclease. Three different properties (optical density, optical rotation, and intrinsic viscosity) were measured and used to compute values of f_{app}. It is clear that all three measurements show the same profile for f_{app} versus T over the entire temperature range examined. Apparently, the ribonuclease transition is two-state under these conditions.

Calorimetric test for intermediates

A second kind of test for intermediates can be made by comparing the enthalpy of the transition as measured calorimetrically with the van't Hoff enthalpy deduced from the temperature dependence of K_{app}. A calorimetric measurement of the heat released in the transition from the fully native to the fully denatured state gives ΔH_D. The apparent heat ΔH_{app} of the reaction obtained from the van't Hoff relationship is given by

$$\partial(\ln K_{app})/\partial T = \Delta H_{app}/RT^2 \qquad (21\text{-}16)$$

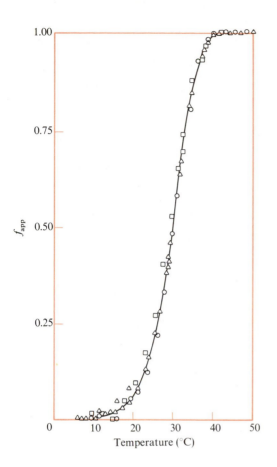

Figure 21-7

Apparent fractional denaturation versus temperature for ribonuclease at pH 2.1. The same curve provides a good fit for experimental points determined by intrinsic viscosity (□), optical rotation at 365 nm (○), and difference spectroscopy (△). [After A. Ginsburg and W. R. Carroll, *Biochemistry* 4:2159 (1965).]

By straightforward differentiation of Equation 21-15, followed by simplification, we obtain

$$\Delta H_{\text{app}} = \Delta H_{\text{D}} \left(1 - \frac{\sum\limits_i d_i (K_i/K_{\text{D}})(1 - \Delta H_i/\Delta H_{\text{D}})}{1 + \sum\limits_i d_i (K_i/K_{\text{D}})} - \frac{\sum\limits_i (1 - d_i)K_i(\Delta H_i/\Delta H_{\text{D}})}{1 + \sum\limits_i (1 - d_i)K_i} \right) \quad (21\text{-}17)$$

where

$$\partial(\ln K_i)/\partial T = \Delta H_i/RT^2 \quad (21\text{-}18)$$

Note that ΔH_i is the enthalpy change for the conversion of N to X_i. If, as before, we assume $0 \leqslant d_i \leqslant 1$ and if we assume $(\Delta H_i/\Delta H_{\text{D}}) \leqslant 1$, then each of the two quotients (which contain summation terms) in Equation 21-17 is greater than or equal to zero, and is less than or equal to unity. Under these conditions, $|\Delta H_{app}| \leqslant |\Delta H_D|$; when there are no intermediates, $\Delta H_{app} = \Delta H_D$.

The reason that the existence of intermediates leads to $|\Delta H_{app}| \leqslant |\Delta H_D|$ is clear. The van't Hoff relation of Equation 21-17 gives a kind of average heat from all of the reactions involved. Because the heat of transition from the folded state to an intermediate state is assumed to be less than that for the conversion to the completely unfolded form, the partially unfolded states weight the average with smaller heat values.

The development of scanning microcalorimetry has made it possible to measure directly and precisely the enthalpy changes for denaturation of a number of proteins. (See Box 21-1 for a discussion of calorimetry.) P. L. Privalov and colleagues in the Soviet Union have been prominent in this work. Figure 21-8 summarizes calorimetric data obtained for the unfolding of ribonuclease, lysozyme, chymotrypsin, and

---

Box 21-1   CALORIMETRY

Differential scanning calorimetry has gained widespread use in recent years and has emerged as an important tool for investigating a diversity of systems. The essential experimental purpose is to measure the heat absorption in a system as the temperature is raised from some initial to some final temperature, where the system under investigation has one or more transformations (such as conformational changes if the system consists of a protein dissolved in aqueous buffer media). Measurements are taken simultaneously with a suitable reference solution, so as to cancel out heat changes due to effects of no interest. Thus, measurements of a protein in buffer would be compared with those of buffer alone.

The resulting data often are plotted as C_p (constant-pressure heat capacity) versus T. For a single transformation, such as a protein conformational change $N \rightleftharpoons D$, a curve such as that shown here might be obtained. The peak in C_p occurs in the transition region as the molecule goes from the N to the D form.

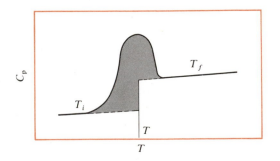

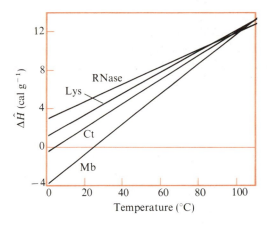

Figure 21-8

Temperature dependence of the enthalpy difference of denatured and native states of globular proteins. The calorimetric $\Delta \hat{H}$ (in cal g^{-1}) of denaturation is plotted against temperature for ribonuclease (RNase), lysozyme (Lys), chymotrypsin (Ct), and myoglobin (Mb). [After P. L. Privalov, *FEBS Letts.* 40:S140 (1974).]

To calculate ΔH for the transition, assume that the temperature is raised from an initial value T_i (below the transition region) to a final value T_f (above the transition region). The total heat change ΔH_{Tot} in going from T_i to T_f is

$$\Delta H_{Tot} = \int_{T_i}^{T_f} C_p \, dT$$

where C_p is the apparent (measured) heat capacity at constant pressure. The ΔH_{Tot} value contains not only the enthalpy change associated with the transition from N to D, but also the "background" heat absorption that comes as a result of heat absorption by N and D molecules that are present in the region between T_i and T_f.

To compute just $\Delta H(T)$—the enthalpy change associated with the transition at temperature T—we must subtract from ΔH_{Tot} the contribution to the total heat absorption that the N species makes between T_i and T and that the D species makes between T and T_f:

$$\Delta H(T) = \Delta H_{Tot} - \int_{T_i}^{T} C_p^N \, dT - \int_{T}^{T_f} C_p^D \, dT$$

where C_p^N and C_p^D are the heat capacities of the N and D forms, respectively. Referring to the figure, we note that $\Delta H(T)$ is given by the shaded area under the curve. The equation can be used to calculate $\Delta H(T)$ at *any* temperature between T_i and T_f.

In some systems, more than one peak may be seen in a plot of C_p versus T. This is evidence for more than one structural transition, occurring at distinctly different temperatures.

For further discussion of this important experimental approach, see J. H. Sturtevant, *Adv. Biophys. Bioeng.* 3:35 (1974), and K. S. Krishnan and J. F. Brandts, *Methods in Enzymol.* 49:3 (1978).

myoglobin. The enthalpy changes are reported on a per-gram basis and are defined as $\Delta \hat{H}$. In all cases, the $\Delta \hat{H}$ values increase with increasing temperature and converge at high temperatures. The van't Hoff enthalpies also have been measured and, in each case, these enthalpies match those for the corresponding protein as determined calorimetrically. Thus, with the calorimetric and van't Hoff tests, no intermediates are discernible in these cases.

21-5 KINETIC STUDIES OF PROTEIN FOLDING

We have seen how equilibrium results on reversible unfolding may be analyzed in order to obtain mechanistic information. However, considerably more information can be derived from kinetic studies, where the time dependence of the unfolding process is analyzed with an eye toward discriminating among possible mechanisms.

The folding and unfolding of protein chains are intramolecular processes. Therefore, the kinetics of these reactions always can be described in terms of first-order rate equations (Chapter 16). This means that the concentration of a given species is described by a sum of one or more exponential terms, $\sum_i A_i e^{-\lambda_i t}$, where the λ_i parameters are characteristic time constants, and the A_i terms are constant amplitude parameters (see Chapter 16 and Eqn. 16-20). A. Ikai and C. Tanford (1971) have made a detailed analysis of protein folding–unfolding kinetics, and our treatment here follows along the lines developed by them.

Kinetics of the two-state model

We consider first the kinetics of the simple two-state model. Assume that initially only native or only denatured protein is present. The protein then is subjected to a rapid temperature jump, a sudden change in solvent, or some other quick change that causes the protein to fold or unfold to a mixture of native and denatured species. For the reaction

$$N \underset{k_{-1}}{\overset{k_1}{\rightleftharpoons}} D \tag{21-19}$$

the rate equation is

$$-d(N)/dt = k_1(N) - k_{-1}(D) \tag{21-20}$$

or

$$-df_N/dt = k_1 f_N - k_{-1} f_D \tag{21-21}$$

where Equation 21-21 is obtained by dividing both sides of Equation 21-20 by the constant sum $(N) + (D)$. Let us write $f_N = \bar{f}_N + \Delta f_N$ and $f_D = \bar{f}_D + \Delta f_D$, where the overbars denote final (and constant) equilibrium fractions, and the Δf terms are deviations from the equilibrium values. Making these substitutions in Equation 21-21, we obtain

$$-d\,\Delta f_N/dt = (k_1 + k_{-1})\Delta f_N \tag{21-22}$$

The relationship $\Delta f_N = -\Delta f_D$ (which follows from the conservation relationship $f_D + f_N = $ constant) and the equilibrium condition $k_1\bar{f}_N = k_{-1}\bar{f}_D$ have also been used in obtaining Equation 21-22. The solution to Equation 21-22 is simply

$$\Delta f_N = \Delta f_N^0 e^{-\lambda t} \tag{21-23}$$

where $\lambda = k_1 + k_{-1}$, and Δf_N^0 is the value of Δf_N at $t = 0$. (Note that our approach to solving the rate equation is identical to that used in Chapter 16 for unimolecular reactions.) Figure 21-9 is a sketch of f_N versus time that illustrates the parameters Δf_N, Δf_N^0, and $\bar{f}_N$.

In general, we measure a physical property y, where

$$y = f_N y_N + f_D y_D \tag{21-24}$$

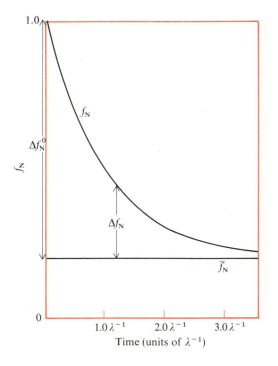

Figure 21-9

Hypothetical plot of f_N versus time to illustrate parameters discussed in the text.

It is convenient to work with the variable Δy, where

$$\Delta y = y - \bar{y} = y_N \Delta f_N + y_D \Delta f_D \qquad (21\text{-}25)$$

$$= (y_N - y_D) \Delta f_N^0 e^{-\lambda t} \qquad (21\text{-}26)$$

$$= \Delta y_0 e^{-\lambda t} \qquad (21\text{-}27)$$

so

$$\Delta y / \Delta y_0 = e^{-\lambda t} \qquad (21\text{-}28)$$

and where Δy_0 is the deviation at $t = 0$ of y from its final equilibrium value $\bar{y}$.

Equation 21-28 indicates that a plot of $\ln(\Delta y / \Delta y_0)$ versus t should be linear, and that the slope is the same regardless of the initial conditions or the property measured, if the two-state mechanism is operative. This is nicely illustrated by the folding and unfolding kinetics of lysozyme, under certain conditions. Figure 21-10a plots y (absorbance at 301 nm) versus time for the denaturation and the renaturation of lysozyme in 2.5 M guanidine hydrochloride. Figure 21-10b shows the corresponding semilogarithmic plots of the same data. It is clear that, in this case, the kinetic data conform very well to a simple two-state scheme with no intermediates.

Consider, however, the case of cytochrome c. Figure 21-11a plots the absorbance at 400 nm versus time, for situations in which both the folding and the unfolding reaction are studied in 2.5 M guanidine hydrochloride. It is clear that the semilogarithmic plots in Figure 21-11b are biphasic, indicating that the simple two-state scheme is not operative. Moreover, note that the exact shape of the semilogarithmic plots is sensitive to the initial conditions.

The biphasic kinetics suggest that two steps occur in the folding–unfolding reaction. (In general, if n phases are observed, there are n independent steps; see Chapter 16.) However, the exact interpretation of the kinetic results in terms of mechanism is rife with pitfalls. Nevertheless, if sufficient care is exercised, the data can be used to eliminate various schemes and to point fairly well to a particular mechanism, as we shall see.

Kinetics of a three-state model

The difficulties encountered with the kinetic approach (as well as its power) are best illustrated by working through a particular example. Let us assume that, in light of the biphasic kinetics, the cytochrome c data can be described by the scheme

$$N \underset{k_{-1}}{\overset{k_1}{\rightleftarrows}} X \underset{k_{-2}}{\overset{k_2}{\rightleftarrows}} D \qquad (21\text{-}29)$$

which involves a single intermediate in the pathway between native and denatured

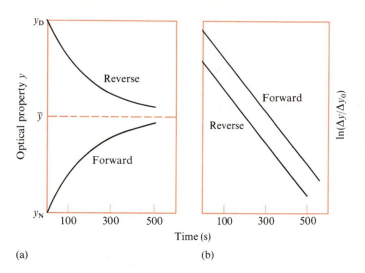

Figure 21-10

Folding–unfolding kinetics of lysozyme. **(a)** The optical density at 301 nm is plotted against time for the denaturation and renaturation of lysozyme in 2.5 M guanidine hydrochloride (pH 2.6) at 25°C. **(b)** A semilogarithmic plot of the data in part a. The curves are displaced from each other by a constant. [After A. Ikai and C. Tanford, *Nature* 230:100 (1971).]

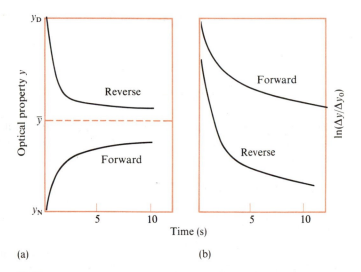

Figure 21-11

Folding–unfolding kinetics of cytochrome c. These plots are analogous to those given in Figure 21–10 for lysozyme. In part b, a constant has been added to the ordinate values of one of the curves in order to displace the two curves for clarity. [After A. Ikai and C. Tanford, *Nature* 230:100 (1971).]

states. The rate equations are

$$-d(N)/dt = k_1(N) - k_{-1}(X) \tag{21-30a}$$

$$-d(D)/dt = -k_2(X) + k_{-2}(D) \tag{21-30b}$$

There are only two independent rate equations, because the time dependence of the third species, $d(X)/dt$, is simply the sum of Equations 21-30a and 21-30b. By making substitutions and manipulations analogous to those employed in solving the one-step mechanism, we can rewrite Equations 21-30a,b as

$$-d\,\Delta f_N/dt = (k_1 + k_{-1})\,\Delta f_N + k_{-1}\,\Delta f_D \tag{21-31a}$$

$$-d\,\Delta f_D/dt = k_2\,\Delta f_N + (k_2 + k_{-2})\,\Delta f_D \tag{21-31b}$$

where the conservation relationship $\Delta f_X = -\Delta f_N - \Delta f_D$ has been used. Equations 21-31a,b are simultaneous linear differential equations, and the standard procedure for solving such equations is to assume that the solutions are of the form

$$\Delta f_N = Ce^{-\lambda t} \tag{21-32a}$$

$$\Delta f_D = C'e^{-\lambda t} \tag{21-32b}$$

where C, C', and λ are constants. Substituting Equations 21-32a,b into Equations 21-31a,b, and noting that $d\,\Delta f_N/dt = -\lambda\,\Delta f_N$ and that $d\Delta f_D/dt = -\lambda\,\Delta f_D$ (from Eqns. 21-32a,b), we obtain

$$0 = (k_1 + k_{-1} - \lambda)\,\Delta f_N + k_{-1}\,\Delta f_D \tag{21-33a}$$

$$0 = k_2\,\Delta f_N + (k_2 + k_{-2} - \lambda)\,\Delta f_D \tag{21-33b}$$

Nontrivial solutions (giving nonzero values for Δf_D and Δf_N) to these simultaneous homogeneous equations are obtained for those values of λ for which the determinant of the coefficients of Δf_N and Δf_D vanishes:

$$\begin{vmatrix} k_1 + k_{-1} - \lambda & k_{-1} \\ k_2 & k_2 + k_{-2} - \lambda \end{vmatrix} = 0 \tag{21-34}$$

or

$$f(\lambda) = \lambda^2 - \lambda(k_1 + k_{-1} + k_2 + k_{-2}) + k_1k_2 + k_1k_{-2} + k_{-1}k_{-2} = 0 \tag{21-35}$$

The two roots to Equation 21-35, λ_1 and λ_2, are obtained by the quadratic formula:

$$\lambda_{1,2} = \{(k_1 + k_{-1} + k_2 + k_{-2}) \pm [(k_1 + k_{-1} + k_2 + k_{-2})^2$$

$$- 4(k_1k_2 + k_1k_{-2} + k_{-1}k_{-2})]^{1/2}\}/2 \tag{21-36}$$

where the positive sign yields λ_1 (the largest root), and the negative sign yields λ_2. [This problem is exactly analogous to finding the eigenvalues of the matrix $\underset{\sim}{\mathbf{M}}$ in Chapter 20 (cf. Eqns. 20-49 through 20-52) and to finding the reciprocal relaxation times for a two-step mechanism (cf. Eqns. 16-80 through 16-82).]

A particular solution to our original differential equations (Eqns. 21-31a,b) is given by Equations 21-32a,b with $\lambda = \lambda_1$. When $\lambda = \lambda_1$, the constants C and C' take on particular values that we will designate C_1 and C_3, respectively. Substituting Equations 21-32a,b into Equation 21-33a, we can solve for C_3 in terms of C_1:

$$(k_1 + k_{-1} - \lambda_1)C_1 + k_{-1}C_3 = 0 \tag{21-37}$$

or

$$C_3 = \gamma_1 C_1 \tag{21-38}$$

where

$$\gamma_1 = (\lambda_1 - k_1 - k_{-1})/k_{-1} \tag{21-39}$$

Similarly, Equation 21-33b could be used to show that $\gamma_1 = k_2/(\lambda_1 - k_2 - k_{-2})$; it is easy to verify that the two expressions for γ_1 are equivalent. When $\lambda = \lambda_2$, the constants C and C' take on the specific values C_2 and C_4; proceeding as before, we obtain

$$C_4 = \gamma_2 C_2 \tag{21-40}$$

where

$$\gamma_2 = (\lambda_2 - k_1 - k_{-1})/k_{-1} \tag{21-41}$$

Thus, the two solutions are $\Delta f_N = C_1 e^{-\lambda_1 t}$ and $\Delta f_D = C_3 e^{-\lambda_1 t}$, and $\Delta f_D = C_2 e^{-\lambda_2 t}$ and $\Delta f_D = C_4 e^{-\lambda_2 t}$. The most general solution is simply the sum of the particular solutions:

$$\Delta f_N = C_1 e^{-\lambda_1 t} + C_2 e^{-\lambda_2 t} \tag{21-42a}$$

$$\Delta f_D = C_3 e^{-\lambda_1 t} + C_4 e^{-\lambda_2 t} = \gamma_1 C_1 e^{-\lambda_1 t} + \gamma_2 C_2 e^{-\lambda_2 t} \tag{21-42b}$$

It is easy to show that these equations for Δf_N and Δf_D satisfy the original differential equations (Eqns. 21-31a,b). C_1 and C_2 are determined by initial conditions.

In actual practice, one follows the time dependence of a physical property rather than that of a particular species concentration. By analogy with the one-intermediate mechanism (cf. Eqns. 21-19 through 21-28), we can obtain an expression for the time dependence of the fractional change $\Delta y/\Delta y_0$ in the physical property y:

$$\Delta y/\Delta y_0 = A_1 e^{-\lambda_1 t} + A_2 e^{-\lambda_2 t} \tag{21-43}$$

where $A_1 + A_2 = 1$ (because $\Delta y/\Delta y_0 = 1$ at $t = 0$).

Equation 21-43 shows that two exponential phases are associated with the two-step mechanism of Equation 21-29. However, the relative magnitude of each phase depends on the amplitude parameters A_1 and A_2. These, in turn, depend on the initial conditions (that is, on whether we start with completely native, completely denatured, or something in between). Indeed, Figure 21-11b shows that the biphasic semilogarithmic plots differ according to whether the forward or reverse reaction is considered.

Using amplitude parameters to discriminate between mechanisms

As shown by A. Ikai and C. Tanford (1971), expressions may be obtained for A_1 and A_2 in terms of rate constants and physical-property parameters. When these amplitude parameters are tested against the data in Figure 21-11, it is clear that the simple scheme of Equation 21-29 cannot be reconciled with the observed values of A_1 and A_2. This observation illustrates a crucial point: even though the biphasic kinetics per se may be consistent with the one-intermediate mechanism of Equation 21-29, the amplitude parameters associated with the kinetic phases may not be consistent with that mechanism. Therefore, the mechanism must be rejected.

A scheme that fits better with the observed amplitude parameters is

$$N \rightleftharpoons D \rightleftharpoons X \qquad (21\text{-}44)$$

in which X is an incorrectly folded form. For this scheme, it turns out that all A_i values are positive, whether the reaction is studied in the forward or in the reverse direction. This agrees with the experimental observations. In actuality, the amplitude parameters for cytochrome c are best accounted for by a mechanism involving both an intermediate state and an improperly folded form:

$$N \rightleftharpoons X_1 \rightleftharpoons D \rightleftharpoons X_2 \qquad (21\text{-}45)$$

This scheme would be expected to give triphasic kinetics, but biphasic kinetics are observed because two of the λ_i parameters are very similar in value.

The preceding discussion indicates the care with which the kinetic analyses must be carried out if certain serious pitfalls are to be avoided. In particular, both the amplitudes and the rates associated with kinetic processes must be examined carefully in order to avoid misinterpretations. If properly carried out, however, the kinetic analysis gives very important mechanistic conclusions that are not readily duplicated through any other approach.

21-6 REPORTER GROUPS (EXTRINSIC PROBES)

Measurements of overall physical-property changes do not, of course, provide information about isolated sections or segments of the protein during the folding process.

For this purpose, a reporter group or molecule (also called an extrinsic probe) may be attached to a specific section of the macromolecule. The reporter group has some physical property (such as its absorption spectrum) that may be studied separately from the corresponding property of the protein itself. In this way, changes at the site of the reporter molecule are followed separately from (and may be compared with) those changes associated with the overall physical properties. It is hoped, of course, that the reporter group does not significantly perturb the structure. This hypothesis may be tested, for example by using an overall property to compare the folding–unfolding reactions of normal and derivatized protein.

Studying ribonuclease with a reporter group

Ribonuclease A provides one example of this approach. The unfolding of this protein may be studied by changes in absorption at around 287 nm, as mentioned earlier. Kinetic studies of the unfolding reaction reveal two processes—one in the millisecond time range, and the other a slow phase of the order of 10 to 100 sec. This is an interesting finding, because the equilibrium data discussed in Section 21-4 indicate that the transition for ribonuclease is two-state, and thus should give only a single kinetic phase. It clearly is of interest to determine whether the same equilibrium and kinetic changes observed at 287 nm are also evident at a single site in the macromolecule, or whether the 287 nm changes represent an aggregate effect peculiar to the six tyrosine residues (intrinsic probes) that absorb in this region. To answer this question, a dinitrophenyl (DNP) reporter group (Fig. 21-12) has been selectively attached to the ε-amino group of Lys^{41}. The optical absorption of the DNP group may be specifically

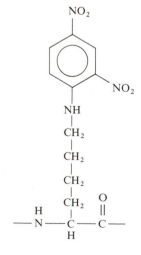

Figure 21-12

The dinitrophenyl (DNP) group attached to the ε-amino group of lysine.

monitored at 358 nm, a wavelength well separated from the protein's absorption bands.

Figure 21-13 compares the equilibrium temperature dependence of the absorbance at 286.5 nm with that at 358 nm, at pH 1.3. In both cases, the curves undergo a reversible transition at about 30°C. The midpoints of the transitions are the same

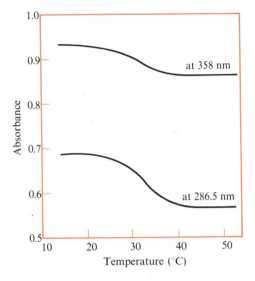

Figure 21-13

Thermal unfolding of 41-*DNP-ribonuclease* at pH 1.3, as measured by absorbance at 286.5 nm and at 358 nm. [After T. Y. Tsong and R. L. Baldwin, *J. Mol. Biol.* 69:149 (1972).]

within experimental error. This suggests that the DNP reporter group monitors the conformational change in exactly the same way as do the tyrosine residues.

Further comparisons have been made by kinetic studies. It turns out that both fast and slow kinetic phases can be monitored by the DNP abosprtion changes, as well as by the absorption changes of the tyrosine residues. Figure 21-14 compares the fast phases for a temperature-jump experiment in which the temperature was raised from 35°C to 40°C in about 10 μsec. The oscilloscope plots light transmission versus time, for $\lambda = 286.5$ nm and $\lambda = 358$ nm. The time constants associated with each of the exponential decay processes are the same. Furthermore, it is found that the rapid and the slow kinetic phases monitored at 286.5 nm are similar in both DNP-containing and unmodified ribonuclease. This similarity is evidence that the modification does not perturb the structure. Hence, both equilibrium and kinetic data support the notion that the overall changes are concerted with those that occur at a specific site.

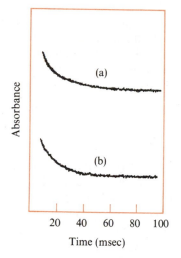

Figure 21-14

Oscilloscope traces from a temperature-jump experiment. Absorbance is shown versus time following a temperature jump from 35°C to 40°C for 41-DNP-ribonuclease at pH 1.3. (**a**) Absorbance at 286.5 nm. (**b**) Absorbance at 358 nm. The curves have been displaced from each other for clarity. [From T. Y. Tsong and R. L. Baldwin, *J. Mol. Biol.* 69:149 (1972).]

21-7 INTERPRETATION OF BIPHASIC KINETICS OF RENATURATION

Further studies by R. L. Baldwin and colleagues have established that the biphasic kinetics associated with the refolding of ribonuclease are due to two different unfolded states that refold to the native structure at different rates. These forms are designated U_1 and U_2. These two states persist when ribonuclease is placed in high concentrations of urea or guanidine hydrochloride, suggesting that neither state contains residual amounts of ordered structure. A scheme that quantitatively fits the kinetic data is

$$U_1 \xrightleftharpoons{\text{slow}} U_2 \xrightleftharpoons{\text{fast}} N \qquad (21\text{-}46)$$

where N is the native form. In this scheme, U_1 slowly converts to another unfolded form, U_2, which rapidly folds up. Although U_1 and U_2 appear to be unfolded chains, it is reasonable to assume that U_2 can more easily form partially ordered structures, which then fold into the native conformation.

Recent data of F. X. Schmid and R. L. Baldwin (1978) suggest that the slow transition is due to the *cis–trans* isomerization of proline imide bonds. Of the four prolines in ribonuclease A, it is believed that at least two must be in the *trans* state in order for the native structure to form; one or both of the others possibly can be accommodated in either the *cis* or *trans* state. It is conceivable that some preliminary folding occurs in the $U_1 \rightarrow U_2$ transition, simultaneous with the imide bond isomerization. Other data on proline *cis–trans* isomerization in protein refolding have been obtained by L.-N. Lin and J. F. Brandts (1978).

21-8   NMR STUDIES

What is ideally desired is a means of simultaneously monitoring various specific regions of the protein during the folding or unfolding process. In this way, one can test directly various mechanisms for folding (see Fig. 21-6). For example, by simultaneously observing several regions during folding, we can easily distinguish a concerted transition from a nonconcerted one. Furthermore, the possibility of folding-up sequentially from the N-terminus can be tested directly. This possibility is of interest because polypeptides are synthesized in vivo from the N-terminus to the C-terminus; if folding occurs during synthesis, we would expect it to occur sequentially from the N-terminus.

One way simultaneously to study several regions of the molecule is to attach extrinsic probes (such as fluorescent probes or spin labels) to several sites on the macromolecule. However, in most cases, the attachment of probes to more than one specific site is not an easy task. Moreover, one always fears that the probe per se perturbs the structure and gives results that do not accurately represent the situation with regard to the underivatized protein. Simultaneous attachment of several probes only exacerbates this problem.

An attractive alternative is to employ a technique that directly monitors changes in the environments of specific sites in the protein, without the introduction of additional groups. The method of choice is nuclear magnetic resonance (NMR) because high-resolution (e.g., 220 MHz) devices are capable of resolving individual amino acid residues, and the resonances themselves are quite sensitive to the local environment. In fact, there is at present no other solution technique with the same resolving power and sensitivity as high-resolution NMR. (See Chapter 9 for a discussion of NMR.)

NMR of lysozyme in the transition zone

A rather detailed study of the 220 MHz NMR spectrum throughout the course of thermal denaturation has been carried out for lysozyme. Figure 21-15 shows regions of the spectrum in D_2O (pD 5) at 35°C, 65°C, and 80°C. (D_2O is used as solvent rather than H_2O because the latter's spectrum overlaps some of the proton resonances of the protein.) The spectra are shown with the low-field region on the left and the high-field

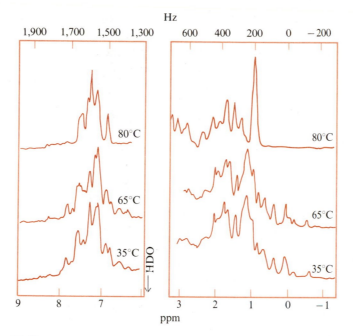

Figure 21-15

Temperature dependence of lysozyme 220 *MHz NMR spectrum* in D_2O at pD 5.0. NH protons are preexchanged. Low-field amplification is greater than that for the high field. Shifts are relative to a DSS standard (see Chapter 9). [After C. C. McDonald et al., *J. Am. Chem. Soc.* 93:235 (1971).]

region on the right. The NH protons, which generally appear in the range of 6 to 11 ppm, are not seen because they have been exchanged by deuterons. The region from 3 to 6 ppm is not shown because it is obscured by the HDO resonance.

This spectrum largely represents various CH resonances. In the denatured form (at 80°C), the spectrum corresponds nicely to the sum of the spectra of the individual amino acids. For example, all 12 alanine methyl groups occur at 1.41 ppm ($\sim$310 Hz), even though each has a different location in the primary structure. Other CH resonances occur in the range of about 200 to 450 Hz for the aliphatic amino acids and about 1,500 to 1,700 Hz for the aromatic CH groups of tryptophan, tyrosine, phenylalanine, and histidine.

The spectrum for the native form (35°C to 65°C) is substantially different from that for the unfolded species. Particularly striking is the appearance of new lines between 100 and −200 Hz. These are ascribed to methyl protons that lie close to the faces of aromatic rings, so that the resonances have been shifted to high fields by ring-current effects (see Chapter 9). Certain of the lines have been tentatively assigned to specific residues, which are distributed throughout the molecule. Because their resonances are very sensitive to ring-current fields of nearby groups, they are excellent probes for sensing structural changes at a variety of sites in the marcromolecule. It has been estimated that a change of a few hundredths of an angstrom in the position of one of these residues could give a detectable shift in the associated resonance.

It is clear from Figure 21-15 that, between 35°C and 65°C, there is very little variation in the spectrum of the native protein. This observation suggests that the native state is a single, uniquely defined structure, and not a family of rapidly "exchange-averaged" structures, which would produce different chemical shifts with varying temperature.

Under these conditions, the enzyme undergoes unfolding as the temperature is raised above 65°C. The question of interest is whether the changes in the NMR spectrum that accompany unfolding indicate a particular mechanism for the unfolding process. Figure 21-16 shows the high-field region as the temperature is varied between 64°C and 77°C. As the transition takes place, the spectrum is accurately represented as a composite of spectra characteristic of the folded and unfolded states. New resonances, indicative of intermediate forms, do not appear. Moreover, the lines characteristic of the denatured state appear simultaneously in a concerted fashion; the disappearance of the "native" resonances also is concerted. These data clearly point, therefore, to a two-state, highly cooperative mechanism.

Spectra have been recorded also at acid pH (pH 3.3) in H_2O; this procedure provides clear resolution of the indole NH groups. At acid pH, the exchange of the indole hydrogen with H_2O is slow enough for the NH resonance to be observed clearly. Figure 21-17 shows some spectra for both the low-field and the high-field

Figure 21-16

Temperature dependence of lysozyme high-field NMR spectrum in D_2O at pD 5.0. Shifts are relative to a DSS standard. [After C. C. McDonald et al., *J. Am. Chem. Soc.* 93:235 (1971).]

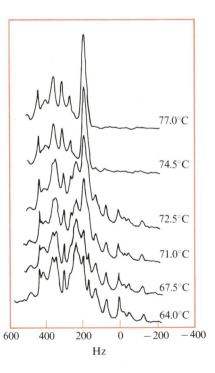

77.0°C

74.5°C

72.5°C

71.0°C

67.5°C

64.0°C

600   400   200   0   − 200   − 400

Hz

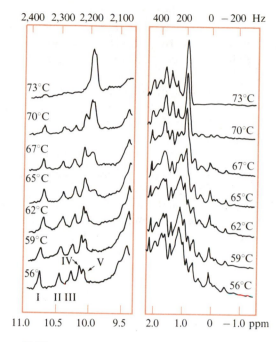

2,400 2,300 2,200 2,100   400   200   0 − 200 Hz

73°C
70°C
67°C
65°C
62°C
59°C
                    IV   V
56°
        I   II III

73°C
70°C
67°C
65°C
62°C
59°C
56°C

11.0   10.5   10.0   9.5   2.0   1.0   0   − 1.0 ppm

Figure 21-17

Temperature dependence of lysozyme NMR spectrum in H_2O at pH 3.3. Shifts are relative to a DSS standard. [After C. C. McDonald et al., *J. Am. Chem. Soc.* 93:235 (1971).]

regions in the range of 56°C to 73°C. At 73°C, the spectrum is characteristic of the denatured state, and the six tryptophans of lysozyme all appear at 2,190 Hz. Upon folding of the protein, however, there is a beautiful splitting of this resonance into separate lines of unit intensity for each residue. Five lines are evident in Figure 21-17 (between 2,200 and 2,400 Hz), with the sixth apparently occurring in the amide NH region (around 1,700 to 2,140 Hz). The splitting is simply a reflection of the nonequivalent environments of these indole groups in the native structure.

Once again, as the folding–unfolding transition occurs, the spectra of both the low-field indole NH region and of the high-field region are simple combinations of the spectra of folded and unfolded species, giving further support to the concerted two-state mechanism for lysozyme.

NMR of ribonuclease histidines at pH 1.3

The imidazole C-2 proton resonances for the four histidine groups in ribonuclease have been studied as a function of temperature by Fourier-transform NMR at 100 MHz resolution. In this experiment, the areas under the distinct peaks for the four histidines are measured; the peaks disappear as the protein denatures, giving rise to a new peak (identical for the four residues) that corresponds to the histidine C-2 protons in the denatured species. By accurately measuring the areas (which are directly proportional to the proton concentrations), we get a picture of the unfolding of distinct segments of the chain.

Figure 21-18 shows an example of the data obtained. The top panel shows that the area of the denatured peak increases in a smooth fashion. However, the peaks corresponding to the C-2 imidazole protons in the native structure do not change in a concerted fashion. In particular, the environment around His[119] appears to denature at a somewhat lower temperature than that around His[48]. These data, and other data on the behavior of the chemical shifts, suggest that unfolding is not a simple two-state process. However, the discrepancies from two-state behavior might represent only minor local structural fluctuations. Regardless of the exact explanation, these data nevertheless illustrate the power of the NMR approach for revealing detailed and specific information on a challenging problem.

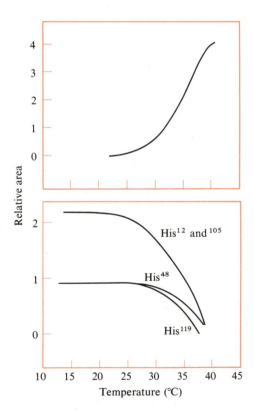

Figure 21-18

Areas of histidine imidazole C-2 proton peaks as a function of temperature. The upper panel shows the behavior of denatured histidine resonances. The lower panel shows the behavior of native resonances of specific histidines. [After D. G. Westmoreland and C. R. Matthews, *Proc. Natl. Acad. Sci. USA* 70:914 (1973).]

21-9 STATUS OF RESEARCH IN PROTEIN FOLDING

In this chapter we have directed attention chiefly to some of the physical chemical approaches to studying the reversible folding–unfolding equilibria of proteins. We have emphasized the question of the existence of intermediates and ways to detect them experimentally. (See also Box 21-2 for a discussion of intermediates in the reoxidation and refolding of reduced proteins.) At this point, it is worth reflecting on what is generally known about the folding–unfolding reaction, and about the conjectures that can be made based on current knowledge.

It is clear that, even though the demonstration of species in other than the native and denatured states often is experimentally elusive, it still is possible with the right methods and systems to detect more than two discrete states if they exist. We have discussed cases in which some kinetic and NMR measurements have given evidence for more than two states. In the case of ribonuclease, it has been shown that there are both fast-refolding and slow-refolding forms. This raises the question of whether the fast-refolding form easily equilibrates with conformations possessing elements of order that might serve as nucleation sites for refolding. In this regard, some data obtained by Christian Anfinsen and associates with staphylococcal nuclease (149 residues) are of interest (see Sachs et al., 1972). Using an immunological approach, they have obtained results showing that an isolated fragment of the enzyme (51 residues) can mimic the conformation of the same fragment contained within the intact enzyme. These data have been interpreted to mean that the isolated piece occasionally adopts a "nativelike" conformation N′, with an equilibrium constant (N′)/(U) (where U represents random-coil forms) of roughly the order of 10^{-4}. These results suggest that some unfolded species may easily equilibrate with forms having elements that are structurally ordered along patterns similar to those of the same elements in the native structure.

If conformations containing small elements of structural order equilibrate with a population of denatured species, we can ask what conformations these elements of order might represent. A good guess is that segments of α-helical and β-sheet structures may occasionally form in a collection of random coils. In this connection, some ultrasonic relaxation measurements by Gordon Hammes and Peter Roberts (1969) on the dynamics of the helix–coil transition of poly-L-ornithine are of interest. These investigators estimated that the relaxation time for the transition is on the order of 10^{-8} sec. This means that helices can form and break many orders of magnitude faster than what has been observed for the overall folding–unfolding rates of proteins. Thus, denatured molecules might maintain a dynamic equilibrium involving some species that contain helical stretches; these regions in turn might serve as nucleation centers for the overall folding reaction.

It is worth noting that, if certain nucleated structures are readily formed and the folding then builds upon these, the final structure that is adopted may not be at a global free energy minimum. This is because the structure of truly lowest free energy

Box 21-2   REOXIDATION AND REFOLDING OF REDUCED PROTEINS

Pathways of refolding can be studied with proteins in which disulfide bonds are reduced. The reoxidation is accompanied by refolding, and intermediates may be detected at stages where oxidation is incomplete. This is done by quenching the reaction when a fraction of the disulfides have formed, and then treating the protein with one or more proteases so as to generate peptide fragments containing S–S linkages; the resulting disulfide-linked peptides may be compared with those obtained from the native protein.

R. R. Hantgan, G. G. Hammes, and H. A. Scheraga (1974) have examined in detail the refolding of reduced bovine ribonuclease, which in the native state has four disulfide bonds. Reoxidation was accomplished with a mixture of reduced and oxidized glutathione. With peptide mapping, it was shown that many intermediates are formed with incorrect, scrambled disulfide bonds. These slowly rearrange to generate the native structure. S. S. Ristow and D. B. Wetlaufer (1973) also have done studies with lysozyme. In this case, it appears that linkages other than native ones can form, although perhaps not all permutations are adopted.

T. E. Creighton (1974) has done some careful studies on the renaturation of reduced bovine tryspin inhibitor, a small protein of 58 amino acid residues with three disulfide bonds. Renaturation was accelerated by oxidized disulfides, such as oxidized dithiothreitol and hydroxyethyl disulfide; these compounds stimulated disulfide interchange reactions. Intermediates with just one disulfide were trapped and characterized. Four different one-disulfide intermediates were found. Of these, the most prominent (representing half of the total "one-disulfide" material) corresponds to a linkage found in the native form, whereas the other three species have bonds not present in the native molecule. It is suggested that the relative proportions of these four species represent a kind of equilibrium distribution of intermediates at the one-disulfide stage. The preponderance of the species with the linkage corresponding to one in the native structure suggests that, even at an early stage in refolding, elements of the native structure are thermodynamically preferred and therefore adopted.

What relevance do these results have to the question of refolding of proteins from a denatured to a native state? First, the *reoxidation* and *refolding* of a *reduced, denatured* protein is not the same issue as the refolding of a denatured protein in which all of the disulfide bonds are already formed and in the correct configuration. The work on disulfide intermediates shows that many different species form in the course of oxidation, and that these species may simply accumulate in amounts dictated by their relative thermodynamic stabilities. However, once the "correct" set of linkages is formed, the protein still must search through a set of conformations to find the native structure. For example, it is clear from the discussion in the text that alternate, denatured conformations are accessible to proteins that have all disulfides intact and correctly registered, and that the refolding of these species to the native conformation is not instantaneous. In vivo, there is evidence for a disulfide interchange enzyme (Epstein et al., 1963). Perhaps the shuffling of disulfide linkages in vivo occurs rapidly until some species have the correct set; these then proceed to fold into the native structure.

may be kinetically "inaccessible"; that is, the easily formed nucleated structures may not lead to the conformation representing the global free energy minimum, and the time required to find the minimum may be too long to be biologically useful. These ideas have been advanced by C. Levinthal (1968); they also are reviewed by D. B. Wetlaufer and S. Ristow (1973). However, for practical purposes, most experimental results are readily interpreted by assuming that the native structure is the thermodynamically preferred one.

We have summarized some of the present thinking in research on protein folding. Clearly, more work is needed in characterizing the pathway from denatured to native species. This research will require both further refinement of existing methods and the development of entirely new approaches.

Summary

The study of folding–unfolding reactions of proteins gives insights into the forces and factors that determine native-structure formation. The reversible denaturation of proteins commonly is a cooperative process. Many thermodynamic measurements can be interpreted in terms of a two-state model involving the native structure and the collection of random coil-like species known as the denatured form. For insight into the mechanism of folding, folding–unfolding transitions are studied by several different physical techniques. Discrepancies in results obtained with different methods can be ascribed to formation of stable intermediates in the folding process. One useful approach is to compare the overall enthalpy of denaturation measured calorimetrically with that obtained from a van't Hoff analysis of the thermal transition measured by a physical parameter (such as optical absorption). When intermediates exist, the magnitude of the former can be greater than that of the latter.

Kinetic studies provide one of the most powerful tools for examining details of the folding process. The existence of more than one kinetic phase indicates that the reaction is not a simple two-state process. Site-specific reporter groups are especially useful for monitoring and comparing conformational changes in different regions to determine, for example, whether local structural changes are concerted or decoupled. In this regard, NMR measurements also are particularly useful, because they permit examination of the resonances of specific residues, such as the histidine C-2 protons.

At present, the picture of the mechanism of protein folding is incomplete. Although some data on certain systems are consistent with a simple two-state model, other results indicate that (at least in some systems) more states may be involved. A plausible scheme is that nucleated structures involving segments of α-helical and/or β-sheet structure form rapidly, and that these structures act as "building blocks" around which the rest of the folding occurs.

Problems

21-1. Derive the expression in Equation 21-17 for ΔH_{app} of a protein that denatures through intermediate states.

21-2. Assume that a protein denatures according to the mechanism $N \rightleftarrows X \rightleftarrows D$. You measure two physical properties at three points in the denaturation where the true values of f_D are 0.2, 0.5, and 0.8; the corresponding values of $K = f_X/f_D$ are 1.0, 0.4, and 0.125. For one physical property, $d_1 = 0.1$; with the other, $d_1 = 0.2$. Will you be able to detect the presence of the intermediate by measuring the physical property? (Assume that the experimental error in each measurment of a physical property is 5%.)

21-3. Assume that the temperature dependence of ΔG for the denaturation of a protein can be represented by an equation of the form

$$\Delta G = a + bT + cT^2$$

where a, b, and c are constants. Calculate ΔH, ΔS, and ΔC_p in terms of a, b, c, and T. Does this protein have a temperature of maximal stability, T_{max}? If so, what is its value (in °C) if $a = 100$ cal mole^{-1}; $b = 1{,}200$ cal K^{-1} mole^{-1}; and $c = -2$ cal K^{-2} mole^{-1}?

21-4. The kinetics of folding and unfolding of a protein is studied under a variety of conditions with an attached reporter-group chromophore that absorbs in the visible range, away from the protein's absorption band. When the absorption at 280 nm of the protein's aromatic groups is monitored, both the folding and unfolding kinetics are strictly monophasic. When the reporter group's visible absorption is simultaneously studied with the same molecules, however, biphasic kinetics are seen in both directions. A control experiment is done with molecules having no attached reporter groups; with these, the kinetics at 280 nm in both directions are the same as the 280 nm kinetics of the molecules containing the attached chromophore. "These data collectively establish that a single $N \rightleftarrows D$ transition occurs, and that the reporter group perturbs the structure," declares a normally shy student. A very self-confident peer says, "This time you are probably right in your interpretation, although you did miss at least one point." Who (if either) is right, and why? If neither, state carefully why not. If you agree with the self-confident peer, explain the subtle point of interpretation. What other data would be useful to have?

21-5. Assume the following mechanism for folding and unfolding of a protein:

$$N \underset{k_{-1}}{\overset{k_1}{\rightleftarrows}} X_1 \underset{k_{-2}}{\overset{k_2}{\rightleftarrows}} X_2 \underset{k_{-3}}{\overset{k_3}{\rightleftarrows}} D$$

with X_3 connected via k_4, k_{-4} and k_5, k_{-5} to X_1 and X_2.

What is the minimum number of rate equations required to describe the kinetics of this system, using Δf_N, Δf_{X_1}, Δf_{X_2}, and Δf_D as variables? In the solution to these equations, how many λ_i parameters are there? Why? Now assume that all other rate constants are much larger than k_1 and k_{-1}. In which direction ($N \rightarrow D$ or $D \rightarrow N$) will the study of the kinetics most clearly reveal more than one kinetic phase? Why?

References

GENERAL

Anfinsen, C. B., and H. A. Scheraga. 1975. Experimental and theoretical aspects of protein folding. *Adv. Protein Chem.* 29:205.

Baldwin, R. L. 1975. Intermediates in protein folding reactions and the mechanism of protein folding. *Ann. Rev. Biochem.* 44:453. [A clearly and thoughtfully written summary; an excellent place to start.]

Némethy, G., and H. A. Scheraga. 1977. Protein folding. *Quart. Rev. Biophys.* 10:239.

Tanford, C. 1968. Protein denaturation, part A: Characterization of the denatured state. Part B: The transition from native to denatured state. *Adv. Protein Chem.* 23:121.

———.1970. Protein denaturation, part C: Theoretical models for the mechanism of denaturation. *Adv. Protein Chem.* 24:1 [This series of articles provides a comprehensive treatment of the protein-denaturation problem and helps to put the problem in concrete physical chemical terms.]

Wetlaufer, D. B., and S. Ristow. 1973. Acquistion of three-dimensional structure of proteins. *Ann. Rev. Biochem.* 42:135. [The authors stress the importance of kinetic factors that may limit acquistion of the global free energy minimum.]

SPECIFIC

Creighton, T. E. 1974. Renaturation of the reduced bovine pancreatic trypsin inhibitor. Intermediates in the refolding of reduced pancreatic trypsin inhibitor. The single-disulphide intermediates in the refolding of reduced pancreatic trypsin inhibitor. *J. Mol. Biol.* 87:563.

Epstein, H. J., R. F. Goldberger, and C. B. Anfinsen. 1963. The genetic control of tertiary protein structure: Studies with model systems. *Cold Spring Harbor Symp. Quant. Biol.* 27:439.

Hammes, G. G., and P. B. Roberts. 1969. Dynamics of the helix–coil transition in poly-L-ornithine. *J. Am. Chem. Soc.* 91:1812.

Hantgan, R. R., G. G. Hammes, and H. A. Scheraga. 1974. Pathways of folding of reduced bovine pancreatic ribonuclease. *Biochemistry* 13:3421.

Ikai, A., and C. Tanford. 1971. Kinetic evidence for incorrectly folded intermediate states in the refolding of denatured proteins. *Nature* 230:100.

Kanehisa, M. I., and T. Y. Tsong. 1978. Mechanisms of the multiphasic kinetics in the folding and unfolding of globular proteins. *J. Mol. Biol.* 124:177.

Karplus, M., and D. L. Weaver. 1979. Diffusion-collision model for protein folding. *Biopolymers* 18:1421.

Levinthal, C. 1968. Are there pathways for protein folding? *J. Chem. Phys.* 65:44.

Levitt, M., and C. Chothia. 1976. Structural patterns in globular proteins. *Nature* 261:552.

Lin, N.-L., and J. F. Brandts. 1978. Further evidence suggesting that the slow phase in protein unfolding and refolding is due to proline isomerization: A kinetic study of carp parvalbumins. *Biochemistry* 17:4102.

Ristow, S. S., and D. B. Wetlaufer. 1973. Evidence for nucleation in the folding of reduced hen egg lysozyme. *Biochem. Biophys. Res. Commun.* 50:544.

Sachs, D. H., A. N. Schechter, A. Eastlake, and C. B. Anfinsen. 1972. An immunological approach to the conformational equilibria of polypeptides. *Proc. Natl. Acad. Sci. USA* 69:3790.

Schmid, F. X., and R. L. Baldwin. 1978. Acid catalysis of the formation of the slow-folding species of RNase A: Evidence that the reaction is proline isomerization. *Proc. Natl. Acad. Sci. USA* 75:4764.

Nucleic acid structural transitions

22-1 NUCLEIC ACID STUDIES

How do nucleic acids behave in aqueous solutions? What kinds of ordered structures do they form, and how can the observed physical and chemical properties be rationalized in terms of these structures? How does a nucleic acid convert from one structural form to another? These questions are the focus of this chapter and the following two. It is clear that they must be answered if one is to claim a real understanding of how nucleic acids carry out their central function in the storage and transmission of genetic information. Earlier chapters have laid the foundations for an explanation of some of the complicated features of nucleic acid behavior. Chapter 3 covers structure and bonding. Chapter 6 deals with the forces that stabilize nucleic acid conformations. Chapters 7 through 12 outline techniques capable of revealing conformational properties in solution. Here we begin to integrate all this information and to examine some of the properties of natural nucleic acids and of various convenient model systems. The focus must be on solution behavior, because that is where nucleic acids function in nature.

Comparison with protein studies

It is instructive to contrast the general problems of protein conformational states and changes, treated in Chapter 21, with the problems faced here.

The average protein of interest is much smaller than a typical nucleic acid. The molecular weights of all but a few nucleic acids are in the millions of daltons. The power and ease of application of most physical techniques diminishes considerably with increasing molecular weight.

Simple informative biological assays exist for most proteins, but very few for nucleic acids. Such assays can provide reassurance that the form being studied is of some biological significance.

It is comparatively easy to focus on a particular region of a protein by exploiting infrequently occurring (or particularly reactive) amino acid side chains, or by using prosthetic groups or substrates. Nucleic acids have few residues with useful selective reactivity or easily resolved physical properties.

The range of protein structures and functions is so wide that comparative studies can add extra dimensions to the investigation, and sometimes can enable a seemingly intractable problem to be circumvented by using a different system. Comparatively few nucleic acid species exist except for the plethora of mRNA species, most of which are present in amounts too small (or are too difficult to purify) for convenient study.

Most complex proteins are composed of subunits. Frequently, these can be reversibly disassembled, making it possible to study the properties of the individual components separately and then to generalize back to the original intact structure. The only subunits of nucleic acids are the separate strands of double-stranded polymers, and studying these individually is not necessarily useful.

However, nucleic acids do have a few distinct advantages over proteins for physical studies. Because there are essentially only four side chains (except in tRNA), the range of fundamental interactions is much more limited. As you will see, most of the interactions occurring in intact nucleic acids also occur, in almost identical form, in the smallest fragments of nucleic acids, in dinucleotides, and in compounds one to two nucleotides longer. Therefore, powerful techniques can be applied to small fragments, and the results can be generalized to infer or rationalize properties of the much larger polymers. In contrast, short peptides are useless as models for proteins and, in fact, are not even very good models for polypeptide secondary structure.

In almost any real protein, the answers to most functionally significant questions are found at the level of the detailed tertiary structure. This poses a strain on most solution techniques, because they do not readily provide the detailed information required. At present, it appears that some aspects of nucleic acid function can be understood at the level of secondary structure. This may simply be an admission of the primitive state of present knowledge. Certainly some nucleic acids, such as tRNA, have tertiary structures as complex and as important as those of a protein. But in many other cases, one need not consider tertiary interactions of side chains to explain currently known properties. As long as this situation continues, it allows an enormous simplification.

22-2 STRUCTURE AND STABILITY OF NUCLEIC ACID SINGLE STRANDS

It is common knowledge that double-stranded nucleic acids form ordered helical structures such as the Watson–Crick helix of B-form DNA. But what happens if you remove one of the two strands of a DNA duplex? Even better, what happens if you synthetically construct a homopolynucleotide such as poly dA, in which all possibilities for hydrogen-bonded base pairs as they exist in DNA are eliminated? Consider two extreme hypotheses. First, the single strand might form a random-coil structure—perhaps stiff and extended because of the limited rotational freedom in the phosphodiester backbone (see Chapter 6), but a coil none the less. At the other extreme, the single strand might form a perfect helix with the same radius and pitch as if it were part of a base-paired double strand.

Neither of these two extreme hypotheses is correct but, as we shall see, the second is closer to the truth than the first in many cases. Nucleic acid single strands are highly ordered and are at least locally helical. Their properties depend strongly on the base composition and sequence of the phosphodiester chain. However, they are easily perturbed by alterations in temperature, pH, and solvent. Although their local properties are not especially sensitive to ionic strength, such long-range properties as hydrodynamic parameters are quite sensitive, as would be expected for any polyelectrolyte with the high charge density of a nucleic acid.

Evidence of ordered structure in homopolynucleotides

The interactions governing the spectroscopic properties of arrays of identical or similar residues are very sensitive to the orientation and distance of nearby groups. These include the optical dipole interactions responsible for CD and hypochromism, and the ring-current effects important in NMR. A true random-coil structure consists of an ensemble of residues with many different conformations. Therefore, the optical properties or NMR of a random coil should closely resemble the properties of the individual monomer residues,[§] because all interactions would tend to be canceled by orientational averaging. Indeed, the absorption spectra of CD of homopolynucleotides such as poly rA and poly rC (also called poly A and poly C) at temperatures near 100°C are quite close to the corresponding properties of mononucleosides or mononucleotides. At lower temperatures, however, the optical properties of homopolynucleotides become dramatically different from the corresponding monomer properties. The CD of poly rA, poly rC, their deoxy analogs, and most other simple-sequence nucleic acids is much more intense at room temperature than is the CD of the corresponding monomer. The signs, positions, and numbers of observable ellipticity bands in the near UV are altered. Figure 22-1 shows CD results for poly rA. These data immediately suggest that poly rA has some kind of ordered structure at neutral pH and room temperature (see Chapter 8).

It is not easy to deduce many structural details of a polymer from CD data alone. The conservative nature of the two nearest UV bands suggests exciton splitting. This splitting indicates that, on the average, neighboring bases in the polymer have

---

[§] Dipolar broadening might be expected in NMR because of possible slow tumbling rates of a large, stiff coil. We ignore this effect in our discussion here.

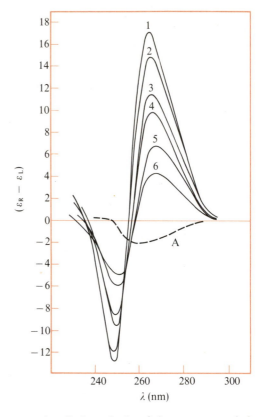

Figure 22-1

Circular dichroism (per residue) of poly rA as a function of temperature. Data were obtained in 0.1 M NaCl (pH 7.4) at (1) −2° to 6°C, (2) 17°C, (3) 34°C, (4) 42°C, (5) 57°C, and (6) 70°C. The CD of the monomer adenosine also is shown (at 0°C); it is essentially temperature independent. [After P. O. P. Ts'o, in *Basic Principles in Nucleic Acid Chemistry*, vol. 2, ed. P. O. P. Ts'o (New York: Academic Press, 1974).]

preferred orientations. Without a much more detailed analysis of the spectrum, it is impossible to be certain whether the order extends over the whole polymer or is only local order. To carry out such analysis, it usually is necessary to have, in advance, more detailed hypotheses about the nature of the structure.

Table 22-1 summarizes CD data on some ribonucleosides, oligoribonucleotides, and polyribonucleotides at room temperature. The results for poly rC are qualitatively similar to those for poly rA, and this is the pattern generally observed. An exception is poly rU, which has an optical activity not much different from that of its monomer. This result alone does not permit the conclusion that poly rU is devoid of an ordered structure, because CD is very sensitive to the angles between neighboring residues. It is possible that an ordered structure could show no optical activity in excess of that for the monomers. For example, a stacked parallel structure—like the teeth of a comb—would have no exciton contribution to the CD if the chromophore had only a single electronic transition in the accessible region of the spectrum.

Absorption spectra provide more clues that most homopolynucleotides are not random coils. Poly rA, poly rC, and many others show substantial hypochromism compared with the spectra of their constituent monomers (Table 22-1). Hypo-

Table 22-1

Optical properties of oligoribonucleotides and polyribonucleotides

Substance	Absorbance			Circular dichroism	
	λ_{max} (nm)	ε_{max} $\times 10^{-4}$ per residue	Hypochromicity at λ_{max}	λ_{max} (nm)	$\theta(\lambda_{max})$ $\times 10^{-4}$ per residue
Adenosine (A)	259.2	1.49	0.00	263.5	−0.38
ApA	257.5	1.36	0.12	271	2.03
Poly rA	256.5	0.90	0.33	264	7.2
Cytidine (C)	271	0.91	0.00	272.5	1.04
CpC	270	0.83	0.07	280.	3.08
Poly rC	269.5	0.62	0.31	278	7.4
Uridine (U)	261.5	1.01	0.00	267.5	0.92
UpU	261	0.98	0.02	271	1.61
Poly rU	260	0.92	0.08	267.5	1.50

SOURCE: Data from the work of I. Tinoco and colleagues.

chromism is indicative of a local stacked arrangement of chromophores. Again, poly rU is an exception to this general pattern. Hypochromism thus supports the implication from CD that poly rU may have little, if any, ordered local structure. Once again, however, it is difficult to interpret the degrees of hypochromism seen for the other polymers in terms of any details of the local structure beyond a preference for stacking. Nor do existing theories of hypochromism allow predictions accurate enough to assess the extent of the local ordering. However, the degree of hypochromism of poly rA and poly rC is considerably less than that seen for double-helical structures such as the DNA B form. This is an indication that the extent of local order in the homopolymers is less than that expected for a quasirigid perfect helix.

The structure assumed by homopolynucleotides must be the result of thermodynamic interactions among monomer residues and between monomers and solvent. Polydeoxyribonucleotides and polyribonucleotides differ only by the presence or absence of the 2′-hydroxyl of the ribose sugar. However, this difference has a profound effect on the preferred puckering of the ribose ring, which in turn affects the statistical chain dimension of single strands. It also must be responsible for the very different double helices adopted by complementary duplexes of RNA or DNA. It is not surprising that the local structures of individual homopolymers seen by optical properties are also affected by the 2′-hydroxyl. For example, the CD spectra of poly dC and poly rC are really quite different in intensity (Fig. 22-2). The electronic states of dC and rC are virtually identical, so one can be quite sure that the spectral differences reflect structural differences. It is the simple presence or absence of the oxygen that causes these structural differences. Substitution of the 2′-OH by 2′-OMe results in polymers that closely resemble the normal polyribonucleotides in most properties.

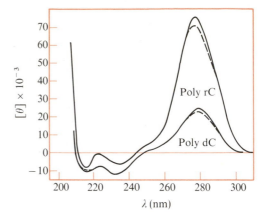

Figure 22-2

Circular dichroism of poly rC and poly dC at pH 10.0. Results are shown in 2 M NaCl (solid curves) and in 10^{-4} M NaCl (dashed curves). [After A. Adler et al., *Biochemistry* 7:3836 (1968).]

Effect of salt on local single-strand structure

Any polymer in a fully ordered structure is at a deep free energy minimum. It usually is safe to assume that small perturbations of experimental conditions will have little effect on properties such as CD. The easiest variables to test for most polymers are pH, ionic strength, temperature, and solvent. Solvent effects are very pronounced with homopolymers. Virtually any organic substance except sucrose, added in moderate quantities, tends to disrupt the ordered forms. The effects of pH are complex; we shall discuss them in Section 22-3.

Individual homopolynucleotides have optical properties relatively insensitive to ionic strength over a very large range, as Figure 22-2 shows for poly rC. Very high salt concentrations (>1 M) will cause structural changes, but here sensitivity to ionic strength is no longer a simple measure of electrostatic effects. The structure of water may be altered enough at these concentrations that the added salt should probably be viewed as a solvent change. The absence of salt effects at lower ionic strengths means that all local structures available to compounds such as poly rC must have similar average separations between near-neighbor phosphates.

The electrostatic potential Φ at a position $\mathbf{r}$ near an array of z charges in a nonionic medium is expressed by Coulomb's law:

$$\Phi(\mathbf{r}) = \sum_{k=1}^{z} e/(\varepsilon|\mathbf{r} - \mathbf{r}_k|) \tag{22-1}$$

where ε is the dielectric constant, e is the charge on one electron, and $\mathbf{r}_k$ is the position of the kth charge. The electrostatic free energy can be calculated as the work needed to place all these charges at their respective positions $\mathbf{r}_l$:

$$G_{\text{el}} = \sum_{l=1}^{z} \int_0^e \Phi(\mathbf{r}_l)\,de \qquad \text{for } l \neq k \tag{22-2}$$

As this equation indicates, the calculation is made by summing the individual work needed to place each charge in its position, assuming that all others are already in place.

Equation 22-1 is not appropriate for charged polymers in aqueous buffer with a supporting electrolyte, because of the screening effects of counterions (discussed in preceding chapters). Instead, a shielded potential must be used in such a case. A convenient approximate form is

$$\Phi(\mathbf{r}) = \sum_{k=1}^{z} \left[e/(\varepsilon|\mathbf{r} - \mathbf{r}_k|) \right] \exp(-\kappa|\mathbf{r} - \mathbf{r}_k|) \qquad (22\text{-}3a)$$

where κ (the Debye–Hückel screening parameter) is a function of the ionic strength I. The parameter κ measures the extent to which electrostatic interactions die off faster than expected because of intervening counterions:

$$\kappa = (8\pi e^2/100\varepsilon kT)^{1/2} I^{1/2} \qquad (22\text{-}3b)$$

The ionic strength I is related to the total concentration of charged species in the solution:

$$I = \sum_i C_i z_i^2/2 \qquad (22\text{-}3c)$$

where C_i is the molar concentration of the ith ionic species with charge z_i. For a 1:1 electrolyte (such as NaCl), I is equal to the molar concentration of cations or anions.

To calculate the electrostatic free energy of a polymer, we now substitute Equation 22-3a into Equation 22-2, and use the definition $r_{kl} = |\mathbf{r}_l - \mathbf{r}_k|$. The electrostatic free energy thus becomes

$$G_{el} = \sum_{l=1}^{z} \int_0^e (e/\varepsilon)\, de \sum_{k \neq l}^{z} \left[\exp(-\kappa r_{kl})\right]/r_{kl}$$

$$= (e^2/2\varepsilon) \sum_{l=1}^{z} \sum_{k \neq l}^{z} \left[\exp(-\kappa r_{kl})\right]/r_{kl} \qquad (22\text{-}4)$$

for a polymer such as poly rA, where all sites have the same charge. This is a very approximate equation. It assumes a macroscopic dielectric constant that is the same at all points in the structure. Also, the accuracy of the form of Equation 22-3a is open to question. It is adequate, however, for the simple argument that follows.

At room temperature in aqueous buffer, κ is 0.33 $I^{1/2}$ Å^{-1}. Thus, at ionic strengths of 0.1 M or higher, κ is large, so contributions to G_{el} are quite small, except for charges in very close proximity ($r_{kl} \to 0$). At low ionic strength, the exponential terms in Equation 22-4 approach a value of one. Then G_{el} not only becomes enormous, but it also becomes very sensitive to long-range effects, because of the r_{kl}^{-1} term.

According to Equation 22-4, the insensitivity of poly C optical properties[§] to ionic strength allows several conclusions. Poly C is unlikely to be a multiple-stranded structure, because in such a case charge repulsion at low ionic strength ought to be so large that strands separate and alter the ionic properties. Poly C is unlikely to have any kind of a structure in which residues distant along the covalent chain are able to come into close proximity and·cause detectable optical interactions. However, the lack of ionic-strength effects leaves open the question of whether poly C has any short-range order.

Effect of temperature on homopolymer properties

The optical properties of poly A, poly C, and most other homopolymers are strongly temperature dependent (see Fig. 22-1). Throughout the whole range of temperatures in which biological molecules usually are studied (the range that corresponds to temperatures capable of supporting life as we know it), CD and absorption spectra vary monotonically. Near 100°C, these changes tend to saturate, as optical properties approach the monomer values. At this extreme, it is clear that local conformation is a random-coil or denatured structure. At very low temperatures (below 0°C in most cases), there usually is some indication of a leveling-out of the temperature-induced effects. This plateau represents the maximal obtainable ordering, but it by no means demands the presence of a perfectly formed structure with long-range order.

At room temperature, poly A and the other homopolymers clearly are in the midst of a structural transition from the high-temperature coil to the low-temperature ordered form. The breadth of this temperature transition suggests that the apparent enthalpy change is rather small compared to typical macromolecular conformational changes. The actual CD spectra as a function of temperature form a family of curves with very similar shapes, mostly differing only in magnitude. This is evidence that, at each temperature, whatever structural regions exist are similar, and only their extents may vary.

Hydrodynamic data best indicate whether the ordered forms of homopoly-nucleotides are rigid. Figure 22-3 shows a summary of the extensive studies by G. Felsenfeld and colleagues on poly A. The degree of local structure (base stacking) was determined as a function of temperature by optical techniques. The rigidity was measured as a function of temperature by hydrodynamic and scattering techniques. The radius of gyration derived from these measurements changes only slightly with

---

[§] The designations "poly C" and "poly rC" are equivalent; references to the deoxy forms are always made explicit (e.g., poly dC). From this point, we shall explicitly indicate the ribo form only when we are comparing ribo and deoxy forms in the immediate context.

Figure 22-3

Dependence on extent of base stacking of the root-mean-square radius of gyration of poly rA. Points shown are experimental values for a sample with an average length of 1,740 residues, corrected for heterogeneity. Curves are the predicted dependence of $\langle R_G^2 \rangle^{1/2}$ upon stacking for this sample, given four possible sets of restrictions on the torsional angles of the unstacked form. [After L. D. Inners and G. Felsenfeld, *J. Mol. Biol.* 50:373 (1970).]

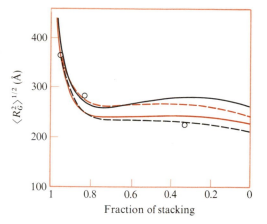

temperature, except at very low temperatures where near-maximal local order is achieved. Combining the optical and hydrodynamic results, one can plot the radius of gyration as a function of the degree of base stacking (Fig. 22-3). Only where the bases are nearly fully stacked does the polymer begin to show a marked increase in rigidity. This is the result of two effects. The coiled forms of polynucleotides already are relatively extended and rigid. Furthermore, at intermediate degrees of stacking, the ordered regions clearly are too short to alter markedly the overall chain configuration (see Chapters 6 and 19).

Poly U at room temperature has a radius of gyration close to that observed for poly A at high temperature, where nearly all local structure has been eliminated. This observation provides strong confirmation that poly U does not form the ordered structures typical of other homopolynucleotides, except in high salt concentrations at very low temperature.

Single-stacked helices

None of the data discussed thus far allow one to conclude rigorously that homopolymers are single-stranded structures (Fig. 22-4a) at neutral pH. It is instructive to examine why this hypothesis is difficult to prove from solution physical data on homopolynucleotides alone.[§] Two kinds of multiple-stranded structures (Fig. 22-4b,c) also must be considered plausible. Both can be formed intramolecularly, so that no change in molecular weight will result. Although Watson–Crick base pairs cannot form in homopolymers, there could be structures (Fig. 22-4b) containing other hydrogen-bonded base pairs known from studies on model systems. The second kind

---

[§] For small molecules, intermolecular associations can always be distinguished by dependence of physical properties on concentration. However, for polymers, intermolecular interactions can be so strong that the dilutions required to counteract them would preclude virtually any type of physical measurements.

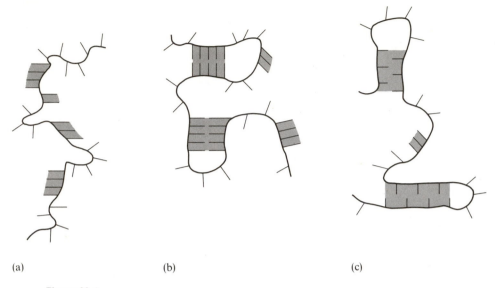

Figure 22-4

Three hypothetical models for partially ordered poly rA (or any other homopolynucleotide).
(**a**) Base stacking.   (**b**) Base pairing.   (**c**) Intercalation.

of possible double-stranded homopolymer structure is formed by intercalating adjacent bases to form a stack that alternates from one strand to the other (Fig. 22-4c). To visualize this, imagine the bases corresponding to the teeth of intermeshed gears.

Base pairs can be excluded for such compounds as poly A and poly C at neutral pH. Infrared spectra are particularly sensitive to hydrogen-bonded interactions, and the spectra of most homopolynucleotides at neutral pH fail to show the shifts expected if any kind of base pairing were present. The same conclusion can be reached from NMR studies, which also should demonstrate characteristic shifts if base pairing occurred (see Chapter 9).

Studies of polymers formed by base analogs offer another argument against base pairing in single homopolymers. Evidence for ordered structures resembling normal homopolymers can be found in analogs in which various possible hydrogen-bond donor or acceptor positions (such as purine N-1, N-6, and N-7) have been blocked by methylation. More indirect evidence comes from studies on chemical modifications. Reagents such as formaldehyde react quite rapidly with the N-6 *exo* amino group of poly A, yet do not react on anything near the same time scale with adenosines known to be involved in base pairing, Watson–Crick or otherwise. Tritium exchange with single homopolymers also is much faster than with known double-stranded systems. However, one must be very cautious about relying too heavily on such data. A double-stranded structure capable of rapid opening and closing could easily show such fast rates.

None of the arguments raised thus far rule out intercalated structures. Even if a fairly ordered form were available, hydrodynamic or small-angle x-ray data would not permit ready discrimination between the intercalated structure and a stacked single strand. For long rigid rods, both of these methods essentially measure the mass per unit length. Various forms—including a stacked single helix (e.g., one strand of the DNA B form) or an intercalated tightly stacked duplex—could each have one base pair every 3.4 Å. Solution studies with small-angle x-rays actually show relatively short ordered regions at low temperature with one base each 3.5 Å. Hydrodynamic studies on poly A at low temperature show an average residue spacing of 3.2 Å.

Probably the strongest evidence presented thus far against the intercalated structure in solution is the insensitivity of the neutral forms of homopolymers to decreasing ionic strength. However, the comparisons of oligonucleotides and polynucleotides described in the following subsection do decisively rule out the intercalated structure in solution.

The ultimate evidence that homopolymers are single-stranded, stacked helices comes from studies of x-ray fiber diffraction. The structure of poly C has been determined in this way (Fig. 22-5). It is evident that the polymer is a single-stranded helix, and that there is extensive stacking between adjacent bases. Note however that the structure of the polymer is very different from that of a single strand of the usual RNA or DNA double helices. Poly C has only six bases per turn, compared to ten or eleven for the double helices.

Oligonucleotides as models for polynucleotide single strands

It is very difficult to extract structural information from even simple homopolymers, thus discouraging the use of homopolymers as convenient models for polynucleotides. However, oligonucleotides of length two to six residues, containing either normal bases or a wide variety of analogs, can be prepared easily. Effects of base composition and sequence thus are readily accessible, as are examinations of characteristics as a function of chain length.

In principle, oligonucleotides offer several advantages over polynucleotides. Their optical properties are simpler to analyze, because there can be only a fixed number of interactions such as exciton coupling. The NMR spectrum of oligonucleotides is much easier to analyze than that of polynucleotides for three reasons. The number of bands must be smaller, facilitating both detection and spectral assignment. The molecular weight is less, enabling studies at high concentrations. Finally, any ordered structures are small enough that tumbling is fast enough to average out dipolar broadening, so that sharp resonances can be observed. The conformational changes in oligonucleotides do not resemble phase transitions in sharpness or in apparent enthalpy. Thus, concentration is a useful thermodynamic variable, and indeed is a natural and convenient variable for distinguishing between intramolecular and intermolecular effects. Oligonucleotides are also small enough for crystallization and subsequent high-resolution x-ray studies.

All these advantages, however, are tempered by some drawbacks. There is no way to use oligonucleotides to study long-range interactions or interactions between points distant in the covalent structure. If these interactions are important, as they

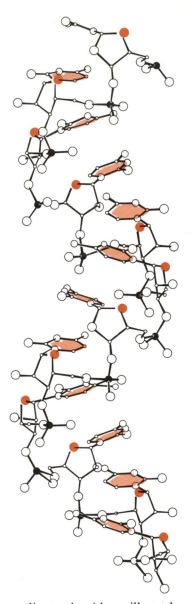

Figure 22-5

Structure of poly C determined by x-ray diffraction. Note the extensive base stacking of adjacent residues; bases are shaded. [After S. Arnott, R. Chandrasekaran, and A. G. W. Leslie, *J. Mol. Biol.* 106:738 (1976).]

must be for globular tertiary structures, oligonucleotides will not be good models. The oligomers must contain all of the short-range thermodynamic interactions that contribute to a polymer, and these interactions often can be sorted out and individually measured in such small model systems. However, there is no reason to expect a priori that these interactions in the oligomer will result in a structure or conformation identical (or even similar) to a polymer. Remember that an oligomer is nothing but a short polymer. If end effects exist, they must be more important in determining an average oligomer structure than they are in an average polymer structure. Great caution must be used in extrapolating from the oligomer structure to the polymer structure. If the properties of a polymer can be predicted correctly by assuming that its structure is similar to that determined for an oligomer, then it usually is safe to conclude that the oligomer really is a valid structural model.

Do oligonucleotides have physical properties similar to those we have discussed for homopolymers? The optical properties of dinucleoside phosphates are quite different from those of the monomers (Table 22-1). Compounds such as ApA and CpC have much more intense CD than do A and C. The spectra of the dimers closely resemble those of poly A and poly C in shape, though the magnitudes of individual bands are smaller. These dimers also are hypochromic like the polymers, but again the degree of hypochromism is less than that of the polymers. The dimer UpU is an exception; its optical properties closely resemble those of the mononucleoside U, but the same is true for poly U. The spectral properties of dimers are sensitive to changes in temperature, solvent, and pH, just as the polymers are. Like the polymers, the dimers show little effect from alterations in ionic strength. All of these parallels suggest that the dimers may be valid models for the polymers, thus justifying more detailed studies of the structure of dimers.

Several crystallographic studies of short oligonucleotides have been completed. Three types of compounds are represented in the available set of data. A few structures were studied at acid pH; these should not be good models for neutral polynucleotides because the state of ionization of the bases is different. Other structures contain self-complementary sequences such as GpC; these form base-paired duplexes in the crystal. Such structures are fine models for double strands. Of more interest here is the third category: neutral non-self-complementary oligonucleotides. The major feature of interest in these structures is that the bases are near one another and are in an approximately stacked array.

Such crystal structures provide a wealth of information about preferred bond angles and rotational torsional angles, information that is essential for the analysis of chain configuration. However, an x-ray structure provides a single static view of molecular conformation. In the case of single-stranded oligonucleotides or polynucleotides, all available solution data show equilibria among more than one conformation. It is not obvious whether the relative base–base geometry seen in x-ray results represents typical stacked geometry in solution or some extreme of the range of structures that are significantly populated in solution. However, the x-ray results are an essential benchmark, and they do indicate that the concepts developed thus far in this chapter must be on the right track.

Optical comparisons of oligonucleotides and polynucleotides

We have already seen evidence of structural differences between the deoxy and ribo forms of polynucleotides. Figure 22-6 shows evidence from optical studies indicating that these differences persist down to the level of dimers (dinucleoside phosphates). This observation suggests that a sufficiently detailed understanding of the structural

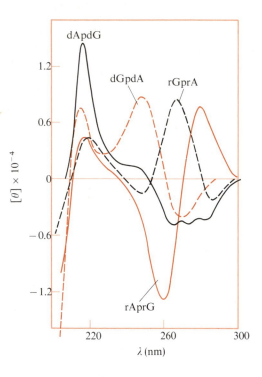

Figure 22-6

Circular dichroism of four dinucleoside phosphates, each containing one adenine and one guanine chromophore, at pH 7. The ultraviolet absorption spectra of these four compounds are nearly identical. [After M. Warshaw and C. R. Cantor, *Biopolymers* 9:1079 (1971).]

differences between the deoxy and ribo forms of the dimers might enable us to rationalize the differing behaviors and structures of DNA and RNA. The absorption spectra of ApG and GpA, as well as those of their deoxy analogs, are virtually superimposable, indicating that the electronic states must be very similar in all four compounds. However, the CD spectra for these dimers are strongly dependent on sequence (Fig. 22-6). The CD of mononucleosides is so weak that virtually all the spectral effects seen in Figure 22-6 must be the result of base–base interactions. The critical aspect of the structure for CD is the relative orientation and proximity of the two chromophores. Thus we conclude that the orientation of the bases must differ between ApG and GpA. Because both compounds show appreciable hypochromism, the bases must be stacked in each. One must look primarily at the angular rotation of the stack to account for the observed differences.

A helical structure comes to mind as an obvious possibility. Consider the molecule ApGpA. If this were a helix, the orientation of transition dipoles between the first A and the G would be different from that between the G and the second A. For example, let α be the angle between the transition dipoles of particular transitions in A and G when the N^9–$C^{1'}$ bonds of the two nucleosides are superimposed. If θ is the helix angle, then the angle between the transition dipoles in ApG will be $\theta - \alpha$, whereas in GpA it will be $\theta + \alpha$. This immediately leads to the prediction of different CD for ApG and GpA, even if the structure of the phosphodiester backbone and the glycosidic torsional angles are the same for the two dimers.

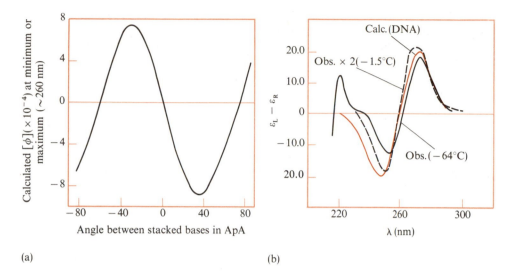

(a)   (b)

Figure 22-7

Optical activity of a dinucleotide, ApA.   (**a**) Calculated long-wavelength maximal or minimal optical rotation as a function of the angle between the bases. A plot of the long-wavelength CD maximum or minimum would be virtually identical. [After C. A. Bush and I. Tinoco, Jr., *J. Mol. Biol.* 25:601 (1967).] (**b**) Calculated and measured CD spectra. The calculations were performed for ApA in the DNA B-form conformation. [After W. C. Johnson and I. Tinoco, Jr., *Biopolymers* 8:715 (1969).]

Attempts to calculate rigorously the CD of GpA and ApG using the methods described in Chapter 8 have been disappointing, because our knowledge about the electronic structure of G is so imprecise. The base A is much better understood; Figure 22-7a shows optical rotation calculated for ApA, essentially by Equation 8-19. In these calculations, the bases were held in parallel planes, but the angle between them was allowed to vary. The sign and intensity of the optical rotation is a sensitive function of this angle. In the range of 30°, an intense negative rotation at the absorption maximum is predicted. This is what is actually observed experimentally. Figure 22-7b shows an actual CD spectrum calculated from this choice of geometry compared with the experimental spectrum. The agreement is impressive. Although the calculations are only approximate and the structures have been modeled by only a single rotation angle rather than explicit variation of allowed degrees of backbone freedom, this result is nonetheless very pleasing. A helix angle of 30° forms a right-handed helix with the same number of residues per turn as the DNA B form. Therefore the result, although just a model, is certainly physically reasonable.

The optical properties of ApA (and virtually all other dimers examined) are invariant over a thousandfold range of concentration. This invariance indicates that only intramolecular effects are contributing to the structure. The similarity of the dimer and polymer spectra immediately suggests that the polymers too must be single-stranded. The structural analogy between dimers and polymers can be tested by attempting to calculate the optical properties of the polymer from the measured properties of the dimer. In such semiempirical calculations, we assume that the structure is independent of chain length and that nearest-neighbor optical interactions are dominant. The methods are described in Chapter 8. For homopolymers and dimers, the equations to use are very simple. For example, the molar rotation per residue of the polymer is

$$[\phi]_{\text{polyN}} = 2[\phi]_{\text{NpN}} - [\phi]_{\text{N}} \qquad (22\text{-}5)$$

where $[\phi]_{\text{NpN}}$ is the molar rotation per residue of NpN, and $[\phi]_{\text{N}}$ is the molar rotation of nucleoside N. Figure 22-8 shows such a curve calculated for poly C compared

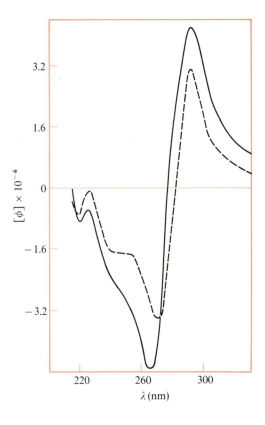

Figure 22-8

Optical rotatory dispersion of poly rC at pH 7. The solid curve shows measured values. The dashed curve is calculated from other measured data by the equation $[\phi]_{\text{polyC}} = 2[\phi]_{\text{CpC}} - [\phi]_{\text{C}}$. [After C. R. Cantor et al., *J. Mol. Biol.* 20:39 (1966).]

with actual experimental results; the agreement is surprisingly good for such a simple approach, strongly suggesting that dimer structural data can be carried over to single-stranded polymers.

Actually, the a priori justification for neglecting optical interactions beyond nearest neighbors is not a good one. We have suggested in Chapter 8 that, for a rigid ordered structure, it is necessary to consider exciton states that are linear combinations of all monomers. However, homopolymers at neutral pH and room temperature are neither rigid nor fully ordered. Thus the extent of disorder will tend to average out the longer-range optical interactions. In other words, dimers are, in practice, good models for single-stranded nucleic acids because they are likely to resemble transient small clusters of stacked residues.

NMR study of the stacked structures

CD is useful for generalization, and it is one of the few techniques that can bridge the gap between large and small structures. For a more detailed picture, however, one must use a technique with higher information content. NMR fits this requirement because it allows simultaneous monitoring of a number of different protons in an oligonucleotide. In practice, one can fairly easily resolve resonances from the H-2 and H-8 of each A, from the H-8 of G, and from the H-1 of each sugar for purine-containing dinucleoside phosphates. With pyrimidines, one can resolve H-5 and H-6 in C and U as well as the methyl protons in T. Figure 22-9 shows an NMR spectrum of ApC and CpA; the major stumbling block in using it is the assignment of resonances to protons. With mononucleosides, a sufficient number of chemical derivatives usually exists so that assignment is easy. In the dimers, however, resonances shift substantially. For example, it is not safe to assume that, because in the monomer H-8 of A is at lower field (more negative ppm with respect to a TMS standard) than H-2 of A, the order of the bands will stay the same in the dimer.

The proton resonances in ApC and CpA have been assigned by the following method. At high temperatures, purine H-8 protons exchange. Therefore, heating a dimer in D_2O will decrease the intensity of the H-8 resonance when the sample is subsequently measured at room temperature. Spin coupling between the spatially close H-5 and H-6 resonances of C splits each into a doublet with a coupling constant of 7.6 cps in the monomer cytidine. Coupling constants are a function of the dihedral angle between the two interacting protons, independent of other magnetic interactions. The geometry of the pyrimidine ring is planar and has no degrees of internal rotational freedom. Therefore the angle between H-5 and H-6 of C is the same in a dimer as in a monomer. The coupling constant J_{5-6} also is the same, and this enables us to distinguish the cytidine H-5 and H-6 protons from all the others. The two protons have resonances at such different positions in the monomer that no ambiguities arise in distinguishing them from each other in the dimer.

The protons yet to be assigned are the H-1' ribose protons of A and C; each of these resonances is split into a doublet by the corresponding H-2' ribose proton. The coupling constants $J_{1'-2'}$ are quite different in monomers and dimers. This is reasonable, because the puckering of the ribose ring can change relatively easily. It is not safe to assume that the adenosine H-1' doublet in a dimer is the one with a coupling constant closest to that seen in adenosine. Instead, we distinguish the H-1' resonances of A and C by chemical shifts. The close proximity of the ring to the H-1' proton ensures that, in a dimer, the chemical shift of H-1' will be dominated by the

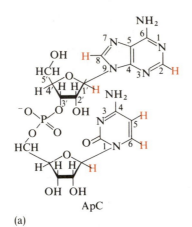

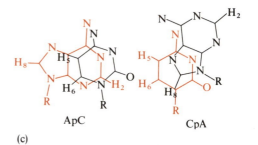

(a)

(c)

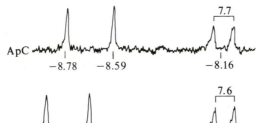

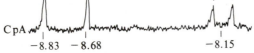

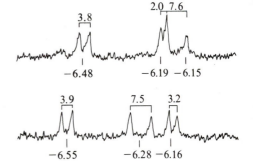

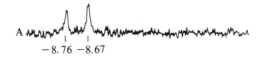

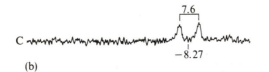

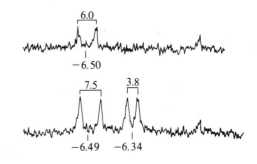

(b)

Figure 22-9

Conformational analysis of ApC and CpA by NMR spectroscopy. (a) Chemical structure of ApC, showing protons detected by NMR in color. (b) Time-average NMR spectra of 0.01 M solutions at neutral pH in D_2O at 29°C. Chemical shifts are indicated below each peak, in ppm; coupling constants are indicated above, in cps. (c) Possible stacked conformations consistent with the NMR data. [After W. Bangerter and S. Chan, *J. Am. Chem. Soc.* 91:3910 (1969).]

nature of the base attached to the same sugar. The ring-current magnetic anisotropy of A is much larger than that of C. This difference causes a larger chemical shift of H-1′ in adenosine than in cytidine. The shifts are downfield because the H-1′ proton is roughly in the same plane as the base. Rotation about the $C^{1′}$–N torsional bond is possible, but none of the preferred positions change the predicted chemical shift very much. Thus, the remaining lower-field doublet in the spectrum of CpA and ApC is assigned to the adenosine H-1′, and the higher-field doublet is assigned to the cytidine H-1′.

We can now analyze the structural information available from the NMR spectra of the dimers shown in Figure 22-9. The most dramatic effects are the large upfield shifts in all three cytidine protons in both ApC and CpA. These shifts must be due at least in part to the large ring-current anisotropy of the adenine base. Cytidine has an almost negligible magnetic anisotropy, and the adenine protons in CpA and ApC show much smaller shifts. Note that the cytidine H-6 shifts relative to cytidine are very similar in both CpA and ApC. In contrast, cytidine H-5 in ApC has a much larger upfield shift than in CpA. This must mean that cytidine H-5 in ApC is closer to the center of the adenine ring (where the magnetic anisotropy is strongest) than it is in CpA.

Ring-current shifts are not the only cause of the observed spectral differences. The charged phosphate group has a strong electric field. It will be felt differently by bases situated 5′ or 3′, because the resulting chemical structure places these at differing distances from the phosphate. All these effects can be combined in at least a semiquantitative way to allow evaluation of proposed structural models for the dinucleoside phosphates (Fig. 22-9c).

The dynamics of stacking interactions occur on a much more rapid time scale than the characteristic times for NMR measurements. As a result, the time-average structure is reflected in an NMR spectrum. In most cases, an NMR spectrum contains insufficient information for complete structural assignment. One cannot get much of a hint about the range of structures over which the temporal average occurs. However, the results of Figure 22-9 do allow us to visualize clear differences between the structures of ApC and CpA. The models shown in Figure 22-9c are consistent with the NMR and with the known stereochemical constraints of the phosphodiester bond. Both structures have stacked bases in a right-handed helical array.

NMR study of ribo and deoxy dimers

The problem of assigning the resonances in homodimers or higher homooligomers is considerably more difficult. There are four aromatic proton resonances and two H-1′ doublets in ApA. Deuterium exchange enables us to distinguish the two H-8 bands from the two purine H-2 bands, and the chemical shifts of the H-1′ protons are sufficiently different that they can be identified. A serious problem remains: we must distinguish between the protons belonging to the 5′-linked adenosine (pA) and those belonging to the 3′-linked adenosine (Ap). The most successful approach has been to add the paramagnetic Mn^{II} ion, which is known to bind primarily to the phosphate groups of oligonucleotides. The paramagnetic ion has a large electron-spin dipole. This dipole couples with nuclear spin dipoles, leading to spectral broadening of the resonances of nearby protons. The effect is very sensitive to distance, because transition probabilities based on the dipole–dipole interaction vary as the inverse sixth power of the distance. Thus, by examining the effect of Mn^{II} on the line width of each resonance, one can estimate relative distances to the phosphate.

Figure 22-10 shows the chemical structure of ApA. In agreement with the discussion of nucleic acid conformation in Chapter 6, both bases have been drawn with *anti* torsional angles. It is easy to see that the H-8 proton of pA is closer to the phosphate than is that of Ap. Thus, the resonance of this proton will be broadened more by Mn^{II}, allowing its assignment. Conversely, the H-1′ proton of pA is farther from the phosphate, so it will be broadened less than H-1′ of Ap. Unfortunately, the two H-2 protons are far from the phosphate and roughly equidistant, so that Mn^{II} binding does not facilitate their assignment. Less direct evidence must be used to distinguish between these two protons. In many cases, the best one can do is to ask which models of the final structure permit a self-consistent assignment of H-2.

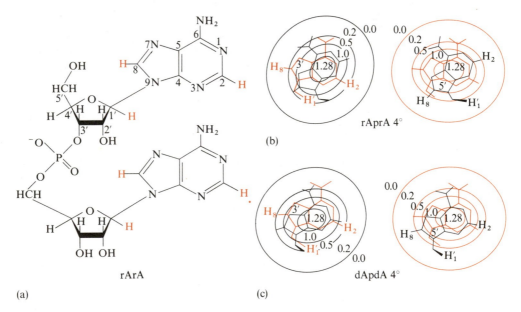

Figure 22-10

NMR analysis of stacking differences between rAprA and dApdA. (a) Chemical structure of rAprA, showing protons monitored by NMR in color. (b) Proposed conformation of rAprA viewed in the 5′ → 3′ direction. Solid lines indicate the top (5′-linked) base and colored lines the bottom base. Also shown are approximate contours calculated for the ring-current magnetic anisotropy in ppm of each base. (c) Results for dApdA, illustrated in the same manner as part b. [After N. S. Kondo et al., *Biochemistry* 11:1992 (1972).]

Table 22-2

Dimerization shifts of the base and H-1′ protons of adenine dinucleoside monophosphates at 4°C (D_2O, pD 7.4)

Dinucleoside	$\Delta\delta_{3'}$			$\Delta\delta_{5'}$		
	H-8	H-2	H-1′	H-8	H-2	H-1′
rAprA	−0.155	−0.315	−0.285	−0.285	−0.11	−0.19
rApdA	−0.15	−0.225	−0.325	−0.23	−0.15	−0.225
dApdA	−0.325	−0.31	−0.42	−0.16	−0.175	−0.275
dAprA	−0.32	−0.28	−0.41	−0.16	−0.21	−0.25

NOTE: $\Delta\delta_{3'} = \delta_{\underline{A}pA} - \delta_{\underline{A}p}$; and $\Delta\delta_{5'} = \delta_{A\underline{p}A} - \delta_{p\underline{A}}$ The underscored residue is the one measured in the dimer.

SOURCE: After P. O. P. Ts'o, *Basic Principles in Nucleic Acid Chemistry*, vol. 2 (New York: Academic Press, 1974).

NMR spectra of homodimers can be used to examine the differences in conformation between deoxy and ribo compounds. Table 22-2 summarizes such NMR data. These data were obtained at 4°C to maximize stacking and thus enhance the stacking-dependent spectral shifts. The results are expressed as dimerization chemical shifts ($\Delta\delta$, resonance location of one nucleoside in a dimer relative to the corresponding mononucleotide): $\Delta\delta_{3'} = \delta_{\underline{A}pA} - \delta_{\underline{A}p}$ and $\Delta\delta_{5'} = \delta_{A\underline{p}A} - \delta_{p\underline{A}}$. This approach tends to compensate for the effect of the phosphate and for variations in the properties of individual deoxy and ribo monomers; it thus highlights changes in the NMR due to stacking. In fact, all the observed $\Delta\delta$ values are negative, reflecting the effect of ring-current magnetic anisotropies in the vertically stacked bases. When ribo and deoxy compounds are compared in this way, an interesting pattern emerges. In dApdA, the H-8 proton of the dAp is much more affected by the stacking than is the pdA proton. Exactly the opposite effect is seen in rAprA. In contrast, the 3′-linked H-2 and H-1′ protons in both compounds appear much more influenced by stacking than does the 5′-linked proton.

Using calculated ring-current shifts for adenine, it is possible to construct detailed models that semiquantitatively account for the observed data. Figure 22-10 shows such models for the time-average structure of dApdA and rAprA. The major differences in structure are predicted to be a greater stacking overlap in the deoxy dimer and a larger helix angle (more rotation from one base to the next) in the ribo dimer. The optical data in Table 22-1 support these conclusions. There is more hypochromism in dApdA than in rAprA; this is consistent with greater base overlap in the stacked conformation of dApdA, although it also could simply indicate more molecules in the stacked (as opposed to an unstacked) conformation.

The optical activity is more intense in rAprA than in dApdA. The optical activity of ApA should get weaker as the angle between the bases is reduced from 40° to 20° or less (Fig. 22-7a). Thus, a smaller helix angle in dApdA could explain its less intense CD, despite the greater stacking implied by the higher hypochromism. This is exactly the conclusion reached from the NMR results. Both optical and NMR techniques have been applied in a similar way to selected trinucleoside diphosphates. In most cases, all results are consistent with the conclusion that—for the normal bases A, U (or T), and G—short oligomers are a homologous series, not just in chemical structure, but also in the actual average conformation attained by neighboring residues. Figure 8-10 shows the kind of optical data that support such a conclusion.

Conformational equilibria in dinucleotides

It should be clear from the preceding discussion that single-stranded oligonucleotides and polynucleotides contain appreciable base-stacked regions, but that they are not rigid or fully ordered structures. Thus, in addition to the average stacked structure, we must know the fraction of time that each base spends as a stack. The rates of stacking and unstacking turn out to be so fast that rates of interconversion are virtually unmeasurable by the standard techniques of rapid kinetics. Thus, accepting a rapid equilibrium among stacked and unstacked forms, we really want to know what the fraction of stacked bases is, and how this fraction depends on environmental parameters.

The simplest model one can imagine for a dinucleotide NpN considers only two states: a fully stacked conformation NpN_{st}, and an unstacked conformation NpN_u. The reaction $NpN_u \rightleftarrows NpN_{st}$ will be characterized by a stacking equilibrium constant:

$$K_{st} = (NpN_{st})/(NpN_u) \tag{22-6a}$$

$$\Delta G_{st}^0 = -RT \ln K_{st} = \Delta H_{st}^0 - T\,\Delta S_{st}^0 \tag{22-6b}$$

If the equilibrium constant can be measured as a function of temperature, the enthalpy of stacking can be calculated from a van't Hoff plot, $\Delta H_{st}^0 = -RT\,d(\ln K_{st})/d(1/T)$, and the entropy of stacking can be computed as $\Delta S_{st}^0 = [\Delta H_{st}^0 - \Delta G_{st}^0(T)]/T$. A plot of $\ln K_{st}$ versus $1/T$ will be linear if the enthalpy of stacking is not a function of temperature, implying that stacking is accompanied by a negligible change in heat capacity.

Chapter 21 discusses the theory of two-state transitions in detail. Two important conclusions are needed here. First, for a two-state transition, any physical parameter that monitors the fraction of unstacked or stacked form should yield the same equilibrium constant and thus the same ΔH_{st}^0. Second, the stacking enthalpy derived from a van't Hoff analysis will be the same ΔH_{st}^0 that would be observed by an actual calorimetric measurement. Most physical properties and all optical properties y are a linear function of the properties of the pure stacked (A_{st}) and unstacked (A_u) states. The observed spectral intensity A (or other physical property) will be

$$A = \chi_{st}A_{st} + \chi_u A_u \tag{22-7}$$

where χ_{st} is the mole fraction of stacked dimer, and χ_u is the mole fraction of unstacked dimer. This relationship allows the stacking equilibrium constant to be written as

$$K_{st} = \chi_{st}/\chi_u = (A - A_u)/(A_{st} - A) \qquad (22\text{-}8)$$

(cf. Eqn. 6-1).

Thus, the equilibrium constant can be determined under any conditions, provided that physical properties of the fully stacked and unstacked forms are available. Herein lies a problem. Oligonucleotide stacking changes very gradually with temperature. In some cases, a fully unstacked form can be reached by 100°C; otherwise, monomer properties can be used to estimate A_u. The only way to determine A_{st} is to produce a fully stacked form. The temperature is lowered until A attains some constant value. In many cases, temperatures as low as $-80°C$ are needed. Such temperatures can be reached by using concentrated aqueous LiCl as a solvent (Fig. 22-11). However, the thermodynamic properties derived for stacking under these conditions must somehow be extrapolated back to dilute salt conditions.

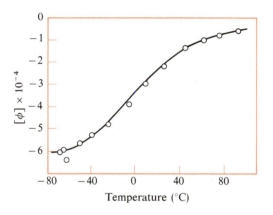

Figure 22-11

Optical rotation of ApA at 262 nm in 25.2% LiCl (pH 7) as a function of temperature. [After R. Davis and I. Tinoco, Jr., *Biopolymers* 6:223 (1968).]

In studying two-state and other conformational transitions, it often is convenient to define a melting[§] temperature T_m by analogy to the sharp transition temperature observed in true phase changes. This is the temperature at which $\Delta G^0_{st} = 0$ and $K_{st} = 1$. Therefore,

$$T_m = \Delta H^0_{st}/\Delta S^0_{st}$$

[§] This temperature T_m might better be regarded as a "midpoint" temperature rather than a "melting" temperature.

It is easy to determine T_m for a sharp transition. The only way to determine T_m for a broad transition is to determine ΔG_{st}^0 as a function of temperature. It is tempting to look at temperature-dependent data such as those shown in Figure 22-11 and to guess that T_m is the temperature of maximal slope, but this is not correct. In fact, T_m is the temperature of maximal slope only when the data are plotted as a function of $1/T$ (see Box 22-1).

Unfortunately, the two-state model does not provide a good fit for data on oligonucleotide stacking. The van't Hoff apparent enthalpies derived from different optical techniques do not agree. Various ORD and CD measurements on rAprA yield ΔH_{st}^0 values between -5.3 and -8.0 kcal mole^{-1}, whereas absorbance and hypochromicity measurements yield values between -8.5 and -10 kcal mole^{-1}. Much of this variation could easily be due to difficulties in choosing the low-temperature limiting value A_{st} of the optical property A. NMR provides even more decisive evidence, however, that a two-state model is not a fully correct description of dimer stacking. For a two-state model, the NMR shifts for each dimer proton and its monomer counterpart should scale as a function of temperature, producing the same melting curve (the same, in fact, as that obtainable from optical data). In practice, different protons show slightly different temperature behaviors, indicating that each measurement is weighting different conditions slightly differently. For example, the van't Hoff ΔH_{st}^0 obtained from the H-2 proton of the 3'-linked A in ApA is -11 kcal mole^{-1}, a value more negative than that seen in any of the optical studies.

In the following discussion, we shall ignore some of these complications and shall use apparent enthalpies from two-state or other oversimplified models to

---

Box 22-1   DETERMINATION OF T_m FOR A BROAD TRANSITION

At $T = T_m$, the value of χ_{st} will be $\frac{1}{2}$. The temperature of maximal slope for a plot of A versus T is found at $d^2A/dT^2 = 0$. From Equation 22-7, this leads to the condition $d^2\chi_{st}/dT^2 = 0$. From Equations 22-6 and 22-9, one can easily show that

$$d\chi_{st}/dT = \chi_{st}(1 - \chi_{st})\Delta H_{st}^0/RT^2$$

Further differentiation leads to the result that, at the temperature of maximal slope, the mole fraction stacked is

$$\chi_{st} = \tfrac{1}{2} - RT/\Delta H_{st}^0$$

Thus only in the limit of very sharp transitions with high ΔH^0 can the temperature of maximal slope be equated with the melting temperature. If data are plotted as A versus T^{-1}, this complication is avoided, because $d\chi_{st}/dT^{-1} = -\chi_{st}(1 - \chi_{st})\Delta H_{st}^0/R$ and (for a two-state transition) the temperature of maximal slope occurs where $\chi_{st} = \frac{1}{2}$ and thus is equal to T_m.

attempt to predict the stability of differing nucleic acid structures. Therefore, it is important to stress the approximate nature of such van't Hoff ΔH^0 measurements. In principle, the problem could be solved by performing actual calorimetric ΔH^0 measurements, but this is not easy for single-stranded oligonucleotides. The ΔH^0 values are small and high concentrations are needed to have measurable heats in convenient ranges. Just a few calorimetric results have been obtained thus far (see Box 6-1).

Stacking equilibria as a function of chain length

The optical changes associated with stacking become more intense as the chain length of oligomers increases. This reasonable because, even for partially stacked structures, the probability of runs of stacked bases is appreciable. A fully stacked polymer has twice as many nearest-neighbor stacking interactions per residue as does a stacked dimer. In fact, for chain lengths of ten residues or more, optical properties are virtually independent of length for single-stranded homooligomers.

How does the thermodynamics of stacking depend on length? The critical question is whether it is energetically more favorable to stack a base next to a pre-existing stack than to start a stack from two unstacked bases. If we describe the former process by an equilibrium constant s, then the constant for beginning a stack can be written as σs, where σ is a nucleation parameter. If $\sigma > 1$, it is harder to continue stacks than to start them; this situation is termed anticooperative. If $\sigma < 1$, it is harder to start a stack than to continue one, and the conformational transition can be called cooperative. The apparent stacking enthalpy from a van't Hoff plot will increase if $\sigma < 1$. The easiest way to see this is to consider the extreme case of an all-or-none transition in a trimer. Then, the only two states are unstacked and fully stacked. The ΔH^0 for this process is the sum of the enthalpies of the two stacking interactions. The resulting melting transition will be sharper than the transition when intermediate stages occur.

This qualitative description can be formulated quantitatively in a manner exactly analogous to the helix–coil statistical mechanics describe for polypeptides (Chapter 20). One must take into account the intrinsic length dependence of optical properties when analyzing actual optical data. Thus, the mechanics of the calculations become somewhat complicated. These calculations are not worth treating in detail here because the actual experimental results are very simple (Figs. 6-12 and 6-13).

There is only very slight cooperativity for oligoribonucleotides such as oligo rA. The melting temperature of poly A is not more than 20°C higher than that of the dimer. Apparent ΔH^0 values are nearly independent of chain length. A detailed analysis yields σ values between 0.8 and 0.5. (A value of 1.0 corresponds to a completely noncooperative conformational change.) There may actually be slight anticooperativity in oligodeoxyadenylates. These results are a far cry from the thermodynamics of α-helix formation or, as we shall see, from the thermodynamics of double-strand formation. However, they are consistent with the general picture we have developed for the nature of single-stranded nucleic acid helices.

22-3   EQUILIBRIA BETWEEN SINGLE STRANDS AND DOUBLE STRANDS

With only a few exceptions, naturally occurring DNAs are fully double-stranded helices formed from two complementary single strands. However, to preserve the information stored in DNA by replication, or to read out that information by transcription, it appears necessary to form locally single-stranded regions by protein-mediated physical disruption of the double helix. Experiments that probe the characteristics of the interconversion of DNA double- and single-stranded conformations thus are of central biological importance. The overwhelming majority of physical studies have involved DNA in the absence of protein. The simplicity of this system is a substantial advantage. These are the experiments we discuss in this chapter. Ultimately, it will be necessary to face the difficult problem of adding back the proteins in order to analyze their role in physical processes.

Most RNA is synthesized in nature as a single strand. However, the driving force for the formation of intramolecular base-paired regions is strong, and most RNAs appear to attain well-defined, ordered secondary structures. Thus, we must understand the thermodynamic forces governing the stability of such structures if we are ever to be able to predict the conformations available to an RNA under a given set of conditions.

Synthetic double-stranded polynucleotides as model systems

Much of the early understanding of the interconversions of double- and single-stranded nucleic acids came from studies on homopolynucleotides or on simple repeating nucleic acid sequences. These substances are easily synthesized enzymatically. They have two intrinsic advantages as model systems. First, their structure and stability are uniform along the whole length of the molecule, so that complicated effects due to heterogeneity are eliminated. Second, in most cases, the individual strands are simple stacked helices, permitting experiments to focus clearly on properties that are functions only of the double-stranded complex.

However, these model systems do have some compensating disadvantages. A preparation of a synthetic polynucleotide usually has a distribution of chain lengths. Duplexes formed from mixtures of complementary homopolymers can be aggregates and have frayed ends or loops (Fig. 22-12). The same is true for most polymers with simple alternating sequences. Self-complementing alternating sequences such as poly $d(AT) \cdot d(AT)$ can form hairpins or loops in addition to perfect duplexes. These complications do not show up in most optical studies, but they can be a real nuisance in hydrodynamic experiments.

Most double-stranded synthetic polynucleotide complexes are formed by mixing separate preparations of pure single strands under conditions where duplex formation is favored overwhelmingly. The RNAs, DNAs, and RNA–DNA hybrids that can be prepared in this way generally follow the same patten of base pairing seen in crystals or in nonaqueous solutions of monomers and known in natural DNAs.

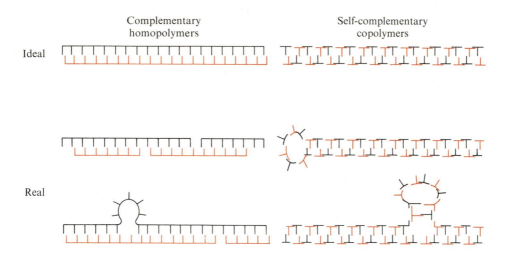

Figure 22-12

Idealized and real base-pairing configurations accessible to complementary homopolymers such as poly rA·rU and to self-complementary copolymers such as poly d(A–T)·d(A–T).

The base G pairs with C, and A with T (or U) in all possible ribo and deoxy combinations. In practice, poly I (inosine) is substituted for poly G in many experiments on model systems. Unlike other homopolymers, poly G is seldom if ever in a single-stranded form, even under rigorous denaturing conditions. The exact form of the poly G self-structured conformation is still not clearly established. However, the monomer G itself forms four-stranded helical hydrogen-bonded complexes in concentrated solution. It is a reasonable guess that poly G does something equivalent. Whatever the actual structure, it is so stable that it is extremely difficult to carry out a reaction such as poly C + poly G → poly C·G, and the duplex must be prepared in indirect fashion. Poly I at high salt concentration and low temperature also forms multiple-stranded self-complexes. Poly rI is a triple strand under these conditions, but it is easily disrupted by a reaction such as poly rC + $\frac{1}{3}$ poly rI_3 → poly rC·rI.

Determining stoichiometry with a mixing curve

How do we know when we have made a double-stranded complex of two polynucleotides? Probably the most direct and simple test is a spectroscopic mixing curve. This technique is illustrated here for UV absorbance, but it is quite a general technique. Consider a mixture of poly A and poly U. If there is no interaction between them, the absorbance A at a particular wavelength using a 1 cm path is given by Beer's law:

$$A = C_A \varepsilon_A + C_U \varepsilon_U \tag{22-10}$$

where C refers to the concentration in units of moles of residue per liter, and ε is the per-residue extinction coefficient. Suppose that some poly A and poly U react to form a complex X. In general, the complex can have an arbitrary number of poly A and poly U strands. The absorbance of the sample is

$$A = C_A \varepsilon_A + C_U \varepsilon_U + C_X \varepsilon_X \qquad (22\text{-}11)$$

where C_X is the concentration of the complex in moles of residue per liter, and ε_X is the per-residue extinction coefficient of the complex.

A classical mixing experiment is performed by varying the ratio of the two components, keeping the total concentration of residues (C_0) constant. Equation 22-11 applies for each mixture and can conveniently be expressed in terms of mole fractions of total residue included in each form (χ_A, χ_U, χ_X), so that

$$A = C_0(\chi_A \varepsilon_A + \chi_U \varepsilon_U + \chi_X \varepsilon_X) \qquad (22\text{-}12)$$

In the general case, the mole fractions of each component in the equilibrium mixture are not known (although they can be found by a multicomponent spectral analysis similar to that described for fitting spectral data to secondary structures in Chapter 8). Only the mole fractions of total A and U residues originally put into the mixture are known directly; we shall call these fractions χ_A^0 and $\chi_U^0 = 1 - \chi_A^0$.

Suppose the complex X contains A and U residues at mole fractions γ and $1 - \gamma$, respectively, Consider the special case in which the stability of the complex is so great that, if A and U residues were mixed at a stoichiometry of $\gamma:(1 - \gamma)$, no free single strands would remain. Suppose in a particular sample that the ratio A:U is greater than $\gamma:(1 - \gamma)$. Then all of the input U will be incorporated into the complex, consuming $\gamma/(1 - \gamma)$ residues of A for each residue of U added. The remaining A stays as free single strands. Thus, Equation 22-12 can be rewritten in terms of the input stoichiometry:

$$A = C_0 \left[\left(\chi_A^0 - \frac{\gamma}{1-\gamma} \chi_U^0 \right) \varepsilon_A + \frac{1}{1-\gamma} \chi_U^0 \varepsilon_X \right] \qquad (22\text{-}13)$$

where the first term describes the absorbance due to the free poly A, and the second term describes the absorbance of the complex. The factor of $1/(1 - \gamma)$ in the second term arises because, for each U residue in the complex, there are $\gamma/(1 - \gamma)$ A residues, making $1/(1 - \gamma)$ residues in all. By substituting $1 - \chi_A^0$ for χ_U^0 in Equation 22-13, we obtain

$$A = C_0 \left(\frac{\varepsilon_X - \gamma \varepsilon_A}{1-\gamma} + \chi_A^0 \frac{\varepsilon_A - \varepsilon_X}{1-\gamma} \right) \qquad (22\text{-}14)$$

If, in another sample, the input ratio A:U is less than $\gamma:(1 - \gamma)$, then all the A is complexed and some free U remains. In this case, the absorbance of the sample from Equation 22-12 is

$$A = C_0\left[\left(\chi_U^0 - \frac{1-\gamma}{\gamma}\chi_A^0\right)\varepsilon_U + \frac{1}{\gamma}\chi_A^0\varepsilon_X\right]$$

$$= C_0\left(\varepsilon_U + \frac{\varepsilon_X - \varepsilon_U}{\gamma}\chi_A^0\right) \tag{22-15}$$

The critical thing to note is that the absorbance of the mixture is a linear function of the input mole fraction of A in both Equations 22-14 and 22-15. However, the slope of A versus χ_A^0 is different on the two sides of the stoichiometric equivalence point, unless the extinction coefficients of the three components are accidentally balanced such that

$$\varepsilon_X = \gamma\varepsilon_A + (1-\gamma)\varepsilon_U \tag{22-16}$$

When a mixing curve of a tight complex is measured, there is a composition at which the slope of an absorbance versus concentration plot abruptly changes. This point is precisely the fractional composition of the complex. If a double helix of 1:1 stoichiometry is formed, $\gamma = 0.5$, and the abrupt change in slope occurs at $\chi_A^0 = 0.5$. Figure 22-13a shows an actual experimental melting curve. This curve verifies that the reaction poly A + poly U → poly A·U occurs.[§] It turns out that, if a complex is

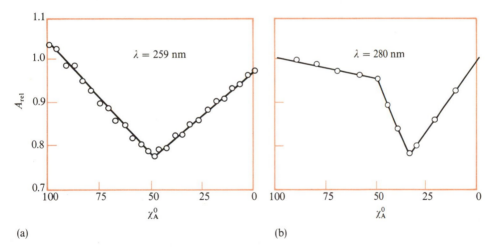

(a)   (b)

Figure 22-13

Absorbance mixing curve for poly A and poly U. Relative absorbance is plotted as a function of the fractional composition of poly A + poly U mixtures at constant total concentration. (a) A_{259} under conditions such that only a 1:1 poly A·U complex forms. (b) A_{280} under conditions where both poly A·2U and poly A·U can form. [After results obtained by Gary Felsenfeld and colleagues.]

[§] We must add one cautionary note about the interpretation of mixing curves. The endpoint gives the stoichiometric ratio of the components, but not the structural ratio. An endpoint of 0.5 could result from A_1U_1, A_2U_2, and so on. Occam's razor dictates that the simplest composition should be considered most plausible unless additional information about the system demands the inclusion of more complex structures.

strong but not infinitely tight, exactly the same arguments hold, in the limits of large excess of either component over the other. Linear extrpolation to intermediate values of absorbance measured near $\chi_A^0 = 1$ or $\chi_A^0 = 0$ will result in two straight lines that intersect at the stoichiometric equivalence point. The actual experimental data will show curvature near that endpoint, because of partial dissociation of the complex. Note that the absorbance A_e at the endpoint (or the extrapolated endpoint) corresponds to pure complex. This allows the per-residue extinction coefficient of the complex to be obtained simply as A_e/C_0.

Mixing-curve evidence for double-stranded and triple-stranded structures

More complicated mixing curves result if complexes with several different stoichiometries can form between two polymers. For example, under some experimental conditions, poly A and poly U show the mixing curve of Figure 22-13b. Two apparent endpoints exist: the expected 1:1 complex at $\chi_A^0 = 0.5$, and a second complex at $\chi_A^0 = 0.33$, which corresponds to the structure poly A·2U. The origin of this mixing curve is explained as follows. As long as A is in mole excess over U, the complex poly A·U forms, and free poly A remains in the solution. Addition of poly U above $\chi_U^0 = 0.5$ results in the reaction poly A·U + poly U $\rightleftarrows$ poly A·2U. At $\chi_U^0 > 0.67$, all of the poly A is converted to the triple helix poly A·2U, and any further U added remains as free poly U. Note that, at $\chi_A^0 = 0.5$, a disproportionation equilibrium should exist:

$$\text{poly A} + \text{poly A·2U} \rightleftarrows 2 \text{ poly A·U}$$

The fact that a sharp endpoint in the mixing titration is seen at $\chi_A^0 = 0.5$ in Figure 22-13b means that the equilibrium under these experimental conditions lies far to the right.

Triple helices such as poly A·2U are fairly common stable forms for homopolymer complexes. Base-paired complexes of ratio 2:1 are commonly seen in crystals of monomers, or when monomer interactions are studied in nonaqueous solutions. Figure 22-14 shows the pattern of hydrogen bonding in several of these interactions. For a long time, the feeling was prevalent that such triple-stranded structures are not very relevant to polymer properties. The constant geometry of sugar attachment for A–T and G–C pairs allows the same regular double helix to be constructed from any arbitrary DNA sequence. However, triple-stranded helical structures are possible only in certain cases. Therefore, only a few simple DNA or RNA sequences will permit formation of extended triple helices such as the poly dA·dT·dT structure shown in Figure 3-16. So long as nucleic acids were viewed simplistically as fully regular structures, there was little place for the short triple helices that might occur at particular sequences. However, the very first nucleic acid tertiary structure determined, yeast tRNA^{Phe}, contains several base triples that play an important role in

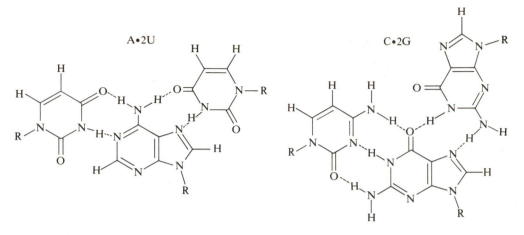

Figure 22-14

Base-pairing schemes for poly A·2U and poly C·2G. Note that other schemes plausible at least in monomer complexes can be generated by rotation of either uracil in A·2U around the N^3–H bond. Similarly, there are other possible C–G multiple base pairs.

establishing a compact, rigid conformation. This discovery has brought a renewed interest in triple-stranded model complexes. Perhaps such interactions will turn out to be much more common in nature than was originally believed.

Thermal stability of homopolymer complexes

Double-stranded polynucleotide complexes in most cases are so stable at room temperature that it is difficult to determine the relative strengths of different base-paired interactions. One must destabilize the double helix to magnify any differences. Increased temperature is by far the most widely used environmental variable for quantitative studies. The absorbance of a homopolymer complex, monitored as a function of temperature, shows a sharp increase centered at a temperature characteristic for the sample (Fig. 22-15a). This is called the melting temperature, T_m. The term is appropriate because the sharpness of the transition approaches that of true phase transitions. The observed melting curve is completely reversible. Note that the absorbance is quite constant at temperatures below the sharp melting region, and in some cases varies only slightly above it. As an isolated observation, this suggests a static structure for both a double helix and the individual single strands. However, as we have seen, the single strands change their properties with temperature. The double-strand T_m for one of the cases shown in Figure 22-15a is so high that the melting reaction is (approximately) double helix → 2 unstacked coils. A duplex that melts at lower temperatures will produce partially stacked single strands and, as the temperature is raised further, will show additional absorbance changes. The lack

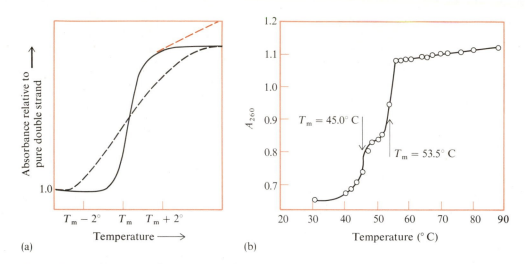

Figure 22-15

Polynucleotide melting curves. (**a**) Schematic results for a natural DNA (*dashed curve*) and for a homopolynucleotide duplex melting either to completely unstacked single strands (*solid curve*) or to partially stacked single strands (*colored curve*). All curves are scaled to have the same T_m and the same total absorbance change. (**b**) Two-step melting of poly A·2U corresponding to formation of poly A·U + poly U at 45°C and all single strands at 53.5°C. [After C. Stevens and G. Felsenfeld, *Biopolymers* 2:293 (1964).]

of absorbance changes below T_m also is deceptive. As we shall see, the DNA duplex is not an invariant static structure. However, absorbance is not a sensitive monitor of the changes that do occur.

The observed melting temperature for any duplex polynucleotide is a sensitive function of environmental conditions such as ionic strength, pH, and solvent conditions. This can be explained by a simple thermodynamic argument. Any small or large molecule M that binds differentially to single strands and double strands will alter the melting temperature. We can write the equilibrium between double helix and single strands as

$$M_n \cdot D \rightleftarrows M_{n'} \cdot S_1 + M_{n''} \cdot S_2 + \Delta n M \qquad (22\text{-}17)$$

where M is the small molecule, S_1 and S_2 are the two single strands, and D is the double helix. According to this oversimplified view, $\Delta n = n - n' - n''$ molecules of M are released upon melting (or are consumed, depending on the relative values of n, n', and n''). In reality, a polymer will have a small-molecule atmosphere (ions, protons, etc.) that requires a description more complicated than simply "bound or free," but this description will suffice for the moment. Using this definition, we can write the equilibrium constant for helix formation as

$$K_h = (M_n \cdot D)/(M_{n'} \cdot S_1)(M_{n''} \cdot S_2)(M)^{\Delta n} = \exp(-\Delta G^0 / RT) \qquad (22\text{-}18)$$

The effect of ligand concentration is most easily seen by calculating

$$\partial(\ln K_h)/\partial[\ln(M)] = -\Delta n \tag{22-19a}$$

Usually, however, what is measured is not K_h but instead T_m. The temperature dependence of K_h is given by the van't Hoff relationship. At T_m this is

$$\frac{\partial(\ln K_h)}{\partial T_m} = \frac{\partial}{\partial T_m}\left(\frac{-\Delta G^0}{RT_m}\right) = \frac{\Delta H^0}{RT_m^2} \tag{22-19b}$$

assuming that the enthalpy change ΔH^0 and the entropy change for the reaction are temperature independent. Using the chain rule for partial derivatives, we can write

$$\frac{\partial(\ln K_h)}{\partial T_m} = \frac{\partial(\ln K_h)}{\partial[\ln(M)]}\frac{\partial[\ln(M)]}{\partial T_m} \tag{22-19c}$$

Substituting Equations 22-19b,c into Equation 22-19a, we obtain a simple result:

$$\frac{\partial T_m}{\partial[\ln(M)]} = -\Delta n \frac{RT_m^2}{\Delta H^0} \tag{22-20a}$$

In Equation 22-20a, ΔH° and Δn refer to the cooperative unit in the melting process. Because this usually is not known precisely, it is more useful to rewrite this result in terms of Δn_p, the ligand release per phosphate, and ΔH_p^0, the enthalpy change per phosphate. With these changes, we have

$$\frac{\partial T_m}{\partial[\ln(M)]} = -\Delta n_p \frac{RT_m^2}{\Delta H_p^0} \tag{22-20b}$$

Note that ΔH_p^0 in Equation 22-20b is the enthalpy of helix formation at T_m; it includes not only the intrinsic heat of helix formation, but also the heat of any small-molecule binding or release. The intrinsic ΔH_p of base pairing is large and negative; in most cases, this dominates ΔH_p^0. Equation 22-20b, then, allows a general conclusion: raising the concentration of a small molecule will raise T_m whenever the small molecule binds in greater numbers to the double helix than to the corresponding single strands ($\Delta n > 0$). The opposite effect on T_m will be seen if the small molecule binds preferentially to the single strands. There are many small molecules, such as ethidium and actinomycin D, that bind preferentially to double-stranded structures. One immediately can make the fairly safe projection that they will raise the T_m of the double strands. Exactly the same kinds of arguments will apply when we measure the effect of small molecules on equilibria between double and triple strands.

Certain proteins show a strong specificity for either single strands or double strands. Again, this can be determined by their effects on T_m, providing that their binding heats are not too large and that the proteins themselves do not denature in the temperature range under study.

Phase diagrams for polynucleotide conformational states

When a sufficient body of experimental data exists on the T_m and stoichiometry of a helical complex as a function of environment, it is possible to construct a phase diagram. This diagram has two useful applications. First, the experimentalist can, by inspection, easily judge the state of the system at any arbitrary set of conditions. Second, the physical effects responsible for the phase transitions sometimes can be assessed easily from the behavior of the phase boundaries. Consider what happens in mixtures of poly rA and poly rU. The three most common experimental variables are temperature, salt concentration, and the composition of the polymer mixture. Four different helix equilibria can occur:

poly A·U $\rightleftarrows$ poly A + poly U	(1)	double-strand melting
2 poly A·U $\rightleftarrows$ poly A·2U + poly A	(2)	double-strand disproportionation
poly A·2U $\rightleftarrows$ poly A + 2 poly U	(3)	triple-strand melting
poly A·2U $\rightleftarrows$ poly A·U + poly U	(4)	release of poly U

This means that, under some conditions, a melting experiment can show two temperature transitions; for example, triple strand can yield double strand by reaction 4, and then (at a different temperature) double strand can melt to single strands by reaction 1; Figure 22-15b shows an actual melting curve corresponding to just this situation.

Figure 22-16 shows a phase diagram constructed for poly A + poly U mixtures. This diagram was assembled by combining absorbance data with a large body of direct calorimetric data on helix transitions. Four discrete phase regions are seen,

Figure 22-16

Phase diagram of the poly A + poly U system, at neutral pH in the absence of divalent ions. Melting transitions are shown as a function of Na⁺ concentration. The effect of the stoichiometry of mixtures in each of the four zones is described in the text. [After H. Krakauer and J. Sturtevant, *Biopolymers* 6:491 (1968).]

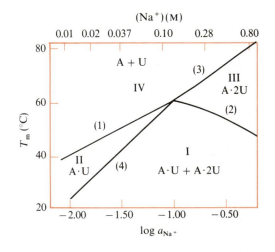

separated by lines corresponding to each of the four reactions indicated above. Only two variables (temperature and Na^+ concentration) are indicated. A third dimension would be needed to describe the effects of the relative concentrations of poly A and poly U. However, these effects are not very pronounced, and we can readily describe them in relation to Figure 22-16.

No double or triple helices are stable in region IV. As one would expect, this region is located at higher temperatures. It becomes more and more dominant at lower salt concentrations. The simple thermodynamic arguments we have given allow the immediate conclusion that all the multiple-stranded structures bind more salt than do the single strands. Another way of looking at this is that formation of a double or triple strand involves bringing together two or more highly negatively charged polymers. A concentrated counterion atmosphere must be created to overcome the electrostatic repulsion. This effect reduces the electrostatic free energy (Eqn. 22-4).

In region II, the triplex helix is unstable with respect to the release of poly U (reaction 4). Therefore, the only multiple helix possible in region II is poly $A \cdot U$. Any stoichiometric excess of poly A or poly U must remain as single strands. The double-stranded helix is unstable with respect to disproportionation into poly $A \cdot 2U$ and poly A in region III (reaction 2). Therefore, the only multiple helix present in this region is poly $A \cdot 2U$, and any excess poly U or poly A beyond this stoichiometry will be present as single strands.

A description of region I is more complicated. Here, both poly $A \cdot U$ and poly $A \cdot 2U$ are stable. The state of the system as a function of composition is described by the mixing curve shown in Figure 22-13b, which was carried out under conditions equivalent to those of region I. If χ_U^0 is less than 0.5, only double-stranded poly $A \cdot U$ and free poly A will be present. For $0.5 \leqslant \chi_U^0 \leqslant 0.67$, a mixture of double and triple strands will form in such proportions that no free single strands exist. At $\chi_U^0 > 0.67$, there will be only triple strands and free poly U. The easiest way to rationalize these observations is to say that any single-stranded forms in region I are essentially infinitely unstable with respect to any multiple-stranded form. Therefore, the set of double and triple helices formed obeys the constraint to minimize the amount of single strand at all costs. Any intrinsic free energy differences between poly $A \cdot 2U$ and poly $A \cdot U$ are negligible compared with the extra energy needed to maintain a free single strand in this region.

Whether an actual melting transition is observed in crossing the phase boundary between regions I and II (T_{m4}, corresponding to reaction 4) or between regions I and III (T_{m2}, corresponding to reaction 2) will depend on composition. If there is any triple strand present, it will melt to a double strand when region II is entered at low ionic strength. Alternatively, any double strand present will disproportionate to triple strand at high ionic strength when region III is entered. The phase boundary between regions I and III is particularly interesting, because $dT_{m2}/d[\ln(Na^+)]$ is negative. This means that a triple strand and a single strand together bind more salt than do two double strands. This is reasonable, because the triple strand will have three phosphate-phosphate interactions at each level in the helix (which must be shielded), whereas each double strand has only one. Consistent with this observation is the greater dependence on ionic strength of T_{m2} and T_{m3} (which each involve triple strands) compared with the simple double-strand melting described by T_{m1}.

Melting temperature and stability

It is tempting to associate these easily measured T_m values with helix stability, and we have been doing this implicitly in the preceding discussion. However, relative T_m values need not correspond rigorously to relative stabilities. The correct measure of stability is the free energy G_s of a given state s. If all components are in their standard states, the standard free energy G_s^0 can be used instead. Whether a given reaction will occur is determined by ΔG^0 (not by ΔH^0, and not by T_m). Let us illustrate this concept for a helix displacement reaction: poly $A \cdot B$ + poly $C \rightleftarrows$ poly A + poly $B \cdot C$. Energy states for this hypothetical system are the following (assuming $G_A^0 = G_B^0 = G_C^0$):

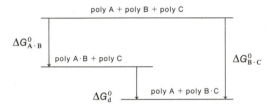

Suppose that we know the T_m values for formation of the helices poly $A \cdot B$ and poly $B \cdot C$ from their components $T_{m(A \cdot B)}$ and $T_{m(B \cdot C)}$. Can this information be used to predict the ΔG_d^0 of the helix-displacement reaction? Because $T_m = \Delta H^0 / \Delta S^0$, it can be used to eliminate one of these two thermodynamic quantities. So, for example,

$$\Delta G_{A \cdot B}^0 = \Delta H_{A \cdot B}^0 - T \Delta S_{A \cdot B}^0 = \Delta H_{A \cdot B}^0 (1 - T/T_{m(A \cdot B)}) \qquad (22\text{-}21)$$

An exactly corresponding equation holds for $\Delta G_{B \cdot C}^0$. Clearly, ΔG_d^0 at temperature T is

$$\Delta G_d^0 = \Delta G_{B \cdot C}^0 - \Delta G_{A \cdot B}^0$$

$$= \Delta H_{B \cdot C}^0 - \Delta H_{A \cdot B}^0 + (\Delta H_{A \cdot B}^0 / T_{m(A \cdot B)} - \Delta H_{B \cdot C}^0 / T_{m(B \cdot C)})T \qquad (22\text{-}22a)$$

We cannot predict ΔG_d^0 from Equation 22-22a unless we know the enthalpy changes as well as the T_m values for the two separate duplex-melting equilibria. In many cases, it may not be a bad assumption that $\Delta H_{B \cdot C}^0 \cong \Delta H_{A \cdot B}^0 = \Delta H^0$. Then Equation 22-22a becomes

$$\Delta G_d^0 = T \Delta H^0 (1/T_{m(A \cdot B)} - 1/T_{m(B \cdot C)}) \qquad (22\text{-}22b)$$

Table 22-3
Observed melting temperatures for duplexes involved in helix-displacement reactions

Reaction	T_m of reactants (°C)	T_m of products (°C)
poly dI·dC + poly rI → poly dI + poly rI·dC	46.1	52.3
poly rI·dC + poly rC → poly dC + poly rI·rC	52.3	60.2
poly dI·rC + poly rI → poly dI + poly rI·rC	35.4	60.2
poly dI·rC + poly dC → poly rC + poly dI·dC	35.4	46.1

SOURCE: Data from M. T. Chamberlin and D. L. Patterson, *J. Mol. Biol.* 12:410 (1965).

Because ΔH^0 is negative for all known duplex-formation reactions, ΔG_d^0 is negative if $T_{m(B \cdot C)} > T_{m(A \cdot B)}$. The double helix with the higher T_m always will be favored in a helix-displacement reaction. This prediction is in good agreement with observed results, such as those shown in Table 22-3. These results also illustrate the general observation that RNA·RNA duplexes are more stable than DNA·DNA duplexes. The relative stability of RNA·DNA hybrids is less systematic.

The approximation we have had to use in order to correlate ΔG_d^0 with the T_m parameters has an interesting consequence. The enthalpy change around a thermodynamic cycle must be zero. Thus if $\Delta H_{A \cdot B}^0 = \Delta H_{B \cdot C}^0$ rigorously, we predict that ΔH_d^0 will be zero, and therefore no melting transition for the strand-displacement reaction will be observed.

If ΔH_d^0 is not zero, an alternative approach is possible. From a measured melting temperature $T_{m(d)}$ for the displacement reaction, we can find the ratio $\Delta H_{B \cdot C}^0 / \Delta H_{A \cdot B}^0$. The temperature at which ΔG_d^0 is zero is $T_{m(d)}$. Therefore, from Equation 22-22a, we have

$$\Delta H_{B \cdot C}^0 / \Delta H_{A \cdot B}^0 = (T_{m(d)}/T_{m(A \cdot B)} - 1)/(T_{m(d)}/T_{m(B \cdot C)} - 1) \qquad (22\text{-}23)$$

Suppose that we know a single enthalpy change for any of the three reactions in the A, B, C reaction scheme. Equation 22-23 permits us to estimate all of the enthalpies (and thus all of the free energies) from T_m determinations. Thus even limited thermodynamic data can be of much use.

Effect of pH on polynucleotide structures

We have concentrated on salt and temperature as environmental variables. The effects of pH and organic solvents are more complicated for a simple reason. The melting transition observed for a polynucleotide duplex is a function of the relative stability of the duplex versus the separate individual single strands. Salt was shown to have little effect on single-strand local structure, and the effect of temperature on single strands is so gradual that it largely can be ignored. Other variables can have more consequential effects on the single strands. We here illustrate such effects for pH.

Increasing the pH of a polynucleotide duplex above pH 7 will destabilize double strands and lower the T_m. The bases U, T, and G lose a proton at alkaline pH (see Chapter 3). A double helix cannot be formed by the deprotonated bases, because the proton lost is one directly involved in base pairing. For example, at a pH where single-stranded poly U is deprotonated, the formation of the double helix poly A·U can be written as

$$\text{poly A·U} \rightleftarrows \text{poly A} + \text{poly U}^{-n} + n\text{H}^+ \tag{22-24}$$

Compare Equation 22-24 with Equation 22-17, and note that the double helix must become ever less stable with increasing pH, as described by Equation 22-20b).[§]

Titration with acid also lowers the melting temperature of polynucleotide duplexes, but in some cases the process is more complicated. Both poly A and poly C can form double-stranded protonated helical structures. Figure 22-17 shows the base pairing involved. The acid form of poly C is particularly interesting because

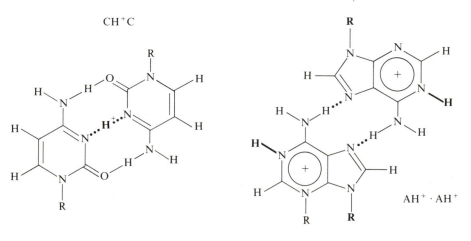

Figure 22-17

Hydrogen bonding between protonated nucleosides.

each C–C base pair shares one proton. The most direct evidence for this is the titration of poly C (Fig. 22-18). At one-half an equivalent of protons bound per C residue, there is an extremely sharp transition to the acid helix poly C—H⁺—C. If more protons are added, this helix becomes destabilized, and further titration eventually

---

[§] Note that, in practice, RNA strands (but not DNA strands) are readily hydrolyzed in alkali, so that the specific example shown by Equation 22-24 may be complicated by the irreversible effects of any chain cleavage.

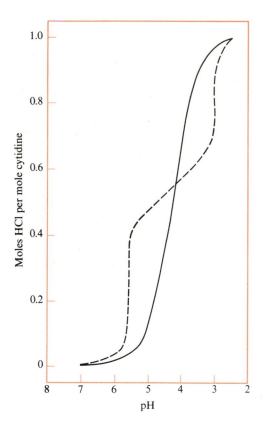

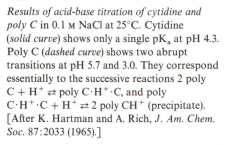

Figure 22-18

Results of acid-base titration of cytidine and poly C in 0.1 M NaCl at 25°C. Cytidine (*solid curve*) shows only a single pK_a at pH 4.3. Poly C (*dashed curve*) shows two abrupt transitions at pH 5.7 and 3.0. They correspond essentially to the successive reactions 2 poly $C + H^+ \rightleftarrows$ poly $C \cdot H^+ \cdot C$, and poly $C \cdot H^+ \cdot C + H^+ \rightleftarrows 2$ poly CH^+ (precipitate). [After K. Hartman and A. Rich, *J. Am. Chem. Soc.* 87:2033 (1965).]

converts it into two single-stranded, fully protonated poly CH^+. The acid double helix of poly A also is not a fully protonated form. However, the protons added to poly A are not directly involved in base pairing. The acid form becomes stable relative to single strands at degrees of proton uptake far less than one per A residue.

Electrostatic attraction between the positively charged bound protons and the negatively charged phosphates should contribute to the stabilization of acid poly A and poly C—H^+—C. The effect of ionic strength on the stability of acid forms at constant pH can be examined to test this. The result is striking. Both poly C and poly A acid double helices show increasing T_m with decreasing salt concentration. This behavior is opposite to that of normal double helices. It indicates that electrostatic free energy favors attraction of acid helices, even though it causes repulsion of normal helices. In fact, poly C and poly A can form acid helices even near pH 7, in the limit of no added salt. The large shift in apparent pK_a values (from monomer values near 4 to polymer values near 7) is explained by coupling of the titration to double-helix formation and by stabilization of the protonated form in the double helix by hydrogen bonding and electrostatic interactions with nearby phosphate residues.

Hydrodynamic studies of duplex melting

When the melting of polynucleotide duplexes is examined by hydrodynamic methods, a more complex picture emerges than is visible from optical measurements. Figure 22-19 shows two examples illustrating quite different phenomena. Poly d(AT)·d(AT) and poly dI·d5BrC melting transitions, which appear from absorbance studies to be fully reversible, demonstrate large hysteresis effects when monitored by viscosity.

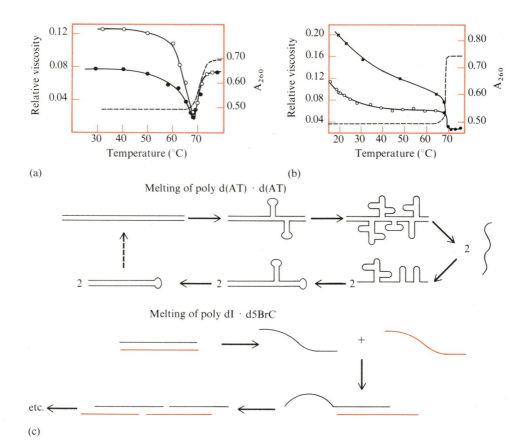

(a)   (b)

Melting of poly d(AT) · d(AT)

Melting of poly dI · d5BrC

etc.

(c)

Figure 22-19

Polynucleotide melting monitored by absorbance (*dashed curves*) and relative viscosity (*solid curves*). (a) Poly d(A–T)·d(A–T). Points indicate viscosity values obtained during original heating (○) and during cooling and second heating (●).   (b) Poly dI·d5BrC. Points indicate viscosity values obtained during heating and cooling at temperatures less than 65°C (○), during heating to 80°C (●), and during cooling from above T_m (■). Note that the absorbance data is completely unchanged by repeated cycles of heating and cooling, whereas viscosity shows dramatic hysteresis effects. [Parts a and b after R. Inman and R. Baldwin, *J. Mol. Biol.* 2:181 (1960) and 8:452 (1964).]   (c) Schematic illustrations of melting and renaturation.

Furthermore, both samples show marked temperature-dependent viscosity changes in regions far outside the narrow absorbance melting region. Figure 22-19c shows a possible origin of these effects. Native synthetic poly d(AT)·d(AT) is a duplex at low temperature. As the temperature is raised, it becomes possible for very small amounts of melting to occur in local regions. The actual amount of base-pair disruption is too small to register as an absorbance change, but the melted regions allow chain slippage to occur and hairpin loops to form. Because these regions contain a few unpaired bases, they are higher in enthalpy than the native duplex. However, there are a large number of energetically equivalent configurations in a structure with several hairpins. The statistical weight of the ensemble of these structures is higher, the entropy is higher, and thus the free energy will be lower.

Viscosity is very sensitive to length changes. The hairpin-containing structures are much shorter, thus accounting for the large drop in viscosity as the melting region is approached. Above the melting temperature, the viscosity actually rises again. Once all base pairs are disrupted, the strands separate, and the relatively globular condensed cluster of hairpins below T_m is converted into two expanded coils. Note that the plateau viscosity reached at high temperature is a sensitive function of ionic strength. The lower the salt concentration, the stronger becomes the electrostatic repulsion of negative phosphates in the coil. This effect causes the coil to expand and stiffen, raising the viscosity. In the limit of very low ionic strength, one expects the coil to approach a rodlike configuration, because this form minimizes the electrostatic free energy of a uniformly charged linear molecule. The resulting viscosity then will be almost the same as the viscosity of the original double helix.

When a poly d(AT) coil is cooled below T_m, intrastrand base pairing is so rapid that there is little chance of two individual coils starting a helix together. The viscosity of the cooled solution is much less than it was before the native poly d(AT)·d(AT) was first melted. The molecular weight of the renatured material is half the starting molecular weight. The two single-strand hairpins are higher in energy than the native form, and should eventually reconvert to it. If this does occur, the rate is too slow to measure. This is reasonable; one can predict a very high activation energy for reconversion, because many base pairs must be disrupted to initiate stable inter-strand pairing.

Poly dI·d5BrC shows much less evidence of "premelting" than poly d(AT)· d(AT), because the individual strands are not self-complementary. Hairpin loops cannot be stabilized by base pairing. When melted poly dI·d5BrC is renatured by cooling below T_m, the viscosity shows much greater temperature dependence, and the final low-temperature sample is much more viscous than the original native material. The origin of these effects is aggregation (Fig. 22-19c). As temperatures below T_m are reached, duplexes form without regard to length. Each of these attempts to maximize the amount of double helix, but single-stranded regions generally remain, the result of any mismatched lengths. At temperatures only slightly below T_m, long single-stranded ends can easily form duplexes by base pairing with others. Shorter ones may not be able to form duplexes at such high temperatures because, as we shall see, the stability of double helices is length dependent. However, as the temperature is lowered, more and more ends can pair, leading to larger and larger aggregates on the average. This causes the continued increase in viscosity, even at temperatures far below T_m.

Correlation of double-stranded DNA properties with base composition

The last few examples show that simple-sequence model systems have certain disadvantages. Their great degree of symmetry or homogeneity may favor phenomena not accessible to a natural system. It is now time we compared the properties of synthetic polynucleotides with those of natural nucleic acids. We start with native DNA. Because this usually is completely double-stranded, it presents a much simpler problem than most RNAs.

Most natural DNA is composed of only the four normal bases A, T, C, and G. It is a good first assumption that any given sample is composed of continuous double helices with no strand breaks or loops, and with only minor imperfections at the ends. In practice, unless extraordinary care is taken, DNA in the test tube contains broken pieces of the original chromosomal DNA. Except for small amounts of DNA with simple repeating sequence common in eukaryotes, the average DNA piece has a very asymmetric sequence. All such DNA, so far as we know, has approximately the structure of the B-form double helix seen in x-ray fiber diffraction at high humidity see Chapter 3).

In comparing the average properties of DNA samples, it is safe to assume that we are starting with fundamentally similar structures, differing only in the details of base composition and sequence. It is natural then to ask if we can understand any variation in observed bulk properties as a function of sequence or composition. Let P be some observed property of a DNA. In the simplest possible case, P will be a function only of individual base pairs. If P_{at} and P_{gc} are intrinsic properties of A–T and G–C pairs, respectively, then

$$P = f(P_{at}, P_{gc}, \chi_{gc}) \tag{22-25}$$

where χ_{gc} is the mole fraction of G–C in the DNA. For optical properties, a linear function is expected if there are no sequence effects:

$$P = \chi_{gc}P_{gc} + (1 - \chi_{gc})P_{at} = P_{at} + (P_{gc} - P_{at})\chi_{gc} \tag{22-26}$$

For other properties, the functional form should be more complex.

Effect of base composition on DNA melting temperature and density

The melting temperature of a double-stranded oligonucleotide is a function of sequence but, for long DNAs of heterogeneous sequence, the detailed sequence effects are almost completely averaged out. The dependence of ΔH^0 and ΔS^0 on mole fraction of G–C pairs must be known in order to compute the effect of base composition on T_m.

Ignoring any sequence differences in vertical stacking enthalpy or entropies, one can express the different enthalpies and entropies of G–C and A–T base–pair formation as

$$\Delta H^0_{gc} = \Delta H^0_{at}(1 + h) \tag{22-27a}$$

$$\Delta S^0_{gc} = \Delta S^0_{at}(1 + y) \tag{22-27b}$$

where the parameters h and y will be small, because observed differences in A–T and G–C thermodynamic parameters are small. Then the total enthalpy and entropy changes on melting, per residue, are found by using Equation 22-26.

$$\Delta H^0 = \Delta H^0_{at}(1 - \chi_{gc}) + \Delta H^0_{at}(1 + h)\chi_{gc} = \Delta H^0_{at}(1 + h\chi_{gc}) \tag{22-28a}$$

$$\Delta S^0 = \Delta S^0_{at}(1 - \chi_{gc}) + \Delta S^0_{at}(1 + y)\chi_{gc} = \Delta S^0_{at}(1 + y\chi_{gc}) \tag{22-28b}$$

From the definition $T_m = \Delta H^0/\Delta S^0$, valid for long polymers, the melting temperature of an arbitrary DNA can be estimated as

$$T_m = \frac{\Delta H^0_{at}}{\Delta S^0_{at}}\left(\frac{1 + h\chi_{gc}}{1 + y\chi_{gc}}\right) \cong T_{m(at)}[1 + (h - y)\chi_{gc}] \tag{22-29}$$

because $1/(1 + y\chi_{gc}) \cong 1 - y\chi_{gc}$ when $y\chi_{gc}$ is small. Equation 22-29 predicts a linear dependence of melting temperature on the G–C content. Experimental results on natural DNAs are in good agreement with this prediction (Fig. 22-20). From a fit to the observed data (in 0.15 M NaCl), we can evaluate the parameters in Equation 22-29:

$$T_m \,(^\circ C) = 69.3 + 41\,\chi_{gc} \tag{22-30}$$

The parameters $T_{m(at)}$ and $h - y$ naturally are functions of other environmental conditions such as salt and pH. It is interesting that conditions can be found where $h - y = 0$. Figure 22-20 shows this for 2.36 M tetraethylammonium chloride. Even higher concentrations of this salt make $dT_m/d\chi_{gc}$ negative, so that DNAs with higher G–C content melt at lower temperatures.

The melting behavior of synthetic DNAs with regularly alternating sequence is not consistent with Equation 22-29. The actual variation of enthalpy or entropy with sequence must be considered in order to explain these discrepancies.

Another frequently measured property of double-stranded DNA is ρ_0, the buoyant density in CsCl. The value of ρ_0 is a function of the partial specific volume and the preferential hydration of the DNA. It has been found empirically (Fig. 22-21) that most natural DNAs can be fit by the equation

$$\rho_0 \,(g\,cm^{-3}) = 1.660 + 0.098\,\chi_{gc} \tag{22-31}$$

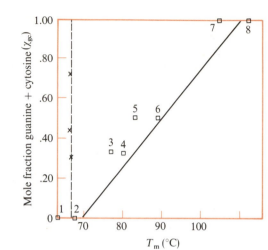

Figure 22-20

Dependence of melting temperature on base composition. T_m is plotted against the mole fraction of G + C for a series of natural DNAs (*solid line*) in 0.15 M NaCl, 0.015 M citrate. Also shown are the T_m values of a number of synthetic polymers (□) under comparable conditions: (1) d(A–T)·d(A–T); (2) dA·dT; (3) d(T–A–C)·d(G–T–A); (4) d(A–T–C)·d(G–A–T); (5) d(T–C)·d(G–A); (6) d(T–G)·d(C–A); (7) dG·dC; (8) d(G–C)·d(G–C). Melting data for three DNAs (×) also are shown under quite different conditions: 2.36 M tetraethyl ammonium chloride; here the dependence of T_m on base composition is abolished (*dashed line*). [After R. Wells et al., *J. Mol. Biol.* 54:465 (1970); and W. B. Melchior and P. von Hippel, *Proc. Natl. Acad. Sci. USA* 70:295 (1973).]

Equation 22-32 indicates that DNAs of differing base compositions can be physically separated by centrifugation in CsCl. The technique is quite useful for analyzing the relative amounts of DNAs of different compositions. It also is useful for purifying DNAs of fairly different base composition, but the small coefficient of χ_{gc} makes it difficult to resolve fully DNAs of very similar base compositions.

A number of DNAs in Figure 22-21 do not obey the relationship of Equation 22-31. These exceptions are interesting. Synthetic DNAs of regularly alternating sequence have anomalous buoyant densities because particular nearest-neighbor pairs are heavily weighted. Some bacteriophage DNAs have buoyant densities that deviate greatly from the predictions of Equation 22-31. This is due to the occurrence of unusual bases in these DNAs, such as glycosylated 5-hydroxymethyl cytidine, which replaces C in certain DNAs. Thus, buoyant density measurements are a sensitive way to spot DNAs with unusual constituents.

Effect of sequence on double-strand properties

The simplest general way to take sequence into account is to assume that all effects are dominated by nearest-neighbor interactions. There are ten such interactions in

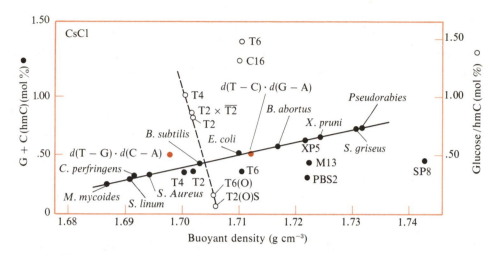

Figure 22-21

Buoyant density of native DNAs in CsCl gradients at 25°C as a function of base composition. The left-hand scale shows mole fraction of G plus C or C analogs. The right-hand scale shows equivalents of glucose per hydroxymethyl cytidine for T-even bacteriophage DNAs (○). [After W. Szybalski, in *Procedures in Nucleic Acid Research*, vol. 2, ed. G. L. Cantoni and D. R. Davies (New York: Harper & Row, 1971), p. 311.]

antiparallel double-stranded DNA:

A–A	A–T	T–A	A–G	G–A	A–C	C–A	G–C	C–G	G–G
. .	. .	. .	. .	. .	. .	. .	. .	. .	. .
T–T	T–A	A–T	T–C	C–T	T–G	G–T	C–G	G–C	C–C

Any observed property P can be analyzed in terms of the intrinsic properties P_i (for $1 \leqslant i \leqslant 10$) associated with each interaction and of the mole fraction χ_i for the occurrence of that interaction in the DNA. In polymers long enough for the ends to be neglected, the ten mole fractions are not independent variables. Two constraints can be developed from the antiparallel base pairing of DNA. Considering both strands together,

$$\chi_{AT} + \chi_{AC} + \chi_{AG} = \chi_{TA} + \chi_{CA} + \chi_{GA} \qquad (22\text{-}32)$$

$$\chi_{GA} + \chi_{GC} + \chi_{GT} = \chi_{AG} + \chi_{CG} + \chi_{TG} \qquad (22\text{-}33)$$

where, for example, χ_{AT} is the mole fraction of the A(3′ → 5′)T sequence. Equation 22-32 simply means that any block of A residues must be bordered on both sides by bases other than A. Equation 22-23 is the equivalent conclusion for blocks of G residues.

These constraints mean that only eight mole fractions χ_i' and eight intrinsic properties P_i' are needed to fit observed data. Or to put it in a more crucial way, only eight independent parameters are derivable from experimental data. This still is a very large number, and it is indeed rare to find any measurable property of a system that justifies the evaluation of eight parameters. An exception is CD spectra, which are remarkably sensitive to sequence. The CD of DNAs is fit moderately well by an equation of the form

$$[\theta] = \sum_{i=1}^{8} \theta_i' \chi_i' \tag{22-34}$$

where θ_i' and χ_i' are linear combinations of the properties and occurrences of the ten fundamental interactions. These combinations can be determined by studies on model DNAs of known nearest-neighbor sequence.

For other properties, it is sufficient to use much more simplified analogs of Equation 22-34. As an example, the change in extinction coefficient on melting a DNA can be described adequately by grouping the ten nearest-neighbor interactions into three sets. Let $\Delta\varepsilon_{AT}$ be the average $\Delta\varepsilon$ for the three pure A–T nearest-neighbor interactions, $\Delta\varepsilon_{GC}$ be the average $\Delta\varepsilon$ for the three pure G–C interactions, and $\Delta\varepsilon_{AG}$ be the average $\Delta\varepsilon$ for the four types of neighboring A–T and G–C base pairs. Furthermore, suppose that the nearest-neighbor frequencies are random. Then the mole fractions of these groups as a function of the mole fraction of G–C are simply $(1 - \chi_{gc})^2$, χ_{gc}^2, and $2\chi_{gc}(1 - \chi_{gc})$. The $\Delta\varepsilon$ for melting is

$$\Delta\varepsilon = (1 - \chi_{gc})^2 \Delta\varepsilon_{AT} + 2\chi_{gc}(1 - \chi_{gc})\Delta\varepsilon_{AG} + \chi_{gc}^2 \Delta\varepsilon_{GC}$$

$$= \Delta\varepsilon_{AT} + 2\chi_{gc}(\Delta\varepsilon_{AG} - \Delta\varepsilon_{AT}) + \chi_{gc}^2(\Delta\varepsilon_{AT} + \Delta\varepsilon_{GC} - 2\Delta\varepsilon_{AG}) \tag{22-35}$$

This equation predicts that the absorbance change will be a quadratic function of the mole fraction of G–C pairs. In practice, at most wavelengths, $\Delta\varepsilon_{AT} + \Delta\varepsilon_{GC} \cong 2\Delta\varepsilon_{AG}$. Then the χ_{gc}^2 term drops out, and Equation 22-35 reduces to

$$\Delta\varepsilon = \Delta\varepsilon_{AT} + \chi_{gc}(\Delta\varepsilon_{GC} - \Delta\varepsilon_{AT}) \tag{22-36}$$

This is just the form predicted by Equation 22-26, where sequence effects were ignored.

Effect of ionic strength on thermal stability

For alkali cations, the effect of ionic strength on the stability of the DNA duplex is essentially independent of base composition or sequence of natural DNAs. Figure 22-22 shows that T_m is linear in the logarithm of the total salt concentration (C_s).

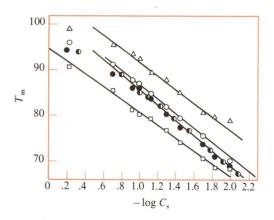

Figure 22-22

Dependence of DNA melting temperature on salt concentration. Points shown are for *D. pneumoniae* DNA in KCl (□), *E. coli* DNA in KCl (○), *Ps. aeruginosa* DNA in KCl (△), *E. coli* DNA in sodium phosphate buffer (●), and *E. coli* DNA in sodium citrate buffer (◐). The deviations from linear behavior at high salt concentrations are real. In fact, at even higher salt concentrations, T_m can decrease with increasing molarity. [After C. Schildkraut and S. Lifson, *Biopolymers* 3:195 (1968).]

The slope of the plot of T_m versus C_s is effectively independent of base composition, so a wide variety of DNA melting data can be fit by the equation

$$T_m = 16.6 \log C_s + 41 \chi_{gc} + 81.5 \qquad (22\text{-}37)$$

It is interesting to compare Equation 22-37 with Equation 22-20b, in which we attempted to explain salt effects by differential binding of cations to the double helix. Equation 22-20b predicts that T_m^{-1} is in the form $a \log C_s + b$, where a and b are constants. The data included in Figure 22-22 actually cover a rather narrow range of absolute temperature and, in this range, a scale linear in T_m^{-1} is close to linear in T_m also. Thus the two equations are consistent.

Although the simple theory that led to Equation 22-20b is sufficient to explain the qualitative effects of ionic strength on polynucleotide melting, it turns out that this theory is not correct quantitatively. In the next few subsections, we develop the outlines of modern polyelectrolyte theory and show how it is applied to the analysis of polynucleotide melting.

Some DNAs of regularly alternating sequence do not behave like normal DNAs in their ionic-strength effects. A linear dependence of T_m on $\log C_s$ still holds, but some DNAs such as poly d(T–G)·d(C–A) and poly d(A–T–C)·d(G–A–T) show a considerably lower sensitivity to salt. The explanation for this behavior is not clear. Perhaps these particular sequences lead to more effective counterion binding in the single strands. Another possibility is that the structure of some of these DNAs is different.[§] Whatever the explanation, this observation reinforces the important fact

---

[§] For example, M. A. Viswamitra recently has solved the crystal structure of pdApdTpdApdT. Here the 5'-AT sequence pairs with the 3'-AT of one tetramer, whereas the 3'-AT pairs with the 5'-AT of another tetramer. The middle T and A of each tetramer are unstacked. Schematically, the structure of one tetramer is ⌐⌐⌐⌐. It is not yet known whether this structure is simply an end effect in a short oligonucleotide or whether it represents some unusual feature of the ATAT sequence.

that, in a natural DNA, most techniques represent a rather crude average over local properties. They can obscure local variations that might well be of extreme biological importance.

● Thermodynamics of linear polyelectrolytes

It is convenient to model a nucleic acid as an infinite linear array of z negative charges of average spacing b. For a rigid double strand, b is measured along the helix axis; for more flexible molecules, it is measured along the contour axis.

To understand the effects of ionic environment on nucleic acids, we must consider two phenomena. One (described in approximate form in Eqn. 22-4) is electrostatic shielding by the counterion atmosphere. The other is direct binding (condensation) of counterions onto the nucleic acid.

We first consider counterion binding and then use the results obtained to describe, in a more accurate and useful form, the counterion atmosphere. The dimensionless parameter ξ governs counterion binding; it is defined as

$$\xi \equiv e^2/\varepsilon kTb \qquad (22\text{-}38)$$

where e is the charge on the electron, ε is the solvent dielectric constant, k is Boltzmann's constant, and T is the absolute temperature. Note that $e^2/\varepsilon kT$ has the value of 7.14 Å in water at 25°C. Gerald Manning showed that when $\xi < 1$ no counterions condense on the polymer, but when $\xi > 1$ counterion condensation occurs. The amount of counterion binding does not depend on the ionic strength. The fraction of a counterion condensed per polymer charge is

$$\theta_{\text{M}} = 1 - \xi^{-1} \qquad (22\text{-}39)$$

For double-strand DNA B, there is one base pair (or two phosphates) per 3.4 Å. Thus b is 1.7 Å, and ξ is 7.14/1.7 = 4.2. This means that counterion condensation occurs, and we can evaluate θ_{M} as 0.76. Experimental measurements of θ_{M} have been obtained through NMR using ^{23}Na. The result, 0.75 ± 0.1, is in excellent agreement with the predictions from polyelectrolyte theory.

After counterion condensation, nucleic acids still are highly charged polyelectrolytes. A net negative charge of $1 - \theta_{\text{M}}$ remains on each phosphate. Because $1 - \theta_{\text{M}} = \xi^{-1} = b/7.14$, there is one net remaining negative charge for each 7.14 Å. These charges are now screened by the counterion atmosphere. We can evaluate this screening by using Equation 22-4, after substituting the net charge $e(1 - \theta_{\text{M}})$ for the electronic charge e. It is most convenient to calculate the electrostatic free energy per mole of phosphate:

$$G_{\text{el}} = \frac{N_0}{z} \frac{e^2(1 - \theta_{\text{M}})^2}{\varepsilon} \sum_{l=1}^{z} \sum_{k \neq l}^{z} \frac{1}{2} \exp[(-\kappa|l - k|b)]/|l - k|b \qquad (22\text{-}40)$$

The double sum can be shown, relatively easily,[§] to be equal to $-(z/b) \ln [1 - \exp(-\kappa b)]$ in the limit of large z. The terms preceding the sum can be rewritten as $(N_0/z)(bkT\xi)\xi^{-2} = bRT\xi^{-1}z^{-1}$. Thus one has the particularly simple result that

$$G_{el} = -RT\xi^{-1} \ln[1 - \exp(-\kappa b)] \qquad (22\text{-}41)$$

In most cases of interest, $\kappa b \ll 1$, and thus the exponential in Equation 22-41 can be expanded to yield an even simpler result:

$$G_{el} = -RT\xi^{-1} \ln \kappa b \qquad (22\text{-}42)$$

As the ionic strength increases, the Debye–Hückel screening parameter κ increases as $(M)^{1/2}$, where (M) is the molar concentration of cations. Thus κb moves closer to one; the $\ln \kappa b$ term becomes less negative; and G_{el} becomes less positive. In effect, electrostatic repulsions are shielded. Note that the magnitude of these repulsions is scaled as ξ^{-1}, the net fractional charge on each phosphate. Thus, screening is more important for single-stranded polynucleotides than for double strands, because the net charge after counterion condensation is much higher for single strands (0.60 per phosphate) than for double strands (0.24 per phosphate).

● Effects of ionic strength on polynucleotide melting

Here we derive a general equation to describe ligand effects on polynucleotide melting. We apply this equation to the particular case of excess monovalent cation M, where Equation 22-20b is not accurate. Let D be a nucleic acid double helix, and S_1 and S_2 be the two single strands. The observed equilibrium constant for double-strand formation is

$$K_{obs} = (D)/(S_1)(S_2) \qquad \text{or} \qquad \Delta G_{obs} = -RT \ln K_{obs} \qquad (22\text{-}43)$$

The true equilibrium constant, in a particular reference state for the melting reaction $S_1^{(0)} + S_2^{(0)} + \Delta n M \rightleftarrows D^{(0)}$, is

$$K_T^{(0)} = (D^{(0)})/(S_1^{(0)})(S_2^{(0)})(M)^{\Delta n} \qquad \text{or} \qquad \Delta G_T^{(0)} = -RT \ln K_T^{(0)} \qquad (22\text{-}44)$$

[§] This is an excellent exercise for the mathematically inclined reader. One starts by recognizing that, among the $z(z-1)/2$ terms in the sum, there are $z - 1$ terms corresponding to nearest-neighbor interactions between charges a distance b apart, $z - 2$ terms for next-nearest neighbors spaced $2b$ apart, and so on. When identical terms are gathered together, the double sum can be rewritten as

$$(1/b)\sum_j (z - j) \exp[(-j\kappa b)]/j$$

where the sum is taken from $j = 1$ to $j = z - 1$. The resulting single sum is evaluated by decomposing it into geometric series or integrals of geometric series.

where the superscript (0) denotes reference state, and Δn is the stoichiometric counterion uptake upon double-strand formation (or, equivalently, the stoichiometric counterion release upon melting). It usually is convenient to pick as a reference state the polymers with their condensed complement of counterions.

The difference between K_{obs} and $K_T^{(0)}$ is due to two factors: the free energy change ΔG_c involved in the uptake of counterions, and the difference ΔG_{el} in ion-atmosphere shielding between the double and single strands. Consider a double helix (or a cooperatively melting double-helical unit) involving z phosphates. From the polyelectrolyte theory in the preceding subsection, the amounts of counterion bound to the double helix and to the two single strands are $z(1 - \xi_D^{-1})$ and $z(1 - \xi_S^{-1})$, respectively. Thus the stoichiometric counterion uptake accompanying double-strand formation is $\Delta n = z(\xi_S^{-1} - \xi_D^{-1})$. If the counterion concentration is (M) during the melting, the resulting free energy change is

$$\Delta G_c/RT = -z(\xi_S^{-1} - \xi_D^{-1})\ln(M) \tag{22-45}$$

The shielding contribution to the melting free energy is given by Equation 22-42. For a duplex with z phosphates,

$$\Delta G_{el}/RT = -z(\xi_D^{-1}\ln \kappa b_D - \xi_S^{-1}\ln \kappa b_S) \tag{22-46}$$

where b_S and b_D are the spacings between phosphates in single and double strands, respectively, and κ is the Debye–Hückel screening parameter. For a monovalent electrolyte, κ is $0.33 (M)^{1/2}$ as described earlier, so that $\Delta G_{el}/RT$ becomes

$$\Delta G_{el}/RT = -(z/2)(\xi_D^{-1} - \xi_S^{-1})\ln(M) - z(\xi_D^{-1}\ln 0.33b_D - \xi_S^{-1}\ln 0.33b_S) \tag{22-47}$$

We can combine $\Delta G_c/RT$ and $\Delta G_{el}/RT$ to write

$$\ln K_{obs} = \ln K_T^{(0)} + \Delta G_c/RT + \Delta G_{el}/RT$$
$$= \ln K_T^{(0)} - (z/2)(\xi_S^{-1} - \xi_D^{-1})\ln(M) - z(\xi_D^{-1}\ln 0.33b_D - \xi_S^{-1}\ln 0.33b_S) \tag{22-48}$$

Thus, the effect of salt concentration on the observed equilibrium constant is

$$\partial(\ln K_{obs})/\partial[\ln(M)] = -(z/2)(\xi_S^{-1} - \xi_D^{-1}) = -\Delta n/2 \tag{22-49}$$

Note that this is exactly one-half the effect predicted by the simpler theory developed earlier (Eqn. 22-19a). A more accurate treatment would include nonideality corrections, but these are fairly small under typical conditions used to study DNA melting. When Equation 22-49 is used to replace Equation 22-19a, we can rewrite Equation 22-20a as

$$\partial T_m/\partial[\ln(M)] = -(\Delta n/2)(RT_m^2/\Delta H^0) \tag{22-50a}$$

Again, the ionic-strength effects are only one-half those predicted by simple theory. The reason is fairly easy to explain physically. By mass action, salt will favor the form that binds more salt (the double helix). But electrostatic shielding will favor this form less because it has less net charge. Thus salt binding and shielding oppose each other, although the former effect remains dominant.

In general, Δn and ΔH^0 refer to the cooperative unit in the melting transition. Because this often is not known precisely, it is convenient to express the melting properties per phosphate. Let ΔH_p^0 be the enthalpy change per phosphate, and recall that $\xi_S^{-1} - \xi_D^{-1}$ is the stoichiometric counterion uptake per phosphate. Defining this uptake as $\Delta \xi^{-1}$, we can write

$$\partial T_m/\partial [\ln(M)] = -(\Delta \xi^{-1}/2)(RT_m^2/\Delta H_p^0) \tag{22-50b}$$

Integration of this expression from some reference melting temperature $T_m^{(0)}$ and salt concentration ($M^{(0)}$) to the experimental values T_m and (M) yields

$$1/T_m = 1/T_m^{(0)} - (R\Delta\xi/2\Delta H_p^0)\,\ln(M^{(0)}) + (R\,\Delta\xi^{-1}/2\Delta H_p^0)\,\ln(M) \tag{22-51}$$

This shows that it is actually $1/T_m$ that should be linear in $\ln(M)$, but experimental data normally cover such a narrow range of absolute temperature that a scale linear in $1/T_m$ is almost linear in T_m also. Thus, polyelectrolyte theory accounts nicely for the observed salt-dependent melting behavior of DNA.

Applications of polyelectrolyte theory

We can use Equation 22-50b in a number of different ways. For example, in the case of T4 DNA, $RT_m^2/\Delta H_p^0$ measured by calorimetry is -50 deg; $\partial T_m/\partial [\ln(Na^+)]$ is 8.9. Thus, from Equation 22-50b, we can evaluate

$$\Delta\xi^{-1} = -2(\Delta H_p^0/RT_m^2)\,\partial T_m/\partial [\ln(Na^+)] = 0.36 \tag{22-52}$$

Because $\xi_D^{-1} = 0.24$ (calculated as shown in the preceding subsection), we can evaluate ξ_S^{-1} as 0.60. This in turn allows us to evaluate the average spacing between phosphates in the single strand. Because $\xi_S = 1.7$, we are still in the range of counterion condensation. From Equation 22-39, the fraction of a charge condensed per phosphate is 0.40. Now, using Equation 22-31, we can calculate that b (the average spacing between phosphates in the single strand) is 4.3 Å.

As a second example, consider the various melting equilibria for the poly A and poly U system summarized in the phase diagram of Figure 22-16. The salt dependence of the various transitions is summarized in Table 22-4. From measured values of $\partial T_m/\partial [\ln(Na^+)]$ and $RT_m^2/\Delta H_p^0$, one can evaluate $\Delta\xi^{-1}$ for each transition.

Because the structures of poly A·U and poly A·2U are known from x-ray fiber diffraction, we can evaluate the phosphate spacings as $b_{A·U} = 1.5$ Å and $b_{A·2U} = 1.0$ Å. Using these values in Equation 22-38, we calculate that $\xi_{A·U}^{-1} = 0.22$ and $\xi_{A·2U}^{-1} = 0.15$.

Table 22-4
Salt dependence of melting transitions

Transition	$\partial T_m / \partial [\ln(Na^+)]$	$RT_m^2 / \Delta H_p^0$	$\Delta \xi^{-1}$
poly A + poly U $\rightleftarrows$ poly A·U	8.5	−55	0.32
poly A + 2 poly U $\rightleftarrows$ poly A·U$_2$	11.3	−54	0.44
poly A·U + poly U $\rightleftarrows$ poly A·U$_2$	15.6	−159	0.20
poly A·U$_2$ + poly A $\rightleftarrows$ 2 poly A·U	−6.5	−230	−0.06

NOTE: ΔH_p^0 is the enthalpy of double-helix formation per phosphate; $\Delta \xi^{-1}$ is the counterion uptake per phosphate.
SOURCE: After M. T. Record et al., *Quart. Rev. Biophys.* 11:103 (1978).

These results can be combined with the data in Table 22-4 to compute net charges and chain dimensions of single-strand poly A and poly U.

Remember that $\Delta \xi^{-1}$ is the stoichiometric ion uptake per phosphate. Thus, for the reaction poly U + poly A·U $\rightleftarrows$ poly A·2U, we can write

$$\Delta \xi^{-1} = (1/3)\xi_U^{-1} + (2/3)\xi_{A \cdot U}^{-1} - \xi_{A \cdot 2U}^{-1} \tag{22-53}$$

This yields $\xi_U^{-1} = 0.61$. Similarly, for the equilibrium poly A + poly U $\rightleftarrows$ poly A·U, we can evaluate ξ_A^{-1} as 0.46. In turn, these results yield $b_A = 3.3$ Å and $b_U = 4.3$ Å. Thus, poly U is very much like a typical single-strand denatured DNA, whereas poly A has a much more condensed structure with an average phosphate spacing not too much larger than that expected for a single strand of a nucleic acid double helix.

Shape of melting transition of natural DNA

The width of the melting transition of a synthetic DNA with a regularly alternating sequence is quite sharp. The fraction of double helix can decrease from 0.75 to 0.25 in 1°C or less. In contrast, the widths of melting of natural DNAs are several times broader (see Fig. 22-15a). Equation 22-30 suggests an explanation for this observation. The melting temperature T_m is sufficiently sensitive to base composition that local fluctuations in χ_{gc} will produce local regions with varying T_m. Because the coupling between these regions is not infinitely strong, individual regions melt independently from one another. Thus, an all-or-none melting model is not valid for DNA.

A clear demonstration of independent melting domains was made by E. Reich and colleagues, who took advantage of the availability of enzymes and base analogs to synthesize DNAs with regions of differing local stability. The ends of DNA duplexes were made single-stranded by exonuclease-III digestion and then resynthesized as

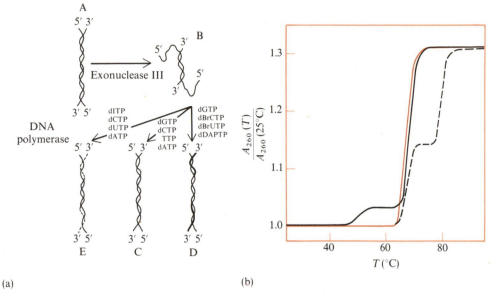

(a)   (b)

Figure 22-23

Melting of DNAs with large regions of differential stability. (**a**) Enzymatic preparation of DNA with extra stable ends (sample D), unstable ends (sample E), and normal ends (sample C). (**b**) Melting behavior of normal DNA (*colored curve*), a sample where the end 25% is extra stable (*dashed curve*), and a sample where the end 6% is unstable (*solid curve*). [After W. Beers et al., *Proc. Natl. Acad. Sci. USA* 58:1624 (1967).]

duplexes using different base analogs (Fig. 22-23a). Substitution of dI for dG destabilizes the duplex because the I–C base pair is weaker than G–C. Weaker stacking interactions in dU compared with dT also lower the T_m. The result is that a DNA with dI and dU selectively incorporated at the two ends shows a two-step transition in which the less-stable ends melt before the more-stable middle of the molecule. A complementary experiment involves making the ends more stable; this is accomplished by substituting 2,6-diaminopurine for A (stronger base pairing) and 5BrU for T and 5BrC for C (stronger stacking). In this case, the center of the DNA melts first, followed by the ends at a much higher temperature (Fig. 22-23b).

Base-composition effects are not as extreme in natural DNAs, but they still are significant. Suppose that a DNA has 50% G–C, and that you arbitrarily break up a long duplex into successive regions of 50 base pairs each. The average region, of course, will have 25 G–C pairs but, if the sequence is random, it will be quite common to find regions with as few as 21 or as many as 29 G–C pairs. This range corresponds to a range in χ_{gc} from 0.42 to 0.58. Equation 22-30 predicts a variation in T_m of 6.5°C.

This variation is somewhat larger than observed widths, indicating that 50 bases is not a long enough region to be considered fully independent in melting.

D. Crothers has performed more realistic calculations of melting. His results show that, as T_m is approached, denatured zones begin in regions of highest A–T content. At slightly higher temperatures, it becomes energetically favorable to disrupt regions between such denatured nuclei, regardless of the intervening G–C content. Such calculations predict widths typical of the observed melting transitions of natural DNAs. However, observed widths are dependent on ionic strength. This is a complicated effect that surely must arise from the electrostatic free energy of the melted nuclei. Because the two single strands in such nuclei are constrained to meet at their ends where duplex structure continues, the G_{el} is neither that of separated strands nor that of duplex. It is a complex function of the length of the nuclei.

If a DNA has certain unique regions of low thermal stability, it should be possible to identify these by trapping the partially melted structure and blocking renaturation. An elegant version of this approach is denaturation mapping in the electron microscope. The technique is most easily applicable when a unique piece of DNA is available. This piece can be either an entire viral DNA or a particular fraction of a eukaryotic DNA, such as repeated ribosomal genes. The DNA is partially melted (in many cases by alkali addition rather than heating) and then allowed to react with formaldehyde. The formaldehyde preferentially couples to the amino groups of single-stranded residues. It blocks renaturation of the melted sections when the DNA is neutralized or cooled and examined in the electron microscope.

Many molecules must be examined to make a clear identification of the preferred locations of melted regions. The melting transitions of individual regions still have a finite pH or temperature width and take a finite time. Reactivity with formaldehyde is not an all-or-none reaction. Thus, each individual molecule examined in the electron microscope will show a different pattern of single-stranded loops (or loops collapsed into globular or branched structures) interspersed with double-stranded segments. Figure 22-24a shows one such molecule, a piece of DNA from *Xenopus laevis*. The locations of structures indicative of melted regions are recorded by measuring their distances from the ends, and this procedure is repeated for a large number of individual molecules. There is no intrinsic way to distinguish one end of the DNA from the other. However, the prominent denatured features usually are asymmetric enough to permit left–right discrimination. Once all of the denatured patterns are aligned, the major consistent features are apparent by inspection (Fig. 22-24b). This pattern is an important signature of the DNA, and it is useful for a variety of biochemical and genetic studies.

A variation on the concept of denaturation mapping is secondary-structure mapping. This technique has found its greatest use thus far in studies of RNA, but it should be equally applicable to single strands of DNA. Rather than focusing on the first regions to melt, secondary-structure mapping highlights the last regions to melt. These appear as hairpins or more complex treelike figures along the otherwise extended single-stranded RNA. Ideally, the most stable regions of RNA secondary structure should correspond to the most stable regions of the DNA duplex that codes for this RNA. Temperature is not a good environmental variable for secondary-structure mapping, because of the cooperativity of the final stretches of thermal melting and because formaldehyde reactivity at temperatures above T_m probably is too vigorous. Organic solvent denaturants have proven to be more useful, and most work has been done with aqueous formamide solutions.

Figure 22-24c shows an example for ribosomal DNA and RNAs from *Xenopus laevis*. By comparing RNA and DNA results, one can assign the region of the DNA coding for the 40S rRNA precursor, and also can assign regions within it that code for the mature 18S and 28S rRNA.

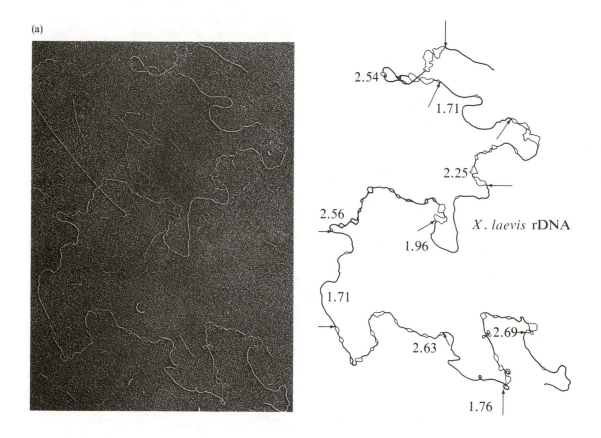

Figure 22-24

Use of electron microscopy in sequence analysis of ribosomal DNA and RNA from *Xenopus laevis*.

(a) Denaturation map of a single rDNA molecule. Alkali treatment was used to cause partial denaturation. After formaldehyde fixation, the molecule is viewed by the Kleinschmidt technique. The tracing clearly shows a repeating pattern of easily melted and stable duplex regions.

1164

Figure 22-24 *(continued)*

(**b**) Denaturation diagram of several DNA molecules; molecule 1 is the one shown in part a. Each line is a single repeat of the stable and unstable regions, and each line begins where the one before it ends. The bars show melted regions. Molecule 2 is the same as molecule 1, but is aligned by eye. This molecule is the most regular of all those examined. Molecule 3 is an average molecule, and molecule 4 is the least regular denaturation pattern seen. Note that a fine structure to the pattern of repeated stable and unstable regions emerges in such an alignment. The stable regions are high-(G–C)-content spacers; the unstable regions are of low (G–C) content, and each is transcribed into a 40S precursor rRNA, which is ultimately trimmed into 18S and 28S rRNAs.

(**c**) Secondary structure maps of rRNAs and rDNA prepared by formamide treatment. The photomicrograph labeled 40S shows a single 40S precursor rRNA. The schematic of the molecule indicates the symbols used to describe single loops (bar) and complex loops (bar with a vertical line). The photomicrographs labeled 18S and 28S show single 18S and 28S rRNA molecules. The photomicrograph labeled rDNA shows a section of a single strand of rDNA. The region between the white arrows is a single transcription unit for one 40S rRNA precursor. The regions outside of the arrows are the (G–C)-rich spacer, which corresponds to the stable regions seen in the repeating pattern in parts a and b. Schematic secondary structure maps are shown for three 18S, three 28S, and three 40S rRNA molecules. Note that the pattern of secondary structure permits assignment of order of the 18S and 28S molecules within the 40S precursor. The schematic secondary structure map of a section of rDNA shows repeating sections of the region coding for 40S rRNA (the first complex loop in the 28S region is indicated by the arrow) and the (G–C)-rich stable region (the complex secondary structure pattern at the left of the repeating unit). [Parts a and b after P. C. Wensink and D. D. Brown, *J. Mol. Biol.* 60:235 (1971); part c provided by P. Wellauer.]

(b)

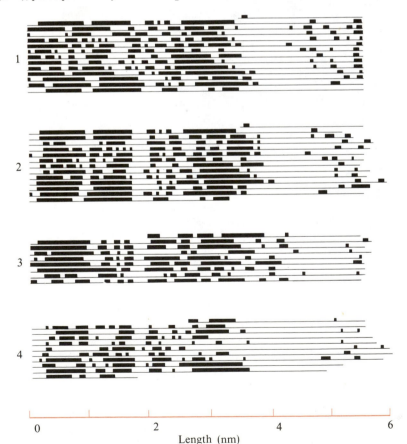

(c)

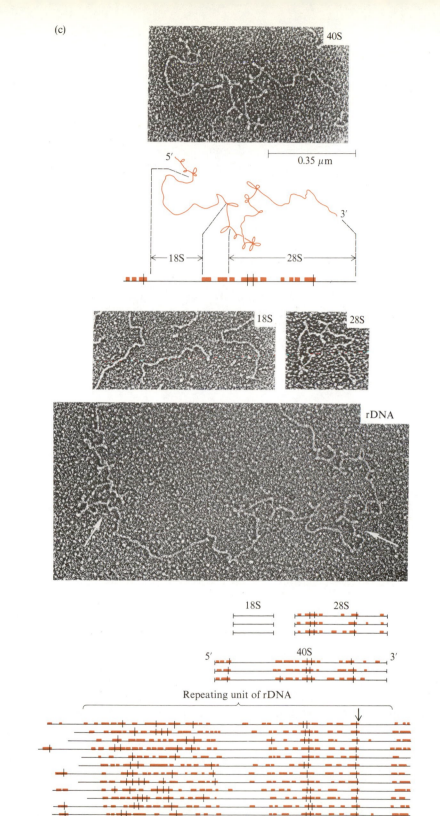

0.35 μm

5′

3′

18S

28S

18S   28S

rDNA

18S   28S

5′   40S   3′

Repeating unit of rDNA

22-4   STRUCTURE FLUCTUATIONS AND CHANGES
IN DOUBLE-STRANDED HELICES

Electron microscopy is a technique of unusual power; it provides a much more detailed picture of DNA melting than do many solution techniques. However, it can provide only a static picture of DNA. One area of interest is any possible tendency for double-stranded regions to open transiently at temperatures far below T_m. After all, what happens to a DNA at temperatures far above 37°C may contain clues about biological function, but the information obtained as not as useful a priori as it would be to know about behavior at biologically significant temperatures.

Chemical reactivity and exchange are effective methods for examining conformational fluctuations. Isotopic exchange is a particularly convenient technique because no significant alteration in structure is induced by the measurements. Here we discuss some of the different uses of tritium exchange. The general idea behind such methods is that more-exposed residues exchange chemically identical protons faster than do less-exposed residues.

Tritium-exchange rates of accessible protons

In principle, one can choose either to measure the rate of ^3H exchange into a macro-molecule initially containing no ^3H, or to preexchange with ^3H and then measure the rate of ^1H displacement of bound ^3H. Because of kinetic isotope effects, these two rates need not be the same. With small molecules, exchange is expected for any protons having a finite probability of being involved in an ionization reaction during the time period of the experiment. For example,

Reaction 1: $R{-}OT + OH^- \xrightleftharpoons{\text{slow}} R{-}O^- + HTO \xrightleftharpoons{\text{fast}} R{-}OH + OT^-$

Reaction 2: $R{-}NHT + H_3O^+ \xrightleftharpoons{\text{slow}} R{-}NH_2T^+ + H_2O \xrightleftharpoons{\text{fast}} R{-}NH_2 + H_2TO^+$

In each case, the rate-determining step involves a proton-transfer reaction of the general type

$$D{-}H^+ + A \underset{k_{-1}}{\overset{k_1}{\rightleftharpoons}} D{\cdots}H^+{\cdots}\dot{A} \underset{k_{-2}}{\overset{k_2}{\rightleftharpoons}} D + H^+{-}A$$

where D is a proton donor, and A is an acceptor. In Reaction 1, D–H is R–OT; in Reaction 2, D–H is H_3O^+.

Such proton-transfer reactions have been extensively studied. In most cases, the kinetics can be fitted to the following kind of model. Collision of the proton donor and acceptor occurs at a diffusion-controlled rate, so $k_1 \cong 10^{10}$ M^{-1} sec^{-1}. In the overall reactions in which we are interested, back exchange is negligible, and so k_{-2} can be ignored. To account for the observed rates, we must determine the fraction of collisions that result in proton transfer. This is done by arguing that proton transfer is so rapid in the $D \cdots H^+ \cdots A$ ternary complex that the proton equilibrates between the two bases D and A. The exchange rate will be determined by the fraction α_{ex} of the complex in which the proton is attached to the acceptor.

The proton binding strengths of donor and acceptor can be described by acid dissociation constants:

$$K_a = (H^+)(A)/(H^+A) \qquad K_d = (H^+)(D)/(DH^+) \qquad (22\text{-}54)$$

We can model the equilibrium in the ternary complex by the equilibrium reaction

$$DH^+ + A \rightleftharpoons D + H^+A$$

Using the results in Equation 22-54, the equilibrium constant for this reaction can be written as

$$(H^+A)(D)/(A)(DH^+) = K_d/K_a \qquad (22\text{-}55)$$

Thus, the fraction α_{ex} of the ternary complex containing a proton on A can be described in terms of this equilibrium constant as

$$\alpha_{ex} = (H^+A)(D)/[(H^+A)(D) + (A)(DH^+)] = (K_d/K_a)/(1 + K_d/K_a) \quad (22\text{-}56)$$

Using the definition $\Delta pK = pK_a - pK_d$, we can rewrite this result as

$$\alpha_{ex} = 10^{\Delta pK}/(1 + 10^{\Delta pK}) \qquad (22\text{-}57)$$

The second-order rate constant for exchange is $k_1 \alpha_{ex}$, the rate for forming the ternary complex times the fraction of that complex leading to exchange. This shows explicitly that the exchange rate will depend on the difference in pK between donor and acceptor.

In practice, because donors are present only in trace radioactive amounts, acceptors are always in such mole excess that their concentration is constant. The exchange kinetics are pseudo first order with an apparent first-order proton transfer-rate constant given by

$$k_{tr} = k_1(A)\alpha_{ex} \qquad (22\text{-}58)$$

For a given proton, pK_d is fixed; α_{ex} increases as pK_a increases, as shown by Equation 22-57. However, at a fixed pH, (A) decreases as the pK_a increases. Thus the observed exchange rate usually will be maximized when $pH > pK_a > pK_d$.

Equation 22-58 has the form of a base-catalyzed reaction. In pure water, the base is OH^-. The expression for α_{ex} in Equation 22-57 implies that donors with $pK_d \cong 14$ will exchange protons in milliseconds at pH 7, because pK_a for water is 14. Donors with $pK_d \cong 20$ will exchange in less than one hour of incubation. An equivalent treatment can be carried out for exchange reactions of the general class shown by Reaction 2; these are acid-catalyzed reactions.

The effect of any catalyst C on the proton-exchange rate of a molecule M can be put in a general form analogous to Equation 22-58:

$$k_{tr} = k_1(C)\alpha_{ex} \tag{22-59a}$$

where
$$\alpha_{ex} = (K_M/K_C)/(1 + K_M/K_C) \tag{22-59b}$$

where K_M is the acid dissociation constant of the site on the molecule containing an exchanging proton, and K_C is the acid dissociation constant of the catalyst. One must be prepared to consider more complex mechanisms involving catalysis by both acid and base, by water molecules, and by any other solute molecules with pK_a values near the pK_a of the proton of interest. In most cases, one particular exchange mechanism will be dominant for a particular reaction, and Equation 22-59 or an analog of it will hold.

Tritium exchange of nucleic acid bases in double strands

Figure 22-25 shows experimental arrangements for the study of tritium exchange in nucleic acids. After equilibration with tritiated water for a long time, the macro-

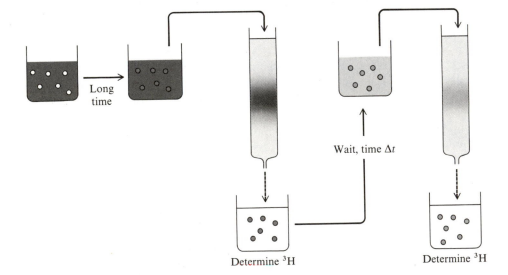

Determine ^{3}H   Determine ^{3}H

Figure 22-25

Schematic tritium-exchange experiment using the two-column technique. The density of shading indicates (roughly logarithmically) the amount of tritium present at various stages. The columns contain a gel exclusion resin (such as Sephadex) that passes macromolecules rapidly but retards the flow of small molecules.

molecular sample is passed rapidly through a short gel exclusion column. This treatment efficiently replaces all tritiated solvent with nonradioactive solvent. Back exchange starts and, after a given time, a second column is run and the difference in tritium remaining in the sample is assayed. Only a single column is used for very fast exchange rates; the sample is run halfway down and, after a pause to allow exchange to occur, is run the rest of the way and analyzed.

In a DNA, the only protons we expect to become fully equilibrated with tritium in the initial exchange are NH_2 and NH protons of the four bases. For an RNA, there is also the 2'-hydroxyl. The pK_a of this proton is around 12 in a nucleoside and, even if it is shifted-up several pH units in an RNA, it is so close to the pK_a of water that Equation 22-58 predicts rapid exchange. The pK_a values of the NH protons of U, T, and G are around 9 in the monomers; even large polyelectrolyte effects would leave them less than 14 in polymers. Therefore, millisecond or faster exchange rates are predicted. The pK_a values for deprotonation of the NH_2 protons of the bases are not known but, by analogy with other amines, they must lie in the range of 20 to 30. This means that base-catalyzed mechanisms of direct proton transfer at these positions are expected to be very slow. Similar arguments show that acid-catalyzed exchange by protonation at the exocyclic amino groups will also be slow.

If a polynucleotide or other sample has several different classes of exchangeable tritium, the exchange kinetics should be a sum of first-order rates:

$$H(t) = \sum_i n_i e^{-k_i t} \qquad (22\text{-}60)$$

where $H(t)$ is the total number of tritium atoms remaining per molecule at time t, n_i is the number of tritium atoms in the ith class, and k_i is the exchange rate for this class. Equation 22-60 arises directly from the fact that the exchange of every proton should be a pseudo first-order reaction, and the exchange of each proton is independent of any of the others.

In the first exchange experiments on calf thymus DNA, the shortest times that could be studied were about 120 sec after the start of back exchange. At this point, 1.9 ^3H atoms were retained per base pair. At times up to about 1,000 sec, exchange kinetics fit a single exponential decay down to 0.2 ^3H atoms per base pair, with $\tau_{1/2} \cong 330$ sec. The rate becomes slower at longer times. Extrapolation of the data back to zero time yielded an apparent initial ^3H content of 2.3 ^3H atoms per base pair. This value is in very good agreement with what would be calculated by assuming that all interstrand hydrogen-bonded protons exchanged slowly and that all others exchanged too rapidly to be observed. There are two hydrogen-bonded protons for each A–T and three for each G–C. From the base composition, 2.42 protons per average base pair are expected.

Improved experimental techniques soon made it possible to obtain data at shorter exchange times. Figure 22-26 shows an example of such an experiment on calf thymus DNA. Now, in addition to the class of protons seen before, a faster class

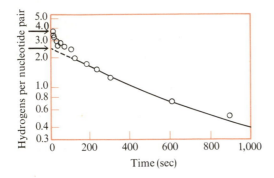

Figure 22-26

Exchange rate of tritium from native calf thymus DNA at 0°C, 0.1 M NaCl, pH 7.6. The arrows on the ordinate show the total exchangeable hydrogens expected per base pair in this DNA (3.84) and the total expected to be involved in interchain hydrogen bonds (2.42). In earlier work, only the long-time data shown by the solid line were available. [After S. W. Englander and P. von Hippel, *J. Mol. Biol.* 63:171 (1972).]

is seen with $\tau_{1/2} \cong 25$ sec. When the total visible exchanging tritium is extrapolated back to zero time, the result is in excellent agreement with the 3.84 ^{3}H atoms per base pair expected for complete exchange of all NH_2 and NH protons.

The original assignment of this faster class of protons was to the non-hydrogen-bonded NH_2 protons of C, A, and G. Contradictions developed when attempts were made to explain quantitatively the observed results using these assignments. It is instructive to trace the steps by which these relationships finally were unraveled. The earliest thinking was that all non-hydrogen-bonded proton exchange would be fast, regardless of the pK_d of the protons involved. This, in view of the discussion in the preceding subsection, is a drastic oversimplification, but it accounted very well for the early data. The emphasis was on explaining why base-paired protons exchange slowly.

A dynamic picture of the DNA double helix

Suppose that a base pair must transiently open in order for exchange to occur. This assumption is quite reasonable if protonation or deprotonation at the exchange position is a prerequisite for exchange. Then the kinetics of exchange should be described by the following scheme:

$$\text{Closed} \underset{k_{cl}}{\overset{k_{op}}{\rightleftharpoons}} \text{Open} \overset{k_{tr}}{\rightleftharpoons} \text{Exchange} \tag{22-61}$$

where k_{op} and k_{cl} are the rates of opening and closing a base pair, and k_{tr} is the chemical exchange rate of the protons of an open base pair. For simple chemical mechanisms, k_{tr} will be given by Equation 22-58. The structural difference between the closed and open forms is vague. By definition, the closed form is one from which tritium exchange cannot occur. The open form is one that allows exchange at the rate expected if there were no base pairing. For nucleic acids, the changes in duplex structure required to permit exchange probably represent much smaller base motions than those seen in actual melting of a single base pair.

The kinetic scheme in Equation 22-61 is readily solved by recognizing that the concentration of open forms is likely to be very small and, applying the steady-state condition, $d(\text{Open})/dt = 0$. Thus we obtain $(\text{Open}) = (\text{Closed})k_{op}/(k_{cl} + k_{tr})$. However, we are interested only in those closed states that still have tritium, so $(\text{Closed}) = H$, where H is the amount of tritium retained by the DNA. Then the rate of tritium loss is just

$$-dH/dt = k_{tr}(\text{Open}) = k_{op}k_{tr}H/(k_{cl} + k_{tr}) \qquad (22\text{-}62)$$

Consider two limits of the exchange process. If $k_{tr} \gg k_{cl}$, open base pairs exchange instantly. The observed exchange rate is just the opening rate: $k_{ex} = k_{op}$. If $k_{tr} \ll k_{cl}$, the open and closed forms will approach a true equilibrium and $k_{ex} = K_{eq}k_{tr}$, where $K_{eq} = k_{op}/k_{cl}$.

It is possible to decide experimentally when exchange is limited by the opening rate, because in that case k_{ex} will not be a function of any catalyst concentration. H. Teitlebaum and S. W. Englander (1975) have carefully examined the tritium exchange of a number of model polynucleotide systems. For poly $rA \cdot rU$, poly $r(A\text{--}U) \cdot r(A\text{--}U)$, and poly $d(A\text{--}T) \cdot d(A\text{--}T)$, they find two exchange classes in each case: one faster proton and two slower protons per base pair. The faster one is insensitive to pH or added catalysts such as imidazole or tris(hydroxymethylamino) methane; it must be limited by helix opening. Thus k_{op} can be calculated from the fast exchange rate: $k_{op} = k_{ex} = 0.06 \text{ sec}^{-1}$. The slower exchange class also is independent of pH, but it is sensitive to catalysts. Furthermore, the catalyst stimulation is virtually identical to that seen for the acid poly A double helix. The only base-paired protons in the acid helix are the adenine NH_2 protons. These must be the slow-exchanging protons in Watson–Crick base-paired helices, whereas the fast-exchanging proton is the pyrimidine NH proton at N-3. This result is the opposite of what is predicted from the simple steric argument that the purine N-1 and pyrimidine N-3 positions are the least accessible in the double helix. However, the above conclusion is in agreement with the fact that the pK_a of this proton is much closer to the pK_a of water than is the pK_a value of the exocyclic amino protons.

Now one must explain why the amino protons, which should open at the same rate as the NH proton, exchange more slowly but not as slowly as would be expected from their pK_a. A reasonable hypothesis is that protonation of adenine, not at N-6 but at N-1, is a necessary intermediate (Fig. 22-27). It is known from model compounds that N-1 protonation lowers the NH_2 pK_a from 20–30 to around 8. Then Equation 22-59 must be modified to recognize that a preequilibrium $A \rightleftarrows H^+_{N-1}\text{--}A$ exists with an ionization constant K_{N-1} and that only the fraction of A opened and protonated participates in the exchange:

$$k_{tr} = \frac{(H^+)/K_{N-1}}{1 + (H^+)/K_{N-1}} (C)k_1\alpha_{ex} = (H^+)K_{N-1}^{-1}k_1\alpha_{ex}(C) \qquad (22\text{-}63)$$

The simplification shown in the final expression of Equation 22-63 is possible because K_{N-1} in adenine polymers is about $10^{-4.3}$ and, at neutral pH, $(H^+)K_{N-1}^{-1} \ll 1$. Equation 22-63 shows that, if the catalyst is hydroxide ion, the chemical exchange rates will be independent of pH because, with $(C) = (OH^-)$ in Equation 22-63, the product

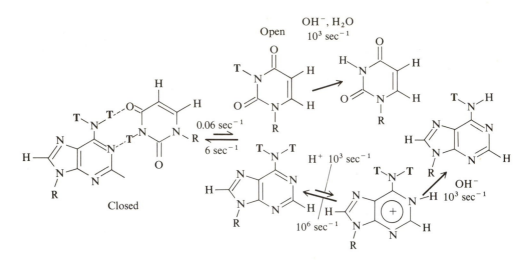

Figure 22-27

Postulated steps in the exchange of the protons of an A–U pair. Observed data are fitted by a model in which opening of the base pair is required for all exchange steps. Base pairs are open 1% of the time. Then the uridine NH proton can exchange by general base catalysis, whereas the exchange of the adenosine NH_2 protons requires a concerted acid–base catalysis in which the A is first protonated at N-1.

$(H^+)(OH^-) = K_w = 10^{-14}$. If a different catalyst is present, the rates will be catalyst sensitive, in agreement with the observed results for the other protons.

Because k_1 and α_{ex} are known from experiments in model systems, Equations 22-62 and 22-63 permit calculation of K_{eq} from the measured amino proton exchange rates. The result is about 0.01 at 0°C. This is a very significant result. For one percent of the time, a base pair in the double helix is open enough to allow protons to eneter, even at such a low temperature. This observation rules out ideas that DNA is a rigid static structure. Combining the equilibrium constant with the opening constant permits prediction of a closing rate of about 6 sec^{-1}; this is really a very slow rate. Figure 22-27 summarizes all the rates and equilibria for A–U proton exchange.

Similar studies have been performed with G–C model polymers. Five protons are monitored: guanine NH and NH_2, and cytosine NH_2. The guanine NH proton is the fastest exchanging; its rate is opening-limited. The cytosine NH_2 proton exchange must be preceded by base-pair opening, but the rate is limited by protonation comparable to that in adenine NH_2 protons. Guanine NH_2 protons exchange about as fast as the cytidine NH_2 protons. However, the pK_a for ring protonation of guanine is two pH units lower than that for cytidine, thus predicting a rate 100 times slower if ring protonation is required for exchange, as it is in adenine. The fact that guanine NH_2 protons exchange so much faster than expected suggests that this exchange, unlike all the others, may occur without base-pair opening. The overall constants derived for G–C base-pair opening rates and equilibria are comparable to the A–U or A–T values.

Studies on the covalent reactions of formaldehyde, hydroxylamine, and Hg(II) with DNA generally agree with the result that DNA base pairs are transiently open about one percent of the time. Thermodynamic analysis of melting data predicts much smaller fractions of open residues. However, the opening equilibrium constant found by protonation or reaction need not be the same as that found by melting. The two processes look at the DNA in somewhat different ways. Melting may require a larger disruption of the structure before it is visible.

Direct study of exchange of particular protons

The lengthy example just given shows that considerable information is potentially available from tritium-exchange experiments. However, the difficulty of resolving kinetic classes becomes great when more-complex systems are examined. The difficulty of assigning these classes to particular sets of protons becomes even greater. Early studies on tRNA showed many more slowly-exchanging protons than could be accounted for by actual hydrogen-bonded protons. At that time, these extra protons were suspected to arise from tertiary structure. Now, as a result of improved understanding of the DNA exchange data, a detailed analysis of the tRNA results seems formidable. The fundamental problem is that so much of the information is lost if only the total exchange rate is monitored.

Clearly, what is needed is a technique that permits the exchange of individual protons, or at least of particular groups of protons, to be monitored independently. NMR is one such approach; it has the resolution to distinguish separate classes and some individual protons quite well, but the problem of assigning resonances is quite difficult. Examples of this were given earlier, in our discussion of the NMR of single-strand nonexchangeable protons. An alternative is to perform direct chemical exchange, trap molecules that have partially exchanged, and analyze the location of individual exchanging residues by fragmenting the nucleic acid and looking at the location of tritium atoms at each point in the sequence. This is not an easy procedure, but the effort is repaid by the large amount of information potentially available.

The trick is to find an exchange process fast enough under some conditions to allow detectable amounts of exchange, yet slow enough under other conditions to permit the time-consuming enzymatic fragmentation and electrophoretic or chromatographic analysis of oligonucleotides without appreciable further exchange. Tritium exchange of the purine C^8–H proton appears to fit these requirements nicely. Table 22-5 summarizes some typical results. The activation energies for C^8–H exchange are quite high. Thus, at 90° to 100°C in mononucleosides, exchange half-times are on the order of 1 to 3 hr. Recall how deuterium exchange under these conditions was used to assist NMR assignments in dimers. At 30°C, exchange half-times lengthen to the order of 1,000 hr. This time is sufficient to allow almost any kind of chemical analysis without significant further exchange, so long as extremes of pH are avoided. In the case of purine C^8–H protons, it is not as obvious that exchange rates should necessarily depend on whether or not the base is paired. After all, the

Table 22-5

Exchange rates of hydrogen–tritium protons linked to carbon atoms in nucleic acid bases

Temperature (°C)	G	A	poly A	poly A·U (A)	Viral RNA (A + G)	DNA (A + G)
100	7.7×10^{-1}	3.7×10^{-1}	——	——	——	5.5×10^{-1}
95	——	2.5×10^{-1}	2.3×10^{-1}	2.5×10^{-1}	——	——
60	2.1×10^{-2}	1.0×10^{-2}	6.2×10^{-3}	2.1×10^{-3}	——	9.1×10^{-3}
37	1.7×10^{-3}	8.3×10^{-4}	2.7×10^{-4}	4.8×10^{-5}	6.1×10^{-5}	4.2×10^{-4}

NOTE: Exchange rates are given in cpm nmole^{-1} hr^{-1} Ci^{-1}.
SOURCE: Data from work of P. Schimmel and colleagues.

C^8–H points into the major groove of the DNA-B helix. Fortunately, C^8–H exchange actually is three times slower in stacked single-stranded polymers than in monomers at 37°C, and is 40 times slower in double-stranded RNAs. Similar, although smaller, effects are seen in DNA. Thus C^8–H exchange is a useful technique for examining secondary structures of nucleic acids where the sequence is known. Examples are shown in Chapter 24.

Changes in the type of secondary structure

All of the methods discussed thus far allow monitoring of the extent and stability of secondary structure in DNA. They do not provide any proof of what that secondary structure actually is in solution. One of the most promising approaches for identifying secondary structure was developed by M. Maestre. He demonstrated that it is possible to measure reliable CD spectra on films of DNA kept at constant relative humidity. In this way, spectral data are obtained under conditions exceedingly close to those used for x-ray diffraction analysis of DNA fibers. A similar approach has been used by W. Peticolas with Raman spectra.

The double-helix structure of DNA fibers changes in salt concentration and humidity from the B form to the A form or C form. Maestre was able to observe dramatic changes in CD with similar environmental perturbations of DNA films (Fig. 22-28a). Lowering the relative humidity of a film of the sodium salt of calf thymus DNA from 92% to 75% changes the CD from a conservative exciton-type spectrum to a single near-UV peak. Such a humidity change in fibers is known to cause the transition B → A. This transition identifies the CD spectra of DNA in the two helical forms. Figure 22-28b compares the film B-form spectrum with the CD of calf thymus DNA in dilute aqueous buffer; they are virtually identical. This is strong evidence that, in dilute salt solutions, DNA is in the B form.

The spectral change in going from the B form to the A form is consistent with the change in structure. W.-C. Johnson and I. Tinoco have calculated the CD of DNA and RNA helices using the theoretical approaches described (in simplified form) in Chapter 8. The RNA-11 structure is very close to the A form of DNA and

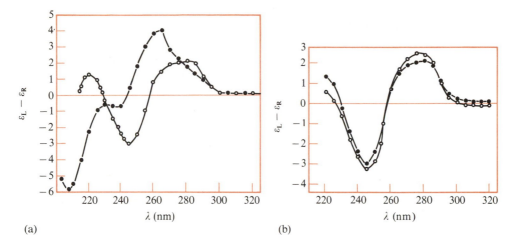

Figure 22-28

Circular dichroism of calf thymus DNA in films at controlled humidity. (**a**) The sodium form at 92%
relative humidity (○) where x-ray diffraction indicates a B-form helix is present, and at 75% relative
humidity (●) where x-ray data show an A-form helix. (**b**) Comparison of the B-form spectrum of part
a with that from an aqueous solution of the same DNA. These results provide strong evidence that DNA
in dilute aqueous solution adopts the B form. [After M. Tunis-Schneider and M. Maestre, *J. Mol. Biol.*
52:521 (1971).]

is expected to have very similar CD. Figure 22-29 compares the results of these
calculations with experimental results. Although agreement is not perfect (not
surprising, considering the imprecision in our knowledge of the electronic properties
of nucleic acid bases), the calculations do correctly predict the much less conservative
spectrum of RNA-11 or DNA-A structures. Also note how similar the CD of double-
stranded RNA is to the CD of the DNA A form (Fig. 22-28).

Perturbation of the environment in solution should be able to cause conversion
of one DNA helical form to another. Addition of ethanol up to 80%, in fact, results
in a spectral change closely akin to the B → A CD change in films. This is reasonable,
because so much ethanol may be equivalent to the drying of a film. However, these
solvent-perturbation experiments are tricky, because one must tread a narrow path
between causing a helix change and causing denaturation or condensation. Much
more complex effects are observed with certain solvents and conditions that clearly
affect not only DNA secondary structure but tertiary structure as well.

Premelting transitions in DNA

The structure of DNA has some complexities even in dilute salt solution that are
just beginning to be understood. The CD of native DNA is strongly temperature

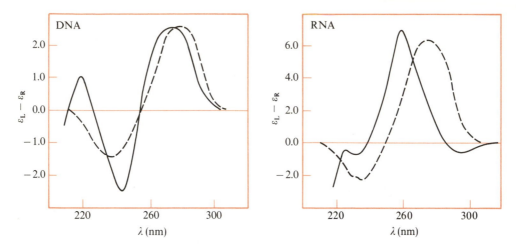

Figure 22-29

Circular dichroism of the DNA B form and of double-strand RNA. The experimental curves (*solid*) were obtained at room temperature in dilute neutral solutions. The calculated curves (*dashed*) come from quantum mechanical considerations of base–base interactions. [After W. C. Johnson and I. Tinoco, Jr., *Biopolymers* 7:727 (1969).]

dependent through the range from T_m down to 0°C. Figure 22-30 shows an example of this for poly d(A–T)·d(A–T). Similar results have been obtained for every DNA examined. The structural change responsible for this premelting is a gradual and partial shift from a DNA-B helix at high temperature to a DNA-C helix at low temperature. Several alternative explanations can be eliminated easily. Premelting seen by CD is not due to local melting of particularly unstable regions, because this melting would have to be enhanced at low salt concentration. However, salt effects below T_m are very small (Fig. 22-30). Premelting effects cannot be an A ↔ B helix interconversion, because that also should be salt sensitive. The A helix is much squatter than the B helix, and phosphate-phosphate distances are much closer in the A form. The G_{el} calculated from Equation 22-4 favors the B form by over 0.1 kcal per mole of phosphate at 10^{-2} M salt concentration, but there is a negligible difference between the G_{el} of the A and B forms at 1 M salt concentration. Analysis of the CD spectra of poly d(A–T)·d(A–T) by comparison of solution and film data clearly suggests that the premelting seen for DNA is a transition from a B-like helix to a C-like helix.

The premelting seen by CD and other techniques shows that the base-pair geometries of DNA in solution are not fixed; they vary with salt concentration and with temperature. This dependence shows up dramatically in studies of circular DNA (Chapter 24). Much more knowledge is needed about the range of DNA double-helix structures before all observed properties can be rationalized.

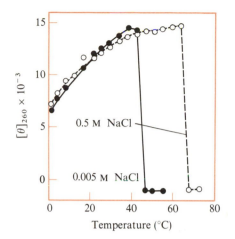

Figure 22-30

Circular dichroism of poly d(A–T)·d(A–T) as a function of temperature at pH 7.0 and two different salt concentrations. Unlike the absorbance for this polymer (Fig. 22-19a), the CD data show significant reversible changes prior to melting. [After R. B. Gennis and C. R. Cantor, *J. Mol. Biol.* 69:381 (1972).]

Collapsed forms of DNA

DNA in aqueous salt solution behaves as a wormlike coil with considerable rigidity and little tendency for interaction between distant portions of the DNA chain (Chapter 19). In short, it has no preferred tertiary structure. However, addition of certain polymers, such as polyethylene oxide, causes a remarkable change in all observed physical properties of DNA solutions. The effects are fully reversible and occur over an exceedingly narrow range of salt and polymer concentrations. This transition to a form called ψ DNA by L. S. Lerman (its discoverer) is accompanied by a 40-fold increase in the CD intensity. In addition, the sedimentation coefficient rises precipitously. For T4 DNA, $s_{20,w}$ increases from 62 S in normal solutions to values ranging from 300 to 600 S.

The only way to explain such an enormous change is a condensation of the DNA into a tightly packed structure. The driving force for this condensation is excluded-volume effects. Coiled polymers extend over such a large volume that much of the solvent lies within the coiled domain of one polymer molecule or another. If the net interaction force between two different polymers is repulsive, however small, then in general a homogeneous solution of a mixture of the two species is not expected to form. Instead, the solution separates into two phases, each one predominantly containing polymers of one species. Presumably, ψ DNA is one such separate phase except that, instead of forming a macroscopically visible phase boundary, each DNA molecule simply becomes individually compacted.

The structure of ψ DNA still is not fully characterized, but the best evidence currently available suggests that it is composed of tightly folded rods packed into a bundle. This structure is very similar to crystals formed by many other linear polymers and also seen in crystals of low-molecular-weight DNA. Figure 22-31 shows two schematic forms for such a ψ particle. One is based on fold lengths seen in crystals,

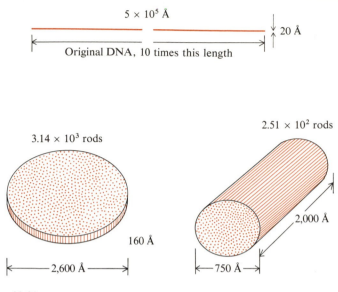

Figure 22-31

Two possible structures for the ψ form of T4 DNA. The striking observation is that
neutral polymers such as polyethylene oxide cause a dramatic condensation of the
DNA. Both forms shown are qualitatively consistent with the extent of DNA
compaction, but the actual details of the folding of ψ DNA are unknown.

the other on electron microscopic observations of rods of T4 DNA prepared from
ethanol. Of course, these are just extremes of various possible models. They lead to
predicted sedimentation coefficients of 469 S and 713 S, respectively, which are
comparable to observed values.

At first glance, polyethylene oxide seems a long way from biology. However,
there is a reasonable chance that the ψ form of DNA or its analogs are very important
in nature. DNA inside the head of a bacteriophage must be very tightly compacted.
For example, the sedimentation velocity of the whole T4 phage is 890 S. This is in the
same range as the ψ form of DNA. In condensed chromatin, such as that found in
metaphase chromosomes, DNA also must be organized into tight structures; the ψ
form may be a useful model for this. However, nature has selected particular molecules
to organize and order the folding of DNA. These include histones and protamines.
They should permit generation of much more precise and complete ordered structures
than that of ψ DNA.

Summary

Nucleic acid single strands are partially stacked helices that gradually melt as the temperature is raised. The properties of dinucleotides closely resemble those of their corresponding polymers. Ionic strength has little effect on single-strand spectroscopic properties. These observations indicate that, although single strands have considerable local ordered structure, they have little long-range structure. The differences in conformation seen between single-stranded RNA and DNA persist all the way down to the level of dinucleotides.

Double-stranded nucleic acids melt to single strands fairly sharply as the temperature is rasied. The melting temperature increases as the logarithm of the ionic strength of the solution, indicating that electrostatic repulsions tend to destabilize the double helix. Double strands bind more counterions than do two corresponding single strands. The stability of double strands increases with the mole fraction of G–C pairs. It decreases as the pH is varied on either side of neutrality. The melting of natural DNAs is more complex than that of synthetic polynucleotides, because of the differential stability of various local regions. Although base composition is the major factor determining local stability, in some cases base sequence also can play a significant role.

Some homopolymers can form triple-stranded as well as double-stranded helices. Studying optical or thermodynamic properties as a function of environmental conditions allows construction of a phase diagram that conveniently summarizes all the properties of the system. It is especially helpful in selecting a set of conditions that allows only one pure conformation.

Nucleic acid double strands show considerable evidence of dynamic behavior and conformational flexibility. Tritium-exchange studies indicate that base pairs can open and close fairly rapidly. Optical studies demonstrate that, as the temperature continues to be lowered below the melting temperature, a gradual change occurs from a B-like to a C-like double helix. Addition of organic solvents can stimulate transitions of B to A or, in some cases, can lead to a dramatic condensation of the DNA into a compact folded structure.

Problems

22-1. Oligonucleotides will form double-helical or triple-helical complexes with polynucleotides. For example, ApApApApApA will form a duplex with poly U, and UpUpUpUpUpU will form a duplex with poly A. Schematically, these complexes are ————————.

  a. What experiment would determine if the oligomers had bound to all available sites on the polymer?

      b. How could you tell if the helical structure of the oligomer·polymer complex was similar or identical to that of the corresponding polymer·polymer complex?

      c. Explain why the oligo A·poly U complex is very much more stable than the oligo U·poly A complex.

22-2. Construct a schematic phase diagram for poly C as a function of temperature (vertical axis) and pH (horizontal axis). Use the following (hypothetical) thermodynamic data for ionization and base-pair formation:

$$C + H^+ \rightleftarrows CH^+ \qquad \Delta H^0 = -2 \text{ kcal/base}$$

$$CH^+ + C \rightleftarrows CH^+\!\!-\!\!C \qquad \Delta H^0 = -5 \text{ kcal/base pair}$$

Are there any pH values at which poly C will show two melting transitions as the temperature is varied? HINT: First, write reactions for the overall formation of acid double helix, and for the dissociation of that helix into two protonated single strands.

22-3. At low temperature in high Mg^{2+} concentrations, poly U shows a sudden increase in hypochromicity and optical activity. Presumably, a helical structure has formed. What kind of experiments would be needed to show if the helix is double-stranded or multiple-stranded? whether the strands are parallel or antiparallel?

22-4. Consider long DNA strands containing only A–T pairs. How many independent variables are needed to describe nearest-neighbor contributors to optical properties such as absorbance or CD? What set of simple repeating-sequence DNAs should be studied to obtain the intrinsic properties needed to fit spectra of DNAs containing only A–T pairs?

22-5. How will a bulge loop imperfection affect the tritium-exchange rates in poly d(A–T)· d(A–T)?

```
                    A—T
                    |  |
                    T  A
                     \ |
              T-A-T-A  T-A-T-A-T-A

              A-T-A-T-A-T-A-T-A-T
```

References

GENERAL

Bloomfield, V., D. Crothers, and I. Tinoco, Jr. 1974. *Physical Chemistry of Nucleic Acids.* New York: Harper & Row.

Felsenfeld, G., and H. T. Miles. 1967. The physical and chemical properties of nucleic acids. *Ann. Rev. Biochem.* 36:407.

Record, M. T., Jr., C. F. Anderson, and T. M. Lohman. 1978. Thermodynamic analysis of ion effects on the binding and conformational equilibria of proteins and nucleic acids: The roles of ion association or release, screening and ion effects on water activity. *Quart. Rev. Biophys.* 11:103.

Ts'o, P. O. P. 1974. Dinucleoside monophosphates, dinucleotides and oligonucleotides. In *Basic Principles in Nucleic Acid Chemistry*, vol. 2, ed. P. O. P. Ts'o (New York: Academic Press), p. 305.

SPECIFIC

Brahms, S., J. Brahms, and K. E. Van Holde. 1976. Nature of conformational changes on poly [d(A–T)·d(A–T)] in the premelting region. *Proc. Natl. Acad. Sci. USA* 73:3453.

Lerman, L. S. 1974. The polymer and salt-induced condensations of DNA. In *Physico-Chemical Properties of Nucleic Acids*. vol. 3, ed. J. Duchesne (London: Academy Press).

Manning, G. R. 1978. The molecular theory of polyelectrolyte solutions with applications to the electrolyte properties of polynucleotides. *Quart. Rev. Biophys.* 11:179.

Palacek, I. 1976. Premelting changes in DNA conformation. In *Progress in Nucleic Acid Research and Molecular Biology*, vol. 18, ed, W. E. Cohn (New York: Academic Press), p. 151.

Sundaralingam, M., and S. T. Rao. 1975. *Structure and Conformations of Nucleic Acids and Protein–Nucleic Acid Interactions*. Baltimore, Md.: University Park Press.

Teitlebaum, H., and S. W. Englander. 1975. Open states in native polynucleotides, I: Hydrogen-exchange study of cytosine-containing double helices. *J. Mol. Biol.* 92:55, 79.

Wellauer, P. K., and I. B. Dawid. 1974. Structure and processing of ribosomal RNA: A comparative electron microscopic study in three animals. *Brookhaven Symp. Biol.* 26:214.

Wells, R. D., J. E. Larson, R. C. Grant, B. E. Shortle, and C. R. Cantor. 1970. Physicochemical studies on polydeoxyribonucleotides containing defined repeating nucleotide sequences. *J. Mol. Biol.* 54:465.

Wensink, P. C., and D. D. Brown. 1971. Denaturation map of the ribosomal DNA of *Xenopus laevis*. *J. Mol. Biol.* 60:235.

23

Statistical mechanics and kinetics
of nucleic acid interactions

23-1 STATISTICAL THERMODYNAMICS
OF DOUBLE-HELIX FORMATION

In this chapter, we take a more detailed look at the equilibrium between single and double strands of nucleic acids. We demonstrate how information obtained from studies on the stability of oligonucleotide complexes can be used directly to predict the secondary structure of nucleic acids. Interpretation of the kinetics of interconversion between single and double strands also is assisted by kinetic studies on oligonucleotides. Finally we look at studies on complexes between ligands and oligonucleotides, using them as tools for the analysis of nucleic acid–ligand interactions.

We begin with a look at the statistical thermodynamics of double-helix formation. The complexities of natural DNAs and RNAs, and even of simple-sequence synthetic polynucleotides, frustrate attempts to sort out all of the thermodynamic interactions responsible for the maintenance of ordered structures. Oligonucleotides provide an excellent system in which to model these interactions, because length and sequence can be varied at will. Figure 23-1 illustrates the kinds of specific questions that can be answered using oligonucleotides. Are there large sequence effects on the stability of double strands? How is the stability of a duplex affected by single-stranded tails, single-stranded loops, single-stranded breaks, or double-stranded breaks? How does the size of a loop affect the stability of a hairpin? In principle, one must examine the influence of specific sequences (or at least of base compositions) for each of these questions. In the ideal case, one would like to assign free energies of formation to each of the ten specific nearest-neighbor duplex interactions, and to reveal how these energies are influenced by the effects of various nonpaired base configurations.

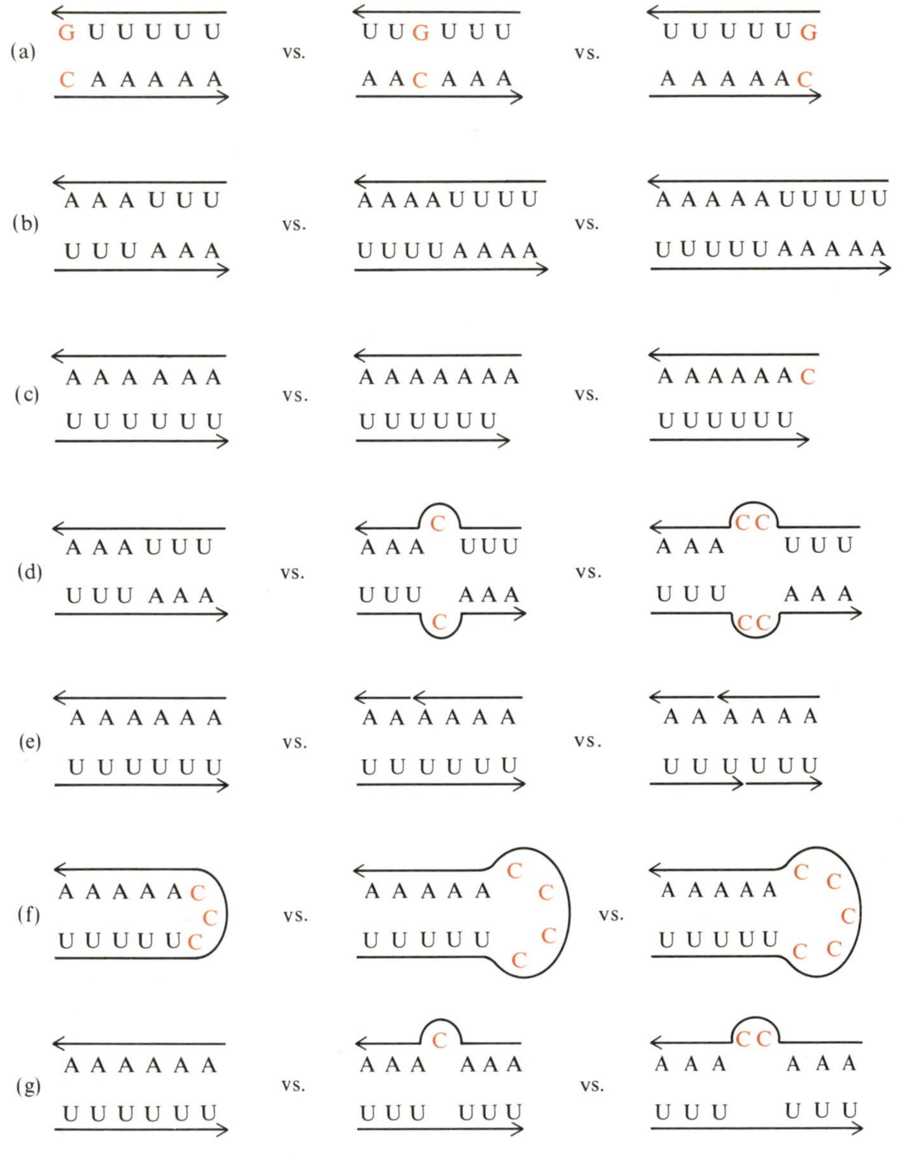

(a)
```
←————————
G U U U U U        U U G U U U        U U U U U G
                                                              
C A A A A A        A A C A A A        A A A A A C
————————→
```
vs. ... vs.

(b)
```
←————————
A A A U U U        A A A A U U U U        A A A A A U U U U U
                                                              
U U U A A A        U U U U A A A A        U U U U U A A A A A
————————→
```
vs. ... vs.

(c)
```
←————————
A A A A A A        A A A A A A A        A A A A A A C
                                                              
U U U U U U        U U U U U U         U U U U U U
————————→
```
vs. ... vs.

(d)
```
                        C                      C C
A A A U U U        A A A   U U U        A A A     U U U
                                                              
U U U A A A        U U U   A A A        U U U     A A A
                        C                      C C
```
vs. ... vs.

(e)
```
A A A A A A        A A A A A A        A A A A A A
                                                              
U U U U U U        U U U U U U        U U U U U U
```
vs. ... vs.

(f)
```
                                  C                    C C
A A A A A C        A A A A A   C        A A A A A     C
          C                  C                      C
U U U U U C        U U U U U   C        U U U U U     C
                                  C                    C C
```
vs. ... vs.

(g)
```
                        C                      C C
A A A A A A        A A A   A A A        A A A     A A A
                                                              
U U U U U U        U U U   U U U        U U U     U U U
```
vs. ... vs.

Figure 23-1

Oligonucleotide comparisons can lead to fundamental information about the interactions stabilizing duplex structures. By measuring relative T_m values, one can examine the effects of (**a**) base sequence, (**b**) duplex length, (**c**) single-strand stacking, (**d**) mispairing (internal loops), (**e**) chain fragmentation, (**f**) hairpin loop length, and (**g**) mispairing (bulge loops).

NMR studies of short oligonucleotide duplexes

It is most convenient to study the shortest oligonucleotides that will form a given structure of interest. Otherwise, the observed conformational free energy is a sum over many different interactions, and a whole series of analogous complexes must be examined before individual contributions can be factored out. Mononucleosides will form base-paired complexes in organic solvents. It is possible to observe indications of base pairing between mononucleosides in water, but only indirectly. Such high concentrations are required for base-pair formation in water that stacked monomer aggregates form, complicating any detailed quantitative analysis. Dinucleoside phosphates can form base-paired complexes at concentrations around 0.04 to 0.08 M. Although these concentrations are too high for most optical techniques, they are convenient for NMR study. The observed proton NMR spectrum of self-complementary dinucleoside phosphates or dinucleotides can be measured directly in water using Fourier-transform techniques that eliminate the strong interfering NMR of the solvent. The results are strongly dependent on concentration.

Figure 23-2 summarizes an NMR study of pdGpdC and pdCpdG. As the concentration is increased, all protons are observed to shift somewhat upfield, except for

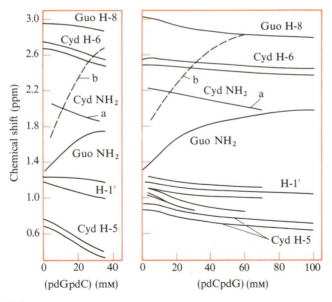

Figure 23-2

Self-association of pdGpdC *and* pdCpdG *monitored by NMR.* The chemical shift of each proton is plotted as a function of molar concentration of dinucleotide. Note that the two Cyd NH$_2$ protons are observed as separate peaks, whereas only a single resonance is seen for the Guo NH$_2$ protons. Shifts to lower ppm with increasing concentration indicate stacking; hydrogen bonding causes shifts to higher ppm. [After M. A. Young and T. R. Krugh, *Biochemistry* 14:4841 (1975).]

one proton assigned to the cytidine NH_2 and one assigned to the guanosine NH_2; these protons show downfield shifts of much larger magnitude. The upfield shifts are characteristic of ring-current effects due to stacking; the downfield shifts are what would be expected for strong hydrogen bonding. The concentrations at which the trends level out are higher in pdCpdG than in pdGpdC; thus, the base pairing is weaker in pdCpdG than in pdGpdC.

A few interesting features of the spectra in Figure 23-2 are worth further comment. The guanosine NH proton is not observed. Its band appears to be broadened by exchange with solvent, even under conditions where a duplex complex is fully formed. Similar exchange effects have been seen in larger oligonucleotide complexes. This is perfectly reasonable because NH exchange rates are much faster than NH_2 exchange rates (Chapter 22). Two cytidine NH_2 protons are seen in each dimer. One shifts downfield with increasing concentration and must be the hydrogen-bonded proton; the other shifts upfield. Apparently, rotation around the Cyd C–NH_2 bond is very slow because of its partial double-bonded character. Therefore, the two NH_2 protons remain NMR nonequivalent in the free dimer and in base-paired complexes. In contrast, the guanosine NH_2 protons show up as only a single resonance. This means that, even in the base-paired dimer, rotation around the Guo C–NH_2 bond is fast enough to render these two protons equivalent. Recall that these protons can exchange in the DNA duplex without opening the helix; perhaps this ease of exchange is related to the ease of their interconversion by rotation.

Data like those in Figure 23-2 can be used to analyze the thermodynamics of oligonucleotide interactions. In the fast-exchange limit for the equilibrium $2M \rightleftarrows M_2$, a single resonance is observed for each proton at a chemical shift δ_{obs}:

$$\delta_{obs} = \chi'_M \delta_M + \chi'_{M_2} \delta_{M_2} = \delta_M + (\delta_{M_2} - \delta_M)\chi'_{M_2} \tag{23-1}$$

where δ_M is the chemical shift in the pure noncomplexed molecules, and δ_{M_2} is the shift in the complex. This equation has the same form as Equation 22-26, except that χ'_{M_2} is the mole fraction of a specific proton in the base-paired complex. Because this complex contains two equivalents of each specific proton, $\chi'_{M_2} = 2(M_2)/[2(M_2) + (M)]$. If K is the formation constant of the complex, and M_0 is the total molar concentration of dinucleotide, a simple equilibrium calculation yields

$$\delta_{obs} = \delta_M + (\delta_{M_2} - \delta_M)[(1 + 4KM_0 - \sqrt{1 + 8KM_0})/4KM_0] \tag{23-2}$$

By fitting experimental data to this equation, one can determine the NMR parameters δ_M and δ_{M_2} and the equilibrium constant K. The NMR of equilibria between different dimers (for example, $GpG + CpC \rightleftarrows$ Duplex) can be analyzed in a similar way. Table 23-1 summarizes results from some of the systems that have been studied. These results show significant sequence dependence for the strength of base pairing. One can also infer that A–T and A–U pairs are considerably weaker than G–C pairs.

Table 23-1

Formation constants of dinucleotide complexes as determined by NMR

Structure	Temperature (°C)	K (M^{-1})
pdGpdC·pdGpdC	2	7.8 ± 0.7
pdCpdG·pdCpdG	2	3.0 ± 0.5
pdGpdG·pdCpdC	2	9.1 ± 1.1
pdGpdC·pdGpdC	25	<2
pdGpdT·pdApdC	2	<3
CpG·CpG	4	14.0 ± 2.5
GpU·ApC	1	$\ll 14$
GpC·GpC	4	$\gtrsim 14$

NOTE: All samples at very low salt concentration.
SOURCE: Data from the work of Thomas Krugh and colleagues.

Optical study of longer oligonucleotide duplexes

Most longer oligonucleotide complexes are stable enough that complex formation can be monitored by optical absorbance changes or by circular dichroism. Only in a few cases have available quantities of sample been large enough for NMR studies of these longer duplexes. Table 23-2 summarizes some of the results of the many optical studies on oligomer complexes. This table is restricted to oligoribonucleotides because a much more extensive body of data is available on these compounds than on their deoxy analogs. Figure 23-3 shows the presumed structures of those complexes from the table that include unpaired bases. A number of qualitative conclusions can be drawn from the properties of the complexes shown in this table (and confirmed by results on many other complexes that have been studied).

For a homologous series (such as $A_n U_n \cdot A_n U_n$), the melting temperature of the complex increases with length. (Compare complexes 1, 7, 14, 25, and 26 in Table 23-2.) In fact, most results are well fit by the relationship

$$1/T_m = A + B/N \qquad (23\text{-}3)$$

where N is the chain length, and A and B are constants having different values for each set of experimental conditions (Fig. 23-4a). An extra unpaired stacked base (e.g., $A_{n+1} U_n \cdot A_{n+1} U_n$) increases the stability slightly, but the effect is much smaller than that of an extra base pair. (Compare complexes 11 and 18 with complexes 7, 14, and 25 in Table 23-2; also see Fig. 23-3.)

For every oligonucleotide duplex, the melting temperature increases with increasing concentration. This is one of the experimental ways to detect the formation

Table 23-2

Melting temperatures of oligoribonucleotide duplexes as determined by optical techniques

No. Complex	Presumed helix length	Melting temperature (°C) at oligonucleotide concentration of		
		10 μM	25 μM	100 μM
1 $A_3U_3 \cdot A_3U_3$	6	—	-14	-9
2 $A_2GCU_2 \cdot A_2GCU_2$	6	20	—	28
3 $A_4G_2 \cdot C_2U_4$	6	$14^§$	—	$23^§$
4 $A_2CGU_2 \cdot A_2CGU_2$	6	11	16	22
5 $U_2CGA_2 \cdot U_2CGA_2$	6	2	—	11
6 $A_4CG \cdot CGU_4$	6	$-14^§$	—	$-1^§$
7 $A_4U_4 \cdot A_4U_4$	8	5	8	12
8 $A_4CU_3 \cdot A_3GU_4$	8	$15^§$	$18^§$	$23^§$
9 $A_3CGU_3 \cdot A_3CGU_3$	8	—	32	35
10 $A_3GCU_3 \cdot A_3GCU_3$	8	35	—	42
11 $A_5U_4 \cdot A_5U_4$	$8^\#$	—	9	14
12 $A_4CU_4 \cdot A_4CU_4$	$8^¶$	—	—	-5
13 $A_4GU_4 \cdot A_4GU_4$	$8^¶$	—	-1	5
14 $A_5U_5 \cdot A_5U_5$	10	18	20	23
15 $A_5CU_4 \cdot A_4GU_5$	10	$26^§$	$29^§$	$33^§$
16 $A_4UAU_4 \cdot A_4UAU_4$	10	22	—	28
17 $A_4CGU_4 \cdot A_4CGU_4$	10	—	39	43
18 $A_6U_5 \cdot A_6U_5$	$10^\#$	—	23	27
19 $A_5CU_5 \cdot A_5CU_5$	$10^¶$	—	5	10
20 $A_5GU_5 \cdot A_5GU_5$	$10^¶$	—	15	18
21 $A_4GCU_4 \cdot A_4GCU_4$	10	41	—	47
22 $A_4GC_2U_4 \cdot A_4GC_2U_4$	$10^¶$	~20	—	—
23 $A_4GC_3U_4 \cdot A_4GC_3U_4$	$10^¶$	11	—	—
24 $A_4GC_4U_4 \cdot A_4GC_4U_4$	$10^¶$	7	—	—
25 $A_6U_6 \cdot A_6U_6$	12	26	29	32
26 $A_7U_7 \cdot A_7U_7$	14	36	37	39

§ These melting temperatures are for twice the listed concentration, because the complex formation reaction is A + B → C rather than 2A → A_2 (see text).

\# Unpaired bases at the ends of the sequence presumably stack adjoining the duplex region (see Fig. 23-3).

¶ Unpaired bases within the sequence presumably form a central loop (see Fig. 23-3).

SOURCE: Data collected from various sources. For example, see F. Martin, O. Uhlenbeck, and P. Doty, *J. Mol. Biol.* 57:201 (1971); O. Uhlenbeck, P. Borer, B. Dengler, and I. Tinoco, Jr., *J. Mol. Biol.* 73:483 (1973); J. Gralla and D. Crothers, *J. Mol. Biol.* 73:497.

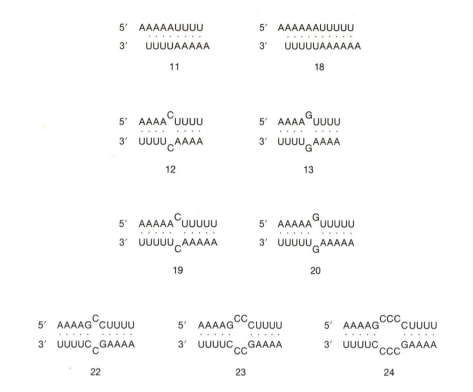

Figure 23-3

Structures of complexes in Table 23-2 that include unpaired bases.

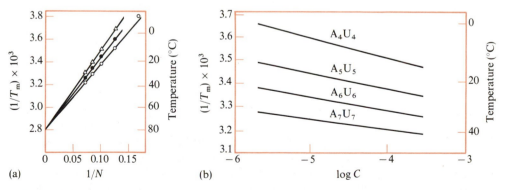

Figure 23-4

Melting behavior of oligonucleotides in the series A_nU_n. **(a)** Plot of $1/T_m$ versus the reciprocal chain length ($N = 2n$). Measurements were made at strand concentrations of 10^{-3} M (○), 10^{-4} M (●), and 10^{-5} M ( ). **(b)** Dependence of T_m on strand concentration for various chain lengths. All results at pH 7.0 in 1 M NaCl. [After F. H. Martin, O. C. Uhlenbeck, and P. Doty, *J. Mol. Biol.* 57:201 (1972).]

of an intermolecular complex. That data are well fit by the relationship

$$1/T_m = A' - B' \ln C \qquad (23\text{-}4)$$

where C is the oligonucleotide concentration, and the constants A' and B' have different values for each oligomer (Fig. 23-4b).

A mismatched base presumably leads to an internal loop, which causes a marked lowering of the stability of the complex. (Compare complexes 12 and 19 with complexes 7, 14, and 25 in Table 23-2.) Furthermore, it is important *which* bases form the loop. (Compare complexes 13 and 20 with complexes 12 and 19.) The size of the internal loop also matters: the larger the loop, the less stable the duplex. (Compare complexes 21 through 24; also see Fig. 23-3.)

The base composition of perfectly complementary duplex structures strongly influences stability. Even a single G–C pair produces a marked increase in the T_m of an A–U duplex. (Compare complexes 8 and 15 with complexes 7 and 14 in Table 23-2.) Two G–C pairs have an even larger effect. (Compare complexes 9, 10, and 17 with complexes 7, 14, and 16.)

The sequence of the duplex also is critically important. Table 23-2 includes five different hexanucleotide duplexes with the base composition $A_4U_4C_2G_2$ (complexes 2 through 6); the T_m values vary over a range of $29°C$ at concentrations around $100 \ \mu M$. Thus, in order to predict the most stable structures of particular RNAs or fragments, one has to consider the details of the sequence.

Table 23-3 shows the melting temperatures of some hairpin loops, formed by intramolecular base pairing. These structures can be distinguished from duplexes by the fact that their optical properties are independent of concentration. Molecular-weight determination by equilibrium ultracentrifugation also has been used to discriminate between duplexes and hairpins. This discrimination is a serious problem, because any sequence that can form a hairpin also can form a duplex with a central loop. For example,

$$2 \begin{array}{l} ^{XX}\!AAAAA \\ \cdots \cdots \\ _{XX}\!UUUUU \end{array} X \rightleftharpoons 2 \ AAAAAXXXXXUUUUU \rightleftharpoons \begin{array}{l} AAAAA^{XXXXX}UUUUU \\ \cdots \cdots \quad\quad \cdots \cdots \\ UUUUU_{XXXXX}AAAAA \end{array}$$

The hairpin must be favored in the limit of low concentration because of simple mass-action considerations. Similarly, the duplex will be favored in the limit of high concentration. In between, the most important factor dictating the relative preference for hairpin or duplex is the size of the loop. Hairpin loops smaller than three residues are not likely stereochemically. The stability of hairpins relative to separated strands reaches a maximum with a loop size of around six or seven residues and then declines for larger loops (Table 23-3). The body of model-compound data is not adequate to assess the importance of sequence or base-composition effects in determining hairpin stability. What fragmentary data do exist indicate that these effects are significant and cannot be neglected. This is reasonable because formation of a loop not only

Table 23-3

Melting temperatures of oligoribonucleotide hairpins as determined by optical techniques

Complex	Presumed helix length	Melting temperature (°C) at oligonucleotide concentration of 10 μM
5′ AAAAG$^{\text{C}}_{}$ C · · · · · C 3′ UUUUC$_{\text{C}}$	10	20
5′ AAAAG^{CC} · · · · · C 3′ UUUUC$_{\text{CC}}$	10	36
5′ AAAAAA$^{\text{C}}_{}$ C · · · · · · C 3′ UUUUUU$_{\text{C}}$	12	2
5′ AAAAAA^{CC} · · · · · · C 3′ UUUUUU$_{\text{CC}}$	12	13
5′ AAAAAA$^{\text{CC}}_{}$ C · · · · · · C 3′ UUUUUU$_{\text{CC}}$	12	21
5′ AAAAAA^{CCC} C · · · · · · C 3′ UUUUUU$_{\text{CCC}}$	12	13

SOURCE: Data collected from various sources (see source note for Table 23-2).

involves alterations in the phosphodiester backbone, but also must change the base-stacking interactions of single strands.

In the following discussion, we attempt to rationalize the data shown in Tables 23-2 and 23-3 and then to show how they can be used to extract the thermodynamic parameters of particular interactions in base pairs or loops. Then we demonstrate how these parameters are combined to analyze the possible conformations of arbitrary sequences. Finally, we show how experimental data on the stability of model oligonucleotide systems can be used to predict the most stable secondary structure of an RNA, given only the sequence.

Models for duplex formation

First, consider the steps in the formation of a duplex from two complementary single strands (Fig. 23-5). Start with the two strands at equal concentration and, for the moment, ignore sequence and base-composition effects on stability. The equilibrium constant κ for initiating a duplex stretch must be considered to be different from the constant s for adding one base pair to a preexisting duplex. The latter constant will contain the enthalpy and entropy of stacking a base pair on top of an existing one, in addition to base-pairing energies. The constant κ must include the entropy change of bringing the two strands together. (The initiation equilibrium constant often is written as a function of s: $\kappa = \sigma s$.)

The equilibrium constant for nucleating a second helical region is not κ, because the strands are already together. Instead, it is some new constant α_l that will be a function of the loop size l between the two initiated regions. The equilibrium constant for forming a given unique configuration with j base pairs is just

$$k_j = \kappa \left(\prod_{l=1}^{\gamma-1} \alpha_l \right) s^{j-\gamma} \qquad (23\text{-}5)$$

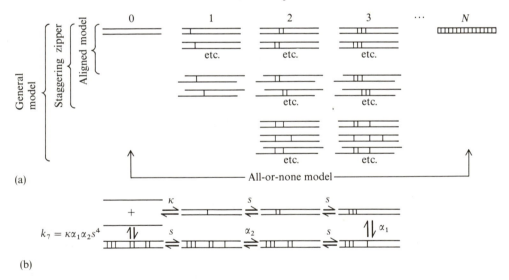

(a)

(b)

Figure 23-5

Duplex states considered by various models of oligonucleotide melting. **(a)** The aligned model includes only states with continuous duplex regions properly in phase. The staggering zipper model adds continuous duplex out-of-phase regions such as might occur whenever homopolymers or repeating sequences are considered. The general model includes also states with discontinuous duplex regions. The all-or-none model considers only the aligned states with zero or N base pairs. **(b)** Some individual steps in the general model, showing equilibrium constants for particular interactions.

where γ is the number of individual initiations required, and $\gamma - 1$ is the total number of loops formed. Figure 23-5b shows an example of a configuration with seven base pairs and two loops.

A complete description of the statistical thermodynamics of the formation of the duplex is contained in the partition function of the system, $q = \sum_i e^{-G_i/RT}$, where G_i is the free energy of the ith configuration (whether single-stranded or double-stranded), and the sum is taken over all configurations. We are interested here in the details of only the duplex, and it is more convenient to work with the conformational partition function. This is derived by considering all energy changes relative to the single strands. Using the fact that $\Delta G_i = -RT \ln K_i$ (where K_i is the equilibrium constant for forming the ith configuration from separated strands), we obtain

$$q_c = \sum_i e^{-\Delta G_i/RT} = \sum_i K_i \qquad (23\text{-}6)$$

As defined in Equation 23-6, q_c is nearly identical in form to the partition function q used in Chapter 20 to describe the polypeptide helix–coil transition (it actually is $q - 1$). The concentration of the strands does not appear in Equation 23-6. To compute the actual amount of duplex formed, one must employ a normal equilibrium expression using q_c as the equilibrium constant, and explicitly take the strand concentrations into account. We shall demonstrate this in the following subsection.

Note the penalty that one pays for using a conformational partition function: the parameters κ, α_l, and s contributing to K_i now are functions of the state of the single strands. In practice, we already know that single-strand properties change little with temperature or salt concentration or length. Thus it will be a safe assumption that values of these properties derived from double strands represent almost entirely the thermodynamic interactions of base pairing and double-strand stacking.

In the general case, it is not possible to reduce q_c to any simple tractable form, although an analytical expression always exists for oligomers of finite length. Instead, it makes sense to use experimental results such as those shown in Table 23-2 to eliminate particular states that clearly will be unstable and will therefore contribute insignificantly to q_c. The least severe approximation is the staggering zipper model of J. Applequist and V. Damle. This model discards all configurations that require more than one initiation. Such an approximation is quite reasonable for short chains. Removing the loop-size-dependent α_l results in an enormous simplification. The partition function now can be written as

$$q_c = \kappa \sum_{j=1}^{N} g_j(N) s^{j-1} \qquad (23\text{-}7)$$

where the sum is taken over all possible numbers of base pairs, $g_j(N)$ is the number of distinguishable states[§] with j base pairs formed, and N is the length of the single-stranded oligomers.

The staggering zipper model is most important for the analysis of homooligomeric complexes; in that case, duplexes not aligned in perfect phase are fairly probable at low degrees of base pairing. As conditions stabilize the duplex more and more, most molecules in the solution attain the maximal number of possible base pairs, which requires a correctly aligned duplex. Figure 23-6 shows some results of the staggering zipper model that illustrate this point. For heterooligomer sequences, staggered configurations become extremely unlikely because they can permit only short helices to form; as you can see in Figure 23-6, this situation is very improbable. Therefore, it is reasonable to make even more severe approximations.

Figure 23-6

Population analysis for the staggering zipper model for the formation of a duplex with 11 potential base pairs. Parameters used in the calculation actually refer to the acid helix $(AH^+)_{11} \cdot (AH^+)_{11}$, but the qualitative results should be general. The equilibrium constant s for adding a base pair to a preexisting duplex is described in the text. Note from these results that the all-or-none approximation should be reasonably accurate. [After J. Applequist and V. Damle, *J. Am. Chem. Soc.* 87:1450 (1965).]

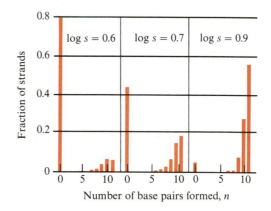

If we allow only perfectly aligned states, the number of distinguishable states with j base pairs for an oligomer of total length N is simple $g_j(N) = N - j + 1$, which you can determine by inspection from Figure 23-5. Then the partition function can be evaluated by explicitly summing the series (as we did in Chapter 20):

$$q_c = \kappa \sum_{j=1}^{N} (N - j + 1)s^{j-1} \tag{23-8a}$$

$$= \kappa \left[(N + 1) \sum_{j=1}^{N} s^{j-1} - \sum_{j=1}^{N} js^{j-1} \right] \tag{23-8b}$$

[§] To be complete and rigorous, one must divide $g_j(N)$ by the symmetry number of the state to account properly for degeneracy in the partition function.

The first sum in Equation 23-8b is just a simple geometric series, and the second can be converted into one by noting that it is equal to $\partial(\sum_j s^j)/\partial s$. The final result is

$$q_{\mathrm{c}} = \kappa[s^{N+1} - (N+1)s + N]/(s-1)^2 \qquad (23\text{-}9)$$

Note that, if $s > 1$ and N is large, Equation 23-9 is approximately of the form

$$\boxed{q_{\mathrm{c}} \cong \kappa s^{N-1}} \qquad (23\text{-}10)$$

Physically, Equation 23-10 corresponds to the most extreme possible simplification of the scheme shown in Figure 23-5. If an all-or-none duplex formation is assumed, the equilibrium constant for formation of the duplex is just $K_{\mathrm{eq}} = \kappa s^{N-1}$. The results in Figure 23-6 show that the all-or-none approximation is not a bad one.

Analysis of experimental data: self-complementary oligomers

Only a solution method such as NMR has the power to distinguish some of the individual configurations that go into the partition function q_{c}. Because at present one must work mostly with optical data, it is necessary to relate q_{c} to the average structural properties that CD or absorbance can monitor. A convenient property with which to work is the fraction of total bases paired, X_{b}. Although X_{b} cannot be measured directly, it can be estimated by measuring such properties as absorbance. The simplest estimate assumes that the observed property A is linear in the amount of base paring. Experimentally, one fits data on a single-strand–double-strand equilibrium to

$$A = \chi' A_{\mathrm{d}} + (1 - \chi')A_{\mathrm{s}} \qquad (23\text{-}11)$$

where χ' is an adjustable parameter, and A_{d} and A_{s} are the properties of the pure fully-formed duplex and of the mixture of single strands, respectively. Rearranging, we obtain the parameter χ' as

$$\boxed{\chi' = (A - A_{\mathrm{s}})/(A_{\mathrm{d}} - A_{\mathrm{s}})} \qquad (23\text{-}12)$$

The parameter χ' will equal X_{b} (the actual fraction of bases paired) when helices are long enough so that any intrinsic dependence of A on length (other than a simple linear dependence on the number of base pairs) is negligible. More elaborate expressions for analyzing observed absorbance can be derived where necessary to take length dependence explicitly into account. Usually, χ' is measured by varying the temperature. Because the melting transition of a short duplex occurs over a broad temperature range, the dependence of A_s and A_d on temperature must be considered in analyzing the data.

Experimentally determined values of X_b permit the computation of q_c. Remember that q_c is just the overall equilibrium constant for formation of all base-paired structures M_2 from a self-complementary oligomer M. Thus, for the reaction $2\,M \rightleftarrows M_2$, we can write

$$q_c = (M_2)/(M)^2 = X/2(1-X)^2 C_T \qquad (23\text{-}13)$$

where $X = 2(M_2)/[2(M_2) + (M)]$ is the fraction of strands that are paired, and C_T is the total concentration of oligonucleotide strands. When the all-or-none approximation is valid, $X = X_b = \chi'$. However, in general $X \neq X_b$, because not all the possible base pairs in a duplex are formed; instead,

$$X_b = Xf_b \qquad (23\text{-}14)$$

where f_b is the fraction of bases actually paired in an average over all duplex structures.

Return to Equation 23-6 and notice that, because the index i runs only over base-paired structures, K_i/q_c is the fraction of *duplexes* with the ith configuration. If we group together configurations with j base pairs, using Equation 23-7, then

$$f_j = \kappa g_j(N)s^{j-1}/q_c \qquad (23\text{-}15)$$

is the fraction of duplexes containing j base pairs. The average number $\langle n \rangle$ of base pairs per duplex thus is

$$\langle n \rangle = \sum_{j=1}^{N} jf_j = (1/q_c) \sum_{j=1}^{N} \kappa g_j(N)js^{j-1}$$

$$= \frac{1}{q_c} \frac{d}{ds} \sum_{j=1}^{N} \kappa g_j(N)s^j$$

$$= (1/q_c)\, d(sq_c)/ds \qquad (23\text{-}16)$$

The fraction of possible base pairs formed is just $f_b = \langle n \rangle/N$, so[§]

$$\chi' \cong X_b = X\langle n \rangle/N = (X/q_c N)\, d(sq_c)/ds \qquad (23\text{-}17a)$$

[§] The advantage of expressing the nucleation step as σs is that, then, Equation 23-17a has the particularly simple form $\chi' \cong X_b = (X/N)d(\ln q_c)/d(\ln s)$. Note also that, in the limit of the all-or-none model, Equation 23-17a leads immediately to $X_b = X$.

For an actual experimental system, N is known, and X can be expressed in terms of q_c and C_T by directly solving the quadratic Equation 23-13. This gives an analytical form for χ' as a function of q_c and C_T:

$$\chi' = \left(\frac{1 + 4q_c C_T - \sqrt{1 + 8q_c C_T}}{4q_c^2 C_T N} \right) \frac{d}{ds} (sq_c) \qquad \text{(23-17b)}$$

One selects either Equation 23-9 or Equation 23-10 for the partition function. Then the observed absorbance of an equilibrating self-complementary oligonucleotide can be calculated by specifying s and κ as a function of conditions. Alternatively, one can attempt to determine s and κ from a set of absorbance measurements.

Analysis of experimental data: complementary pairs and hairpins

The data in Table 23-2 encompass another type of equilibrium not governed by Equation 23-13. For a complex between two complementary nonidentical oligomers at equal concentrations (such as $A_4G_2 + C_2U_4 \rightleftarrows A_4G_2 \cdot C_2U_4$), we can write the reaction as $A + B \rightleftarrows C$. Then

$$q_c = (C)/(A)(B) = 2X/(1 - X)^2 C_T \qquad \text{(23-18)}$$

where $X = 2(C)/[(A) + (B) + 2(C)]$. Equation 23-18 can be checked by substituting this definition of X and realizing that, because the strands are at equal concentration, $(A) = (B)$, and $1/4[(A) + (B)]^2 = (A)(B)$.

When the all-or-none approximation is accurate, the observed T_m occurs at $X = 0.5$. Equations 23-13 and 23-18 then become

$$q_c(T_m) = 1/C_T \qquad \text{for self-complementary oligomers} \qquad \text{(23-19a)}$$

$$q_c(T_m) = 4/C_T \qquad \text{for complementary oligomer pair} \qquad \text{(23-19b)}$$

These expressions indicate that, to compare the melting temperatures of two such duplexes, we should compare the self-complementary oligomer with studies of the complementary pair made at fourfold greater concentration.

However, it turns out that the correct comparison is at only twice the concentration. A self-complementary oligomer complex has a twofold rotational symmetry not present in either single strand or in any non-self-complementary oligonucleotide complexes. This symmetry makes duplex formation appear less favorable by a factor of two. Equation 23-19a becomes

$$q_c(T_m) = 2/C_T \qquad \text{for self-complementary oligomers} \qquad (23\text{-}19c)$$

(The same phenomenon raises the observed macroscopic dissociation constant of a proton in a dibasic acid by a factor of two over the individual microscopic dissociation constant for one site.) For example, there are two ways in which the complex $G_3C_3 \cdot G_3C_3$ can lose one base pair to produce identical structures:

```
    GGGCCC          GGGCCC              GGGCCC
    · · · · · · ─────→ · · · · ·    or   · · · · ·
    CCCGGG          CCCGGG              CCCGGG
```

We are interested in using measured q_c values to compute interactions in systems that are almost inevitably non-self-complementary. To remove the complications of symmetry, imagine that only one of the two dissociations shown above can actually occur. To obtain the equilibrium constant governing it, the observed q_c must be multiplied by a factor of two. Thus, self-complementary oligomers should be studied at a concentration of $C_T/2$; then differences in stability relate directly to differences in individual microscopic equilibrium constants. This is why the T_m values of some samples in Table 23-2 are shown at twice the concentration of the other samples.

The data in Table 23-3 describe the intramolecular equilibrium between a single strand and a hairpin loop formed from it. Equations 23-8 through 23-10 and 23-16 and 23-17 were derived for intermolecular complexes, but exactly the same equations should hold for intramolecular complexes. By using the conformational partition function, we removed from the partition function all concentration dependence. Therefore, to describe hairpin-loop formation ($S \rightleftarrows H$, where S represents the single strand and H represents the hairpin loop), all we have to do is write the equilibrium in terms of the fraction of hairpin, $X = (H)/[(S) + (H)]$:

$$q_c = X/(1 - X) \qquad (23\text{-}20)$$

All of the other equations for the quantities X, X_b, and χ' apply for this case as well. However, in computing q_c, one must remember that κ_l (the nucleation parameter for hairpin-loop formation) will be a function of loop size l and is entirely different from κ for duplex formation.

Intuition can be used to simplify the conformational partition functions of hairpins. For example, a sequence such as $A_4GC_6U_4$ is very unlikely to form a loop in which A–U pairs were made but the G–C pair was left open. Let the initiation parameter κ_5 refer to closing a loop of 5 residues by a G–C pair. The equilibrium constant s_{AG} for stacking an A–U atop a G–C is different from s_{AA} for stacking an

A–U atop another A–U. Because the hairpin will nucleate from the G–C pair, only configurations containing the G–C pair and continuous adjacent duplex need be considered. Starting from Equation 23-7, and recognizing that $g_j(N) = 1$ for all j because we have fixed nucleation size and growth direction, we can closely approximate the conformational partition function as

$$q_c = \kappa_5 \left(1 + s_{GA} \sum_{j=0}^{3} s_{AA}^j \right) \tag{23-21}$$

If s_{AA} and s_{GA} are independently known from measurements on linear duplexes, determining the stability of the hairpin permits computation of the nucleation parameter κ_5.

Calculating free energies for individual duplex interactions

There are a number of different ways to analyze the melting temperature of oligonucleotides to extract thermodynamic parameters of interest for individual interactions. Let us here restrict attention to model complexes with only one favorable alignment of base-paired sequences. Then the free energy of duplex formation ΔG_T is simply

$$\Delta G_T = \sum \Delta G_{nuc} + \sum \Delta G_{gr} \tag{23-22}$$

where sums are taken over all individual nucleation steps (ΔG_{nuc}, governed by κ or α_l values) and growth steps (ΔG_{gr}, governed by s values) necessary to complete the final structure. Enthalpies and entropies for each of the interactions and for the total duplex are defined analogously.

To obtain a value for a particular ΔG_{gr} interaction, the best procedure is to compare the melting of two oligomer complexes (1 and 2) as identical as possible except for a difference in the desired interaction. If both are self-complementary, then Equation 23-19c implies that, for each oligomer complex at its T_m, $q_c(T_m)C_T = 2$. Adjust or extrapolate the experimental conditions until the melting temperatures of the two complexes are the same. Then

$$q_c^{(1)} C_T^{(1)} = q_c^{(2)} C_T^{(2)} \qquad \text{at } T_m^{(1)} = T_m^{(2)} = T \tag{23-23}$$

Because the free energy ΔG_T is simply $-RT \ln q_c$, the difference in free energies is simply

$$\begin{aligned} \Delta\Delta G_T = \Delta G_T^{(1)} - \Delta G_T^{(2)} &= -RT \ln q_c^{(1)} + RT \ln q_c^{(2)} \\ &= RT(\ln C_T^{(1)} - \ln C_T^{(2)}) \end{aligned} \tag{23-24}$$

Figure 23-7

Determination of the free energy of adding an A–U pair to a preexisting duplex. The T_m of two different oligomers is measured as a function of concentration. Then, at constant T_m, we have (from Eqn. 23-24) $\Delta\Delta G_T = \Delta G_T^{(1)} - \Delta G_T^{(2)} = RT(\ln C_T^{(2)} - \ln C_T^{(1)})$. Because the A_6U_6 duplex has two extra A–U pairs, the equilibrium constant for adding just one of them is $s_{AA} = \exp(-\Delta\Delta G_T/2RT)$ (from Eqns. 23-26, 23-27, and 23-28). [After J. Gralla and D. M. Crothers, *J. Mol. Biol.* 73:497 (1973).]

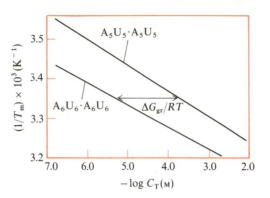

Figure 23-7 shows an example of the application of Equation 23-24. What facilitates the use of this approach is the linearity of $1/T_m$ versus $\ln C_T$. It is easy to explain this linearity. For an oligomer complex at T_m, from Equation 23-19b,c,

$$\Delta G_T/RT_m = -\ln q_c = \ln(\gamma'' C_T) = (\Delta H_T/RT_m) - \Delta S_T/R \qquad (23\text{-}25)$$

where ΔG_T, ΔH_T, and ΔS_T represent the formation of the entire complex from its components. The constant γ'' is 2 for self-complementary oligomers and is 4 for complementary pairs. Note that Equation 23-25 is identical in form to the empirically derived Equation 23-4, where $A' = \Delta S_T/\Delta H_T$, and $B' = -R/\Delta H_T$.

After $\Delta\Delta G_T$ has been measured experimentally by Equation 23-24, it must be correlated with the microscopic equilibrium constants for individual duplex interactions. For two structures with the same nucleation step(s) in the limit of the all-or-none reaction, Equation 23-22 becomes

$$\Delta\Delta G_T = \Delta\sum \Delta G_{gr} \qquad (23\text{-}26)$$

However, each ΔG_{gr} is a term of the form $-RT \ln s_i$, where s_i is the equilibrium constant for placing the ith base pair next to an adjacent pair. Thus $\sum \Delta G_{gr}$ is $-RT \ln \prod_i s_i$ (where the product is taken over $i = 1$ to $i = N$) for an oligomer duplex with $N + 1$ base pairs. So, when two duplexes of lengths N_1 and N_2 are compared by Equation 23-24,

$$\Delta\Delta G_T = -RT \ln(q_c^{(1)}/q_c^{(2)}) = -RT \ln\left(\prod_{j=1}^{N_1} s_j^{(1)} \middle/ \prod_{k=1}^{N_2} s_k^{(2)} \right) \qquad (23\text{-}27)$$

Let us see how Equation 23-26 applies in practice. For the two oligomers in Figure 23-7, the 12-mer (complex 1) has two extra AA stackings, each with a formation Constant s_{AA}. Therefore,

$$\Delta\sum\Delta G_{gr} = -RT\ln s^2_{AA} \tag{23-28}$$

Suppose we compare the duplexes $A_2GCU_2 \cdot A_2GCU_2$ and $A_2CGU_2 \cdot A_2CGU_2$:

$$\Delta\sum\Delta G_{gr} = -RT\ln(s_{AG}s_{GC}s_{CU}/s_{AC}s_{CG}s_{GU}) \tag{23-29}$$

If data on more oligomers are available, a set of equations can be developed for each of the ten possible duplex neighbor interactions. A full resolution of all these is not yet justified by existing results. However, current progress is quite rapid, and Table 23-4 summarizes the best available parameters for individual interactions, expressed as $\Delta G_i^0 = -RT\ln s_i$. Note that ΔG_{gr} is dependent on salt concentration and temperature, and data must be extrapolated to the same conditions in order to compare two oligomers by Equation 23-26. Thus, in practice, a standard set of conditions (25°C, 1 M Na^+) has been chosen for such comparisons.

The ΔG_{gr} terms are the free energy of stacking one base pair on top of another; the corresponding s_i terms are stacking equilibrium constants. Thus the formalism contained in Equations 23-27 through 23-29 demonstrates the dominance of stacking over simple hydrogen bonding in determining the energetics of nucleic acid interactions.

Estimating enthalpies of oligonucleotide interactions

Where sufficient data exist, ΔH_T values for an oligonucleotide complex can be obtained by writing Equation 23-25 in the form

$$\Delta H_T/R = d(\ln C_T)/d(1/T_m) \tag{23-30}$$

where it is assumed that $d\,\Delta H_T/d(1/T_m)$ and $d\,\Delta S_T/d(1/T_m) \cong 0$. The ΔH_T obtained will be $\Delta H_{nuc} + \sum\Delta H_{gr}$ for all interactions (where "nuc" represents nucleation, and "gr" represents growth). Solving linear equations derived from various oligomer complexes permits sorting out of ΔH_{gr} values for individual interactions. From ΔH_{gr} and ΔG_{gr}, one can then compute ΔS_{gr}.

Thermodynamic parameters can be obtained also by the direct analysis of the melting curves of individual oligomer complexes. Because $\Delta G_T = -RT\ln q_c = \Delta H_T - T\,\Delta S_T$, a simple van't Hoff relationship applies to the total enthalpy change:

$$[\partial(\ln q_c)/\partial T]_{C_T} = (1/q_c)(\partial q_c/\partial X)_{C_T}(\partial X/\partial T)_{C_T} = \Delta H_T/RT^2 \tag{23-31}$$

The quantity $(1/q_c)(\partial q_c/\partial X)_{C_T}$ can be evaluated at T_m because $X \cong 0.5$ at T_m. From Equations 23-13, 23-18, and 23-20, it is easy to show that (at T_m)

$$(\partial X/\partial T)_{C_T} = \Delta H_T/\gamma' R T_m^2 \qquad (23\text{-}32)$$

where $\gamma' = 4$ for intramolecular hairpin formation, and $\gamma' = 6$ for either self-complementary or complementary-pair duplex formation. The ΔH_T here is the apparent total enthalpy of the duplex or the hairpin. In the all-or-none limit, there is only a single nucleation step, so $\Delta H_T = \Delta H_{nuc} + \sum \Delta H_{gr}$. In the special case where nucleation enthalpy is zero and all the base-paired interactions are the same, $\Delta H_T = (N-1)\Delta H_{gr}$. In practice, enthalpies from Equations 23-32 and 23-30 do not agree all that well. Equation 23-32 gives lower values, which probably are less reliable.

A final way to extract thermodynamic parameters is to extrapolate results to the melting of infinite homopolymers. For the formation of intramolecular hairpins, $\Delta G_T(T_m) = 0$, and the melting temperature T_m^∞ of a long hairpin polymer (to a very good approximation) is

$$T_m^\infty = N \Delta H_{gr}/N \Delta S_{gr} = \Delta H_{gr}/\Delta S_{gr} \qquad (23\text{-}33)$$

because nucleation contributions are negligible. For a homooligomer hairpin of duplex length N,

$$T_m = [(N-1)\Delta H_{gr} + \Delta H_{nuc}]/[(N-1)\Delta S_{gr} + \Delta S_{nuc}] \qquad (23\text{-}34)$$

Combining these two equations and rearranging, we can write

$$1/T_m = (1/T_m^\infty) + (\Delta S_{nuc} - \Delta H_{nuc}/T_m^\infty)/[(N-1)\Delta H_{gr} + \Delta H_{nuc}] \qquad (23\text{-}35)$$

An equivalent expression can be derived for the intermolecular complexes. This equation correctly predicts that a plot of $1/T_m$ versus $1/N$ will be approximately linear for a series of homologous oligomers (Fig. 23-4a). Note that Equation 23-35 is similar in form to the empirically derived Equation 23-3 (and becomes almost identical if ΔH_{nuc} is very small).

Nucleation of duplexes; small loops and hairpins

In order actually to calculate the T_m or ΔG_T of a duplex or hairpin, we must account for the free energy of nucleation and growth. The latter parameters are measurable as described in the preceding subsection. However, the nucleation parameters are difficult to determine for two reasons. The form of the partition function means that, in simple sequences, errors in s will multiply and lead to large errors in κ if one attempts to determine it from the observed ΔG^0 for the whole duplex. Also, the conformational partition function defines κ at unit activity, which leads to standard conditions of 1 M single-strand concentrations. Actual data are obtained at much lower concentrations, and a long extrapolation is required.

For duplex formation from separate strands, values of κ_{AU} and κ_{GC} have been estimated as 4×10^{-5} M^{-1} and 2.5×10^{-4} M^{-1}, respectively, corresponding to ΔG_{nuc} values of $+6$ kcal mole^{-1} and $+5$ kcal mole^{-1}, respectively. The former value applies only to an oligomer with no G–C pairs, because a single G–C pair will presumably always be a preferred nucleation site.

To obtain enthalpies and entropies of duplex formation, it has been assumed that $\Delta H_{nuc} = 0$ because hydrogen bonding (the only major interaction in nucleation) has negligible enthalpy in water. Thus, $\Delta G_{nuc} = -T\Delta S_{nuc}$. The thermodynamic parameters shown in Table 23-4 thus are somewhat uncertain because of these and other approximations involved in their derivation. However, these parameters work rather well in predicting the T_m of RNA duplexes not included in the library of data used to establish the table. Combining Equations 23-22 and 23-25, and using the fact that $\Delta G_{gr} = \Delta H_{gr} - T\Delta S_{gr}$, one can easily show that

$$(1/T_m)(\textstyle\sum \Delta H_{gr}) = \Delta S_{nuc} + \sum \Delta S_{gr} + R \ln \gamma'' C_T \tag{23-36}$$

This expression allows calculation of the T_m (in Kelvin) of a duplex.

Free energies for loop and hairpin formation can be handled in much the same way as duplex formation. For a hairpin, as before, $\Delta G_T = \Delta G_{nuc} + \sum \Delta G_{gr}$. However, $\Delta G_T = 0$ at the melting temperature because hairpin formation is an intramolecular process. The ΔG_{gr} energies are known from linear duplexes, so ΔG_{nuc} at T_m can be determined. Table 23-5 shows the values available to date. Internal double-stranded loops can be treated in much the same way. For example, compare $A_4U_4 \cdot A_4U_4$ and $A_4CU_4 \cdot A_4CU_4$. The difference in ΔG_T should be just $\Delta G_{AU} - \Delta G_{loop}$, assuming that the central AU–AU stacking interaction is lost when the loop forms. Table 23-5 also lists some values for internal loops. Single-stranded bulge loops will arise in cases such as $A_4U_4 \cdot A_4CU_4$. These are more of a problem because little actual experimental data exist on such complexes. Bulge-loop parameters have been estimated from the thermal stabilities of imperfectly paired polymers, such as those prepared by random chemical modification or by incorporation of an occasional noncomplementary base.

There are two problems with the loop data in Table 23-5. Although different experimental determinations of loop free energies agree, it is not clear how these divide into enthalpy and entropy. Too few data are available to test assumptions used in analyzing the available results. These various assumptions have resulted in variations of ΔH_{loop} from 0 to rather large positive enthalpies. Thus, although Tables 23-4 and 23-5 together permit the estimation of $\Delta G^0(25°C)$ for specific looped structures, they do not permit calculations of T_m. A second problem is how to handle large loops. This is an important factor in predicting the most stable structures of partially melted RNAs. There are few experimental data on the stability of such large loops.

Table 23-4

Thermodynamics of adding a base pair to a double-stranded helix

Reaction	ΔH^0 (kcal)	ΔS^0 (kcal deg^{-1})	ΔG^0 (kcal) at 25°C
—A / A over —U \ U → —A—A / over —U—U \	−8.2	−0.0235	−1.2
—A / U over —U \ A → —A—U / over —U—A \ or —U / A over —A \ U → —U—A / over —A—U \	−6.5	−0.0164	−1.6
—A / C over —U \ G → —A—C / over —U—G \ or —A / G over —U \ C → —A—G / over —U—C \ —C / A over —G \ U → —C—A / over —G—U \ or —G / A over —C \ U → —G—A / over —C—U \	−5.9	−0.0127	−2.1
—C / G over —G \ C → —C—G / over —G—C \	−13.0	−0.0335	−3.0
—G / C over —C \ G → —G—C / over —C—G \	−14.7	−0.0349	−4.3
—C / C over —G \ G → —C—C / over —G—G \	−13.7	−0.0298	−4.8
—G / U over —U \ G → —G—U / over —U—G \	——	——	0.3
—X / G over —Y \ U → —X—G / over —Y—U \ or —G / X over —U \ Y → —G—X / over —U—Y \	——	——	0.0

NOTE: The assumed values for the standard (strand concentration of 1 M) free energy of initiation at 25°C are +6.0 kcal for an A–U base pair and +5.0 kcal for a G–C base pair. In the last reaction, X and Y are any Watson–Crick complementary base pair.

SOURCE: After P. Borer et al., *J. Mol. Biol.* 86:843 (1974).

Table 23-5
Loop free energies at 25°C, 1 M Na$^+$

Loop structure (N = number of unpaired bases)	ΔG^0 (kcal mole^{-1})
Hairpin	
Closed by A–U (add 1 kcal mole^{-1} if no U in loop)	
$N = 3$	$+7$
$N = 4$ to 5	$+6$
$N = 6$ to 7	$+5$
$N = 8$ to 9	$+6$
$N \geqslant 10$	$7 + 0.9 \ln(N/10)$
Closed by G–C	
$N = 3$	$+8$
$N = 4$ to 5	$+5$
$N = 6$ to 7	$+4$
$N = 8$ to 9	$+5$
$N \geqslant 10$	$6 + 0.9 \ln(N/10)$
Internal loop	

$$-G \diagup^{\text{X}}\diagdown C-$$
$$\cdot \qquad \cdot$$
$$-C \diagdown_{\text{X}}\diagup G-$$

	0
$N = 2$ to 6 (except structure above)	$+2$
$N = 7$	$+3$
$N \geqslant 8$	$3 + 0.9 \ln(N/7)$
Single-stranded bulge loop	
$N = 1$	$+3$
$N = 2$ to 3	$+4$
$N = 4$ to 7	$+5$
$N \geqslant 8$	$6 + 0.9 \ln(N/8)$

SOURCE: After V. Bloomfield, D. Crothers, and I. Tinoco, Jr., *Physical Chemistry of Nucleic Acids* (New York: Harper & Row, 1974).

Calculating the nucleation entropy for a large loop

It seems reasonable that the length dependence of the free energy of formation for a large loop should show up entirely as entropy. Any interaction enthalpies involved in loop initiation or closure should be constant once the loop has reached a certain size. To calculate the entropy change in closing a double strand or a hairpin loop, consider the schematic illustration in Figure 23-8. Closure can be modeled as the process of bringing the two ends of a single nucleic acid strand together to within some volume V_i. The internal entropy per mole of a chain is related to the number

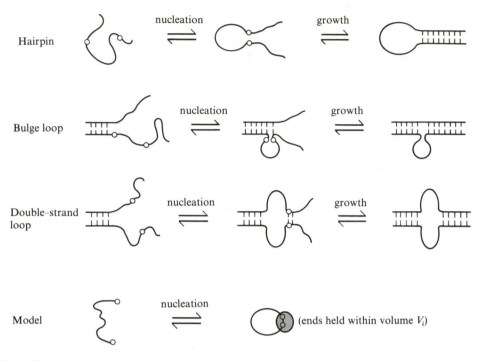

Figure 23-8

Three types of loops, and the way in which the chain configurational statistics of a linear single strand are used to model the free energy of loop formation.

Ω of available conformations by

$$S = R \ln \Omega \qquad (23\text{-}37)$$

so the entropy change for cyclization is

$$\Delta S_{\text{loop}} = R \ln(\Omega_{\text{cl}}/\Omega) \qquad (23\text{-}38)$$

where Ω_{cl} represents the number of conformations available with the ends constrained to lie within the volume V_i.

The simplest polymer chain to consider is an unrestricted random coil. The normalized distribution function of end-to-end distances for such a coil (Chapter 18) is

$$W_N(r)\,dr = 4\pi r^2 (3/2\pi Nl^2)^{3/2} \exp(-3r^2/2Nl^2)\,dr \qquad (23\text{-}39)$$

where N is the number of monomer units in the coil, and l is the length of one monomer. Equation 23-39 cannot actually apply to a real polynucleotide, but it can be used at least to examine the form of ΔS_{loop}.

The fraction of configurations permitting the ends of the chain to approach within a distance a is

$$\Omega_{cl}/\Omega = \int_0^a W_N(r)\,dr \tag{23-40}$$

If N is large and if one is interested only in small r (corresponding to ends very close together), the exponential in Equation 23-39 can be approximated by unity. Then Equation 23-40 becomes

$$\Omega_{cl}/\Omega = (3/2\pi N l^2)^{3/2} \int_0^a 4\pi r^2\,dr = (3/2\pi N l^2)^{3/2} V_i \tag{23-41}$$

where the approach volume V_i is just $4\pi a^3/3$. This expression allows us to write the entropy change as the Jacobson–Stockmayer equation:

$$\Delta S_{loop} = -(3/2)R \ln N + R \ln[(3/2\pi l^2)^{3/2} V_i] \tag{23-42}$$

The length dependence is contained entirely in the first term of Equation 23-42, whereas the second term should be a constant for all loops. The free energy of closing a large loop is just $-T\Delta S_{loop}$. The constant term in Equation 23-42 is not easy to determine because of the rather vague quantity V_i. Instead, the loop free energies of large loops are compared with the longest one known experimentally. If that one is of length n, then

$$\Delta G_{loop}(N) = \Delta G_{loop}(n) + (3/2)RT \ln(N/n) \tag{23-43}$$

This expression allows any loop enthalpy to be included in the term $\Delta G_{loop}(n)$. Table 23-5 demonstrates values from Equation 23-43 for the various types of oligonucleotide loops.

The preceding treatment of loop entropies is rather crude because a polynucleotide single strand is not well approximated by an unrestricted Gaussian chain. Much more exact treatments are available under theta conditions (Chapter 18). However, melting experiments on oligomers rarely are done under theta conditions. One still can expect that the loop-closure entropy of a real chain under any conditions will follow a form analogous to Equation 23-42:

$$\Delta S_{loop} = -\alpha_c R \ln N + \text{Constant} \tag{23-44}$$

The constant α_c is greater than 3/2 and probably is around 1.75 in good solvents for stiff chains with excluded volume.

Admitting the approximate nature of Equation 23-43, let us see what it tells us qualitatively about the relative stabilities of various configurations. Consider the equilibrium shown in Figure 23-9. Can the two A–U pairs form and close the big

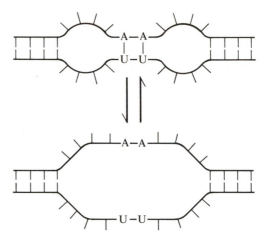

Figure 23-9

An example of loop fusion. See text for a quantitative discussion of the factors that decide which form is favored.

interior loop ($N = 20$) to two small loops ($N = 8$)? From Tables 23-4 and 23-5, the the relative free energies of formation of the two structures at 25°C are

$$\text{big loop:} \quad \Delta G_T = \Delta G_{\text{loop}}(20) = 3 + 0.9 \ln(20/7) = 3.9 \text{ kcal mole}^{-1} \quad (23\text{-}45)$$

$$\text{small loops:} \quad \Delta G_T = \Delta G_{AA}^0 + 2\,\Delta G_{\text{loop}}(8) = -1.2 + 2 \times [3 + 0.9 \ln(8/7)]$$

$$= 5.0 \text{ kcal mole}^{-1} \quad (23\text{-}46)$$

The small loops are significantly higher in energy, and so the A–U base pairs will be broken most of the time. Note that, if there were two G–C pairs in the center of the loop, they could close (because ΔG_{CG} is much more negative than ΔG_{AA}).

It is instructive to extend the example of Figure 23-9 to larger loops. The ΔG_{fus} for the fusion of two loops of size N separated by m base pairs to make one loop of size $2(N + m)$ is

$$\Delta G_{\text{fus}} = \Delta G_{\text{loop}}(2N + 2m) - 2\,\Delta G_{\text{loop}}(N) - (m - 1)\Delta \bar{G}_{\text{gr}}$$

$$= 3 + 0.9 \ln[2(N + m)/7] - 6 - (2 \times 0.9) \ln(N/7) - (m - 1)\Delta \bar{G}_{\text{gr}} \quad (23\text{-}47)$$

where $\Delta \bar{G}_{\text{gr}}$ is the average of the stacking interactions shown in Table 23-4, or about

-2.5 kcal mole^{-1}. Thus, if $m \ll N$,

$$\Delta G_{\text{fus}} \cong -4.9 - 0.9 \ln(N/7) + 2.5m \qquad \text{(kcal mole}^{-1}) \qquad (23\text{-}48)$$

It takes more and more base pairs to keep two loops from merging as the loop sizes get bigger. Furthermore, the loop free energies are more-or-less temperature independent, because they are dominated by entropy. In contrast, $\Delta \bar{G}_{\text{gr}}$ is strongly temperature dependent and becomes rather small near T_{m}. This explains why DNA melting is so cooperative despite sequence heterogeneity. Once loops form near T_{m}, there is a strong likelihood that they will merge.

Predicting RNA structure

All of the efforts outlined thus far in Section 23-1 have unfolded over the past decade. The result is that the approximate parameters in Tables 23-4 and 23-5 permit the prediction of the most stable secondary structures of specific RNA sequences. The current accuracy of these predictions is uncertain, but the results have been fairly good where they can be tested. More important, they are sure to improve as more model-compound data become available and permit refinement of the individual contributions to free energy.

Figure 23-10 shows an example of the use of the predictive scheme on a real sequence. For any hypothetical secondary structure, the free energy of formation from single strands is

$$\Delta G_{\text{T}} = \sum \Delta G_{\text{nuc}} + \sum \Delta G_{\text{loop}} + \sum \Delta G_{\text{gr}} \qquad (23\text{-}49)$$

where ΔG_{gr} represents contributions from stacking interactions. Keep in mind that, powerful as this approach is, all it permits is the computation of the free energy surface of the molecule point by point. Unless all possible structures are tested, there is no guarantee that one has found the structure corresponding to a minimal free energy. This procedure is impractical for large sequences, and one must return to the formidable problem of approximate searches for free energy minima (a problem discussed elsewhere in this book). Also, one must be aware that nowhere in the formalism of this chapter have any tertiary structure interactions been included. The observed structure of a nucleic acid will correspond, if it is in thermodynamic equilibrium, to a minimal total free energy, and this need not correspond to a minimum in the secondary-structure free energy. The hope in nucleic acid structure prediction is that secondary-structure interactions are so large that they will strongly constrain possible tertiary structures. The future will reveal whether this hope is justified.

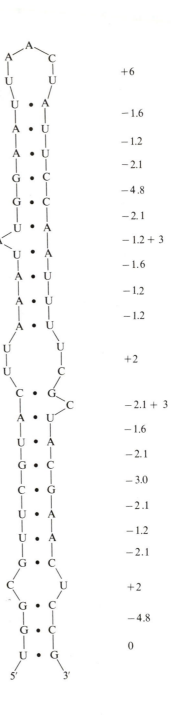

Figure 23-10

Calculating the stability of a particular secondary structure model, using data from Tables 23-4 and 23-5. The sequence is a 55-nucleotide fragment of R17 viral RNA. The values shown are the contributions to the total free energy at 25°C from each loop and stacking interaction. The overall structure is predicted to be stable with respect to single strand by -20.0 kcal mole^{-1}. For this particular sequence, it is unlikely that a more stable secondary structure model could be found. [After I. Tinoco, Jr., et al., *Nature New Biol.* 246:40 (1973).]

23-2   KINETICS OF CONFORMATIONAL CHANGES

An understanding of the mechanism of a biochemical process almost inevitably requires some knowledge about the kinetics of individual reaction steps. Alterations in nucleic acid secondary structure certainly are involved in DNA replication and transcription, and they are very likely to be involved in protein synthesis as well as in numerous levels of regulation of gene expression. If the rates of these processes are intrinsically fast enough, a biological system may be able to afford to wait until they occur spontaneously and then use the desired state. If the intrinsic rate is too slow, the system must provide a specific catalyst (in most cases a protein), which interacts directly with an existing secondary structure and speeds the rate of conversion to the desired structure. These two mechanisms pose very different challenges to an organism. It is important to understand the intrinsic rates of conformational changes of nucleic acids in order to acquire some insights into the biochemical function of nucleic acids.

Here we concentrate on the kinetics of interconversions between single strands and double strands in simple model oligonucleotide complexes. Then we show how this information can help us analyze the kinetics of denaturation and renaturation of DNA. In the process, we develop the framework needed to understand quantitative nucleic acid hybridization, one of the most frequently used techniques in current eukaryotic molecular biology.

Individual base-stacking interactions

The two fundamental interactions that must dominate nucleic acid kinetics are base stacking and base pairing, the very same interactions that determine structure. In many cases, kinetic examination of such interactions can lead to a more detailed understanding of the processes involved. Base stacking in an oligonucleotide is an intramolecular reaction; therefore, it can be studied kinetically only by perturbing the system. A logical choice of perturbing variable is temperature change, because this change is known to alter the extent of stacking. When microsecond temperature jumps are applied to an oligonucleotide in solution, unstacking occurs and is accompanied by a decrease in hypochromism. However, the rates are too fast to monitor conveniently, even by the sophisticated technology conventionally used in relaxation kinetics. See, however, Dewey and Turner (1979).

One of the few pieces of kinetic information available on intramolecular stacking comes from fluorescence studies on flavin ethenoadenine dinucleotide (FεAD), an analog of the dinucleotide coenzyme FAD (flavin adenine dinucleotide) in which one nucleoside is flavin mononucleoside (FMN) and the other is ethenoadenosine (εA). The bonding in FAD or in the analog is not the same as in normal dinucleotides, and stacking rates need not correspond (Fig. 23-11a). However, the method by which the rates were measured is instructive, and the results should at least be indicative of what

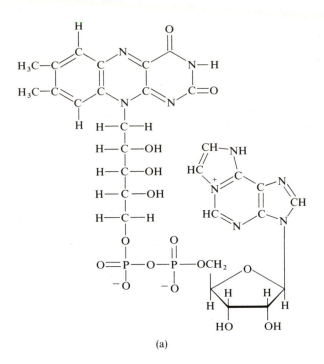

(a)

(b)

10%

k_s ‖ k_{-s}

+ hν

→ F ⓔA*

Energy transfer

→ F* ⓔA —hν→ F ⓔA
       5%
(Fluorescence)

k_s^* ↓ 5%

F* ⓔA Dynamic quenching

      Nonfluorescent forms

+ hν → F* ⓔA Static quenching

90%
complex

Figure 23-11

Kinetics of stacking in a dinucleotidelike compound, FεAD. (**a**) Structure of FεAD. (**b**) Schematic of the states needed to explain observed fluorescence data. The ground state is 90% stacked. The stacked form is quenched essentially instantaneously. The rate of conversion of the excited unstacked form to stacked form (k_s^*) is fast enough to compete with fluorescence. [After J. R. Barrio et al., *Proc. Natl. Acad. Sci. USA* 70:941 (1973).]

can be expected in nucleic acid single strands. Figure 23-11b shows how fluorescence can be used to measure a stacking rate.

The chromophore εA is intensely fluorescent, but it is quenched a thousandfold in FεAD by the singlet-singlet energy transfer and cannot be detected. In FεAD, the static fluorescence of the flavin, $F_{\text{F}\varepsilon\text{AD}}$, is quenched about 18-fold relative to FMN, but it still is intense enough to study. The ground state of FεAD is an equilibrium between stacked and unstacked forms. Two extreme kinetic models can be considered. If the rate of stacking or unstacking were much slower than the excited-state lifetime of flavin in the dimer, then the fluorescence decay of the flavin should simply be the sum of two one-exponential terms, corresponding to emission of flavin in the stacked and unstacked forms. The latter decay should have the same lifetime as that of FMN. Alternatively, if stacking rates were much faster than decay, only a single fluorescence lifetime should be seen. Recalling that τ^{-1} (the total rate of decay) is the weighted sum of the decay rates of the two forms, we expect this single decay rate to be

$$\tau^{-1} = \chi_s^* \tau_s^{-1} + (1 - \chi_s^*)\tau_u^{-1} \tag{23-50}$$

where χ_s^* is the mole fraction stacked in the excited state, and τ_s and τ_u are the fluorescence lifetimes of the stacked and unstacked states, respectively. We expect τ_u to be equal to τ_{FMN}, the lifetime observed in FMN alone, which is 4.7 nsec. For this model to be consistent with observed static flavin fluorescence, τ/τ_u must be equal to $F_{\text{F}\varepsilon\text{AD}}/F_{\text{FMN}}$, so that τ is predicted to be

$$\tau = \tau_u F_{\text{F}\varepsilon\text{AD}}/F_{\text{FMN}} = (4.7 \text{ nsec})(1/18) = 0.26 \text{ nsec} \tag{23-51}$$

The observed results are consistent with neither extreme kinetic model. Only a single fluorescence lifetime is observed, thus eliminating the first model. However, this lifetime is 2.6 nsec, tenfold larger than the value predicted by the second model. Thus, we must consider a more accurate and realistic kinetic scheme. The only easy way to resolve the discrepancy between static and lifetime data is to postulate that the stacked form is quenched extremely rapidly, so that all observed emission comes from the unstacked form. However, the rate of stacking is fast enough to cause some quenching of any excited unstacked forms (Fig. 23-11). Otherwise the lifetime of FεAD would be the same as that of FMN. If k_s^* is the first-order stacking rate constant of the excited state of FεAD, then

$$1/\tau_{\text{F}\varepsilon\text{AD}} = k_s^* + k_F + k_e \tag{23-52a}$$

$$1/\tau_{\text{FMN}} = k_F + k_e \tag{23-52b}$$

where k_F is the rate of fluorescence, and k_e is the sum of all nonradiative decay rates. Therefore, k_s^* can be determined by

$$k_s^* = \tau_{\text{F}\varepsilon\text{AD}}^{-1} - \tau_{\text{FMN}}^{-1} \tag{23-52c}$$

The result is $1.7 \times 10^8 \text{ sec}^{-1}$.

The equilibrium constant for stacking can be determined also from the fluorescence data. Only the unstacked form fluoresces. If it were the only form present, then $F_{\text{F}\varepsilon\text{AD}}/F_{\text{FMN}}$ should equal $\tau_{\text{F}\varepsilon\text{AD}}/\tau_{\text{FMN}}$. Because some F$\varepsilon$AD is already stacked prior to excitation, the static fluorescence is lower:

$$F_{\text{F}\varepsilon\text{AD}}/F_{\text{FMN}} = (1 - \chi_s)\tau_{\text{F}\varepsilon\text{AD}}/\tau_{\text{FMN}} \tag{23-53}$$

From Equation 23-53 we can determine χ_s, the mole fraction of stacking; we can then calculate the ground-state equilibrium constant as $K_s = \chi_s/(1 - \chi_s) = 9$. If the stacking rates were the same for the ground and excited states, then the rate of unstacking would be $k_{-s} = k_s^*/K_s = 1.9 \times 10^7 \text{ sec}^{-1}$. Thus, the relaxation time that would be measured by temperature-jump or equivalent perturbation techniques (Chapter 16) is $\tau = (k_s^* + k_{-s})^{-1} = 5.3$ nsec. It is much too fast to be seen by standard relaxation techniques. This means that we can pretty much neglect single-strand stacking in the analysis of other nucleic acid conformational kinetics. Rate-limiting processes or relaxation processes observed in double-strand systems must correspond to other elementary steps.

Individual base-pairing interactions

As a model for base-pairing kinetics, one can examine the relaxation rates of hydrogen-bonded complexes of individual bases or their derivatives. These are too unstable to be studied in water, but they can be studied in nonaqueous solvents. Relaxation kinetics of base-pair equilibria can be monitored by placing the sample in a strong electric field ($\sim 200{,}000 \text{ V cm}^{-1}$) and applying a second high-frequency modulating field. Because the base-pairing reaction results in a changed dipole moment, the equilibrium is perturbed by field changes. It can be detected by methods equivalent to the dielectric dispersion technique (Chapter 12). The reaction is

$$\text{A} + \text{B} \underset{k_{-1}}{\overset{k_1}{\rightleftharpoons}} \text{A} - \text{B}$$

so the relaxation time (Chapter 16) is

$$\tau^{-1} = k_1(\bar{A} + \bar{B}) + k_{-1} \tag{23-54b}$$

where the overbars denote equilibrium concentrations. Therefore, both forward and backward rates can be measured by observing the effect of concentration on the relaxation time.

The results show that $k_1 > 10^9 \text{ sec}^{-1} \text{ M}^{-1}$. This is so fast that the reaction is diffusion controlled. That is, it is limited only by the rate at which bases encounter each other through random collisions in solution. All base pairs show about the same formation rates. Dissociation rates differ, depending on the stability of a complex, with the least stable complexes naturally having the fastest dissociation rates. In general, the dissociation rates also are quite fast, in the range of 10^6 to 10^8 sec^{-1}. Once again, then, the most basic nucleic acid interactions occur on a very short time scale, Of course, the collision rate of base pairs is slowed at very low concentration, as must occur for any bimolecular kinetic process. At 10^{-6} M concentration, base pairs have an intrinsic formation rate of only 10^3 sec^{-1}. This is slow enough that it should be a significant factor when oligonucleotide or nucleic acid kinetics are measured in dilute solution.

Oligonucleotide duplex formation: experimental results

Consider the kinetics of melting of a self-complementary or complementary oligonucleotide duplex. To simplify things, assume that only states containing in-phase continuous runs of base pairs contribute and that a unique nucleation site exists. Figure 23-12 shows the most general possible model consistent with this restriction It allows for the possibility that the forward and backward rate constants for the formation of successive base pairs could be different. Such a model has $n - 1$ different relaxation times for a system with n base pairs.

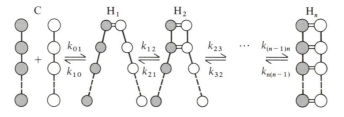

Figure 23-12

Model for the kinetics of oligonucleotide duplex formation. It is assumed that only aligned duplexes form, and that a unique nucleation site exists. In principle, all of the rate constants shown could have different values.

Experimental analyses of such kinetics are hopelessly complicated for $n > 4$. However, the thermodynamics of oligomer melting (discussed earlier) suggests that simpler models should be sufficient to describe melting. A reasonable approximation is to make the steady-state assumption for each kinetic intermediate (H_1, H_2, ..., H_{n-1}) except for the final states C and H_n (Fig. 23-12). This approximation has been used in a number of cases to analyze data for conformational changes (Chapter 21). However, for the mechanism we must consider here, the results would be quite complex. It turns out that an even simpler mechanism, the all-or-none reaction, suffices to explain most of the kinetic data.

The all-or-none mechanism was used earlier in treating the statistical thermodynamics of oligomer melting. For kinetic analysis, we consider that only the initial and final states are populated to significant extents, and therefore that all optical measurements simply monitor the reaction

$$2C \underset{k_{-1}}{\overset{k_1}{\rightleftharpoons}} H_n \qquad (23\text{-}55a)$$

for the melting of a self-complementary oligonucleotide, which yields a relaxation time (Chapter 16):

$$1/\tau = 4k_1 \, \bar{C} + k_{-1} \qquad (23\text{-}55b)$$

In the case of complementary oligomer pairs, the monitored reaction is assumed to be

$$C_A + C_B \underset{k_{-1}}{\overset{k_1}{\rightleftharpoons}} H_n \qquad (23\text{-}56a)$$

which yields a relaxation time:

$$1/\tau = k_1(\bar{C}_A + \bar{C}_B) + k_{-1} \qquad (23\text{-}56b)$$

In making such a drastic approximation, one must realize that the rate constants derived are ambiguous, even if the data fit the results of the model. The constant k_1 will not necessarily refer to the combination of two separated strands. It corresponds to an aggregate of rate constants.

A number of oligomer systems have been studied by temperature-jump relaxation kinetics. In general, only a single relaxation time is observed, and the experimental data can be fit by Equation 23-55b or 23-56b, whichever is appropriate. However, the most simple-minded interpretation of the data is ruled out: k_1 is not the rate of collision of two separate single strands. Table 23-6 shows two factors that lead to this conclusion. The rate constants themselves are of the order 10^5 to 10^7 M^{-1} sec^{-1}; this is much slower than the rate of base pairing seen for monomer complexes, and it is too slow to be diffusion-controlled collision. Even more convincing is the effect of temperature on the forward rate k_1. For a wide variety of systems, k_1 fits the Arrhenius equation:

$$d(\ln k_1)/d(1/T) = -E_a/R \qquad (23\text{-}57)$$

where E_a is the apparent activation energy. However, the rate constant is observed to decrease with increasing temperature for oligomers having no G–C pairs (Fig. 23-13). Therefore, the apparent activation energy E_a is negative, ranging from -4 to -9 kcal mole^{-1}. The activation energies for k_{-1} are large and positive. This is reasonable, because base pairs must be broken to disrupt the helix in any mechanism.

Table 23-6

Relaxation kinetics of oligonucleotides (21° to 23°C)

Ordered form	k_1 (M^{-1} sec^{-1})	E_a (kcal mole^{-1})	v (see note)	k_{-1} (sec^{-1})	E_a (kcal mole^{-1})
$A_9 \cdot U_9$	5.3×10^5	-8	3	640	30
$A_{10} \cdot U_{10}$	6.2×10^5	-14	3	175	45
$A_{11} \cdot U_{11}$	5.0×10^5	-12	3	28	53
$A_{14} \cdot U_{14}$	7.2×10^5	-17.5	3	1	75
$A_4 U_4 \cdot A_4 U_4$	1.0×10^6	-6	2–3	3,000	37
$A_5 U_5 \cdot A_5 U_5$	1.8×10^6	-4	2–3	150	50
$A_6 U_6 \cdot A_6 U_6$	1.5×10^6	-3	2–3	8	60
$A_7 U_7 \cdot A_7 U_7$	8.0×10^5	$+5$	2–3	0.8	65
$A_2 GCU_2 \cdot A_2 GCU_2$	1.6×10^6	$+3$	1–2	450	33
$A_3 GCU_3 \cdot A_3 GCU_3$	7.5×10^5	$+7$	1–2	3	50
$A_4 GCU_4 \cdot A_4 GCU_4$	1.3×10^5	$+8$	1–2	1.5	26
$A_5 G_2 \cdot C_2 U_5$	4.4×10^6	$+7$	1–2	340	43
$A_4 G_3 \cdot C_3 U_4$	4.2×10^6	$+9$	1–2	5	44

NOTE: The parameter v is the length of duplex that corresponds to the rate-limiting step in helix formation.

SOURCE: After D. Riesner and R. Römer, in *Physico-Chemical Properties of Nucleic Acids*, vol. 2, ed. J. Duchesne (London: Academic Press, 1973).

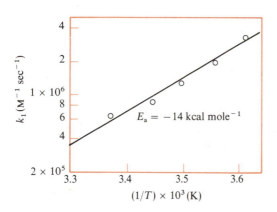

Figure 23-13

Arrhenius plot of the overall rate constant k_1 for duplex formation as a function of temperature. The positive slope implies that the apparent activation energy for the reaction is negative. [After D. Pörschke, unpublished Ph.D. thesis, Univ. of Braunschweig (1968).]

Analysis of oligonucleotide base-pairing kinetics

A simple elementary kinetic step cannot have an activation energy less than zero. Therefore, k_1 (and hence also k_{-1}) must represent composites of rate constants for individual steps. The simplest kind of model to explain this result allows a rapid preequilibrium of single strands to form a partial helix with v base pairs. The reaction from $v \to v + 1$ base pairs then is set as the rate-determining step. When we discussed equilibrium melting earlier, we defined κ as the formation constant of the first base pair, and allowed each subsequent base pair to have the formation constant s. However, it is easy to see that this simple assignment is not consistent with the kinetic results.

The formation of the first base pair is ruled out as rate-limiting by the negative activation energy. The first step that could be rate-limiting is the formation of the second base pair. If this is the case, and formation of the first base pair is a rapid equilibrium, then the observed forward rate should be[§]

$$k_1 = (k_{01}/k_{10})k_{12} = \kappa k_{12} \tag{23-58a}$$

The activation energy then will be

$$d(\ln k_1)/d(1/T) = d(\ln \kappa)/d(1/T) + d(\ln k_{12})/d(1/T) = \Delta H_{nuc} + E_{a,k_{12}} \tag{23-58b}$$

where ΔH_{nuc} is the enthalpy of nucleation of the first base pair, and $E_{a,k_{12}}$ is the activation energy for forming the second base pair. As described earlier, the nucleation enthalpy is at most a small negative value. The activation energy for step k_{12} cannot be less than zero and probably is slightly positive, as is true for most elementary reactions. Therefore, the kinetic scheme implied by Equation 23-58a cannot account for the large negative activation enthalpy seen experimentally for k_1.

The next scheme one can consider allows the rate-determining step to be the formation of the third base pair. One describes the kinetics by allowing a fast equilibrium to form the state with two base pairs. Then,

$$k_1 = (k_{01}/k_{10})(k_{12}/k_{21})k_{23} = \kappa s k_{23} \tag{23-58c}$$

In this case, the activation energy is $\Delta H_{nuc} + \Delta H_{gr} + E_{a,k_2}$. Because ΔH_{gr} is a large negative enthalpy, Equation 23-58c is sufficient to account for the observed negative activation energy. However, the form for Equation 23-58c leads to a contradiction with any all-or-none equilibrium thermodynamics.

If each step after the original nucleation is governed by the same equilibrium constant s, known to be very large from measured free energies of stacking, then

$$s = k_{12}/k_{21} = k_{23}/k_{32} \gg 1 \tag{23-59}$$

Now, if all the base pairs after the first were equivalent, one should also have $k_{12} = k_{23} = k_{34} = \cdots$, and therefore $k_{23} \gg k_{21}$, $k_{34} \gg k_{32}$, etc. But this would mean that the state with two or three base pairs cannot be rate-limiting, because it would rapidly zip up to form more base pairs. The contradiction can be eliminated if one allows

---

[§] More exact treatments yield $k_1 = \kappa k_{12}(1 - k_{21}/k_{12})$. The extra term is necessary to ensure that at T_m the observed forward rate must be zero. See D. M. Crothers, N. Davidson, and N. R. Kallenbach, *J. Am. Chem. Soc.* 90:3560 (1968).

individual nucleation parameters for at least the first two base pairs, κ_1 and κ_2. Then,

$$k_1 = \kappa_1 \kappa_2 k_{23} \qquad (23\text{-}60)$$

This would mean that the nucleation parameter determined by melting-equilibrium measurements actually is $\kappa_1 \kappa_2 / s$. This illustrates the common result that kinetic measurements may reveal information about intermediate steps that was not visible in equilibrium studies.

Oligomers containing G–C pairs show positive activation energies for the bimolecular relaxation step described by k_1. Therefore, one is not so constrained in this case about postulating complex initiation processes. Nevertheless, a detailed consideration of the data indicates that it is unlikely for the formation of the first base pair to be rate-limiting. It seems more plausible, from the actual rates and activation energies involved, that the first base pair exists in a rapid preequilibrium, that the rate-limiting step is the formation of the second base pair. It is fair to ask how one knows where to stop. Why not postulate that formation of the fourth, fifth, or any other base pair is rate-limiting? Here, the activation energies help out because, the longer the preequilibrium region, the more negative the total E_a must become. For oligomers containing A–U, a rate-limiting step of formation of the fourth base pair would result in activation energies too negative to be consistent with available thermodynamic data.

General expressions for k_1 and k_{-1} where the nucleation must be spread over many base pairs have been given by M. E. Craig et al. (1971). We shall not reproduce these results here. However, one general conclusion is that, for the all-or-none model, forward rates for homologous series of oligomers are predicted to increase linearly with the total length. Available data are not consistent with this prediction, although much of the experimental work may not be accurate enough for a valid test. It will not be a surprise if the all-or-none model for oligonucleotides has to be discarded in the future.

The rates and activation energies observed for helix formation (Table 23-6) are largely independent of chain length. In contrast, the rate of helix disruption decreases dramatically with increasing length for most of the oligonucleotide samples studied thus far. The activation energy for melting also is strongly dependent on chain length, and is around 5 to 6 kcal per base pair for almost all of the data in Table 23-6. This value is slightly smaller than the known enthalpy for base-pair breaking. Thus, one can conclude that the rate-limiting step in dissociation of a short helix is the breaking of most of the base pairs, presumably to generate the fast-equilibrating nucleation structure seen in strand combination. Oligonucleotide relaxation kinetic measurements typically are performed at total strand concentrations of about 10^{-6} M. Thus, as you can see from Table 23-6 and Equation 23-55b or 23-56b, the helix dissociation rate dominates the kinetics. The observed relaxation times fall in the range of 1 msec to 1 sec.

Hairpin formation

It is interesting to compare the preceding results with the intramolecular kinetics of hairpin-loop formation. This process will have a concentration-independent relaxation time. If we write the reaction as

$$S \underset{k_{-1}}{\overset{k_1}{\rightleftharpoons}} H \qquad (23\text{-}61a)$$

where S is the single-strand form, and H is the hairpin, then the relaxation time is determined by

$$1/\tau = k_1 + k_{-1} \qquad (23\text{-}61b)$$

Relaxation times at temperatures near T_m are found to be in the range of 10^{-5} sec for a variety of hairpins. This is about two orders of magnitude faster than equilibria between complementary single strands. You can see from the form of Equation 23-61b that it is impossible to determine both k_1 and k_{-1} independently by kinetic measurements alone. The equilibrium constant under the conditions of the relaxation measurement, if also known, yields k_1/k_{-1}. The individual rate constants can be calculated by combining both sets of data. However, the situation is especially simple at T_m, where $k_1 = k_{-1}$, so that the relaxation measurement yields both rate constants.

For the hairpin $A_4GC_5U_5$, one finds $k_1 = k_{-1} \cong 5 \times 10^4$ sec^{-1} at T_m. Similar values are found for a few hairpin loops isolated from natural tRNAs. We can rationalize these results from the previous data on linear complexes. Because the base-paired section of a typical loop is much shorter than a typical model duplex, the faster dissociation rate is a natural extrapolation of the trends shown in Table 23-6. We can estimate the hairpin helix-formation rate by assuming that the limiting step is the same for hairpins and duplexes. Ignoring detailed considerations about the energetics of bending the backbone into a loop, we can imagine the rapid preequilibrium between the two ends of the hairpin prior to the rate-limiting step. A crude estimate shows that the concentration of one end of the oligomer in the presence of the other is on the order of 0.1 м. Therefore, a rough estimate of the intramolecular helix-formation rate constant is $0.1\,k_1$ sec^{-1}, where k_1 is the second-order rate constant of intermolecular helix formation. Using the values shown in Table 23-6, one predicts an intramolecular k_1 of about 10^4 to 10^5 sec^{-1}. This predicted value falls just in the range of observed values. Thus, there is no mystery why the intramolecular relaxation rates are so much faster than intermolecular rates; both reflect the same fundamental kinetic steps.

DNA melting

The thermodynamics of melting of natural DNAs is much more complex than that of model oligonucleotides. It follows that the kinetics of DNA denaturation and

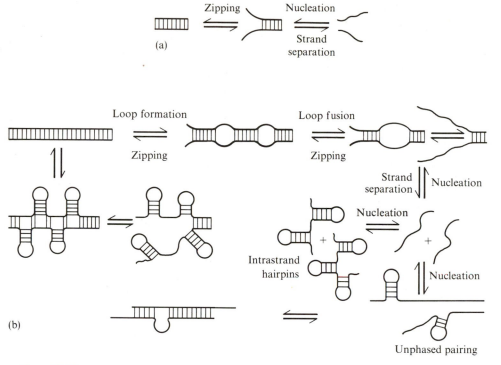

Figure 23-14

States that must be considered during denaturation and renaturation. (**a**) For a short polynucleotide duplex. (**b**) For a natural double-stranded DNA.

renaturation is also more complex. In fact, some aspects still are not well understood, despite intensive study of the phenomena for more than 15 years. Figure 23-14 provides a glimpse into the kind of complications expected. In a short oligomer, nucleation and helix zipping (or unzipping) are sufficient descriptions of most of the observed results. For the denaturation of a DNA, one must also consider loop formation and loop fusion, the results of long-range base-composition heterogeneity. With a short oligomer, melting is strictly reversible. This is not true for model poly-nucleotides. Their optical properties show reversible melting, but their hydrodynamics show hysteresis (Chapter 22).

For most natural DNAs, the problem is even worse because even an optical melting experiment, as typically performed, shows marked hysteresis (Fig. 23-15). This hysteresis is due to kinetic effects on the renaturation step. Figure 23-14 shows some of the causes. When separated DNA strands originally at a temperature above T_m are cooled below T_m, strand recombination can occur. This is a slow process because it must be a second-order reaction, and the concentrations of the individual strands are minute for high-molecular-weight DNA. A competing process is the formation of intramolecular hairpins within each of the separated single strands.

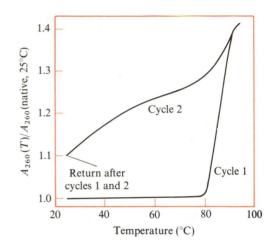

Figure 23-15

Hysteresis in the melting of calf thymus DNA in 0.15 M NaCl, pH 7. The first heating cycle shows a sharp melting. Rapid cooling fails to recover all of the original hypochromism. The amount recovered can be reproduced in subsequent cycles, providing that the cooling rate is the same. The broad remelting shows the presence of many imperfectly paired regions in the once-denatured DNA. [After R. F. Steiner and R. F. Beers, *Polynucleotides* (New York: American Elsevier, 1961).

These hairpins are not thermodynamically as stable as the native DNA duplex, but the rates of the intramolecular reactions at low concentration are orders of magnitude faster than interstrand nucleation. Once intrastrand hairpins form, the single-strand DNA coil is partially collapsed, and it can be very difficult for two such collapsed structures to interpenetrate. Thus, even though the two separated strands are capable of base pairing and forming a nucleus that eventually zips back to the native form, the rate of renaturation can be very slow.

The extent of renaturation observed in practice will be a function of the experimental protocol and of the nature of the DNA. If the temperature is dropped suddenly from far above T_m to far below, the separated complementary strands have no chance of finding one another. Intrastrand secondary structure, once formed, is very stable; thus, reformation of native DNA is effectively blocked completely by this procedure. Such samples traditionally are called denatured DNA. Do not fall into the common trap of considering such samples to be single-stranded. It is their double-stranded sections that prevent renaturation. If, instead of fast cooling, the DNA is slowly annealed by lowering the temperature gradually or by holding it near T_m for a long time, then intrastrand hairpins do form, but these are not very stable. Eventually, correct nuclei form, and these zip back to the native structure as the temperature is further lowered.

Additional complications are possible, however (Fig. 23-14). Suppose the DNA has repeated sequences, as is common for most eukaryotic genomes. Then a nucleus can form near T_m that is every bit as stable as the correct nucleus, except that the two strands are not correctly in phase. As cooling proceeds, a duplex will form containing single-stranded loops and ends. This structure should eventually reconvert to the native form but, if the repeated section is long, the activation energy for breaking it may be so high that complete renaturation takes forever.

From the discussion thus far, it is clear that a complete relaxation kinetic description of DNA melting is a formidable task. We shall not attempt it here. Instead, we focus on two different aspects of DNA melting kinetics: the kinetics of

DNA helix disruption prior to strand separation, and the kinetics of nucleation (that is, initial strand recombination) at temperatures not too far below T_m. Each of these problems should prove reasonably tractable in the right limiting cases.

DNA melting prior to strand separation

First, we need a criterion to distinguish whether or not the strands have separated in a given sample. Hydrodynamics would be the definitive method, but it is easier to use absorbance, based on the kinetic observations just discussed. If a DNA duplex is heated to a temperature near T_m or only slightly above, the strands should not separate, because a few regions of high G–C content will remain unmelted. Therefore, upon quick recooling, rapid zipping between the correctly placed single strands will occur, and all the original hypochromicity should be regained. Higher incubation temperatures start to break these last duplex regions, and strands can separate. Then, rapid recooling can produce little or no hypochromicity.

This "irreversible melting" is not significant for a typical DNA until more than three-fourths of the reversible melting has occurred (Fig. 23-16). It is a good simple

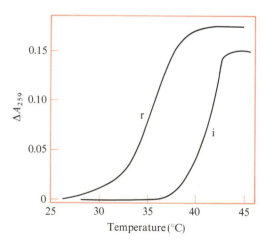

Figure 23-16

Absorbance assay of strand separation upon melting of T2 DNA in 7.2 M NaClO$_4$. Curve r shows the loss in hyperchromism upon heating DNA to the temperature indicated. Curve i shows the irreversible loss in hyperchromism upon heating DNA from 25°C to the indicated temperature and then rapidly cooling to 25°C. [After E. P. Geiduschek, *J. Mol. Biol.* 4:467 (1962).]

assay for strand separation. In principle, the strand-separation temperature should not be reached until all available hypochromism in the reversible melting curve is eliminated. This is because even one short, uniquely stable duplex region could hold the strands in place and yet make only a minute contribution to the absorbance. However, most DNA samples contain single-strand breaks. These arise as artifacts of handling in some DNAs, and they appear naturally in at least a few native DNAs. Because of these breaks, DNA pieces are released progressively during melting. This phenomenon probably accounts for the observed breadth of the irreversible melting curve.

The results shown in Figure 23-16 suggest that double-helix unzipping should be studied by a rapid temperature jump that brings the sample to a final equilibrium loss of hypochromism up to (but no more than) 75%. If this were done for an oligomer duplex with strand separation prevented (consider a hairpin as a reasonable model), it should show a single concentration-independent relaxation time. With native DNA, however, the kinetics of unzipping prove to be surprisingly complicated (Fig. 23-17). First of all, note how slow the rates of unwinding are. For intramolecular hairpin melting near T_m, the relaxation times are around 10^{-5} sec. In contrast, with a typical DNA sample, although the data cannot be fit to a single relaxation time, the time scale over which the reaction occurs is 10^2 to 10^3 sec. We can make a rough estimate from the model-oligomer data of how long it should take a DNA to melt. Suppose that the same helix-unzipping reaction is rate-limiting, and suppose that all unzipping must occur from one single loop. This will provide a maximal unwinding time for DNA, because in practice we know that more than one loop must originally form. A large viral DNA such as T2 DNA has 2×10^5 base pairs. If the original loop were to propagate through the entire DNA irreversibly at the oligomer rate, 10^{-5} sec per five base pairs, it should take only 10^{-1} sec to melt the DNA. With multiple nucleation, the time could be one to two orders of magnitude smaller. Thus we must account for a discrepancy of 10^3-fold to 10^5-fold between predicted and observed melting times.

M. T. Record and B. H. Zimm (1972) have suggested that we view the complex kinetics of Figure 23-17 as being governed by an unwinding rate parameter that varies with time, rather than make what could be a rather arbitrary fit to a number

Figure 23-17

Kinetics of T2 DNA denaturation in 0.08 M Na$^+$, 80% formamide. Three samples were heated to produce final hypochromism losses of 25% (□), 53% (●), and 95% (○) of the total hypochromism of the native DNA. [After M. T. Record and B. H. Zimm, *Biopolymers* 11:1435 (1972).]

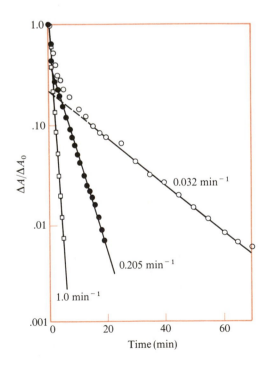

of single-exponential decays. This rate parameter has the following properties. It becomes a constant, k_∞, at long times. Then, presumably, loop nucleation is over so that rates should reflect primarily the unzipping reaction. The long-time rate k_∞ is about 20-fold slower than the initial rates. It is much slower for a temperature jump that results in mostly denatured DNA than for a small temperature perturbation causing only 25% to 50% denaturation. Furthermore, for a fixed final degree of unwinding, k_∞ is inversely proportional to the molecular weight of the DNA. These last two results are most critical, because they imply that the intrinsic rate of breaking a base pair somehow is strongly dependent on the length of the DNA and on whether other sections of the DNA are double helix or coil.

Frictional effects on DNA melting

The physical processes that must be involved in DNA denaturation provide the clues to understanding all of these strange kinetics. Most loop nucleation will occur in the middle of the DNA, because only rarely would you expect high-A–T regions to be at the ends. In order for a loop to grow, base pairs must be broken. Because DNA is a double helix, this can occur only if one end of the growing loop rotates relative to the other (Fig. 23-18). The rotation that accompanies unzipping must

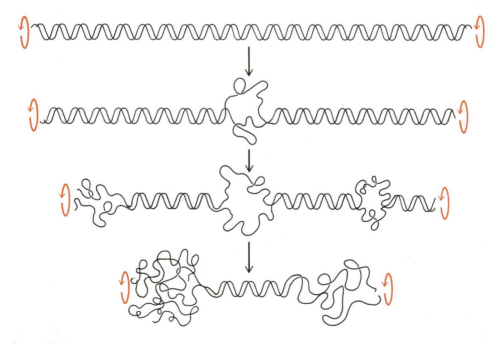

Figure 23-18

Friction-limited unwinding of DNA. As denaturation proceeds, coiled regions form. These coiled regions have greater resistance to rotation around the helix axis of the remaining duplex regions.

occur around the helix axis. At early times (or early degrees in the melting process), rotation should be relatively easy. The driving force is the ΔG for the base-pair breaking. The resisting force is the rotational frictional drag on the DNA, but this is small for a long rod rotating around its long axis. Initially, it is estimated that helix-rotation rates can be as fast as 10^4 rpm. However, as melting proceeds, bulky coils replace the rather streamlined helix. Their rotational friction is much larger, and so the rate of unzipping must slow down. The increasingly large friction effects will be the rate-limiting step throughout the rest of the course of denaturation. As a result, rates will slow until a final terminal value k_∞ is reached for the last small fraction of base pairs to melt. In parallel experiments, this k_∞ will decrease as the final extent of melting increases, because more and larger coils will be present. The marked decrease in k_∞ with increasing molecular weight can be explained in the same way. One piece of evidence supporting the concept of the frictional rate-limiting step is that most workers find unwinding rates to be viscosity dependent.

If a DNA is subjected to a sufficiently strong denaturing force (such as high pH), all base pairs are broken very rapidly, in less than 10^{-2} sec. However, strand separation and unwinding have not occurred in this short time, because lowering the pH still leads to rapid renaturation. Times longer by orders of magnitude are needed for the strands to unwind. One can guess that the initial alkali-denatured form contains tightly twisted nucleic acid single strands. Friction-limited rotation still must occur in order for these strands to unwind and separate. The interested reader can look elsewhere for the details of how the frictional effects are used to predict the total time it takes to unzip a helix (see references cited by Record and Zimm, 1972). A number of different approaches have been used, and all predict total unwinding times that increase with at least the square of the molecular weight. Experimental values range from a linear to a square dependence.

This unwinding time would appear to pose a problem for the replication of large DNAs, because that process also must be accompanied by unwinding. There is evidence that nature has solved this problem by creating nicking–closing enzymes and gyrases. These enzymes act by binding to the DNA and transiently breaking the phosphodiester bond, allowing rapid independent motion of a local region (nicking–closing enzymes) or an energy-driven unwinding (gyrases) before the bond closes again. The necessity for such drastic solutions will become even more apparent when we consider the special properties of double-stranded DNA circles.

Renaturation of complementary DNA strands

Now, having seen how hard it is to take DNA strands apart, we focus on the even harder problem of getting them back together. There are three formidable obstacles: the frictional considerations cited above, intrastrand secondary structure, and the complexities of two coils interpenetrating and finding each other's complementary sequences for nucleation. Experimentally, renaturation rates are found to be maximal about 25°C below T_m. If the DNA is kept within this temperature range, the kinetic effects of intramolecular hairpin formation are minimized because most short hairpins are unstable at such temperatures. The hydrodynamic complications can be effectively eliminated if the DNA is broken into small pieces. This can be done enzymatically, by sonication or other shear forces, or by severe alkali treatment. Generally, it is assumed that these pieces are all of the same length. If necessary, it is possible to modify the following treatment to account for distributions of lengths (or for even more complex situations).

For relatively short DNA pieces (say, 10^3 base pairs) at very low concentrations, it is reasonable to suppose that the nucleation step is rate-limiting in DNA renaturation. We shall assume that, once nucleated, each duplex zips to completion instantly (Fig. 23-19). Suppose that A and B are separated complementary strands at concentrations C_A and C_B. They are mixed at time $t = 0$. The kinetics of production of double-helical A·B complexes will be second-order:

$$dC_d/dt = k_2 C_A C_B \tag{23-62}$$

where C_d is the concentration of double-stranded complexes.

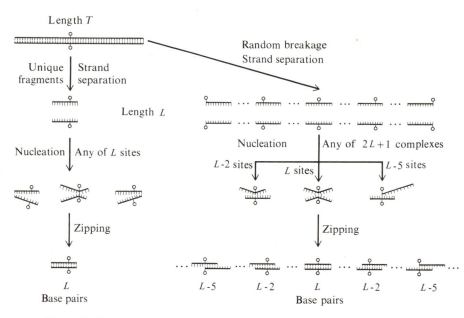

Figure 23-19

Renaturation of small DNA fragments. If fragmentation occurred at unique sites, a given base pair (*flagged*) would always renature into completely paired duplexes. Random fragmentation means that a given base pair can be located at any position in a fragment of length L. This fragment can be involved in any of $2L + 1$ different nucleation complexes, but only one of these complexes is capable of forming all L base pairs. See Box 23-1.

We first explore the experimental implications of Equation 23-62 and then return to make some molecular interpretations of k_2. Suppose that A and B come from an original DNA (T base pairs long), sheared to give unique small pieces including A and B. The original DNA had $2T$ nucleotides. One can easily measure C_0, the total concentration of nucleotides in a solution of this DNA. Therefore, the concentration of the original DNA molecules is $C_0/2T$. This is also the concentration of each unique sheared double-stranded fragment of the DNA—providing that there are no repeated sequences and no heterogeneity, and that the shearing produced unique, nonoverlapping fragments. In general, however, the concentration of particular double-stranded fragments will be $C_{Tot} = nC_0/2T$, where n measures heterogeneity (<1) and repeated sequences (>1) in the original DNA.

Equation 23-62 can be simplified by explicitly including mass conservation for each particular fragment. At any time, the fragment can be either intact duplex or separated single strands. If $C_s(t)$ is the concentration of one of the single strands, and $C_d(t)$ is the double-strand concentration, then

$$C_{Tot} = nC_0/2T = C_s(t) + C_d(t) \tag{23-63}$$

Because we started with separated complementary strands at equal concentrations,

$$C_s(t) = C_A = C_B = nC_0/2T - C_d(t) \tag{23-64}$$

Using this relationship, we can rewrite Equation 23-62 as

$$-dC_s/dt = k_2 C_s^2 \tag{23-65}$$

Integration from time 0 to time t yields

$$[1/C_s(t)] - [1/C_s(0)] = k_2 t \tag{23-66a}$$

or

$$C_s(t)/C_s(0) = 1/[1 + k_2 t C_s(0)] \tag{23-66b}$$

With the boundary condition that there is no helix at $t = 0$, from Equation 23-63 we find $C_s(0) = nC_0/2T$. Now recognizing that $C_s(t)/C_s(0)$ is just the fraction f_s of strands remaining single as a function of time, we have

$$f_s(t) = C_s(t)/C_{Tot} = 1/(1 + nC_0 k_2 t/2T) \tag{23-67}$$

Note that the time at which $f_s = 0.5$ (half-"$C_0 t$") is $(C_0 t)_{1/2} = 2T/nk_2$.

Genome size and repeating sequences

The kinetic equation for renaturation (Eqn. 23-67) has a number of experimental implications. Suppose that there is no heterogeneity and no repeated sequences. Then $(C_0t)_{1/2}$ is directly proportional to the molecular weight of the original unbroken DNA. In fact, even if the original DNA had been in several pieces, it would not affect the final result, because all samples are sheared to small lengths before renaturation anyway. This means that a measurement of $(C_0t)_{1/2}$ can yield the molecular weight of a genome, provided that the isolated sample contains the whole genome and is homogeneous. In practice, genome sizes vary over an enormous range, from less than 10^4 base pairs for a small virus to more than 10^9 base pairs for a eukaryote. This means that renaturation times at constant nucleotide concentration vary 10^5-fold.

Suppose that a sample contained a mixture of DNAs a and b (of two very different molecular weights) before shearing. If they had no sequences in common, each would renature independently. The observed $f_s(t)$ would simply be

$$f_s(t) = \chi_a/(1 + \chi_a C_0 n_a k_{2a} t/2T_a) + \chi_b/(1 + \chi_b C_0 n_b k_{2b} t/2T_b) \qquad (23\text{-}68)$$

where χ_a and χ_b are the mole fractions of nucleotides present from the two DNAs, and where we have allowed for the possibility that the intrinsic rate constants are different.[§] If T_a and T_b are very different, the renaturation of the two DNAs proceeds on very different time scales. Each can be observed separately in the mixture and, as a result, χ_a and χ_b can be determined experimentally, even if the parameters n, k_2, and T are not known, This is a result of the very narrow time range over which most of a second-order reaction actually occurs. See, for example, the typical renaturation data in Figure 23-20.

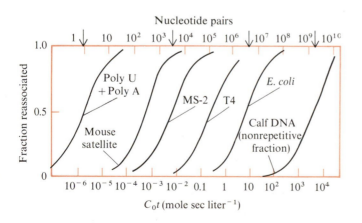

Figure 23-20

Renaturation kinetics of DNAs from different sources. Each sample was sheared to fragments averaging 400 nucleotides in single strands. The abscissa is the total concentration of nucleotides multiplied by the renaturation time. The scale at top shows the complexity X_c of some of the DNA samples. [After R. J. Britten and D. E. Kohne, *Science* 161:529 (1968).]

[§] The intrinsic rate constants would be different, for example, if the two DNAs had different G–C contents. In practice, such effects are small, and corrections for them are easy.

If a DNA contains repeated sequences, these are present at higher concentration than would be expected from fragmentation of a unique sequence. Therefore, renaturation of the repeated sequences proceeds more rapidly, as described by the factor n in Equation 23-67. Notice in Figure 23-20 how much more rapidly poly $A \cdot U$ and a mouse satellite DNA reassociate than do other DNAs. If a homogeneous DNA contains a distribution of different extents of repeated sequences, it is convenient to define a quantity called the kinetic complexity, X_c. This is the length of DNA needed to contain one copy of the entire sequence. For a sequence in which all repeats are of a similar nature, $X_c = T/n$. Equation 23-67 permits experimental determination of X_c, providing that a calibration has been done to evaluate k_2. For a homopolymer pair, $X_c = 1$; for an alternating sequence such as $d(AT) \cdot d(AT)$, $X_c = 2$; for a unique sequence with no repeats, $X_c = T$. Therefore, a renaturation kinetics measurement yields an indication of the average degree of sequence repeats if the genome size is known.

In practice, when the renaturation of a eukaryotic DNA is studied, several kinetic classes usually are observed. Figure 23-21 shows a typical experimental example. A small amount of the DNA renatures so fast that the kinetic course cannot

Figure 23-21

Renaturation kinetics of E. coli DNA (solid curve) and total calf thymus DNA (dashed curve). The latter shows evidence of three classes: unique sequences with C_0t near 10^3; middle-repetitive sequences with C_0t near 10^{-1}; and a small highly repetitive palindromic class with $C_0t < 10^{-3}$. It is the presence of repeated sequences that enables parts of calf thymus DNA to renature more rapidly than E. coli DNA despite the much larger size of the total genome. [After R. J. Britten and D. E. Kohne, Science 161:529 (1968).]

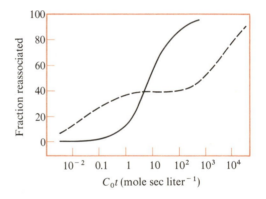

easily be followed. This result corresponds to a very small kinetic complexity ($X_c < 60$) and must represent simple repeating sequences such as satellite DNAs. Also included in this class are DNAs with palindromic sequences (capable of self-complementary loop formation) that always can renature intramolecularly. Then there is a large class of sequences with a complexity of 10^3 to 10^7; this is called middle repetitive and must exist in tens to ten-thousands of copies per eukaryotic genome. Finally, there is unique DNA, where X_c is equal to the genome size. The amounts of each of the classes and the complexities of each are determined by using Equation 23-68 or its multicomponent analogs. Table 23-7 shows some typical results. The dissection into specific classes is not always trivial experimentally. In principle, it should be possible only when the range of complexities represented within one class is far less than the difference in complexities of DNAs in adjacent classes.

Table 23-7

Extent of repeated sequences in some animal genomes

Species and DNA sequence class	Copies per genome	Percentage of genome	X_c
Calf			
Unique	1	55	1.5×10^9
Middle-repetitive	60,000	38	1.7×10^4
Highly repetitive	1,000,000	2	60
Highly repetitive and/or palindromes	——	3	——
Toad			
Unique	1	54	1.6×10^9
Middle-repetitive			
slow	20	6	1.5×10^7
intermediate	1,600	31	6.0×10^5
fast	32,000	6	6.0×10^3
Highly repetitive and/or palindromes	——	3	——
Sea urchin			
Unique	1	38	3.0×10^8
Middle-repetitive			
slow	20–50	25	1.0×10^7
intermediate	250	27	1.0×10^6
fast	6,000	7	1.3×10^4
Highly repetitive and/or palindromes	——	3	——

SOURCE: After E. Davidson and R. Britten, *Quart. Rev. Biol.* 48:565 (1973).

● Significance of the renaturation constant k_2

Now we try to interpret the observed second-order renaturation rate constant k_2. Oligonucleotide studies described earlier indicate that more than one base pair must be formed to nucleate double-helix formation. The total number of nucleation sites on a DNA strand of length T is βT, where β is the nucleation site density—that is, the average number of sites per nucleotide (so that $\beta \leqslant 1$). The total number of distinct nucleation sites is βX_c for DNA with repeating sequences. For separated DNA strands uniquely broken into pieces of length L, at total nucleotide concentration C_0, the original concentration of any one unique nucleation site is $\beta n C_0/2T = \beta C_0/2X_c$. (For randomly broken DNA, see Box 23-1.)

Consider nucleation as the collision of two complementary nucleation sites. The rate of nucleation at one unique site is just $k_n(\beta C_0/2X_c)^2$, where k_n is the nucleation rate constant. However, there are a total of βX_c unique sites. Furthermore, each nucleation leads to the formation of L base pairs, where L is the length of each DNA piece. Therefore, the total initial rate of base-pair formation observed will be

$$dC_d/dt = k_n \beta^3 L (C_0/2X_c)^2 X_c \qquad (23\text{-}69)$$

If you knew nothing about nucleation complexity or length, you would write the initial rate as

$$dC_d/dt = k_2 (C_0/2)^2 \qquad (23\text{-}70)$$

where $C_0/2$ is just the concentration of nucleotides on one strand. Therefore, the observed k_2 is

$$k_2 = k_n \beta^3 L / X_c \qquad (23\text{-}71)$$

Box 23-1   RENATURATION KINETICS OF RANDOMLY BROKEN DNA

The model for renaturation kinetics derived in Equations 23-63 through 23-67 is realistic only for a DNA fragmented by a specific restriction nuclease. All other breakage techniques must give approximately random breaks and, therefore, the solution will contain a population of fragments with overlapping sequences. Figure 23-19 shows the effect of this random breaking on the kinetics.

   With *unique* cuts, each single-strand piece has a concentration $C_0/2T$. Each piece L nucleotides long has L possible nucleation sites, assuming that one base pair can serve to nucleate (just to keep things simple). Suppose that collisions between complementary nucleation sites are characterized by a bimolecular rate constant k_n. The total initial rate of nucleation by a given piece then is $k_n(C_0/2T)^2 L$. Each original DNA strand was cut into T/L fragments. The net rate of nucleation is $k_n(C_0/2T)^2 T$. Each nucleation is followed by rapid chain zipping to produce L base pairs. Therefore, the initial rate of base-pair formation from a particular strand fragment is $k_n(C_0/2T)^2 L^2$. In the whole solution, the net rate of base-pair production is $k_n(C_0/2T)^2 LT$.

   Suppose now that the DNA is cut *randomly* into pieces of length L, and these pieces are separated into single strands. Each particular base of the original sequence can occur on any one of L different single-strand fragments, each of which has a concentration in the solution of $(1/L)(C_0/2T)$. Each particular piece can nucleate with any one of L pieces deriving from the complementary strand. Thus, a total of $2L - 1$ nucleation complexes can be formed (Fig. 23-19). The rate of nucleation for one of these complexes is proportional to n, the number of nucleation sites, where $1 < n < L$; by the same arguments used earlier, the nucleation rate is $k_n(C_0/2LT)^2 n$. The total rate of nucleation for one fragment is thus a sum over all rates of nucleation of the $2L - 1$ possible complexes:

$$k_n(1/L^2)(C_0/2T)^2 \left(L + 2 \sum_{n=1}^{L-1} n \right) = k_n(C_0/2T)^2$$

This expression correctly predicts that the observed initial rate is inversely proportional to the complexity. The reader should be cautious in trying to compare Equations 23-62 through 23-67 with Equations 23-69 through 23-71. Earlier we assumed that all sites have the same nucleation rate and then corrected for multiple sites with the factor n. Now n appears implicitly in our experimental rate constant through the use of the complexity X_c. The nucleation constant for a unique site, k_n, should no longer contain any dependence on complexity, but it could still be length dependent.

Earlier, with oligonucleotide-complex formation, we found that k_n varied somewhat with length, but the data were not accurate enough to pin down this relationship, nor was a sufficient range of lengths available for detailed study. However, J. G. Wetmur and N. Davidson (1968) were able to examine k_2 over a broad range of lengths with sheared DNA. They found that $k_2 \propto L^{1/2}$. This proportionality, combined with Equation 23-71, must mean that $k_n \propto L^{-1/2}$. The dominant reason for this

---

Each nucleation of sequences with n nucleation sites in common rapidly produces n base pairs by zipping. Therefore, the rate of base-pair formation by a particular fragment is

$$k_n(1/L^2)(C_0/2T)^2\left(L^2 + 2\sum_{n=1}^{L-1} n^2\right) = k_n(C_0/2T)^2(2L/3 + 1/3L)$$

Within the entire solution, there are $T - L \cong T$ (for $T \gg L$) different fragments from each original DNA strand. Therefore, the observed net rate of nucleation is $k_n(C_0/2T)^2 T$. The observed net rate of base-pair production should be

$$k_n(C_0/2T)^2 T(2L/3 + 1/3L)$$

You can see that random breakage has no effect at all on the total initial rate of nucleation. That is reasonable because the total concentration of nuclei has not been altered. However, random breakage slows the observed rate of base-pair formation (for large lengths) by the factor

$$\frac{k_n(C_0/2T)^2 T(2L/3 + 1/3L)}{k_n(C_0/2T)^2 LT} = 2/3 + 1/3L^2 \cong 2/3$$

Therefore, Equation 23-67 still holds in the case of random breakage, and renaturation is still predicted to be a simple second-order reaction, but the measured rate constant k_2 is expected to be only 2/3 the value predicted for unique cuts. The decrease occurs because nucleations are less effective at matching long base-paired sequences in the randomly broken fragments.

relationship appears to be excluded-volume effects. To nucleate, two coils must interpenetrate so that complementary base pairs can be juxtaposed. As the lengths become larger, more and more of the base pairs of the single strand become inaccessible for nucleation, effectively cutting down the number of available nucleation sites. Wetmur and Davidson show that this effect can account for the proportionality of k_n to $L^{-1/2}$.

Another effect that must be considered is the rate of zipping. As lengths become larger, the rate of zipping no longer can be considered negligible compared with the nucleation rate. Therefore, our assumption that each nucleation produces L base pairs is wrong for long pieces. Per unit of time, less than L base pairs will be formed, because zipping will lag behind each nucleation. Therefore, even if excluded-volume effects are ignored, k_2 still would depend on less than the first power of the fragment length.

In actuality, sheared DNA is not of a single length L, but is a distribution of lengths. Therefore, the arguments we have just used indicate that we should not observe pure second-order kinetics of base-pair formation, because the rate constant is not a constant. Particularly serious will be any fragments long enough to have appreciable intrastrand secondary structure. This puts us right back amidst all the complications in Figure 23-14. Happily, there is a way out. The complications are real, and a technique (such as absorbance) that monitors total base-pair formation does not yield simple second-order kinetics, even for pure unique-sequence DNA. The solution is to measure nucleation rather than base-pair formation. This technique eliminates concerns about the helix growth rate. It also eliminates the explicit length dependence of k_2, as we have seen. In practice, some length dependence of k_n may still remain.

Separation procedures

The most popular technique for nucleation measurements (indeed, for all renaturation measurements) is hydroxyapatite chromatography. For reasons that remain largely mysterious, DNAs with even a short duplex region (50 base pairs and perhaps considerably less) will stick to hydroxyapatite under conditions where single strands are completely eluted. Therefore, simply measuring the amount of DNA phosphate retained yields the amount of DNA present in a nucleated duplex. These measurements are found to obey pure second-order kinetics, in agreement with the equations derived in the preceding discussion.

The use of hydroxyapatite (or other physical separation procedures) opens up a whole new range of renaturation experiments. Absorbance can detect only the total amount of renaturation; any very minor species can make no detectable contribution. However, if radioactively labeled DNA is used, separation procedures permit study of the renaturation of trace components. A trace amount of a known pure sequence can be added as a probe of a complex system, or an excess of a known pure sequence can be added to drive the renaturation of a trace amount of some component.

Applications of renaturation kinetics

A beautiful example of the power of renaturation kinetics is its application in the search for fragments of a viral genome present in viral-transformed tissue culture cells. The question of interest was whether the entire adenovirus genome is present in these tissue cells. A sample of ^{32}P-labeled adenovirus DNA was prepared and cut with a restriction nuclease into six unique, nonoverlapping fragments. Each of these pieces was purified, mixed separately with whole transformed rat DNA, and sheared; then the renaturation kinetics were measured. From Equation 23-67, the observed half-life $t_{p,1/2}$ for renaturation of a purified viral DNA fragment alone at a total mono-nucleotide concentration C_p is $t_{p,1/2} = 1/K_0 C_p$, where K_0 is a function of the length and complexity of that particular fragment. Equation 23-67 for the fraction of single strand remaining at time t can then be rewritten as

$$f_s^{-1} = 1 + t/t_{p,1/2} \tag{23-72}$$

When the probe is mixed with whole cell DNA, if any copies of the probe DNA are present in the cellular DNA, the total concentration is $C_T = (1 + G)C_p$, where G is the molar ratio of viral sequences in the cellular DNA to those in the added probe. Then the time course of renaturation is (from Eqn. 23-67)

$$f_s^{-1} = 1 + (1 + G)t/t_{p,1/2} \tag{23-73}$$

Thus, G can be determined. If the total amount of cell DNA is known, then G can be converted to the number of copies of viral DNA fragment per cell genome. Figure 23-22 shows some typical data. You can see that fragment B is not present in the cell DNA, whereas fragment C is present. When all the restriction fragments were examined, three of the six were found to exist in about 1.5 copies each per host genome; one was present in less than one copy per genome, and two were completely absent. Experiments such as this should lead to identication of the specific viral DNA regions that carry the information responsible for transformation.

Another typical application of renaturation kinetics is the exploration of the arrangement of repeated sequences in the DNA genome. Consider two extreme models (Fig. 23-23a). Each contains the same ratio of repeated and single-copy DNA sequences. In one case, short regions of the two classes are interspersed. In the other, the two classes are completely segregated. Suppose that renaturation is carried out on samples sheared to lengths much smaller than the length of one repeat. Conditions are selected so that $C_0 t$ is large enough that most of the repeated sequences have reassociated, but almost all the unique sequences are still single-stranded. Both models for the sequence distribution predict the same amount of DNA retained on hydroxy-apatite. Suppose, instead, that the DNA is sheared to lengths quite a bit longer than the length of the repeating sequences. Then, if the interspersed model is correct, much single-copy DNA will be on fragments containing repeated sequences. It thus will renature with the much faster kinetics of that class.

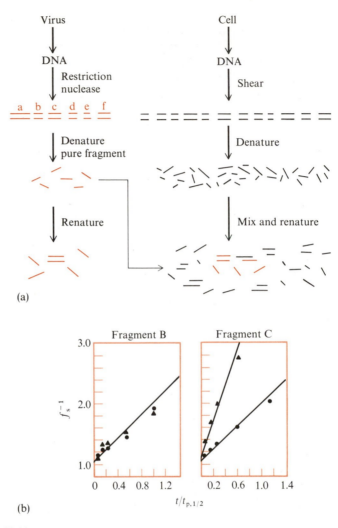

(a)

(b)

Figure 23-22

Use of renaturation kinetics to assay the presence of adenovirus 2 DNA sequences in eukaryotic cells. (**a**) Schematic illustration of the technique. A vast excess of cell DNA is mixed with a small aliquot of a radio-labeled viral DNA restriction nuclease fragment. If viral sequences are present in the cell DNA, the overall concentration of sequences complementary to the unique fragment is increased. Therefore, radio-label is observed to renature more rapidly. In essence, cold DNA drives the renaturation of the labeled DNA. (**b**) Experimental results for two adenovirus DNA fragments mixed with viral transformed cells (▲) or untransformed control cells (●). [After S. J. Flint et al., *J. Mol. Biol.* 96:47 (1975).]

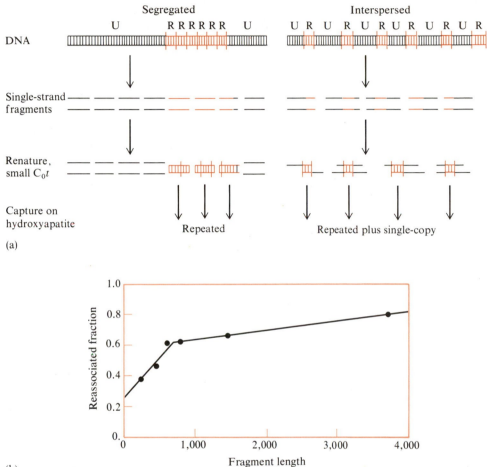

(a)

(b)

Figure 23-23

Use of renaturation kinetics to examine the pattern of unique and repeated sequences in the eukaryotic genome. (**a**) Schematic random fragmentation and reassociation for two extreme models of sequence organization. Repeated sequences are shown in color. At a renaturation C_0t such that only repeated-sequence duplexes can form, the interspersed model predicts that renatured duplexes will contain most or all sequences present in the genome when the size of the DNA fragments becomes somewhat larger than the repeat length. The segregated model predicts that few unique sequences will be found as renatured duplex under these conditions. (**b**) A fraction of *Xenopus laevis* DNA was reassociated at $C_0t = 50$; the reassociated fraction is plotted as a function of fragment length. These results support the interspersed model. [After E. Davidson and R. Britten, *Quart. Rev. Biol.* 48:565 (1973).]

A careful analysis of the amount of DNA renatured as a function of fragment length actually can yield the average lengths of the repeated and single-copy regions. For example, the results in Figure 23-23b show that about 65% of the single-copy DNA contains repeating sequences spaced on the average about 700 nucleotides apart. Another 20% of the DNA has repeated sequences spaced about 4,000 nucleotides apart. This type of information is proving extremely useful in the analysis of eukaryotic genomes. It is a fine example of how work that begins as fairly pure physical chemistry can rapidly become of great importance in attacking biological problems.

Branch migration

Thus far, we have considered only the simplest kinetic steps involving base pairing and base stacking. More complex structure interconversions are known, and one set of processes that has received considerable attention is branch migration (Fig. 23-24).

Single-strand branch migration presumably is involved in the replacement of newly synthesized RNA by a strand of DNA during transcription. It has been observed experimentally in renatured nicked circular DNAs formed by molecules with circularly permuted sequences. The location of a branch will execute a random walk until strands separate or until some particularly stable intermediate structure is formed. The rate of migration is estimated to be faster than 1,000 base pairs per second.

Double-strand branch migration is a more elaborate process, because two

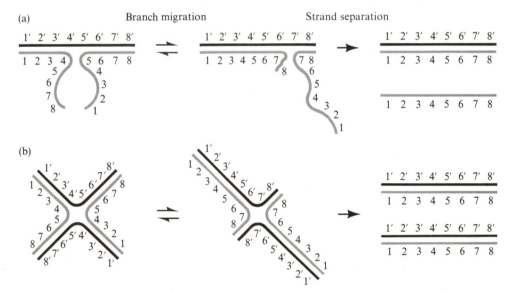

Figure 23-24

Schematic models of branch migration.   (a) Single-strand model.   (b) Double-strand model.   Numbers and primed numbers represent complementary sequences.

separate base pairs (e.g., 5–5 and 5′–5′ in Fig. 23-24b) must simultaneously break before others can reform (e.g., 5′–5 and 5–5′). The rate of double-strand branch migration recently has been measured by direct electron microscopic observation of the disappearance of X-shaped intermediates, prepared at lower temperature and then allowed to equilibrate at higher temperatures. The location of the branch point fluctuates according to a random walk until it reaches the end of the duplexes, where irreversible separation of the duplexes occurs. Although only the latter process can actually be measured reliably, the rate of the migration can be estimated by a computer simulation of the random walk. The result is a rate of branch-point motion of 6,000 base pairs per second at 37°C, and about 1 base pair per second at 0°C. Double-strand branch migration is believed to play a role in genetic recombination.

23-3 BINDING OF SMALLER MOLECULES TO NUCLEIC ACIDS

In Chapter 22 we use thermodynamic arguments to show how the binding of small molecules can alter the equilibrium between single- and double-stranded conformations. Quite a few different classes of small molecules profoundly affect the biological properties of DNA or RNA, even when they are present at concentrations too low to cause overall changes in strandedness. These small molecules include mutagens and carcinogens (such as aminoacridines and polycyclic aromatic hydrocarbons), antibiotics (such as actinomycin D), a variety of planar dyes (some of which, such as ethidium, find use as drugs), and numerous heavy metals (some of which bind to phosphates and some to the bases). Figure 23-25 shows the chemical structures of a few of the best-studied examples.

In addition to small molecules, there are many individual macromolecules that can bind to nucleic acids. These include proteins responsible for packaging DNA into chromatin or RNA into ribosomes, proteins involved in the replication or transcription of DNA, and proteins involved in the regulation of nucleic acid function. Table 23-8 summarizes the properties of some of these proteins; they have surprisingly few common features. Perhaps only a portion of each is involved in nucleic acid binding. Once we know the structures of these binding portions, generalizations about nucleic acid interaction regions may be more evident.

Here we want to formulate the general principles and approaches needed to describe the interaction of a set of identical or equivalent small molecules with one linear region of a nucleic acid. The detailed treatment we shall restrict to molecules that interact with only a few bases at a time. However, the questions of interest and the phenomena expected should apply to larger ligands.

When a small molecule binds to a nucleic acid, some of the things one would like to know are the following.

1. Is there preferential binding to RNA or DNA? to double strands or single strands?

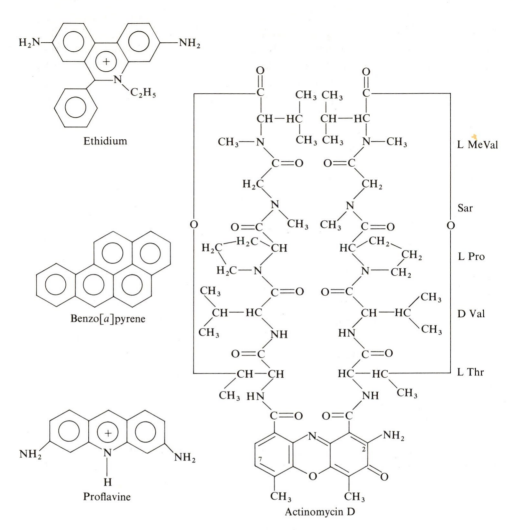

Figure 23-25

Structures of four of the many small molecules believed to bind to double-stranded DNA by intercalation. Sar = sarcosine; MeVal = methylvaline.

Table 23-8
Properties of some proteins known to bind to nucleic acids

Protein	Molecular weight (d)	Subunits	Isoelectric point	Strand specificity§	Sequence specificity	Length of region bound	Association constant (M^{-1}, at ~0.05 M Na^+)	Cooperativity (ζ in Fig. 23-27)
lac Repressor	150,000	α_4	~6	d.s. DNA	Very high	13	10^{14}	1
λ Repressor	26,000–104,000	$\alpha_1 \rightleftarrows \alpha_2 \rightleftarrows \alpha_4$	—	d.s. DNA	Very high	~17	—	?
T4 gene-32 protein	35,000	1	5.5	s.s. DNA	None	~10	10^6	1,000
E. coli RNA polymerase	480,000	$\alpha_2\beta\beta'\sigma$	—	d.s. DNA best	Moderately high	~40	10^{13}	?
Histones	11,000–22,000	1	~14	—	None	~20	—	?
Ribosomal protein S4	25,000	1	10.4	rRNA	Very high	>200	—	1
Aminoacyl tRNA synthases	70,000–200,000	α, α_2, α_4 or $\alpha_2\beta_2$	—	tRNA	Moderately high	40?	~10^7	?
Ribosomal protein S1	68,000	1	5.0	Site I: s.s. DNA or RNA	None	4	10^6	1
				Site II: s.s. RNA	Pyrimidines	10	10^6	30

§ s.s. = single-stranded; d.s. = double-stranded.
# 1 = noncooperative.
SOURCE: Based largely on data provided by Peter Von Hippel.

2. How can one measure the strength of binding and the number of binding sites?

3. Can the ligand bind anywhere on the nucleic acid, or are some base pairs or sequences preferred or excluded?

4. Does the small molecule bind to the outside of the helix? If so, is the structure left largely unperturbed?

5. Does the small molecule penetrate into the helix? Intercalation, in which a planar ligand slips between neighboring base pairs, is a very common mode of binding. How does this affect the physical properties of the nucleic acid?

6. Does binding at one site along the nucleic acid helix alter the probability of binding to neighboring sites? There could be cooperative effects in which binding nearby is enhanced, or anticooperative effects in which it is inhibited.

Experimental detection of ligand binding

Studies of binding inevitably begin with a search for a method of determining what fraction of the total ligand concentration $(M)_0$ is bound. By conservation of mass, if the amount bound is known, then one can determine the molar concentration (m) of free ligand, and vice versa.

$$(M)_0 = (m) + \bar{v}(NA) \tag{23-74}$$

where $\bar{v}$ is the number of bound small molecules per nucleic acid base pair, and (NA) is the total molar concentration of base pairs. This nomenclature is biased in favor of ligands that bind to double strands.

Spectroscopic methods commonly are used to monitor binding. For techniques where the characteristic measuring time is faster than any binding rates, the observed small-molecule property is always a simple average of the properties of free and bound species, so long as there is only one spectroscopically distinguishable type of binding site:

$$A = (m)A_{fr} + \bar{v}(NA)A_{bd} \tag{23-75}$$

where A is the observed optical property of the sample, and A_{fr} and A_{bd} are the corresponding molar properties of free ligand and completely bound ligand, respectively; all of these properties are measured at the same wavelength. If there are several spectroscopically distinguishable classes of binding sites, the second term in Equation 23-75 can be replaced by a sum over these classes of bound ligand.

Equation 23-75 and its analogs are valid for absorbance, fluorescence intensity or anisotropy, CD, and ORD.[§] In general, A_{fr} is found by measurements on the pure small molecule; A_{bd} is found by extrapolating measured A to the limit of infinite (NA)-to-ligand ratios, where presumably all ligand is bound. Equations 23-74 and 23-75 represent two linear equations in two unknowns, (m) and $\tilde{v}$. Therefore, the amounts bound and free can be calculated easily.

Some binding techniques look only at the free ligand. For example, in a potentiometric titration, the solution pH gives the free (H^+). Then the amount of bound protons is available from Equation 23-74, because the amount added to the solution is known. Equivalent techniques for metal ions are conductimetric titrations, or the use of specific ion electrodes. Many binding techniques look only at the bound ligand. These techniques include any assayable biochemical effects, or binding-induced changes in physical properties of the macromolecule (such as apparent molecular weight, buoyant density, viscosity, and macromolecular spectroscopic properties). Such techniques can be used to monitor binding equilibria, but only if there is some way to calibrate the observed macromolecular property with the number of bound ligands. In many cases it is a linear function, and then the problem is trivial.

The simplest binding model involves B binding sites, each of which is thermodynamically and spectroscopically identical to the others. Binding at each site, and the spectroscopic properties that result, are completely independent of what is happening at any other site. For such a model, a normal intrinsic macroscopic binding constant K_b describes the binding equilibrium at each site:

$$K_b = (\text{Filled site})/(m)(\text{Empty site}) \qquad (23\text{-}76)$$

and Equation 23-75 gives the number of filled sites. In an independent-site model, the statistical weights of states of intermediate ligand saturation need not be considered explicitly (Chapter 15). Because the sites are noninteracting, the equilibria will be identical to the case in which we have a solution of B times the macromolecule concentration, where each macromolecule has only a single binding site.

If we use B_0 to represent the number of binding sites per base pair, we can substitute the quantities defined by Equation 23-74 into the equilibrium expression of Equation 23-76:

$$K_b = \tilde{v}(\text{NA})/(m)[B_0(\text{NA}) - \tilde{v}(\text{NA})] \qquad (23\text{-}77)$$

The concentration of base pairs drops out, and Equation 23-77 can be rearranged in the form

$$\tilde{v}/(m) = K_b(B_0 - \tilde{v}) \qquad (23\text{-}78)$$

[§] For techniques with characteristic measuring times slower than binding rates, the form of Equation 23-75 can be altered, but the data analysis frequently is handled in a similar way. For example, Equation 23-75 holds for NMR chemical shifts in the fast-exchange limit, but not for spectral intensities at a particular frequency.

This very useful equation is the basis of the well-known Scatchard plot. It can be used to determine both the apparent binding constant K_b and the number of binding sites per base pair, B_0. A plot of $\tilde{v}/(m)$ versus $\tilde{v}$ will be linear. The slope is K_b, and B_0 is the value of $\tilde{v}$ at the point where $\tilde{v}/(m)$ extrapolates to zero. Curve A in Figure 23-26 is an example of a linear Scatchard plot.

Figure 23-26

Scatchard plots expected for three models of ligand binding to DNA. The parameter $\tilde{v}$ is the number of ligands bound per base pair; (m) is the concentration of free ligand. Curve A is for the model in which all sites are independent. Curve B is for the cooperative-binding model—for example, occupation of a site adjacent to a filled site is favored. Curve C is for the anticooperative-binding model—for example, a bound site renders occupancy of neighboring sites less favorable.

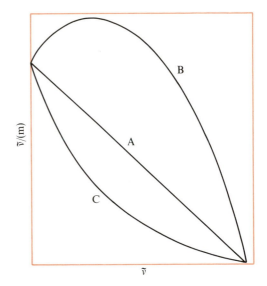

The next-simplest kind of binding equilibrium is a model with two different classes of noninteracting, independently binding sites occurring at B_{10} and B_{20} per base pair, with binding constants K_{1b} and K_{2b}. Equation 23-78 holds for each class separately and, if there is some way to measure the binding fractions $\tilde{v}_1$ and $\tilde{v}_2$ separately, then K_{1b}, K_{2b}, B_{10}, and B_{20} can all be determined. In many cases, however, all that is accessible experimentally is the total binding fraction, $\tilde{v} = \tilde{v}_1 + \tilde{v}_2$. Starting from a separate Equation 23-78 for each type of site, you should be able to show that this fraction has the following form:

$$\tilde{v}/(m) = K_{1b}B_{10} + K_{2b}B_{20} - K_{1b}\tilde{v} + (K_{1b} - K_{2b})K_{2b}B_{20}/[K_{2b} + 1/(m)] \quad (23\text{-}79)$$

Because of the last term, a plot of $\tilde{v}/(m)$ versus $\tilde{v}$ no longer is linear. However, if data of sufficient accuracy exist, Equation 23-79 will permit determination of the four binding parameters. This is illustrated in Section 15-4.

Interactions between binding sites lead to a higher order of complexity. Attractive or cooperative interactions can have several origins. Neighboring ligands bound on the outside of the helix might be able to interact directly. For example, phosphate-bound planar cationic ligands could stack on each other in the same way that the bases stack inside the helix. A binding reaction that disrupts helical structure could facilitate neighboring-site binding if the disruption extends beyond the actual binding domain. Think, for example, of a ligand that stabilizes a loop. Initiation of the loop is difficult, but extension by additional binding could be easier.

Repulsive or anticooperative interactions between bound ligands also can arise from several different mechanisms. A large ligand might contact only one or two base pairs, but sterically exclude the approach of other ligands to nearby sites. Neighboring bound ligands might have unfavorable, direct thermodynamic interactions. As an example of the kinds of complexities that must be present if one looks in sufficient detail, consider the binding of any cationic ligand. It will tend to repel like-charged ligands attempting to bind nearby, but it also will alter the local counterion atmosphere, which means that the electrostatic free energy of the whole polymer becomes a variable in the analysis of the binding experiment.

Describing site–site or ligand–ligand interactions by statistical methods

One must resort to the additional power of statistical thermodynamic descriptions to handle the complexities introduced by ligand–ligand interactions. Here we restrict our attention just to nearest-neighbor ligand interactions, but the treatment described is completely general and is easily extended to more complex cases. Other techniques for treating ligand–ligand interactions are described in Chapter 15.

Allow all binding sites to be identical. The equilibrium constant for binding to a site when the neighbors are empty is $k = \exp(-\Delta G_b^0/RT)$. When two ligands bind at adjacent sites, the total energy change in the system will be $2\Delta G_b^0 + \Delta G_c$, where ΔG_c describes the neighbor interaction. Therefore, the equilibrium constant for producing this state from unliganded nucleic acid is $k_{11} = k^2\zeta$, where $\zeta = \exp(-\Delta G_c/RT)$. For cooperative interactions, $\zeta > 1$; for anticooperative interactions, $\zeta < 1$. Figure 23-27 shows all the possible states of a system with three binding sites. It becomes quite complicated to write down an expression for the amount of ligand bound at a given free-ligand concentration, even for a system with only three sites.

The free energy change ΔG_i that accompanies the binding of n_i ligands at free-ligand concentration (m) to form one particular configuration (such as one of those shown in Fig. 23-27) is

$$\Delta G_i = \Delta G_i^0 - n_i RT \ln(m) \qquad (23\text{-}80)$$

We choose the standard molar free energy change ΔG_i^0 as $-RT \ln K_i$, where K_i is the equilibrium formation constant of that configuration from completely unliganded nucleic acid. This definition of ΔG_i^0 results in setting the free energy of the unliganded nucleic acid at zero, because ΔG_i^0 and n_i for that configuration both equal zero.

A useful parameter for describing complex equilibria is the statistical weight W_i introduced in Section 20-1. The statistical weight of a state is defined as

$$W_i = e^{-\Delta G_i/RT} \qquad (23\text{-}81)$$

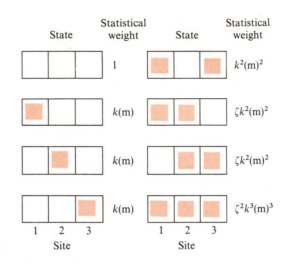

Figure 23-27

Possible states of occupancy of three adjacent ligand binding sites on a polynucleotide duplex. For intercalation, the three sites are formed by four adjacent base pairs. Shown are the (unnormalized) probabilities of occurrence of each state, where k is the binding constant to an unoccupied duplex, ζ describes nearest-neighbor interaction between occupied sites, and (m) is the concentration of free ligand.

Using Equation 23-80 and the definition of ΔG_i^0, we can write the statistical weight as

$$W_i = K_i(m)^{n_i} \tag{23-82}$$

Figure 23-27 shows the statistical weights of each of the states of a system with three sites. Inspection of this figure reveals that K_i contains a factor of k for each occupied site and an additional factor of ζ for each pair of adjacent occupied sites. Therefore, for a system with any number of sites, W_i can be constructed by examining the configuration from left to right and multiplying by a factor of

1 for each unoccupied site, or

$k(m)$ for each occupied site on the right side of an unoccupied site, or

$\zeta k(m)$ for each occupied site on the right side of an occupied site.

The statistical weight of a fully unliganded nucleic acid is one. Thus, the statistical weights defined by Equation 23-82 reveal the relative probability of finding a particular ligand configuration. The absolute probability P_i is computed by comparing the relative probability of one configuration to the sum of the relative probabilities of all possible states:

$$P_i = W_i / \sum_i W_i \equiv K_i(\mathrm{m})^{n_i} / Q \qquad (23\text{-}83)$$

where Q, the configurational partition function of the system, is defined as

$$Q \equiv \sum_i e^{-\Delta G_i / RT} = \sum_i K_i(\mathrm{m})^{n_i} \qquad (23\text{-}84)$$

Any property of the system is computed by averaging over all the i configurations. The property we are interested in here is the average number $\langle n \rangle$ of bound ligands:

$$\langle n \rangle = \sum_i n_i P_i = (1/Q) \sum_i n_i K_i(\mathrm{m})^{n_i} \qquad (23\text{-}85)$$

Equation 23-85 can be rewritten in the following form:

$$\langle n \rangle = \frac{(\mathrm{m})}{Q} \frac{d}{d(\mathrm{m})} \left(\sum_i K_i(\mathrm{m})^{n_i} \right) = \frac{(\mathrm{m})}{Q} \frac{dQ}{d(\mathrm{m})} \qquad (23\text{-}86)$$

If the polymer has N base pairs, the binding fraction $\tilde{v}$ is

$$\tilde{v} = \langle n \rangle / N = N^{-1} d(\ln Q) / d[\ln(\mathrm{m})] \qquad (23\text{-}87)$$

Thus, if we can calculate Q, we will able to calculate $\tilde{v}$ for any model as a function of (m).

● Matrix methods of calculating the partition function

In Chapter 20, we use the Zimm–Bragg matrix method to generate all possible helix–coil configurations for a polypeptide. Here we use the same method to generate the sum over all the possible terms in Equation 23-84 so that we can compute Q. Suppose that $\delta + 1$ consecutive binding sites must be considered to account for all site–site interactions. One starts by describing the configurations of the first δ binding sites. The number of possible configurations of δ sites is 2^δ, because each site can be empty or full. We can describe the statistical weights of all these configurations by a row vector $\mathbf{a}_\delta$ with 2^δ elements. Each element is the statistical weight of one of the possible configurations of the δ sites. The partition function for the first δ sites is the sum of the statistical weights of all the configurations. Thus it is the sum of all the elements of $\mathbf{a}_\delta$. This sum is computed by $\mathbf{a}_\delta \cdot \mathbf{e}$, where $\mathbf{e}$ is the unit column vector:

$$\mathbf{e} = \begin{pmatrix} 1 \\ 1 \\ \vdots \\ 1 \end{pmatrix}$$

For example, the simple nearest-neighbor interaction scheme described in Figure 23-27 requires us to consider only two sites at a time. The first site can be either empty or full. The vector $\mathbf{a}_1$ describing this site is

$$\mathbf{a}_1 = (1, k(\mathrm{m})) \qquad \text{site 1:} \quad \text{(empty, full)} \tag{23-88}$$

where the $2^1 = 2$ elements, 1 and $k(\mathrm{m})$, give the statistical weights of the two possible states for the single site. The partition function for the first site is

$$Q_1 = \mathbf{a}_1 \cdot \mathbf{e} = (1, k(\mathrm{m})) \begin{pmatrix} 1 \\ 1 \end{pmatrix} = 1 + k(\mathrm{m}) \tag{23-89}$$

To generate the partition function for the first two sites, we must consider all states of the first site and how each statistical weight changes when the second site is included. This can be done with a matrix $\underline{\mathbf{M}}$. Each element M_{ij} describes the change in statistical weight when the second site in state j is added to the first site in state i. If we let the index 1 represent empty, and 2 represent full, then the matrix $\underline{\mathbf{M}}_{ij}$ is

$$\begin{array}{cc} & \text{site 2} \\ & \begin{array}{cc} \text{empty} & \text{full} \end{array} \\ \text{site 1} \begin{array}{c} \text{empty} \\ \text{full} \end{array} & \begin{pmatrix} M_{11} & M_{12} \\ M_{21} & M_{22} \end{pmatrix} \end{array} \qquad \underline{\mathbf{M}} = \begin{pmatrix} 1 & k(\mathrm{m}) \\ 1 & \zeta k(\mathrm{m}) \end{pmatrix} \tag{23-90}$$

For example, $M_{12} = k(\mathrm{m})$ is the statistical weight for filling the second site when the first site is empty.

A vector that describes the statistical weights of all states of the two sites is

$$\mathbf{a}_1 \cdot \underline{\mathbf{M}} = (1 + k(\mathrm{m}), k(\mathrm{m}) + \zeta k^2(\mathrm{m}^2)) \tag{23-91}$$

The partition function for the first two sites then is

$$Q_2 = \mathbf{a}_1 \cdot \underline{\mathbf{M}} \cdot \mathbf{e} = 1 + 2k(\mathrm{m}) + \zeta k^2(\mathrm{m})^2 \tag{23-92}$$

The procedure used to generate Q_2 from Q_1 can be repeated as many times as necessary to describe the whole polymer:

$$Q_3 = \mathbf{a}_1 \cdot \underset{\sim}{\mathbf{M}}^2 \cdot \mathbf{e} \tag{23-93}$$

$$\boxed{Q_N = \mathbf{a}_1 \cdot \underset{\sim}{\mathbf{M}}^{N-1} \cdot \mathbf{e}} \tag{23-94}$$

In general, the starting vector $\mathbf{a}_\delta$ describes the first δ sites and has 2^δ elements. The matrix $\underset{\sim}{\mathbf{M}}$ describes the changes in statistical weight M going from sites 1 through δ to sites 2 through $\delta + 1$; thus $\underset{\sim}{\mathbf{M}}$ is a $2^\delta \times 2^\delta$ matrix. The partition function for the polymer becomes

$$Q_N = \mathbf{a}_\delta \cdot \underset{\sim}{\mathbf{M}}^{N-\delta} \cdot \mathbf{e} \tag{23-95}$$

Both $\mathbf{a}_\delta$ and $\underset{\sim}{\mathbf{M}}$ usually can be written by inspection, even for rather complex cases (see Problem 23-1).

● **Computation of binding curves**

Although Equation 23-95 permits exact calculation of Q, in most cases this equation is not very helpful because the final form is quite complex. For large chain lengths, N, the partition function can be approximated as $Q = \lambda_{max}^n$, where λ_{max} is the largest eigenvalue of the matrix $\underset{\sim}{\mathbf{M}}$, and n is the exponent of $\underset{\sim}{\mathbf{M}}$ in Equation 23-95. In the general case, we want to consider ligands that may occupy more than one adjacent site. If B_0 is a fraction denoting the number of sites per base pair, then the partition function will be $Q = \lambda_{max}^{NB_0}$. Then, from Equation 23-87, the binding fraction $\tilde{v}$ is simply

$$\tilde{v} = B_0 \, d(\ln \lambda_{max})/d[\ln(m)] \tag{23-96}$$

We first demonstrate this relationship for a case of completely noncooperative binding, where we can set $\zeta = 1$ in Equation 23-90. The eigenvalues of $\underset{\sim}{\mathbf{M}}$ are found by

$$\begin{vmatrix} 1 - \lambda & k(m) \\ 1 & k(m) - \lambda \end{vmatrix} = 0 = \lambda^2 - \lambda[k(m) + 1] \tag{23-97}$$

The largest eigenvalue clearly is $k(m) + 1$ (the other is zero), and so $\tilde{v}$ is simply

$$\tilde{v} = B_0 \frac{d\{\ln[k(m) + 1]\}}{d[\ln(m)]} = (m)B_0 k/[k(m) + 1] \tag{23-98}$$

If this is rearranged, it is identical to Equation 23-78; thus we have rederived the Scatchard plot, albeit by a much more general method.

For the nearest-neighbor interaction model of Figure 23-27, we can take B_0 as one because all sites potentially can be filled. The eigenvalues of the matrix M in Equation 23-90 are found by

$$\begin{vmatrix} 1 - \lambda & k(m) \\ 1 & \zeta k(m) - \lambda \end{vmatrix} = 0 = \lambda^2 - \lambda[1 + \zeta k(m)] + k(m)(\zeta - 1) \qquad (23\text{-}99)$$

The largest root is

$$\lambda_{max} = (1/2)\{1 + \zeta k(m) + \sqrt{[1 + \zeta k(m)]^2 - 4k(m)(\zeta - 1)}\} \qquad (23\text{-}100)$$

This expression, together with Equation 23-96, allows us to calculate $\tilde{v}$, and thus to compute Scatchard plots of $\tilde{v}/(m)$ versus $\tilde{v}$ as a function of various values of the parameters ζ and k.

The solutions to the eigenvalue equations corresponding to Equations 23-97 and 23-99 become very cumbersome for more complex binding schemes, because cubic or higher-order terms in λ appear. A general way to handle the results is to compute $d(\ln \lambda)/d[\ln(m)]$ directly from the eigenvalue equation. For example, Equation 23-99 can be arranged in the form

$$(\lambda^2 - \lambda)/(1 - \zeta + \lambda\zeta) = k(m) \qquad (23\text{-}101)$$

Then, using the fact that $B_0 = 1$ for the models we have considered, it is easy to show from Equation 23-96 that

$$\tilde{v} = \frac{d(\ln \lambda)}{d[\ln(m)]} = \frac{(m)\, d\lambda}{\lambda d(m)} = \frac{\zeta\lambda^2 + (1 - 2\zeta)\lambda + \zeta - 1}{\zeta\lambda^2 + 2(1 - \zeta)\lambda + \zeta - 1} \qquad (23\text{-}102)$$

Equations 23-101 and 23-102 can be combined to compute $\tilde{v}/(m)$ in terms of λ, ζ, and k. In this treatment, λ is a variable parameter that must be specified before (m) and $\tilde{v}$ can be computed. To be consistent with Equation 23-96, we must use values of λ large enough to correspond to the range of λ_{max}. How can these values be chosen? Inspection of the correct expression for λ_{max} (Eqn. 23-100) shows that $\lambda_{max} > 1$. Therefore, it suffices to rewrite Equations 23-101 and 23-102 in terms of the parameter $y = \lambda^{-1}$ and then choose y in the range $0 \leqslant y < 1$.

Figure 23-26 shows binding isotherms computed by Equations 23-101 and 23-102 for three values of ζ. The choice of $\zeta = 1$ yields a simple linear Scatchard plot. A choice of $\zeta > 1$ usually will produce a Scatchard plot with a maximal value of $\tilde{v}/(m)$ at some intermediate value of $\tilde{v}$. This is a reflection of the cooperativity of binding that leads to more ligand bound at intermediate ligand concentrations than in the absence of ligand–ligand interactions.

Excluded-site model for intercalative binding

A binding case of special interest is that where $\zeta = 0$ (Fig. 23-26). Here, binding at one site absolutely excludes the possibility of binding at the nearest neighboring sites. The resulting Scatchard plot is concave and extrapolates to a maximal value of $\tilde{\nu} = 0.5$, corresponding to a double helix in which filled and empty binding sites alternate (so that every other base pair has a filled site). This extreme of anti-cooperative-neighbor binding is called the excluded-site model. With $\zeta = 0$, Equation 23-102 simplifies to

$$\tilde{\nu} = (\lambda - 1)/(2\lambda - 1) \qquad (23\text{-}103)$$

With $\zeta = 0$, Equation 23-99 is easily solved to yield $\lambda_{\max} = (1/2)[1 + \sqrt{1 + 4k(m)}]$, and so

$$\tilde{\nu} = (1/2)[1 - 1/\sqrt{1 + 4k(m)}] \qquad (23\text{-}104)$$

As $(m) \to \infty$, the correct limit $\tilde{\nu} = 1/2$ is reached. The Scatchard plot produced by Equation 23-104 is concave upward (Fig. 23-28). Note that such a plot could easily

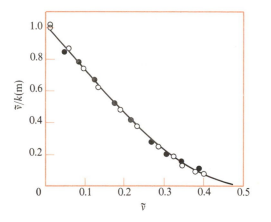

Figure 23-28

Scatchard plot for the binding of ethidium bromide to two DNAs: calf thymus DNA ($\bullet$) and *M. luteus* DNA ($\bigcirc$). The smooth curve shows a fit of the data to the excluded-site binding model ($\zeta = O$). The ordinate has been normalized by dividing by the binding constant. [After J. Bresloff, unpublished Ph.D. thesis, Yale University, 1974.]

be mistaken for one resulting from two classes of binding sites, one strong and one weak. Detailed information on the mechanism of binding must be obtained to distinguish these two different physical binding models. However, the excluded-site model very accurately accounts for the binding of a number of ligands to DNA. Figure 23-28 demonstrates this for the case of ethidium binding.

A fit of data to the excluded-site model involves the adjustment of only a single experimental parameter, the binding constant k. In practice, it seems unlikely that most substances binding to DNA will be completely insensitive to base sequence or base composition. The excellent fit to ethidium–DNA binding in Figure 23-28 suggests that these effects must largely cancel out, just as sequence effects on optical properties become largely invisible in the polymer. Thus, it usually is necessary to resort to studies on the interaction of ligands with small oligonucleotides of known sequence in order to learn the details of sequence-selective interactions.

Physical measurements on ligand–polynucleotide interactions can lead to some important general conclusions. As shown earlier, an increase in T_m upon binding indicates that the double strand is the preferred site. If binding of a ligand increases the length of a DNA with no significant change in diameter, it is easy to show from the hydrodynamic equations in Chapters 11 and 12 that the sedimentation coefficient will decrease, whereas the intrinsic viscosity will increase. The simplest binding mode that increases length is intercalation (Fig. 23-29). Ethidium is the classic example of a ligand that binds by intercalation. Because, according to the excluded-site model, ethidium can fill one site for every other base pair, a fully saturated DNA will increase in length by 50% (the thickness of ethidium is the same as the thickness of a base pair).

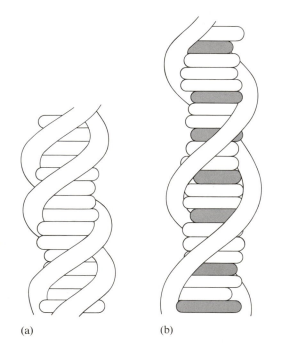

(a)   (b)

Figure 23-29

Schematic illustration of DNA duplex lengthening and unwinding caused by intercalative binding. (a) The original DNA helix. (b) The helix with intercalative binding of ligands. [After M. Waring, *Nature* 219:1320 (1968), as modified by S. J. Lippard, *Accts. Chem. Res.* 11:211 (1978).]

One prediction of the intercalation model is that the plane of the bound ligand should be roughly perpendicular to the helix axis, Linear dichroism provides a straightforward test of this prediction. Because the major electronic transitions of ethidium are polarized in the plane of the dye, bound ethidium will preferentially absorb light polarized perpendicular to the helix axis. Thus, the linear dichroism will be negative.

All of these physical studies on ligands bound to DNA led originally to the hypothesis that ethidium, acriflavine, and other planar ligands bind to DNA by intercalation. Some x-ray fiber results supported this model, but fibers produced in the presence of bound ligands generally are not very well ordered, and so the amount of detail available from these studies is limited. A series of elegant studies on the effect of ethidium bromide binding on the hydrodynamic properties of closed circular DNA produced results not only consistent with the intercalation model, but also providing detailed estimates of the angle through which the double helix would have to unwind to allow binding. However, none of these results could actually prove that intercalation is the predominant mode of binding. Only recently have fibers suitable for high-resolution analysis been provided. These studies employed a platinum-containing intercalation reagent. Analysis of the x-ray scattering provides strong support for a neighbor-exclusion intercalation-binding model.

Crystalline complexes between ethidium and oligonucleotides

Recently, x-ray crystallography has revealed the three-dimensional structures of several crystalline complexes of suspected intercalators and mononucleoside or dinucleoside phosphates. Figures 23-30 and 23-31 show two of these, an ethidium complex with 5-iodoUpA, and an actinomycin D complex with deoxyguanosine. The ethidium complex is typical of many simple intercalation complexes. We will discuss it in considerable detail. However, the reader should be aware that not all known intercalation complexes behave like this ethidium complex.

The ethidium complex contains hydrogen-bonded 5-iodoUpA duplexes with ethidium intercalated between the two A–U base pairs and also stacked on the ends of the duplex. The overall stoichiometry is one ethidium to one 5-iodoUpA, but this may be somewhat misleading. The crystals were prepared at low salt concentrations, so ethidium must act as the cation for each phosphate. The 1:1 stoichiometry is needed to maintain overall electroneutrality. In a solution of DNA–ethidium complex, there is always sufficient supporting electrolyte to provide counterions as needed.

The ethidium–iodoUpA crystal structure provides clear evidence that intercalation occurs. It also provides an interesting indication of the origin of the excluded-site effect. The sugar rings in the B-form double helix are puckered in the 2'-*endo* configuration (Chapter 3). The sugar rings in the A-form DNA and in the RNA-11 helices are puckered in the 3'-*endo* configuration. In the ethidium–5-iodoUpA complex (Fig. 23-30a), however, the sugar puckering is 3'-*endo* for each 3'-linked nucleoside,

Figure 23-30

Model for binding of ethidium bromide to DNA. (a) X-ray structure of a 1:1 complex of ethidium and iodoUpA. Ethidiums are intercalated (*colored*) between the base pairs of the dinucleotide phosphate duplex, and also are stacked at the ends. [After C. C. Tsai et al. *Proc. Natl. Acad. Sci. USA* 72:628 (1975).] (b) Schematic representation of the structure in part a extrapolated to an excluded-site saturated DNA duplex. Note that alternating sugar puckers destroy the dyad axis of the DNA but leave a C_2 axis between each set of base pairs, as shown. The pattern of puckering also leads to a regular alternation of two classes of potential binding sites. (c) Two views of a more realistic model for the ethidium–DNA complex, showing a kink and a displacement of the helix axis (the vertical bars indicate the local helix axis). [From H. M. Sobell et al., *J. Mol. Biol.* 114:333 (1977).]

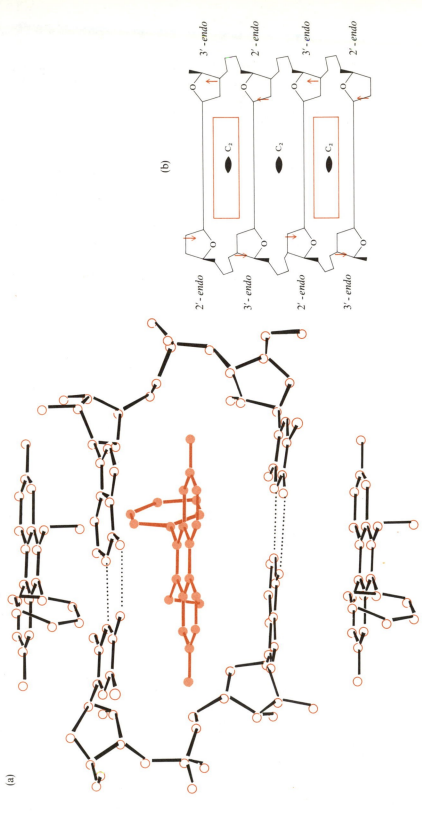

(a)

(b)

2'-endo   3'-endo

3'-endo   2'-endo

2'-endo   3'-endo

3'-endo   2'-endo

C_2

C_2

C_2

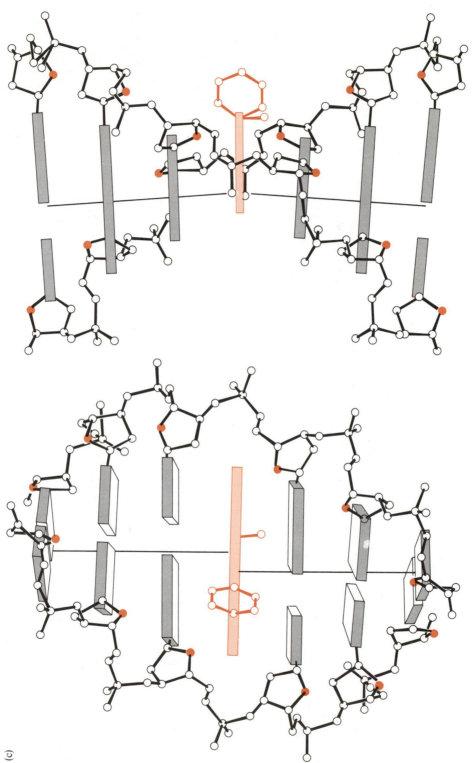

(c)

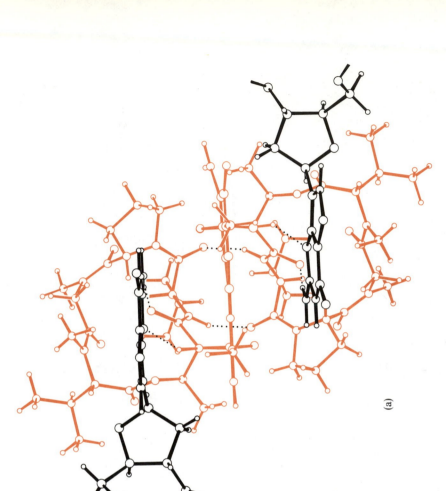

(a)

Figure 23-31

Binding of actinomycin D to DNA. **(a)** X-ray structure of a 2:1 complex of deoxyguanosine (*black*) and actinomycin D. **(b)** Extrapolation of the dinucleoside complex structure to a model for actinomycin D intercalation into duplex DNA. [Parts a and b after H. M. Sobell and S. C. Jain, *J. Mol. Biol.* 68:21 (1972).] **(c)** Revised model for the actinomycin–DNA complex, showing a kink. The vertical bars indicate the local helix axis at each level. [From H. M. Sobell et al., *J. Mol. Biol.* 114:333 (1977).]

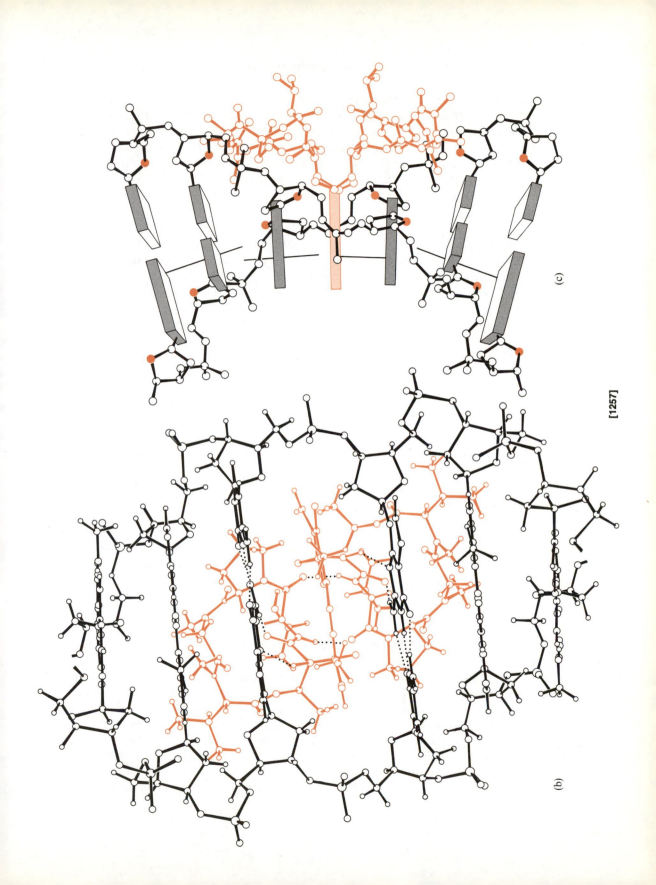

(c)

(b)

and 2'-endo for each 5'-linked nucleoside (that is, 2'-endo-iodoUp–3'-endo-A). Although this structure has been seen only in dinucleotides and could be some kind of end effect, an interest result develops if it is extrapolated to polymers. Each strand could alternate 2'- and 3'-endo sugars (Fig. 23-30b). This observation has two implications. Because of the alternating sugar pucker, the dyad axis in the plane of the base pairs of the double helix is disrupted. However, a C_2 axis between adjacent base-pair planes could be maintained, and now this axis would pass through the center of the bound ethidium ring.

The potential binding sites above and below the one occupied by ethidium also could have C_2 local symmetry. However, their structures are different from that of the ethidium site, because the sugar puckering is opposite. Therefore, if the extrapolation of the dinucleoside-intercalation structure to polymers is correct, it predicts that the double helix will be cast into a conformation in which alternating inter-base-pair regions have different structures. All one need do is postulate that only one of the two structures is energetically favorable for intercalation, and the excluded-site model can be rationalized. More hard structural data will be needed before this attractive hypothesis can be tested thoroughly. Note that an alternating sugar-ring pucker can explain excluded-site binding in both DNA and RNA. In the DNA-B helix, sugars normally are 2'-endo so, to form intercalation complexes, a change from 2'-endo to 3'-endo puckering of every other residue is required. In RNA helices, sugars normally are 3'-endo, so a change from 3'-endo to 2'-endo puckering of every other residue is required.

When detailed models of the ethidium–DNA complex are constructed, a slightly more complicated picture emerges. Figure 23-30c shows two views of an idealized structure of an ethidium intercalated into DNA B. Intercalated base pairs are tilted with respect to one another. This results in a kink of 8° when DNA-B helix is allowed to extend on both sides of the ethidium. In addition, the helix axes above and below the ethidium are displaced about 1 Å in a manner similar to the kinks shown in Chapter 3. Because of the kink, a DNA fully saturated with ethidium will be forced into a superhelical structure. Each ethidium leads to a net angular unwinding of the B helix of $-26°$, a value in good agreement with the results of hydrodynamic studies of ethidium binding to DNA (Chapter 24).

A complex between actinomycin D and deoxyguanosine

Local symmetry plays an important role also in the interpretation of the actinomycin D structure shown in Figure 23-31a. Here the stoichiometry of the actinomycin–deoxyguanosine complex is two deoxyguanosines for each actinomycin. The actual structure contains an approximate C_2 axis that relates one deoxyguanosine to the other, and also relates one actinomycin cyclic peptide to the other. The two bases and the planar actinomycin ring have a tantalizing resemblance to an intercalated structure. All three aromatic moieties are in parallel planes and, if cytidines were brought up and base-paired to the two guanines, the overall structure would have almost the precise geometry of a duplex dinucleotide containing an intercalated ligand. The C_2 axis found in the structure of the complex is the same axis expected to be present in intercalated DNA, such as that seen in the ethidium–DNA complex. Figure 23-31b illustrates this axis in the extrapolated structure for an actinomycin–DNA complex.

An interesting feature of the actually-measured deoxyguanosine structure is the presence of specific hydrogen bonds between the cyclic peptide rings and the two nucleic acid bases. The steric fit is so good that it suggests the possibility that actinomycin D might exhibit some binding preference for such sequences as GpC·GpC. More detailed model-building studies of the actinomycin–DNA complex have been carried out using the results of the simpler ethidium case as a guide. These studies predict a structure in which an exact C_2 axis is no longer present. This effect arises from the inherent asymmetry of the phenoxazone ring of actinomycin and small but significant departures in the symmetry that relates the two polypeptide chains.

The revised model (Fig. 23-31c) has a number of very interesting features. The sugar pucker alternates C-3′-endo(3′ → 5′)C-2′-endo, just as it does in the ethidium complexes. The hydrogen bonds between the 2-amino groups of guanine and the peptide chains (threonine oxygens) are not of equal length. The DNA kinks at the intercalation site, and additional bendings occur on both sides of the site due to the steric bulk of the peptide chains. This results in an overall kink of 40°. The unwinding of the DNA resulting from one bound actinomycin is predicted as $-28°$, in good agreement with experimental data. The revised model also can explain why actinomycin can bind almost as strongly to duplex sequences of the type d(GpX) (where X is G, A, or T) as it can to the sequence with perfect C_2 symmetry, d(GpC).

Solution studies of intercalation complexes with oligonucleotides

NMR and other solution physical studies have been performed on complexes between dinucleoside phosphates and actinomycin D or ethidium to examine possible sequence specificity. Earlier we mentioned that duplexes formed between between complementary dimers are not very stable in aqueous solution, and that physical measurements are rather difficult. However, because ethidium and actinomycin D preferentially bind to duplexes, they increase the stability of the complexes considerably.

Different deoxyribodinucleotides have shown quite varied complexing behavior with actinomycin D. Initial studies monitored changes in the visible absorption band of actinomycin D, induced by added dinucleotides (Fig. 23-32a). It is difficult to draw more than qualitative conclusions from this data alone, because the lack of saturation of the absorbance change makes it hard to calculate binding constants. However, with more complete data, and assuming either 1:1 or 2:1 complexes, it was possible to decipher the differences in complex strength and stoichiometry through least-square fits.

It was found that d(pGpC) forms a strong 2:1 complex with actinomycin D.

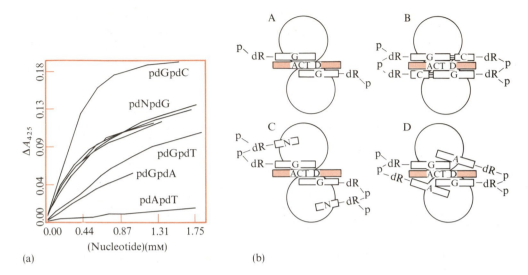

Figure 23-32

Complex formation between actinomycin D and dinucleotides. (**a**) Absorbance changes in actinomycin are used to monitor the fraction of complex as a function of the concentration of dinucleotide added. The pdN–dG set of curves represents all four normal dinucleotides ending in dG, as well as 5′-dGMP. (**b**) Schematic illustration of the types of complexes formed, as inferred from the data in part a and additional studies. When G is the 5′-nucleotide, complexes formed are the most stable if the sequence permits base pairing; if the sequence does not permit base pairing, then these complexes are the least stable because of steric interference. When G is the 3′–nucleotide, the complexes essentially mimic monomer complexes. [After T. R. Krugh and J. W. Neely, *Biochemistry* 12:4418 (1973).]

Furthermore, this complex forms cooperatively, indicating that the two d(pGpC) strands are likely to be base-paired in the complex. All other dinucleotides examined showed the same final stoichiometry, but the complexes formed noncooperatively. This suggests that no base pairing occurs even with self-complementary dinucleotides like d(pCpG).

The binding constant for the sequence isomer d(pCpG) (when corrected for its noncooperative binding mode) is about a factor of three weaker than that of d(pGpC). In fact, d(pCpG) forms complexes no stronger than those between actinomycin D and non-self-complementary dimers such as d(pApG) and d(pTpG). This observation led to the tentative binding models for these complexes shown schematically in Figure 23-32b. Subsequent NMR studies have shown that the notions expressed in these models are correct; they support the idea, seen in the crystal structure of the actinomycin–deoxyguanosine complex, that this drug should selectively recognize the d(pGpC) sequence and other d(pGpN) sequences. In these cases, an actual intercalation complex occurs, as in structure B in Figure 23-32b.

In contrast, with d(pNpG) sequences, the N bases are found to stack on top of the guanine, rather than on the phenoxazone ring of the actinomycin D, as in structure C of Figure 23-32b. It appears that there is a strong preference for one side of the guanine, the 3′-linked side, to stack on the actinomycin D. Clear indication of such specificity has been observed in studies of the binding of actinomycin D to synthetic DNAs of regularly alternating sequence.

Similar oligonucleotide model studies have been performed with ethidium com-complexes. Here, fluorescence changes as well as absorbance and NMR can be used to monitor complex formation, assign stoichiometry, and deduce some structural details. For example, NMR studies confirm that the C-3′-*endo*(3′ → 5′)C-2′-*endo* sugar pucker seen in most crystalline intercalation complexes also occurs in solution. The results on solutions contain one interesting surprise. The ethidium complexes with UpA·UpA, CpG·CpG, and UpG·CpA are all much more stable than complexes with the corresponding sequence isomers. Double strands must have formed in the complexes, because mixtures of noncomplementary dinucleoside phosphates show little or no evidence of interaction with ethidium. These results indicate that ethidium preferentially interacts with pyrimidine(3′ → 5′)purine double-stranded sequences. There is no absolute specificity, and the preference observed has not yet been quanti-tated. However, it is quite significant that a molecule as simple as ethidium appears to recognize some sequence aspects of nucleic acids preferentially. This makes it easy to see how proteins could easily have evolved to have much more dramatic sequence specificity in their interactions with DNA and RNA. It must be added, however, that the physical origin of the specificity of ethidium interaction is not yet clearly es-tablished.

The study of ligand binding to nucleic acids is an extremely active area, and many other interesting ligands exist that have not as yet been studied as completely as have actinomycin D and ethidium. Summaries of this broad area are readily available in the literature. Rather than attempt to summarize these, we turn instead in the next chapter to general studies on circular DNA. Here ligand binding has played a major role in unraveling some remarkable properties of certain natural and synthetic circular DNAs.

Summary

It is possible to predict the stability of specific base-paired structures of RNAs. The stability is a function of the nearest-neighbor base pairs in duplex regions and of the size and nature of the single-strand loops between base-paired regions. The thermo-dynamic interactions needed for the predictions can be derived from studies on the stability of complexes formed between short oligonucleotide pairs and on the stability of synthetic RNA hairpins. An all-or-none model for duplex formation fits the observed results for short helices quite well. This model greatly simplifies the statistical thermodynamic analysis required. Several general conclusions have emerged from a large volume of model compound studies. Adjacent G–C pairs provide far more stability than do sequences containing A–U pairs. Hairpin loops of six or seven bases are the least destabilizing. It is difficult to hold nearby loops separate with only a few base pairs. This observation helps to explain the cooperativity of DNA melting.

Kinetic processes in nucleic acid conformational changes span a wide range of time scales. Individual base-pairing and base-stacking interactions occur in nanoseconds. Short duplexes can form with rate constants of about 10^6 M^{-1} sec^{-1}. A transient base-paired nucleus forms rapidly. The rate-limiting step appears to be the formation of the second or third base pair; then, continued zipping of the helix is rapid. Denaturation kinetics of short duplexes are quite rapid, although activation energies are high. In contrast, the kinetics of DNA melting are quite complex and can be very slow. Friction-limited unwinding of the partially denatured duplex is the reason for this.

The kinetics of nucleation for the renaturation of short DNA duplexes is very convenient to study and analyze. It is a second-order process, and the rate depends on the size of the genome from which the DNA was isolated and on the extent of repeated sequences.

Nucleic acids can interact with a host of small- and large-molecular-weight ligands. A matrix method can be used to compute ligand binding curves. This procedure allows cooperative and anticooperative effects to be treated very conveniently. Intercalators such as ethidium bind to DNA in an excluded-site (anticooperative) fashion. Their binding lengthens the helix, causes unwinding, and may also cause kinking. Binding appears to be accompanied by a change in the puckering of sugar rings in the adjacent nucleosides.

Problems

23-1. Consider a model for binding of a ligand to a polynucleotide in which the ligand excludes binding to adjacent sites, but ligands bound with an empty site between them show cooperative attraction. Let L represent a filled site and E an empty site; for example, LELE represents a sequence of four sites with the first and third sites filled by ligands.
  a. Justify the statistical weights shown for the following configurations: LEEE, $k(m)$; LLEE, zero; LELE, $\zeta k^2(m)^2$.
  b. It is necessary to consider the occupancy of two adjacent sites to predict the binding curve of the ligand. Show that the vector describing the initial state is

$$\text{EE} \quad \text{LE} \quad \text{EL} \quad \text{LL}$$
$$(1, k(m), k(m), 0)$$

  c. Show that the matrix needed to generate the partition function is

		Sites 2 and 3			
		EE	LE	EL	LL
	EE	1	0	$k(m)$	0
Sites 1 and 2	LE	1	0	$\zeta k(m)$	0
	EL	0	1	0	0
	LL	0	1	0	0

d. Compute the eigenvalues of the matrix in part c. (Leave them as a cubic equation.) Then show that, in the limit of $\zeta = 1$ for this model (no cooperativity), the binding curve predicted using Equation 23-102 is identical to that worked out in the text for the excluded-site model in Equation 23-103.

23-2. Estimate the equilibrium constant at 25°C in 1 M salt for the following conformational change:

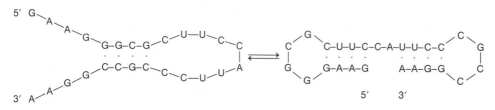

23-3. Consider a DNA that contains regularly alternating blocks of a 300-base-pair constant sequence C and 300-base-pair unique sequences $U_1, U_2, U_3, \ldots$:

$$\cdots CU_1 CU_2 CU_3 C \cdots$$

Suppose the genome size is 3×10^6 base pairs. Make schematic plots of the renaturation kinetics expected for random-cut DNA pieces 60 base pairs long, 300 base pairs long, 600 base pairs long, and 1,800 base pairs long.

23-4. Two tRNA molecules with complementary anticodon triplets form complexes up to 10^6-fold stronger than would be predicted from the data in Tables 23-4 and 23-5. Discuss some of the factors that may contribute to this extra stability, and outline experiments with model compounds that could be used to examine these factors individually.

23-5. Consider the following hypothetical data showing the total oligonucleotide concentration required to obtain a T_m of 37°C for various complexes in the presence of 1 M salt: AATAA·TTATT, 0.031 M; ATATA·TATAT, 0.010 M; AAAAT·ATTTT, 0.120 M. The ΔG^0 of nucleation of an A–T pair (at 1 M strand concentration 1 M salt) is 4 kcal mole^{-1}. Using the all-or-none approximation, predict the ΔG of formation of TAAAA·TTTTA at 1 M strand concentration, 37°C, and 1 M salt concentration.

References

GENERAL

Bloomfield, V., D. Crothers, and I. Tinoco, Jr. 1974. *Physical Chemistry of Nucleic Acids*. New York: Harper & Row.

Riesner, D., and R. Römer. 1973. Thermodynamics and kinetics of conformational transitions in oligonucleotides and tRNA. In *Physico-Chemical Properties of Nucleic Acids*, vol. 2, ed. J. Duchesne (London: Academic Press), p. 237.

Ts'o, P. O. P. 1974. Dinucleoside monophosphates, dinucleotides and oligonucleotides. In *Basic Principles in Nucleic Acid Chemistry*, vol. 2, ed. P. O. P. Ts'o (New York: Academic Press), p. 305.

SPECIFIC

Craig, M. E., D. M. Crothers, and P. Doty. 1971. Relaxation kinetics of dimer formation by self-complementary oligonucleotides. *J. Mol. Biol.* 62:383.

Dewey, T. G., and D. H. Turner. 1979. Laser temperature-jump study of stacking in adenylic acid polymers. *Biochemistry* 18:5757.

Hood, L. E., J. H. Wilson, and W. B. Wood. 1975. *Molecular Biology of Eucaryotic Cells.* Menlo Park, Calif.: Benjamin.

Krugh, T. R., E. S. Mooberry, and Y. C. C. Chiao. 1977. Proton magnetic resonance studies of actinomycin D complexes with mixtures of nucleotides as models for the binding of the drug to DNA. *Biochemistry* 16:740. [See also p. 747.]

Lee, C. S., R. W. Davis, and N. Davidson. 1970. A physical study by electron microscopy of the terminally repetitious circularly permuted DNA from the coliphage particles of *Escherichia coli* 15. *J. Mol. Biol.* 48:1. [Describes single-strand branch migration.]

Patel, D. J., and C. Shen. 1978. Sugar pucker geometries of the intercalation site of propidium diiodide into miniature RNA and DNA duplexes in solution. *Proc. Natl. Acad. Sci. USA* 75:2553.

Record, M. T., and B. H. Zimm. 1972. Kinetics of the helix–coil transition in DNA. *Biopolymers* 11:1435.

Scheffler, I. E., E. C. Elson, and R. L. Baldwin. 1968. Helix formation by dAT oligomers, I: Hairpin and straight chain helices. *J. Mol. Biol.* 36:291. [see also the accompanying paper, and *J. Mol. Biol.* 54:401 (1970).]

Thompson, B. J., M. N. Camien, and R. C. Warner. 1976. Kinetics of branch migration in double-stranded DNA. *Proc. Natl. Acad. Sci. USA* 73:2299.

Wetmur, J. G., and N. Davidson. 1968. Kinetics of renaturation of DNA *J. Mol. Biol.* 31:349.

24

Tertiary structure of nucleic acids

24-1 CIRCULAR DNA

Tertiary structure in nucleic acids includes possible kinks and superhelices in linear duplexes, multiple helices and loops formed intramolecularly by single strands, and topological features of circular DNA (such as knots, supercoils, and catenanes). In this chapter we focus attention on just two aspects: circular DNA or tRNA. Our emphasis is on comparison of various methods that reveal information about nucleic acid tertiary structure in solution.

Virtually all early physical studies on DNA dealt with relatively short linear molecules. Autoradiographic examination indicated that some bacterial DNAs are circular, but more gentle preparative procedures had to be developed before such samples could be isolated without fragmentation due to shear or enzymatic degradation. The first solution evidence for circular DNA came from the bacteriophage ϕX174. The base composition of this DNA is unusual in that $\chi_A \neq \chi_T$, and $\chi_G \neq \chi_C$. Furthermore, the DNA proved resistant to exonuclease digestion. This resistance suggested that ϕX174 DNA is a single-stranded circle and, indeed, electron microscopic examination of samples treated with formaldehyde to open up any intrastrand secondary structure revealed a circular molecule.

Subsequent work with *E. coli* infected by ϕX174 uncovered a second form of ϕX174 DNA. This form has normal base-composition ratios, and it also appears circular in the electron microscope. Single- and double-stranded DNA can be distinguished in the electron microscope when examined by the Kleinschmidt technique. (The film of cytochrome *c* surrounding double-stranded DNA appears slightly thicker, and the double-stranded molecule is not quite as kinky.) Thus, this second form of ϕX174 DNA (the replicative form) could be identified as a double-stranded circle.

Linear and circular species of λ DNA

A second kind of circular DNA was first seen with bacteriophage λ. When examined by ultracentrifugation, this DNA appears as multiple species with varying sedimentation constants. If one pure species is isolated by zonal centrifugation and allowed to stand, it slowly reequilibrates back to a mixture of components. Enzymatic and electron microscopic studies indicate that the two slowest-sedimenting forms are a 21 S double-stranded linear molecule and a 24 S double-stranded circle. The faster-sedimenting forms are aggregates of these, such as double-length linear or circular duplexes. To explain the ability of λ DNA to undergo all of these interconversions, it was postulated that λ has short single-stranded complementary ends (Fig. 24-1).

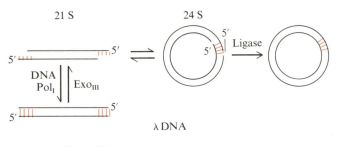

Figure 24-1

Some of the forms of DNA from bacteriophage λ.

This hypothesis was confirmed by enzymatic treatment. In the presence of deoxynucleoside triphosphates, *E. coli* DNA polymerase I can extend a free 3'-end along a complementary strand. Treatment with DNA polymerase I converts the 21 S linear form of λ DNA into a molecule with identical properties except that it will no longer equilibrate to the other forms. In addition, treatment of the 24 S circular form with DNA ligase results in a circular duplex containing no single-strand breaks. Therefore, it cannot equilibrate back to linear forms.

The length of the single-stranded ends of λ DNA can be estimated by studying the equilibrium between 21 S and 24 S forms at concentrations low enough to inhibit the formation of any higher aggregates. The equilibrium constant for circle formation is known once the fractions of linear and circle (χ_c) are measured: $K_c = \chi_c/(1 - \chi_c)$. From the dependence of K_c on temperature, a van't Hoff analysis yields an apparent enthalpy of circle formation of -90 kcal mole^{-1}. It is reasonable to assume that essentially all of this enthalpy change is due to the heat of formation of the extra base pairs between the single-stranded ends of the duplex. A double-stranded DNA of the length of λ DNA is flexible enough that there should not be major enthalpy changes in bending it into a circle. If we use an average enthalpy of -8 kcal mole^{-1} for base-pair formation, we predict the length of extra duplex in the circle to be 11 base pairs. Subsequent sequence studies showed that the actual length of the single-stranded ends of λ DNA is 12 base pairs, in fairly good agreement with the thermodynamic prediction.

The free energy of forming 12 base pairs can be estimated roughly[§] from Table 23-4 as about -36 kcal mole^{-1} at 25°C. At any reasonable concentration in solution, two free complementary dodecanucleotides will exist almost totally as duplex. However, when λ DNA cyclizes, the number of accessible configurations of the DNA is dramatically reduced. This effect leads to a large entropy decrease, which should be proportional to the logarithm of the $-3/2$ power of the length of DNA in the circle (Chapter 23). The entropy of cyclization mostly compensates for the favorable ΔG of base pairing between single-stranded ends. The larger the circle, the larger the single-stranded ends must be to maintain a significant fraction of cyclized molecules. For very short circles (say, less than 500 base pairs), DNA will no longer be flexible enough to cyclize without some kind of energy-requiring bending or distortion.

One can imagine two discrete stages in the kinetics of formation of a λ circle from the linear form. The two ends of the chain must diffuse into each other's neighborhood; then the extra section of double helix can nucleate and zip up. J. Wang and N. Davidson studied the first-order rate of λ cyclization as a function of temperature and viscosity to explore which of these two stages is rate limiting. A diffusion-limited rate will be proportional to the diffusion constant of the colliding species; hence, $k_f \propto T/\eta$. We know from model systems that the rate of helix nucleation from nearby single strands is governed by the formation of the initial two or three base pairs (Chapter 23). There can be an appreciable negative or positive apparent activation energy, and no dependence on viscosity should be observed.

Experiments show that there is a high positive apparent activation energy for cyclization, and that $k_f \eta / T$ increases markedly with temperature. These results clearly show that helix formation, and not diffusion, is rate limiting. This observation illustrates that molecular motions one might intuitively consider quite cumbersome, such as the approach of two ends of a long strand, in fact occur at an appreciable rate on the laboratory and biological time scales.

DNA cyclization reactions are very important in the replication in vivo of some viruses and in various recombinant DNA techniques in vitro. Box 24-1 demonstrates how length and concentration affect the tendency of a DNA to cyclize.

Topological constraints on closed circular DNA duplexes

A large number of circular DNAs have been found in nature, including many animal and bacterial-viral DNAs, plasmids, and mitochondrial DNAs. Most of these are

---

[§] The values in Table 23-4 are for RNA, but DNA results should be qualitatively similar.

neither single-stranded (like ϕX174) nor nicked and capable of equilibrating to linear forms (like λ). Instead, they are circular DNA duplexes with no single-stranded breaks. Such structures are called closed duplexes; any circle with one or more single-stranded nicks is called an open duplex. It is very easy experimentally to distinguish closed from open structures (Fig. 24-2). Suppose the DNA is placed in an alkaline solution. In

---

Box 24-1   DNA CYCLIZATION VERSUS INTERMOLECULAR ASSOCIATION

Any DNA with complementary single-stranded ends, such as bacteriophage λ DNA, is capable of cyclization or self-association into concatomeric linear or circular forms. The relative amount of these forms will depend on the length of the DNA and on the concentration of DNA in the solution. Here we consider only the simple case of competition between formation of linear dimers and cyclic monomers from linear monomers (see figure).

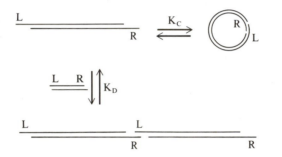

The dimerization reaction, governed by K_D, consists of the complexation of the left end (L) of one molecule with the right end (R) of another. We can write this as

$$2K_D = \text{(Dimer)}/\text{(Monomer)}^2$$

since monomers can combine in two ways to yield the same dimer.

Formally, the cyclization equilibrium is

$$K_C = \text{(Circle)}/\text{(Linear)}$$

We can relate K_D and K_C by realizing that the actual local reaction is the same, but the intramolecular reaction is dependent on the concentration of one end of the molecule in the neighborhood of the other end. Calling this concentration j, we can write

$$K_C = K_D j$$

To calculate the intramolecular end concentration j, we can use Equation 23-41, which shows that (for flexible molecules) the fraction of molecules with both ends located within a any open structure, the high pH will disrupt the base-paired duplex. Even a single nick will allow one strand to unwind about the other, and the two single strands eventually will separate. Alkaline ultracentrifugal analysis of a single-nicked circular duplex will reveal a linear strand and a single-stranded circle. The high pH also can eliminate base-pairing interactions in a closed duplex, but the two single-stranded

volume V_i is

$$\Omega_{cl}/\Omega = (3/2\pi lN)^{3/2} V_i$$

where N is the number of segments, and l is the length of one segment. Because V_i has the dimensions of volume, the concentration j must be

$$j = (3/2\pi lN)^{3/2} \text{ molecules cm}^{-3}$$

The parameters l and N are really statistical segment lengths and segment numbers, and their exact values will depend on the hydrodynamic properties of the particular sample. However, for pure double-stranded DNA, l is constant and N is proportional to molecular weight. For λ DNA with a molecular weight of 30.8×10^6 d, the measured value of j is 3.5×10^{11} molecules cm^{-3} = 5.8×10^{-10} M. For any other DNA of molecular weight M, we can write

$$j = 5.8 \times 10^{-10} (30.8 \times 10^6/M)^{3/2} = 99/M^{3/2}$$

Combining the preceding equations, we can calculate the ratio of cyclic to linear molecules as

$$(\text{Circle})/(\text{Dimer}) = j/2(\text{Linear})$$

If we express the concentration of linear DNA in moles of residue per liter, (C), we have

$$(\text{Linear}) = 330(C)/M$$

Substituting for j and (Linear), we finally can write

$$(\text{Circle})/(\text{Dimer}) = 0.15\ M^{-1/2}/(C)$$

For DNA of a given molecular weight, this final equation shows how the total concentration of linear DNA should be adjusted to favor circles or dimers. This result has enormous practical importance in constructing recombinant DNAs in vitro. Here the usual goal is to favor intermolecular complexes between target and vector DNA molecules, while minimizing unwanted cyclization or higher-aggregate formation. For more details about cyclization, see Wang and Davidson (1966); for experimental examples of the case we have considered, see Dugaiczyk et al. (1975).

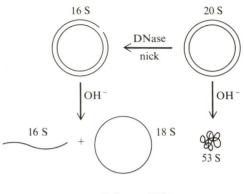

Figure 24-2

Some of the forms of DNA from polyoma virus.
Note the dramatic differences in the changes
caused by alkali treatment of closed versus
nicked circles.

Polyoma DNA

circles formed cannot separate in this case. In fact, as we shall see, they remain so
interwined that a tangled ball of single-stranded DNA results. This very compact
structure has a high sedimentation constant and is easily distinguished from any free
single-stranded forms.

The topological constraints on a closed circular duplex are responsible for the
collapse of the structure at alkaline pH. They also cause the remarkable properties
of closed circles that are the main focus of this section. First, let us show why the two
strands of a closed duplex could not be separated. It is clear that two interlocked rings
cannot be separated unless one is broken. Interwinding two circles once or many
times always results in structures that are inseparable (Fig. 24-3a). Topologically, a
closed circular duplex is equivalent to two circles interwound as many times as there
are turns in the helix.

When simple interlocked rings are viewed with one ring held in a plane, the
linking number α is defined as the number of times one ring passes through the other.
Figure 24-3a,b,c shows that α also is one-half the number of times the two rings appear
to cross each other when they are viewed with one ring in a plane. Now consider the
double-stranded duplex with β turns shown in Figure 24-3d. Suppose we start with
one strand in place and follow the steps needed to thread the other strand in to form
the duplex. For each helix turn, the second strand must pass through the circle
formed by the first. This leads to one interwind per turn. Alternatively, consider the
double helix shown schematically in Figure 24-3e. For each helix turn, the two strands
cross twice when viewed perpendicular to the helix axis. You can see from either
description that $\alpha = \beta$ for a perfect circular duplex with β helical turns.

Now it should be clear why alkali cannot cause strand separation of high-
molecular-weight closed circular duplex DNAs. With a linking number in the
thousands, any attempt to pull one circle free of the other leads to a tangled knot.
The linking number is a constant for any closed duplex. It can be changed only by
breaking one of the two circular strands.

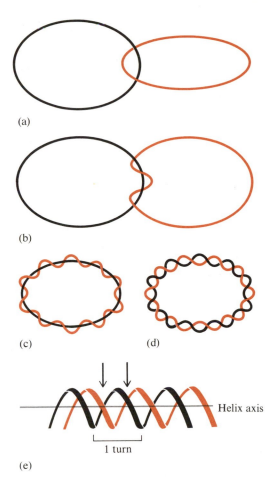

Figure 24-3

Topological properties of interlocked circles.
(**a**) Circles interwound once.   (**b**) Circles interwound twice.   (**c**) Circles interwound ten times.   (**d**) A closed circular duplex helix with ten turns. This structure is topologically equivalent to that in part c.   (**e**) An expanded view of the double helix. Note that the strands cross twice in each turn. From parts a and b, note that each pair of crossings is equivalent to one interwind.   [Parts a–d after W. Bauer and J. Vinograd, in *Basic Principles in Nucleic Acid Chemistry*, vol. 2, ed. P. O. P. Ts'o (New York: Academic Press, 1974).]

Superhelical turns in closed circular duplexes

In addition to helical or duplex turns, a closed circular duplex could contain tertiary (superhelical) turns. To see what these involve, carefully study the thought experiment shown in Figure 24-4a. The duplex turns of the DNA double helix are right-handed. As you travel in either direction along the double helix, the strands rotate clockwise. Suppose you start with a linear duplex β turns long and unwind one turn. The linking number now is $\beta - 1$ and, if you close the DNA into a circle, it will be a circle with one duplex turn unwound. Instead, suppose the linear DNA is grasped firmly and twisted counterclockwise one turn without unwinding any duplex turns. The DNA will kink into a loop. It does this because of topological constraints imposed

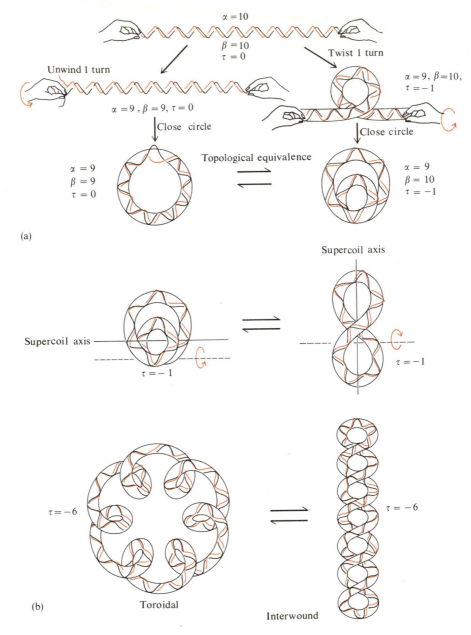

Figure 24-4

DNA supercoils. **(a)** Two schematic ways of introducing one supercoil into a DNA with ten duplex turns. The linking number α, duplex winding number β, and supercoil number τ are defined in the text. For right-handed turns, α and β are positive. The sign of τ is positive for right-handed toroidal turns, negative for right-handed interwound turns. Note that introduction of one negative toroidal turn is equivalent to loss of one duplex turn. **(b)** Interconversion of toroidal and interwound turns. A left-handed toroidal turn is converted to a right-handed interwound turn simply by folding along the dotted line. It is more difficult to visualize the transitions between forms with larger numbers of supercoils.

when you hold it; when your wrists turn to untwist the DNA, it responds by coiling up in the same direction. (The skeptical reader is strongly advised to try this with a piece of clothesline or rubber tubing.) Now, if the ends of the DNA are joined into a circle without allowing it to untwist, the result is a closed duplex with a linking number of $\beta - 1$. There are, however, β duplex turns and one superhelical turn.

From our previous discussion, we see that the two forms of DNA in Figure 24-4a have the same linking number and therefore should be interconvertible without breaking either strand. A real DNA in such a situation must exist as an equilibrium among these two forms and many others with the same linking number. The major biophysical challenge is to discover which forms are thermodynamically favored and then try to rationalize this observation.

The thought experiment illustrated in Figure 24-4a is the consequence of a topological theorem that relates the linking number α to the helix winding number β and the number τ of supercoils or tertiary turns:

$$\alpha = \beta + \tau \tag{24-1}$$

Box 24-2 gives a more rigorous and general definition of the quantities in Equation 24-1.

Because α is constant for any DNA closed circular duplex, any change in β must result in a compensating change in τ. The supercoil shown in Figure 24-4a is called a toroidal turn. A molecule containing many such turns would resemble a doughnut or torus, as shown in Figure 24-4b. The handedness of such a turn can be determined by placing an axis in the plane of the doughnut. A left-handed toroidal turn will result in a 360° counterclockwise rotation of the DNA about this axis. A sign convention must be selected for superhelical turns. It is customary to call right-handed duplex turns positive. Then the left-handed superhelical turn of Figure 24-4a must correspond to $\tau = -1$ in order to maintain $\alpha = \beta - 1$.

Toroidal turns are topologically equivalent to an alternative form of tertiary turn call interwound (Fig. 24-4b). It is easiest to see the equivalence in the case with just one superhelical turn ($\tau = -1$). One structure can be deformed into the other just by a rotation (or folding) along the dotted axis. By convention, the handedness of interwound supercoils is determined by examining the direction of twist around the interwinding axis. Again, a left-handed supercoil will result in a 360° clockwise rotation about this axis. Note carefully that the axes used to define toroidal and interwound turns are not the same. The single left-handed toroidal coil shown in Figure 24-4b is topologically equivalent to the single right-handed interwound coil. Thus we see that, for a self-consistent sign convention, we must take τ as negative for right-handed interwound superhelical turns, even though it is positive for right-handed toroidal superhelical turns. This problem is avoided by the treatment developed in Box 24-2.

Box 24-2   TOPOLOGICAL PROPERTIES OF CLOSED CIRCULAR DUPLEXES

In the text, we define three quantities to describe particular types of double-strand topologies: the linking number α, the helix winding number β, and the supercoiling number τ. Here we seek a more general description of the topology of circular duplexes. The treatment that follows was developed by F. B. Fuller (1971); its applicability to DNA was elucidated by F. H. C. Crick (1976).

    In order to concentrate on the overall topology of the DNA, we model the normal B-form helix as a flat ribbon constructed by joining every tenth phosphate on one strand to form one edge of the ribbon, and every tenth phosphate on the other strand to form the other edge (Fig. a). The edges of the ribbon are indicated with arrows or colors to keep track of the opposite polarity of the two DNA strands. Closed circular DNA with a perfect B-form helix and no supercoils then is described as the closed ribbon shown in Figure b.

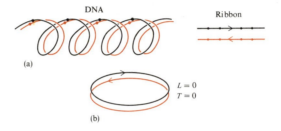

(a)

(b) $L = 0$ $T = 0$

Linking Number

The linking number L of the ribbon is defined as the number of times the two edges (viewed as separate closed lines) are linked in space. For example, the deformed ribbon shown in Figure c has a linking number of $+1$. The positive sign is chosen because the twists are right-handed (clockwise). The mirror image of the structure in Figure c would have a linking number of -1;

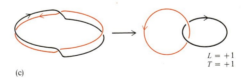

(c) $L = +1$ $T = +1$

in general, reflection in a mirror always causes a change in the sign of L. The linking number of a closed curve can never be changed unless the curve is torn or broken.

    Note that, by modeling DNA as a flat ribbon, we are ignoring any linking due to the interwinding of the two strands of the double helix itself. The relationship between L and α is

$$L = \alpha - \beta_0$$

where β_0 is the number of duplex turns expected for the DNA in a normal unconstrained B-form helix.

The linking number of a closed ribbon must always be an integer. This interger L can be calculated for any closed ribbon in a simple way. View the ribbon from a distance and note the number of times the colored curve crosses *in front of* the black (or vice versa, but not both). Assign a value of $+1$ or -1 to each crossover, depending on whether it is a right-handed or a left-handed twist. Then sum these values to obtain L. Thus, for the example in Figure c, if we look at the black curve crossing the colored curve, we see 3 crossovers: 2 right-handed, and 1 left-handed. Thus, $L = 2 - 1 = +1$. If we look at the colored curve crossing over the black, we see only one crossover and it is right-handed, so again $L = +1$.

Twist Number

The twist is another property of a ribbon. To define it, we first construct an axis in the plane of the ribbon and halfway between the edges. At any point along the ribbon, let $\mathbf{x}$ be the vector tangent to the ribbon. Define a second vector $\mathbf{u}$ perpendicular to $\mathbf{x}$ and intersecting both edges of the ribbon (Fig. d). The twist, as one moves along the ribbon, is measured by the extent to which $\mathbf{u}$ rotates about $\mathbf{x}$. The total twist number T is the integrated twist along the entire ribbon divided by 2π. If ω is the angular rate per unit arc length at which $\mathbf{u}$ rotates about $\mathbf{x}$, and ds is the increment in arc length, then T is the line integral

$$T = (1/2\pi) \oint \omega \, ds$$

The twist number need not be an integer, and it is defined for open as well as closed curves. Reflection in a mirror changes the sign of the total twist number. For the flat-ribbon model of DNA, the total twist number is related to the helix winding number by

$$T = \beta - \beta_0$$

A ribbon can be bent without twisting (Fig. e) or twisted without bending (Fig. f). Also note that $T = 0$ for Figure b, and $T = +1$ for Figure c. The twist can be calculated separately for individual sections of a ribbon. The total twist is then computed by summing the twist numbers of the sections. (L cannot be calculated in this way.)

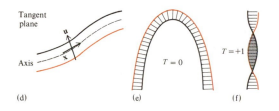

(d) (e) (f)

Box 24-2 (*continued*)

To model supercoils of various types, it is convenient to consider a ribbon wound flat in a cylinder to form a helix with N turns and a constant pitch angle θ. A helix can be projected onto a cylinder and the cylinder unwrapped to give a flat sheet (Fig. g). In the unwrapped form, the vector **u** does not wind around the vector **x** at all. Therefore, to compute T we need only calculate the number of times that **u** winds around **x** when we wrap the cylinder back up to make the helix. It is easier to do this if we resolve **u** into components parallel ($\mathbf{u_1}$) and perpendicular ($\mathbf{u_t}$) to the long axis of the cylinder (Fig. h).

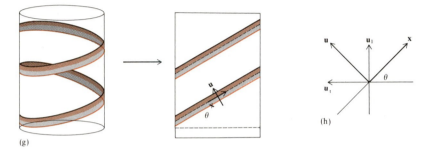

(g)   (h)

Wrapping up the sheet to a cylinder has no effect on $\mathbf{u_1}$. However, for each turn of the helix, the wrapping causes $\mathbf{u_t}$ to rotate about **x** by 360°. Thus, the rotation experienced by **u** per helix turn is the projection on $\mathbf{u_t}$ given by $\cos(90° - \theta) = \sin\theta$. For a helix of N turns, therefore, the total twisting number is

$$T = N\sin\theta$$

Writing Number

Fuller defined the writhing number W of a ribbon as

$$W = L - T$$

Because L and T change sign upon mirror reflection, so must W. It turns out that W has a number of interesting properties. It is a function only of the path of the axis of the ribbon in space. Thus, given a closed curve, any way of placing a ribbon on it results in the same writhing number. Essentially, W is a measure of the extent to which the axis coils and folds in three dimensions. The value of W is zero for any closed curve lying on a sphere or a plane. For such a curve, the twist number can easily be determined by using the simple rule stated earlier to compute the linking number, and then setting $T = L$.

For a DNA helix, using equations given earlier, we can write

$$W = \alpha - \beta$$

Comparing this expression with Equation 24-1, we see that $W = \tau$. However, there is a subtle difference. The discussion in the text defines τ only in relation to arbitrary axes, but W is a totally general property.

The use of W, L, and T

Consider the six cases of a closed ribbon wrapped around a cylinder shown in Figures i through n. Figure i is a model for a toroidal supercoiled molecule. Two complete helix turns are wrapped around the cylinder and connected into a closed curve by a twistless link. The black edge crosses the colored edge in 2 right-handed twists; therefore, $L = +2$. The twist number is $2 \sin \theta$; therefore, $W = 2(1 - \sin \theta)$. In other words, the number of supercoils will depend on the helix pitch angle θ.

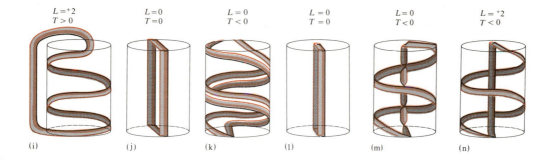

$L = {}^{+}2$	$L = 0$	$L = 0$	$L = 0$	$L = 0$	$L = {}^{+}2$
$T > 0$	$T = 0$	$T < 0$	$T = 0$	$T < 0$	$T < 0$
(i)	(j)	(k)	(l)	(m)	(n)

In Figure j, obviously $L = 0$ and $T = 0$. Figure k can be generated from Figure j by holding the bottom surface of the cylinder fixed and rotating the top surface 720° (two complete turns) clockwise. After this twist, L must still be zero. For each of the two-turn-left-handed helices, $T = -2 \sin \theta$, so the total twist number is $-4 \sin \theta$, and the writhing number W is $+4 \sin \theta$.

In Figure l, again $L = 0$ and $T = 0$. Figure m can be generated from Figure l in exactly the same way that Figure k was generated from Figure j. Again, L must be zero. The twist of the central portion is -2, and the outer helix contributes $-2 \sin \theta$, so the total twist number is $T = -2(1 + \sin \theta)$. Therefore, the writhing number is $2(1 + \sin \theta)$.

Figure n is like Figure m, except that the twists of the central portion have been removed. Therefore, $T = -2 \sin \theta$. However, the ribbon axes in Figures m and n are identical; therefore, W in each case is $2(1 + \sin \theta)$. Thus we can calculate the linking number for Figure n as $L = T + W = +2$. This is the same value of L as in Figure i. Thus, these two forms can be interconverted without breaking the chain. An advantage of the writhing-number approach is that it lets us compute the properties of a form such as Figure n with relative ease. It also shows that the linking number depends critically on how one connects the ends of a coil to make a closed curve.

For any given structure, W and T can vary continuously, whereas L is fixed. Therefore, experimental measurements that reflect size and shape will be averages over various W and T values, but the measurements should differ for species that have different L values. Thus we see that, in the case of supercoiled DNA, what matters critically is not really τ, but α.

Experimental observation of supercoils

Let us use Equation 24-1 to analyze the alkaline denaturation of a simple closed duplex DNA. If the base pairs are disrupted by deprotonation, the two strands tend to repel each other. For the strands to separate locally, some helix unwinding must occur. This unwinding decreases β, but that means that τ must become more positive to maintain constant α. If τ was zero before alkali treatment, then the increase in τ will correspond to right-handed toroidal or left-handed interwound turns. As these accumulate, the resulting structure will wind tighter (as in Fig. 24-4b). The collapsed tertiary structure will lead to a greatly increased sedimentation coefficient.

Equation 24-1 shows how supercoils can be created in the laboratory by varying experimental conditions so that β changes. However, supercoils also exist in natural DNA. Discovery of these natural supercoils originally called attention to the topological aspects of closed DNA. J. Vinograd and colleagues had been studying the DNA of polyoma virus (Fig. 24-2). They observed two forms at neutral pH: form I sedimenting at 20 S, and form II at 16 S. In alkali, they observed three DNA forms: 16 S and 18 S structures (which would be consistent with linear strands and single-stranded circles), and a very fast 53 S, which must have come from a closed duplex. Reasoning that the two neutral forms might differ by only one single-strand nick, they added minute amounts of DNase. This treatment caused a parallel decrease in concentrations of neutral form I and alkaline 53 S material, and a compensating rise in concentrations of neutral form II and the alkaline 16 S and 18 S materials. Thus, form I is closed duplex, and form II has only a single nick. Why then should the neutral sedimentation velocities of these two forms be so different? When both forms were examined by electron microscopy, form I appeared much more twisted or tangled (Fig. 24-5). This observation suggested the presence of supercoils. The challenge was to prove their presence and to determine the number and handedness of the supercoils.

A general strategy has been used in a number of different ways to determine the sign of τ and to estimate its magnitude. Singly-nicked DNA (such as form II polyoma) is used as the reference state for a closed DNA of the same species presumed to have supercoils. Experimental conditions are adjusted to vary β for both forms while measuring a physical parameter such as sedimentation velocity, viscosity, amount of ligand bound, or buoyant density. For example, deprotonation and ethidium binding are expected to unwind the helix, resulting in a decrease in β. Figure 24-6 shows a typical experimental result. As increasing amounts of ethidium are bound to DNA, the sedimentation constants of forms I and II approach the same value and then diverge again. At the point where both forms have the same physical properties, they should have identical shapes and contain the same number of bound ligands. We shall call the amount of ethidium bound per base pair at this point $\bar{v}_c$, the critical saturation ratio.

At any ethidium concentration, the nicked form II should have a chain configuration corresponding to an untwisted duplex circle. Therefore, by binding $\bar{v}_c$ ethidium molecules per base pair, the closed duplex form must have been converted

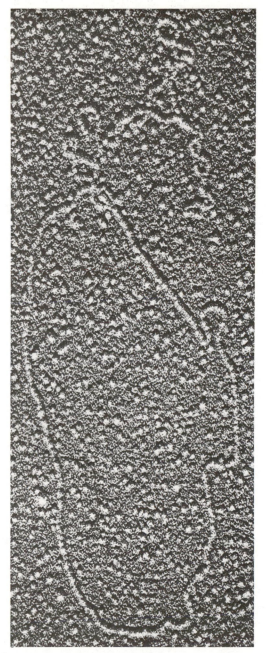

Figure 24-5

Relaxed and supercoiled circular DNA. This electron micrograph shows a DNA catenane containing one relaxed and one supercoiled circular DNA. The samples were prepared by the Kleinschmidt technique. [Courtesy of Mavis Shure.]

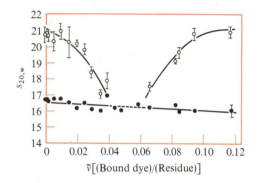

Figure 24-6

Sedimentation velocity of SV40 DNA as a function of bound ethidium. Form I DNA is closed circular duplex (○); form II is nicked circular duplex (●). Measurements were made in 1.0 M NaCl. [After W. Bauer and J. Vinograd, *J. Mol. Biol.* 33:141(1968).]

to a structure with τ very close to zero. Because ethidium binding reduces β, Equation 24-1 indicates that the binding must cause a compensating increase in τ. Therefore, the original τ of the native form I must have been negative, and the supercoils must be left-handed toroidal or right-handed interwound. Adding more ethidium raises the binding ratio above $\tilde{v}_c$. This reduces β further, and now τ must take on positive values in accordance with Equation 24-1 (Fig. 24-7). Ethidium first removes the original supercoils and then introduces new ones of the opposite handedness. At sufficiently high values of bound ethidium per base pair, the sedimentation changes level off. There are two possible reasons for this. It should be difficult to wind more than a certain number of supercoils into the DNA. Furthermore, the excluded-site binding model places a limit on the number of potential ethidium binding sites.

All natural DNAs that contain supercoils behave in qualitatively the same manner as the sample shown in Figure 24-6. Therefore, all have supercoils of the same handedness in the absence of ethidium. If an alkaline titration is used instead of the ethidium treatment, the same general pattern of behavior is observed, thus providing an independent confirmation of the negative values assigned to τ.

Effect of dye binding on the number of supercoils

To estimate the magnitude of τ, one must know the change in β induced by the binding of an ethidium or any other perturbant. Because binding data usually are expressed in ratios independent of the molecular weight of the DNA, it will be convenient to develop a compatible representation for supercoiling. Divide each term in Equation 24-1 by β_0, the number of duplex turns expected for the entire DNA molecule of interest if it were in a perfect B form. With β_0 equal to simply one-tenth the number of base pairs, Equation 24-1 becomes

$$\alpha/\beta_0 = \beta/\beta_0 + \sigma_h \qquad (24\text{-}2)$$

where σ_h is defined as the superhelix density, τ/β_0. It is in units of supercoils per helix turn.

Suppose that the binding of one ethidium or other ligand causes a net change in helix winding of ϕ degrees. A negative ϕ will correspond to unwinding. Binding of one ethidium will change β by $\phi/360$ turns. There is good experimental support for assuming that ϕ is not a function of $\tilde{v}$, the amount of dye bound per base pair. Therefore, binding of $\tilde{v}$ ethidiums per base pair will change β by $\tilde{v}\phi/360$ turns per base pair. The corresponding change in β/β_0 will be $10\tilde{v}\phi/360$ turns per turn. The factor of 10 enters because $\tilde{v}$ is expressed in units of dyes per base pair, and there are 10 base pairs per turn.

Equation 24-2 demands a compensating change in supercoils. If the original unperturbed DNA has a superhelix density σ_0, then ethidium or other ligand binding will change it to

$$\sigma_h = \sigma_0 - \tilde{v}\phi/36 \qquad (24\text{-}3)$$

This equation has a number of diverse and important applications. If ϕ and $\tilde{v}$ are known as a function of the amount of ligand added, then when the DNA is fully relaxed, $\tilde{v} = \tilde{v}_c$, $\sigma_h = 0$, and σ_0 is determined. Alternatively, if σ_0 is known, then ϕ can be measured if $\tilde{v}$ is already available, and vice versa.

An easier experiment is to compute the relative values of ϕ for different ligands, or the relative values of σ_0 for different DNAs. In general, when two samples are compared in titrations that reduce the number of supercoils to zero,

$$\sigma_{0,1}/\sigma_{0,2} = (\phi_1/\phi_2)(\tilde{v}_{c1}, \tilde{v}_{c2}) \qquad (24\text{-}4)$$

Suppose that the DNA is the same and the ligands are different. Then $\sigma_{0,1} = \sigma_{0,2}$, so that $\phi_1 \tilde{v}_{c1} = \phi_2 \tilde{v}_{c2}$. Thus, by measuring the ratio of the critical saturation ratios of two ligands, one can determine their relative unwinding angles. In this way, it has been shown that a whole set of planar molecules believed to bind by intercalation have very similar unwinding angles. Another important application of Equation 24-4 is to compare the superhelix densities of different DNAs. Even if ϕ is unknown, it must be constant so long as the same intercalating dye is used. Then a measurement of the ratio $\tilde{v}_{c1}/\tilde{v}_{c2}$ yields relative values of σ_0.

In early work, a value of $\phi \cong -12°$ for ethidium was estimated from the results of intercalation studies using model building and x-ray fiber diffraction. For alkaline titrations, it was assumed that each G and T must be deprotonated to cause unwinding of one helical turn. The resulting values of τ were in good agreement with one another. However, in practice, this agreement may have been largely fortuitous, because more recent estimates of ϕ are about twice the earlier value. There are several difficulties to the use of Equation 24-3 in addition to uncertainty in ϕ. As you can see from Figure 24-6, the sedimentation rates of forms I and II coincide over a fairly broad range, and this makes precise determination of $\tilde{v}_c$ difficult. The binding measurements that must be done to determine $\tilde{v}_c$ are more complex for supercoiled molecules than for nicked circles or linear duplexes, as we shall see. Finally, all we can measure is the value of $\tilde{v}$ at which τ is a minimum. It is not clear that τ actually is zero at the point we are experimentally defining as $\tilde{v}_c$. Depending on the energetics of supercoiling, relaxed molecules are expected still to have a distribution of τ values and, if this distribution is not symmetrical around zero, the minimum in measured average sedimentation need not correspond to the amount of dye bound expected by Equation 24-3.

Energetics of supercoiling

It clearly is worthwhile to examine the energetics of supercoiling and how this affects the binding of ligands. First, note that the closed and open *relaxed* forms (by definition) must have the same free energy and the same amount of bound ligand. This means that opening a nick or sealing it (Fig. 24-7) must proceed with little change in $\bar{\nu}$ or in conformational free energy.

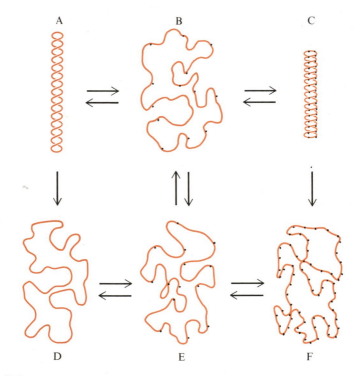

Figure 24-7

Schematic explanation of the experiment shown in Figure 24-6. Structures A through C are form I DNA; structures D through F are form II. In A, the original DNA has 14 left-handed interwound turns. The open forms B and E contain the same number of bound ethidiums. The final form I (C), saturated with ethidium, has 14 right-handed interwound turns. (Actually, it is not known whether the supercoils are toroidal or interwound.) [After W. Bauer and J. Vinograd, *J. Mol. Biol.* 33:141 (1968).]

In the absence of any ligand, superhelices are higher in free energy than are open duplexes. The proof is that introduction of a nick leads to spontaneous removal of the superhelices, resulting in a relaxed form. The origins of this more positive free energy are the entropy loss in constraining the DNA to a more compact ordered supercoil, and any unfavorable enthalpy of strain or distortion. Because the native superhelix is higher in energy than otherwise equivalent nicked DNA at $\tilde{v} = 0$, and their free energies are equal at $\tilde{v} = \tilde{v}_c$, it necessarily follows that below $\tilde{v}_c$ the free energy of binding of any ligand that can cause removal of the supercoils must be more negative to the supercoil. This means that ethidium should bind more strongly to native ($\tau < 0$) supercoils than to normal DNA duplexes. Above $\tilde{v}_c$, it is more difficult to bind additional dye to the closed DNA because now each dye bound introduces high-energy supercoils. Thus, just from ethidium-binding measurements alone, $\tilde{v}_c$ can be determined as the point at which the binding isotherms of forms I and II DNA cross. Figure 24-8 illustrates a sample experiment of this kind.

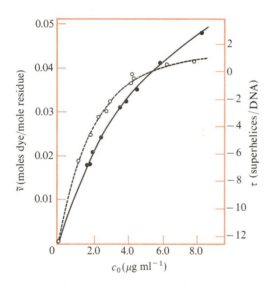

Figure 24-8

Binding of ethidium to SV40 DNA as a function of the total amount of ethidium added to solutions of form I (*dashed curve*) and form II (*solid curve*) with equal DNA contents. Note that the binding isotherms cross at the point where there are no supercoils. [After W. Bauer and J. Vinograd, *J. Mol. Biol.* 33:141 (1968).]

The apparent binding constant for ethidium is a function of τ or σ_h. Several different workers have found that the free energy of supercoiling, which is responsible for this dependence, is well approximated by $\Delta G_\tau = B\tau^2$, where B is a constant. W. Bauer and J. Vinograd estimated, from a detailed analysis of the binding curves, that native form I SV40 DNA (which has 26 supercoils) has a free energy of supercoiling of 100 kcal mole^{-1}. More recent estimates by T. Hsieh and J. Wang suggest a somewhat smaller value, but the qualitative picture remains the same. If the supercoiling free energy is distributed evenly over the molecule, it amounts to only 13 to 25 cal per base pair, which is not large enough to yield any local structural changes sufficient to alter substantially any chemical or physical properties. If, however, some of the superhelix energy can concentrate in one or more particular local regions, it could lead to structural distortion or even to unwinding of the duplex (Fig. 24-4). In fact, some supercoiled DNAs appear to have regions of heightened reactivity or accessibility. Whether these regions are of biological significance is not yet known.

Measuring dye binding by buoyant density shifts

The difficulty with all the determinations thus far discussed remains the actual measurement of the number of dye molecules bound to supercoiled DNA. Many supercoiled DNA samples are quite difficult to obtain, and standard spectrophotometric binding techniques (Chapter 23) require large amounts of sample. An alternative, first used by J. Vinograd and colleagues, is to exploit the shifts in buoyant density that accompany ethidium binding. Ethidium has a density lower than that of nuclecic acids and, furthermore, it should displace Cs^+ counterions when it binds. The most straightforward approach would be simply to titrate DNAs with ethidium and determine the point at which open and closed forms of the same molecule have the same buoyant density. This approach, however, would involve many tedious experiments.

The ease and accuracy of detecting ethidium-induced changes in buoyant density suggests an alternative approach. Suppose that enough ethidium is added to unwind all the native supercoils, and then to wind some of them in the opposite sense. Figure 24-9 shows the effect of this treatment for three molecules of different original superhelix densities (σ_0). In each case, the final extent of ethidium binding for open and closed forms will be compared. It is convenient to divide the binding into two stages. First, a certain amount of ethidium will be consumed to convert each closed DNA into a relaxed circle; this amount increases linearly with the superhelix density. Then more binding of ethidium will occur, but this binding is subject to two restrictions. For a DNA of high σ_0, many of the ethidium binding sites potentially available were filled just to relax the native supercoils. Thus at $\tilde{v} > \tilde{v}_c$, both open and closed forms bind relatively similar amounts of ethidium, because filling the few available sites remaining cannot introduce many new supercoils. In contrast, a closed DNA with low σ_0 will bind a total of much less ethidium than will its open form. Although the amounts bound at $\tilde{v} = \tilde{v}_c$ are equal, at $\tilde{v} > \tilde{v}_c$ it will cost much more free energy to bind to the closed form, because of the increase in supercoils that must accompany the binding. Because more ethidium is bound, the buoyant density of closed DNA in excess ethidium decreases with increasing values of the original superhelix density in the absence of dye. This dye buoyant-density technique is very useful as a preparative technique for separating DNAs as a function of their superhelix densities. Bauer and Vinograd showed that, when two closed DNAs of different σ_0 are compared, the distance of separation in a CsCl gradient is a linear function of the difference in σ_0.[§] Thus, once a calibration is established, σ_0 can be determined by a single experiment

---

[§] This observation is consistent with what can be predicted by treating the four-component (dye, CsCl, H_2O, DNA) thermodynamics of the system.

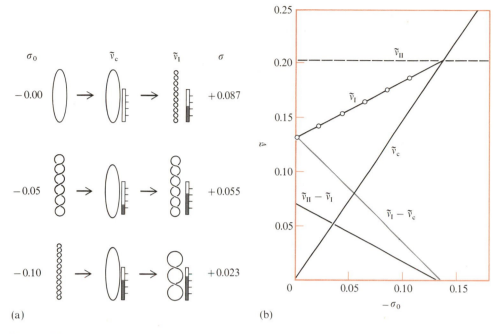

Figure 24-9

Effect of superhelix density on ethidium binding: a schematic experiment. The same amount of ethidium is added to solutions of three DNAs with different initial superhelix densities σ_0. **(a)** The amount of binding needed to relax the original supercoils is $\tilde{v}_c$. The fraction of sites occupied is shown by a vertical bar to the right of each molecule. The final binding fraction is $\tilde{v}_I$, producing a final superhelix density σ. From measurements of $\tilde{v}_I$, it is possible to compute the values of σ_0. **(b)** Binding ratios plotted against initial superhelix density. The horizontal curve $\tilde{v}_{II}$ represents the binding ratio of ethidium to SV40 DNA form II at 330 μg dye ml^{-1}. The curve for $\tilde{v}_{II} - \tilde{v}_I$ represents the difference in binding ratios responsible for the separation between the closed and open DNAs. [After W. Bauer and J. Vinograd, *J. Mol. Biol.* 54:281 (1970).]

in the ultracentrifuge. However, the calibration still requires a knowledge of the unwinding angle.

A technique that has greatly assisted such measurements is synthesizing DNAs of desired σ_0, rather than using DNAs isolated from natural sources. We mentioned earlier that DNA ligase can seal a nicked circular DNA to a closed circle. Suppose that, before ligase treatment, $\tilde{v}$ moles of ethidium per mole of base pairs are bound to the open DNA. After sealing, the ethidium is removed. Because of the topological constraint, left-handed toroidal (right-handed interwound) supercoils must form. Equation 24-3 shows that removal of all the ethidium will produce a superhelical density of $\tilde{v}\phi/36$. Thus, by adjusting $\tilde{v}$ prior to ligase treatment, one can make closed molecules with any desired negative superhelix density in the absence of dyes. When large amounts of original sample are available, the amount of dye bound prior to closing can be determined by direct spectrophotometric measurements and, because the DNA is open, there are no complexities due to supercoiling effects on the binding. Unfortunately, no comparable technique yet exists to produce dye-free positive superhelix densities.

Determining the number of supercoils

All of the methods thus far discussed for computing σ_0 require knowledge of the unwinding angle of bound ethidium. As mentioned earlier, the original estimate of $\phi \cong -12°$ was obtained by accepting rough estimates from model building and showing consistency with unwinding caused by alkaline titration. These latter measurements, however, were quite difficult. They were monitored by sedimentation or by buoyant density, but in either case it was difficult to locate the endpoint of the titration corresponding to the relaxation of the supercoils. Recently, Wang was able to obtain a clearer endpoint by using synthetic supercoils that require titration by the order of 100 or more hydroxide ions per molecule in order to relax the supercoils. Comparing alkaline buoyant-density measurements with ethidium titrations produced a new value, $\phi = -26°$. This revision changes all previous estimates of supercoil numbers roughly by a factor of two.

Several confirmations exist of the new larger ϕ values. More accurate model building, based on the crystal structures of dye–oligonucleotide complexes, yields values of $-26°$ to $-28°$ for ϕ (Chapter 23). The use of covalent but reversible chemical modifications with water-soluble carbodiimides simplifies the problem of determining the amount of reagent bound in unwinding titrations. Radiolabeled reagent allows an unequivocal determination of the amount of attached reagent. It seems reasonable that mild treatment with such a reagent should completely melt out a few susceptible duplex regions, leaving the rest of the DNA unaltered. If a supercoil so treated is then relaxed by nicking and then resealed, subsequent removal of the covalent reagent produces single-stranded loops that will spontaneously close. Supercoils will appear because of the topological constraint. Thus, this approach is essentially identical to the thought experiment in Figure 24-4. It leads to estimates of $\phi \cong -33°$.

Another way of performing the same manipulation would be to anneal complementary single strands to a few particular regions of a circular duplex. Nature has provided one such model system in the mitochondrial D loop. This intermediate in replication consists of a supercoil in which a stretch of 400 to 500 base pairs has been disrupted, and one strand displaced with a single strand of DNA (Fig. 24-10). Because removal of the extra single strand must result in winding of the duplex, the process can be used as a reference for the sign of supercoiling.

Of course, all the data on supercoil numbers we have presented are indirect, and a skeptical observer might still be uneasy about all the assumptions that enter into the determination of τ or σ. A direct approach would be to arrange somehow to count the supercoils. In principle, this should be possible with electron microscopy. That

Figure 24-10

Pictorial representation of a mitochondrial DNA D-loop. A short region of one strand has been replicated, displacing a section of parent single strand. The unwinding necessary to accomplish this displacement reduces the number of supercoils, proving that the original τ was less than zero. [After W. Bauer and J. Vinograd, in *Basic Principles in Nucleic Acid Chemistry*, vol. 2, ed. P. O. P. Ts'o (New York: Academic Press, 1974).]

technique certainly has permitted determination of the handedness of the supercoils, but quantitation is difficult. How can one be sure that β and τ have not been altered while constraining the molecules to lie on a grid in a dry state, and while complexing with protein to allow visualization?

However, there is an alternative to microscopy. W. Keller discovered that DNAs with the same molecular weight but different τ values separate cleanly on agarose gel electrophoresis (Fig. 24-11a,b). Some of these samples were generated by treating native supercoiled SV40 DNA with a nicking-and-closing enzyme that can relax supercoils (Fig. 24-12). The detailed mechanism of this enzyme is unknown, but clearly it must break one or both strands and then reseal the break. To relax supercoils, the enzyme must allow either random rotational diffusion around a nick or a single turn prior to reclosing. The idea that the enzyme would allow only a fixed number of turns greater than one seems most implausible.[§] Whichever mechanism operates, it should produce a series of DNAs spaced by increments of one supercoil. If one naively counts the bands in the partially relaxed sample in Figure 24-11a and assumes that the extremes represent zero and maximal native supercoils, one finds $\tau_0 = -26 \pm 1$.

The real explanation of Figure 24-11 is slightly more complicated, but the result is the same. The native sample (Fig. 24-11a) is heterogeneous in τ, and the most probable form has several fewer supercoils than the maximal-supercoiling peak observed. The gel conditions used in Figure 24-11a have high resolution at high superhelical densities. These allow good resolution of the distribution of supercoiled molecules in the native sample, but not in the relaxed sample. When run under different gel conditions, the fully relaxed sample (Fig. 24-11b) also shows a distribution of supercoiled forms. This is not an artifact of the nicking-and-closing enzyme, because DNase nicking and ligase resealing produce the same results. The relaxed sample should contain molecules with $\tau = 0, \pm 1, \pm 2, \ldots$ (Fig. 24-12). Because the supercoil free energy is $B\tau^2$, a Boltzmann distribution of products will be

$$f_\tau = A \exp(-G/RT) = A \exp(-B\tau^2/RT) \qquad (24\text{-}5)$$

where f_τ is the fraction of species with τ supercoils, and A is a normalization constant. The actual data fit the expected Gaussian form moderately well.

[§] See, however, P. O. Brown and N. R. Cozzarelli, *Science* 206:1081 (1979).

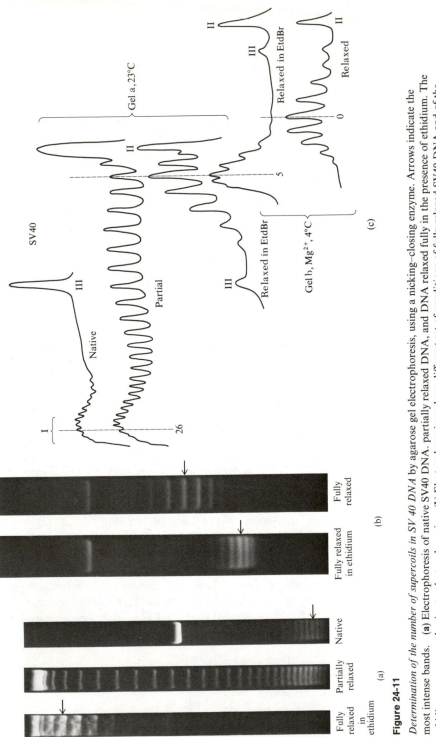

Figure 24-11

Determination of the number of supercoils in SV 40 DNA by agarose gel electrophoresis, using a nicking–closing enzyme. Arrows indicate the most intense bands. (**a**) Electrophoresis of native SV40 DNA, partially relaxed DNA, and DNA relaxed fully in the presence of ethidium. The ethidium was removed prior to electrophoresis. (**b**) Electrophoresis under a different set of conditions of fully relaxed SV40 DNA and of the same sample of DNA relaxed in the presence of ethidium. (**c**) Alignment of the results of parts a and b using the one sample common to both sets of electrophoresis conditions permits counting of all the bands. A densitometric scan of the gel is shown here. The labels I, II, and III refer to native supercoiled, nicked circular, and linear DNA, respectively. [After M. Shure and J. Vinograd, *Cell* 8:215 (1976).]

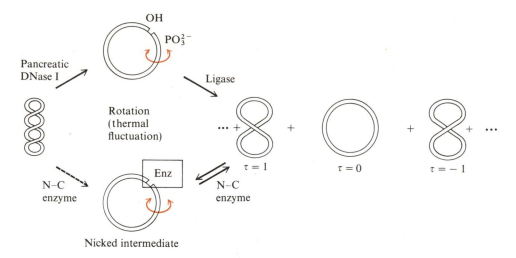

Figure 24-12

Schematic experiment for relaxation of supercoiled DNA by DNase or by nicking-and-closing enzyme. The final equilibrium mixture is a distribution of discrete species. [After D. Pullyblank et al., *Proc. Natl. Acad. Sci. USA* 72:4280 (1975).]

Equation 24-5 allows us to locate the band corresponding to $\tau = 0$ in the relaxed sample. Unfortunately, the gel conditions used in Figure 24-11b do not give good resolution of native SV40 DNA. To bridge the gap between the two gel systems, a sample of DNA was fully relaxed in the presence of ethidium. Then the ethidium was removed, generating a distribution of supercoiled molecules. This sample, on both gel systems, overlaps native and relaxed SV40 DNA samples sufficiently to allow accurate assignment of all the bands. Then counting the bands (Fig. 24-11c) permits the most probable number of supercoils in native SV40 DNA to be set at 26. This value in turn can be combined with the results of dye-binding studies, using Equation 24-3 to compute the ethidium unwinding angle. The result is $\phi = -23°$, in fairly good agreement with other recent estimates. Equation 24-5 can be used also to determine the energy parameter B, and the values obtained are in quite good agreement with those from ethidium-binding studies. This agreement is highly encouraging, considering the enormous difference in the nature of these gel experiments and simple thermodynamic measurements.

Closed duplexes represent an exquisitely sensitive opportunity to monitor changes induced in DNA when a ligand binds. This is applicable to macromolecules as well as to small ligands. One example is the study of the binding RNA polymerase to λ DNA. Samples were closed with ligase in the presence or absence of varying amounts of bound polymerase. Then, after removal of the enzyme, sedimentation differences were measured. These differences can be related to the number of induced supercoils and thus to the unwinding angle, as discussed earlier. The result is an unwinding angle of about 260° per RNA polymerase. This result places very strong constraints on the nature of the enzyme–DNA complex.

Topoisomerases

Are supercoils just a laboratory curiosity, or do they play significant roles in the function of DNA in vivo? The helix winding number β is known to be a fairly sensitive function of experimental conditions. Thus, the observation that many closed circular DNAs are supercoiled as isolated in vitro could, in principle, just reflect the differences in β between conditions in vivo and those in vitro. This hypothesis leads to the prediction that all closed circular DNAs isolated from the same organism should have approximately equal superhelix densities, independent of the molecular weight of the DNA. In practice, superhelix densities (corrected to $37°$ C, and 0.2 M NaCl) vary over a significant range. For example, closed circular bacteriophage λ isolated from infected *E. coli* cells shows σ_h values from -0.032 to -0.044, whereas the plasmids pSM1 and pSM5 have σ_h values of -0.063 and -0.092, respectively.

The strongest evidence that supercoiling plays biological roles is the existence of numerous enzymes that affect the superhelical density, and the demonstration that antibiotics interfering with these enzymes (or bacterial strains lacking these enzymes) result in abnormal nucleic acid function. Enzymes that alter the superhelix density are called by the general term topoisomerases. Four types of such enzymatic activity can be imagined: relaxation of positive or negative supercoils, and introduction of positive or negative supercoils. Various enzymes can be expected to combine one or more of these activities. Among the topoisomerases that have been isolated and characterized thus far are the following.

1. *E. coli* ω protein: this enzyme catalyzes the relaxation of negatively super-coiled DNA specifically.

2. *E. coli* *Nal*A gene product, and KB cell nicking—closing enzymes: these substances catalyze the relaxation of both positively and negatively super-coiled DNA.

3. *E. coli* gyrase: this protein, discovered by M. Gellert and colleagues, catalyzes the introduction of negative supercoils, using the hydrolysis of ATP as an energy source. Gyrase contains two subunits: an ATPase sensitive to the antibiotics coumeromycin and novobiocin, and the *Nal*A gene product. From studies on the effects of various antibiotic inhibitors, we know that gyrase is involved in the replication of various circular and linear DNAs, and that it also is involved in the transcription of certain operons. Gyrases presumably are responsible for the superhelical density seen in various native DNAs.

The role of supercoils in biological function still is not known. However, it is clear that the presence of supercoils affects the ability of DNA to serve as a substrate for various enzymes in vitro.

24-2 TERTIARY STRUCTURE IN RNA

The supercoils we have discussed thus far are largely in the domain of tertiary rather than secondary structure. The conformational equilibria of supercoiling represent the coupling of both levels of structure. There also is clear evidence for the existence of tertiary structures in various RNA molecules. Here we briefly discuss some of the solution techniques that have revealed information about the structure of tRNA, the best-studied RNA. It is of particular interest to try to understand how a nucleic acid with a tertiary structure undergoes conformation changes, such as thermal unfolding.

Evidence of tRNA tertiary structure in solution

The universality of the cloverleaf pattern of base pairing tRNA suggests that it surely must be a structural features of tRNAs. However, a number of lines of evidence indicate that the cloverleaf is not a complete description of the ordered structure of tRNA in aqueous solution. A cloverleaf such as that in Figure 24-13a (or Fig. 24-17) leads to a number of predictions inconsistent with experimental data. Hydrodynamically, a free cloverleaf either should behave like a flexible molecule or, if rigid, should resemble an oblate ellipsoid. However, hydrodynamic measurements are not in accord with such a structure.

Flexibility of tRNA (at least in a nanosecond time scale) is for the most part ruled out by fluorescence polarization measurements. These data yield a rotational relaxation time τ_D of about 26 nsec. For a rigid spherical tRNA, one would predict that τ_D should be only about 12 nsec. Flexibility can only lower this value, and oblate ellipsoids behave rotationally quite similarly to spheres. Thus the large measured τ_D suggests that tRNA is more accurately modeled as a prolate shape, and that it must be fairly rigid. This conclusion is supported by a variety of other measurements, including intrinsic viscosity (~ 6 cm^3 g^{-1}) and sedimentation (~ 4.2 S), both of which are inconsistent with spherical tRNA and argue instead for an elongated structure. Additional evidence comes from small-angle x-ray scattering, which is best fit by rigid elongated structures with lengths of 85 to 92 Å and widths and thicknesses of only 22 to 35 Å.

One of the strongest early indications of some kind of ordered tertiary strucutre in tRNA was the discovery of native and denatured forms. These two forms have relatively similar extents of base pairing as monitored spectroscopically. Therefore, the denatured form cannot be a random coil. The overall hydrodynamic properties of the two forms are quite different. Denatured forms can be created by removal of Mg^{2+} ions. They can be converted to native forms by incubation in the presence of magnesium ions. These observations indicate that divalent cations play a critical role in the stabilization of the native structure. The existence of denatured forms suggests that some kind of ordered folding of the cloverleaf exists in the native struc-

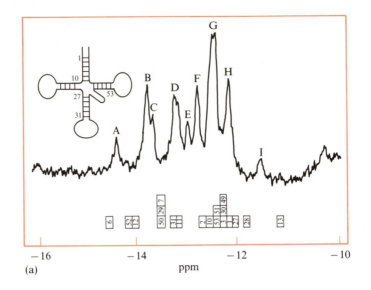

(a)

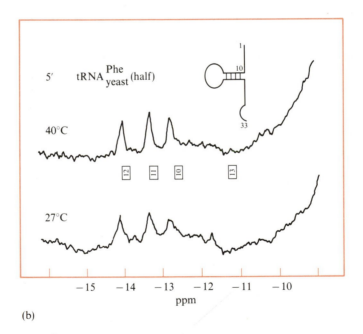

(b)

Figure 24-13

Proton NMR spectra of tRNA at 300 MHz in H_2O. (**a**) Intact yeast tRNA[Phe] at two temperatures. The predicted locations of resonances from secondary-structure base pairs are shown in boxes. (**b**) A fragment of yeast tRNA[Phe]. [After D. R. Lightfoot et al., *J. Mol. Biol.* 78:71 (1973).]

ture; this folding is disrupted upon Mg^{2+} removal. After slight alterations in secondary structure, the conformation can rearrange to the denatured form. This form is metastable relative to the native form once Mg^{2+} is returned to the solution. However, the rate of interconversion can be quite slow.

Another set of experiments strongly indicating a tertiary structure involved chemical modification. If native tRNA were a simple cloverleaf one might expect two distinct classes of reactivity with reagents such as formaldehyde, kethoxal, and perphthalic acid that are known to modify base residues. Part of the tRNA should behave as a single strand, and the rest should behave as pure double-stranded duplex. However, experimental results showed that some residues predicted by the cloverleaf pattern to be single-stranded are inaccessible to reagents capable of reacting with isolated single strands (see Chapter 3). This result can be rationalized if, in native tRNA, the arms of the cloverleaf are folded up to make an ordered tertiary structure. Then experiments must be designed to examine the overall geometry of folding and to discover what specific interactions are responsible for the maintenance of the tertiary structure.

The pattern of chemical reactivity of native tRNA suggested that the anticodon loop and the 3'-end are relatively exposed, whereas other potentially single-stranded regions (such as the TψC sequence) are not accessible. Without *a priori* knowledge of the tertiary structure, it is very difficult to explain the origin of particular alterations in reactivity. Furthermore, it is rather treacherous to attempt to deduce structural information from reactivity data alone, because very subtle alterations in structure can sometimes lead to large reactivity differences. However, the x-ray structural data currently available on yeast tRNA^{Phe} (Chapter 3) do permit rationalization of much of the chemical modification pattern.

A major danger in chemical modification experiments is that structural alterations may be introduced in the process of forming covalent derivatives. Even when biological activity is maintained after the modification, one cannot be absolutely sure that the structure is identical. The advantage of physical studies in solution is that many of these entail no risk of structural alteration.

Estimating tRNA base pairing in solution

One of the first goals in the study of an RNA structure is to estimate the fraction of bases involved in base pairing. When the optical methods outlined in Chapters 7 and 8 were applied to tRNA, it was found that simply modeling cloverleaf as a set of short duplexes and linear single strands cannot account for the observed CD or hypochromism. The discrepancies are in the direction expected if tRNA contains significantly more base pairing than is demanded by a simple cloverleaf model. This observation immediately suggested that the tRNA tertiary structure might be organized around a set of additional base pairs. However, there was no way to discern from optical studies which base pairs these might be.

Inspection of the available tRNA sequences revealed a few potential tertiary base pairs, but it did not indicate an obvious pattern containing a sufficient number of tertiary base pairs that could be common to all tRNAs in the same way as the universal cloverleaf applies to all. Furthermore, one could not rely on the quantitative accuracy of the optical analyses, because it was difficult to take into account any special optical interactions arising from particular loop conformations, unusual nucleotides, and any possible differences between the structure of double helices in folded tRNA and the structure of linear duplexes. Rapid tritium-exchange studies on tRNA were consistent with optical studies in pointing to more base–base interactions in the native strucutre than predicted by the cloverleaf model. However, once again, uncertainties in the interpretation of the data precluded their use for rigorous testing of various proposed tRNA models.

The fundamental problem with optical and rapid tritium-exchange studies is that they provide far too few pieces of information about the system. NMR is potentially vastly more powerful, because the individual contributions of each residue to the observed spectrum are (in principle) resolvable. Given the instruments presently available, most of the NMR spectrum of a tRNA is not easy to interpret. Resonances from bases stand out, but the rest is largely a blur. However, a striking exception involves the NH protons of hydrogen-bonded Urd and Guo and their strange-base analogs. As we have seen in earlier chapters, base pairing leads to large downfield shifts and, as a result, these resonances appear in the region of -11 to -15 ppm with respect to a dimethylsilapentane sulfonate (DSS) standard. In this region of the spectrum, only NH resonances are present. Because only a single proton is involved for each base pair, one can expect that a simple cloverleaf would show only 20 bands in the low-field spectral region. Even allowing for the possibility of a few tertiary base pairs, an experimental spectrum could be simple enough to analyze.

The low-field region of the NMR spectrum must be studied with H_2O as the solvent. Free NH protons exchange rapidly. In base-paired regions, the exchange is slowed (Chapter 22). When the exchange rate is significantly slower than the characteristic NMR times, individual NH proton resonances will be visible. However, actual experimental NMR measurement for a tRNA can take hours because of the extensive signal averaging required. If D_2O is used as a solvent, the resonances disappear once the NH protons exchange. In H_2O, exchange does not matter, as long as it is slow. There always are protons resident on the tRNA, and (on the average) each spends a long enough time bound to be detected before it is replaced with another. The difficulty in working with H_2O is that there is a large background of absorption due to the solvent, which has 10^4- to 10^5-fold greater proton concentration than an individual tRNA NH resonance. This background and the resonances of other tRNA protons are centered at frequencies away from the -11 to -15 ppm region, but they contribute to the baseline there and make quantitative measurements of the intensities of the NH proton resonances rather difficult.

Figure 24-13a shows a typical low-field tRNA NMR spectrum. Nine distinctly resolved NH bands are visible in yeast tRNA^{Phe}. Not all are of equal intensity, and this result suggests that some represent the contribution of two or more protons. Integration of the spectrum (compared with reference standards) originally suggested that 18 ± 1 protons are being observed. Because the yeast tRNA^{Phe} cloverleaf contains 20 base pairs (not counting a single G–U), it is a reasonable first assumption that all the resonances seen are due to base pairs forming the cloverleaf secondary structure.

Assigning base-paired residues in NMR spectra of tRNA

To confirm the suggestion that the resonances in the range of -11 to -15 ppm are due to base pairs, it is necessary to assign the observed resonances to particular residues. A general approach to assignment is to study the NMR spectra of a vast array of model oligonucleotide duplexes. Then the NMR contribution of a base pair in a given sequence (say, the central A–U of ApUpG·CpApU) can be assigned by solving sets of simultaneous equations, much as we have done for optical and thermodynamic data (Chapters 22 and 23). In practice, NMR requires much larger samples of material than do these other techniques, and it has not yet been possible to accumulate a sufficient library of experimental reference spectra.

An alternative approach to assignment, facilitated by the much better resolution of the NMR technique, is to study specific fragments of the tRNA structure. Halves or quarters of a tRNA molecule should contain only 4 to 7 base pairs. Thus there is a reasonable chance that each NH resonance will be cleanly resolved from all the others. This prediction is confirmed very well, as shown for the spectrum of the 5'-half of yeast tRNA^{Phe} in Figure 24-13b; this spectrum has only four low-field resonances. To assign these individual resonances, one resorts to theoretical calculations. An isolated single A–U or G–C NH hydrogen-bonded proton should have some particular fixed resonance frequency, shifted downfield from that of the free molecules by hydrogen bonding. Imbedding any base pair within a double helix will lead to additional shifts upfield, due to the ring currents of bases stacked above or below it. The ring-current magnetic anisotropy of each of the nucleic acid bases has been calculated from quantum mechanics, using some of the best available approximate wavefunctions.

In Chapter 22, we show how these anisotropies can be used to estimate the stacking geometry from the observed structure. For tRNA analysis, it was originally assumed that the geometry of the double-stranded regions is a normal RNA-11 helix. Then the anisotropies can be used to predict the resonance positions as a function of sequence. Because the intrinsic NMR intensities of the isolated base pairs in H_2O are unknown, these values are allowed to be variables and are adjusted to give the best fit for a wide variety of data from tRNAs, fragments, and model oligonucleotide duplexes. The results are chemical shifts for isolated A–U $(-14.8$ ppm$)$, G–C $(-13.7$ ppm$)$, and A–ψ $(\sim -13.5$ ppm$)$. The effect of stacking by neighboring base pairs can only shift these resonances upfield. Thus, in early studies of tRNA, the approximation was made that all resonances downfield from -13.7 ppm must belong to A–U base pairs.[§]

Table 24-1 shows the calculated upfield shifts for various nearest-neighbor base pairs. Two general features of these results can be noted. First, purine ring-current

Table 24-1

Calculated ring-current shifts for base pairs stacked adjacent to an existing base pair (see Note)

5′	3′	5′	3′
U = 0.0	A = 1.3	U = 0.0	A = 1.3
C = 0.0	G = 0.6	C = 0.0	G = 0.7
G = 0.0	C = 0.1	G = 0.0	C = 0.2
A = 0.1	U = 0.0	A = 0.0	U = 0.1
U–A		C–G	
U = 0.1	A = 0.0	U = 0.1	A = 0.0
C = 0.2	G = 0.0	C = 0.25	G = 0.0
G = 0.6	C = 0.0	G = 0.7	C = 0.0
A = 0.7	U = 0.0	A = 1.2	U = 0.0
3′	5′	3′	5′

NOTE: The existing base pair is shown at the center of the table. The upfield shifts are shown (in ppm) for each stacked base attached toward the end of the chain indicated at top and bottom of the table. For example, consider the duplex 5′UpC3′·5′GpA3′. The C is stacked toward the 3′-end of the chain from the U, so its effect is listed in the lower left-hand quadrant of the table: an upfield shift of 0.2 ppm. The G is stacked toward the 5′-end of the chain from the A, so its effect is read in the same quadrant as 0.0 ppm. Thus, the total effect is an upfield shift of 0.2 ppm in the U–A resonance. Similarly, the effects of the U and A bases on the G–C resonances are read in the upper righthand quadrant: the U has no effect, and the A produces an upfield shift of 1.3 ppm.
SOURCE: After R. Shulman et al., *J. Mol. Biol.* 78:57 (1973).

effects are much stronger than those of pyrimidines. Second, the 5′-linked base (3′-end base) on each strand is almost totally responsible for the shifts induced by the base pair. This second result is reasonable. In the RNA-11 helix, the 5′-linked base is partially stacked over the N–H–N hydrogen bond of the adjacent base pair, whereas the 3′-linked base shows little overlap.

Certain assumptions are needed to use the results in Table 24-1. One must take it on faith that the actual helical parameters in the tRNA duplex regions are close to

---

[§] More recent studies have calculated the NMR of base-paired protons in yeast tRNA[Phe] using the molecular coordinates determined by x-ray crystallography. These calculated values fit the observed spectrum very well. However, they indicate that the isolated A—U pair should have a chemical shift of -14.3 ppm, and that reverse Hoogsteen A–U base pairs may account for the very-low-field resonances (see Robillard et al., 1977). As a result, some of the assignments we describe for tRNA[Phe] may have to be revised.

those seen in fibers of long duplex RNAs. The base pairs at the ends of helices or at helix–loop interfaces are especially troublesome. It certainly is reasonable to guess that some helix distortion may occur at those places. Even more serious is the question of whether or not single-strand bases adjacent to the duplex stack on it and, if so, with what geometry. In general, stacking of single-strand residues adjacent to helices has been used and, for intact tRNA, it has been assumed that the CCA stem and TψC stems are stacked to make a continuous helix. The spectra are sensitive to such assumptions and, therefore, tertiary-structure information is (in principle) present in the observed resonances. However, the sensitivity is not all that large, and the data must be better calibrated before they can be exploited.

Figure 24-13b shows proton assignments for the base pairs in the 5′-half of yeast tRNA^{Phe}. Three sharp resonances are in good agreement with the predicted values. The fourth predicted resonance (for G^{22}–C^{13}) is observed not as a sharp band, but rather as a somewhat diffuse band at slightly lower field position than predicted. This result may be an indication that the structure is heterogeneous in the 5′-half; if so, G^{22}–C^{13} might be especially affected because it borders on a loop. The observed results can be explained by assuming that the orientation of A^{14} in the loop is quite variable. In a similar way, assignments were made for the resonances of other tRNA fragments. Then, to treat the whole molecule, one assumes that no major shifts occur in the resonances of individual protons compared with their positions in the fragments. In fact, the predicted resonance positions do not align all that well with the observed peaks in the intact molecule, or even in the fragments (Fig. 24-13b). However, the assignments do fairly well explain the overall qualitative pattern of the spectrum.

Improved analysis of NMR spectra of base-paired residues is possible by using a more empirical approach to band assignment. As we shall see, particular helical regions of a tRNA can sometimes be observed to melt individually. If these regions contain unique methylated bases, a clear assignment of the identity of the melting region often can be made by examining the methyl resonances. This assignment then assists the assignment of hydrogen-bonded proton bands because, in such a melting, all the bands due to one helical region should disappear simultaneously. Figure 24-14 shows an example of the extent to which secondary-structure resonances can be used to fit the NMR spectrum of a tRNA in which most of the bands have been assigned. Here a sample of yeast tRNA^{Asp} was studied at 65°C, a temperature at which no tertiary structure exists. At this temperature, the pseudouridine stem is intact, and the anticodon and amino acid stems are partially melted. After the extent of melting of each region is taken into account, a computer is used to simulate the observed spectrum. Clearly, the agreement between calculation and experiment is excellent.

Effect of tertiary structure on the NMR of tRNA

The success in analyzing the NMR of tRNA assures that, beyond resonable doubt, the cloverleaf secondary structure is present in tRNA in aqueous solution. However, there are quite a few complications. For tRNA^{Phe}, the original integration yielded

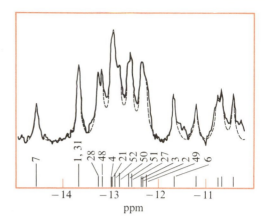

Figure 24-14

NMR spectrum of yeast tRNA^{Asp} *at 65°C.* The observed spectrum (*solid curve*) and computer-simulated spectrum (*dashed curve*) are shown. Numbered vertical bars indicate positions assigned to individual residues. [After G. T. Robillard et al., *Biochemistry* 15:1883 (1976).]

only 18 ± 1 resonances, whereas the cloverleaf contains 20 normal base pairs. The detailed assignment located 19 resonances; one A–U appeared to be missing. Tentatively, this missing band was assigned to A^5–U^{68}, on the argument that the G–U may cause disruption of a relatively weak adjacent A–U. However, as we now know from the x-ray structure, tRNA contains many tertiary base pairs; these NH resonances should be visible in the NMR. More recent attempts to assign the NMR spectrum of tRNA have grappled with this problem.

A major factor that determines whether or not a resonance will be visible is the proton exchange rate. When the exchange rate approaches the NMR characteristic time for the two states of a proton, the resonance will broaden until, at a critical rate, the spectrum becomes infinitely broad and is unobservable. At even faster exchange rates for tRNA in H_2O, the resonance remains invisible because its position is heavily weighted by the fraction of time that each proton spends as water rather than as base pair. Thus the resonance disappears into the huge water background. At present, it is not clear which types of secondary- or tertiary-structure resonances in an arbitrary conformation necessarily will have fast or slow exchange times. Thus, it now appears likely that some of the resonances originally assigned to secondary base pairs may actually belong to tertiary base pairs.

Improved normalization procedures have increased the number of protons that must be assigned to the integrated spectrum for most tRNAs. *E. coli* tRNA$_f^{Met}$ and tRNA$_1^{Val}$ show 27 and 26 low-field protons, respectively, whereas only 20 are demanded by a cloverleaf. Such results increase the difficulty of spectral assignment but are encouraging nonetheless, because solution of the assignment problem will make information on tertiary structure accessible directly from the NMR spectrum. Tertiary NH resonances are unlikely to be assigned from model-compound or oligonucleotide-duplex studies, because the tertiary interactions are absent in these systems. Instead, one must resort to perturbations of the intact tRNA.

Figure 24-15 shows an example of one very promising way to assign resonances in tRNA. A nitroxide free radical was covalently attached to the 4-thiouridine at

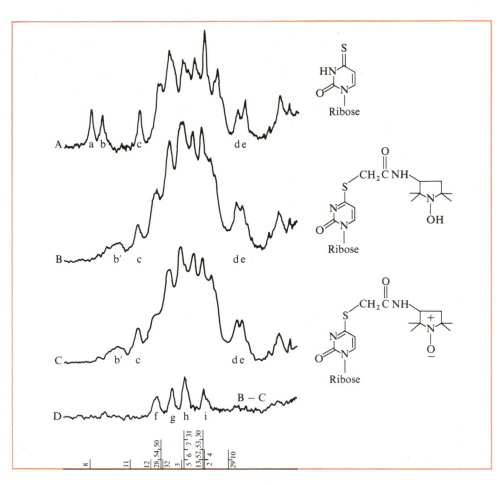

Figure 24-15

Proton NMR spectra of E. coli tRNA$_f^{Met}$ *at 220 MHz in H$_2$O at 30°C. Curve A is the spectrum of native tRNA. Curve B is the spectrum of ascorbate-reduced tRNA spin-labeled at the 4-thiouridine in position 8. Curve C is the spectrum of tRNA spin-labeled at the 4-thiouridine in position 8. Curve D is curve B minus curve C. Predicted positions of secondary-structure resonances are shown at the bottom. [After W. E. Daniel and M. Cohn,* Proc. Natl. Acad. Sci. USA *72:2582 (1975).]*

position 8 of *E. coli* tRNA$_f^{Met}$. This paramagnetic species can broaden the resonances of NH protons located nearby in the structure. Any direct environmental alterations due to introduction of the nitroxide can be examined separately by reducing this species with ascorbate. Compared to unlabeled tRNA$_f^{Met}$, the reduced spin-labeled tRNA derivative is missing a single resonance ("a" in Fig. 24-15), which therefore can be assigned to the proton normally hydrogen bonded between 4-thiouridine and A^{14}. When the difference spectrum caused by reduction of tRNA is computed, four discrete resonances appear. These can be assigned only tentatively at present. By combining expected chemical shifts and the known tRNA tertiary structure, one concludes that these four resonances apparently represent a G–C pair in the acceptor stem, a G–C pair in the dihydrouridine stem, and two G–C tertiary hydrogen bonds.

The NMR methods we have just described represent what is probably the most powerful solution technique currently available for the study of nucleic acid structure. They also are valuable in the dissection of individual steps in nucleic acid conformation changes, as we shall see shortly. NMR methods have been extended to examine slightly larger RNAs, such as 5 S rRNA. It seems likely that serious difficulties will arise in the study of RNAs larger than this. The number of base-pair resonances keeps increasing, and larger molecular weights almost inevitably lead to longer rotational correlation times, where dipolar broadening destroys the spectral resolution.

Slow tritium exchange

The power of NMR is its ability to look at individual residues—if not one at a time, at least no more than a few at a time. Another technique with this potential is slow tritium exchange of C^8-hydrogen (Chapter 22). Here we show the result of its application to tRNA by R. C. Gamble and P. R. Schimmel.

Any particular purine should exchange with tritium according to first-order kinetics:

$$\Delta(T)_i = \Delta(T)_{0i}\, e^{-t/\tau_i} \tag{24-6}$$

where i refers to the ith purine, $\Delta(T)$ is the deviation from the equilibrium solvent tritium concentration at time t, $\Delta(T)_0$ is the deviation at time zero, and τ is the decay time. In actual practice, exchange measurements are carried out for only a small fraction of one decay time. Therefore Equation 24-6 becomes, to a very good approximation, $\Delta(T)_i = \Delta(T)_{0i}(1 - t/\tau_i)$. For a polymer or oligomer containing multiple purines, the total amount of purine labeling will be

$$\Delta(T) = \sum_i \Delta(T)_{0i}(1 - t/\tau_i) \tag{24-7}$$

where the sum is over all purines in the molecule.

Equation 24-7 can be simplified by realizing that $\Delta(T)_{0i}$ must be essentially the same for all purines. This value $\Delta(T)_0$ will depend on the specific activity of the tritiated water used, on isotope effects that shift the final equilibrium extent of the exchange, and on the starting tritium content in each purine (which is zero for typical experimental arrangements). Then, for a polymer with N purines,

$$\Delta(T) = N\,\Delta(T)_0(1 - t/\bar{\tau}) \tag{24-8}$$

where $\bar{\tau}$, the average exchange time, is given by

$$1/\bar{\tau} = N^{-1} \sum_i (1/\tau_i) \qquad (24\text{-}9)$$

If the total exchange into an oligomer or polymer is measured, $\bar{\tau}$ can be computed.

To highlight the effects of secondary or tertiary structure on the exchange, one should compare the results either with monomer exchange rates (τ_A^0 and τ_G^0) or with the rates of a pure single-stranded oligomer. If the former comparison is used, one can compute an expected single-strand exchange time $\bar{\tau}_0$ from Equation 24-9 as

$$1/\bar{\tau}_0 = \chi_A/\tau_A^0 + \chi_G/\tau_G^0 \qquad (24\text{-}10)$$

where χ_A and χ_G are the mole fractions of these residues in the sample. Results usually are expressed as the retardation of tritium exchange:

$$R = \bar{\tau}/\bar{\tau}_0 \qquad (24\text{-}11)$$

In the use of Equation 24-11, $\bar{\tau}$ always is the measured time for a particular structure but, depending on the reference method used, $\bar{\tau}_0$ either is calculated by Equation 24-10 or is measured.

Two levels of resolution can be employed for tritium-exchange studies on tRNA. Easiest experimentally is to exchange tritium in an intact tRNA and then to cool and to carry out a total enzymatic digestion with an endonuclease such as T_1 RNase, which cleaves specifically after guanine residues. This procedure provides a distinct nonoverlapping set of fragments, which are separated in so far as possible by chromatography or electrophoresis. Then radioactive counting on each fragment permits calculation of its $\bar{\tau}$. Even more informative in many cases is further digestion of each separated oligonucleotide fragment into mononucleotides prior to tritium analysis. In this way, the exchange into adenine and guanine residues can be determined separately, and any effects of pyrimidine exchange can be corrected for. In principle, it should be possible to carry the method even further. The use of more than one specific nuclease would permit separate monitoring of the tritium exchange into each individual purine.

Figure 24-16 shows some quite informative results from slow tritium exchange on yeast tRNA[phe]. The general agreement with the known crystal structure of this tRNA is excellent. Regions that definitely are single-stranded (such as the CCA terminus and the anticodon loop) have values of R less than 5, indicating that exchange is only slightly impeded by conformation. Double-strand tRNA purines show values of R greater than 15, which is expected because purines in pure double-stranded RNAs have similar values. A few specific residues that are single-stranded in the cloverleaf representation still show R values between 15 and 30. Among these are the residues A^{14}, G^{15}, and G^{57}. These results are reasonable because these

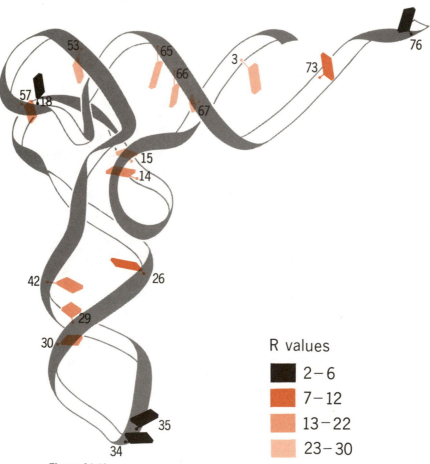

Figure 24-16

Kinetics of purine C^8-hydrogen tritium exchange in yeast tRNA[Phe], superimposed on a schematic drawing of the tertiary structure of this tRNA. Shown is each of the purines for which exchange data at 37°C are currently available. The color indicates the relative exchange rate at each position. Except for residue G^{18}, all purines implicated in secondary or tertiary structure in the crystal show substantially reduced exchange rates. [Drawing by Irving Geis.]

residues either are involved in tertiary-structure base pairing, or else at least appear to be tightly folded into the structure (Fig. 24-16). The only real discrepancy is the exchange rate of G^{18}, whose R value of only 4 suggests that it is exposed or uninvolved in tertiary interactions. Perhaps, at 37°C (the temperature at which tritium-exchange measurements are carried out), some tertiary features seen in the crystal at low temperature are melted, or at least are weakened sufficiently to show rapid tritium exchange.

Oligonucleotide binding

A third high-resolution technique for studying the structure of tRNA in solution is the oligonucleotide-binding method. This method is particularly suited for identifying exposed single-stranded residues. Oligonucleotides are selected to be complementary to short stretches of the known tRNA sequence. Their binding strength to the RNA then is measured by equilibrium dialysis. A high concentration of tRNA is placed in a chamber on one side of a membrane impermeable to tRNA but permeable to smaller molecules. Radioactive oligonucleotide at low concentration is placed in a chamber of equal volume on the other side of the membrane. At equilibrium, the concentration of unbound oligonucleotide must be the same in the two chambers. Let K_b be the binding constant for an oligonucleotide to a single site in the tRNA:

$$K_b = (\text{Oligo} \cdot \text{tRNA})/(\text{Oligo})(\text{tRNA}) \tag{24-12}$$

Because tRNA is present in great excess, its free concentration can be taken as equal to the input concentration. Then, just by rearranging Equation 24-12, we have the equilibrium ratio of total oligomer in the two dialysis chambers:

$$r_d = [(\text{Oligo} \cdot \text{tRNA}) + (\text{Oligo})]/(\text{Oligo}) = 1 + K_b(\text{tRNA}) \tag{24-13}$$

Thus, K_b can be measured in a very simple way.

   If there are multiple binding sites, indexed by i, then we have the following expression for r_d in the limit of tRNA excess:

$$r_d = 1 + \sum_{i=1}^{n} K_{bi}(\text{tRNA}) \tag{24-14a}$$

If all n sites have comparable binding constants, then

$$r_d \cong 1 + nK_b(\text{tRNA}) \tag{24-14b}$$

The result is simply to increase the apparent value of K_b in Equation 24-13; therefore, the presence of multiple sites cannot be detected in the limit of tRNA excess. However, from known tRNA sequences, it is rare that more than two sites exist complementary to a given trimer, and rare that there are even two sites complementary to any particular tetramer. So, if one restricts attention to only large differences in apparent K_b for different oligonucleotides, the effect of numbers of binding sites on an interpretation of data will not be that consequential. How long should the oligonucleotides used for this purpose be? It is comparatively difficult to synthesize compounds of length four or more, and the number of specific oligonucleotides of length n increases as 4^n. However, longer oligonucleotides have greater specificity and higher binding constants. In practice, a common compromise is to work with lengths of three and four.

An effective strategy is to compare the binding strength of a tetramer with the strengths of its two constituent trimers. Consider a tRNA containing the sequence $3' \cdots \text{UpCpGpC} \cdots 5'$. We compare the binding strength to this sequence of the tetramer $5'\text{ApGpCpG}3'$ versus the trimers ApGpC and GpCpG. If all four bases on the tRNA are in a single-stranded section available for simultaneous binding to the four bases of the tetramer AGCG, its association constant should be much larger than that of either trimer. There will be three double-strand stacking interactions in the tetramer complex, but only two for each of the trimers. You can estimate from Table 23-4 that the third stacking interaction is worth about -3 kcal mole^{-1} in free energy on the average; this would correspond to an increase in K_b by a factor of $\exp(3,000/RT)$, or 148 at 25°C. If the complementary tRNA sequence is in a loop, the extra tetramer-binding affinity is slightly more difficult to estimate, but it still should be significant. If only three tRNA bases of the four are accessible simultaneously, then K_b for the tetramer should be quite similar to one of the trimer values. (One might expect the tetramer K_b to be up to 10-fold stronger than the trimer K_b if the extra base is able to stack at the end of the duplex.) If both sets of three tRNA bases are separately available and nonoverlapping, the tRNA site just corresponds to two independent trimer sites; the K_b for the tetramer will be twice the trimer value.

In a similar way, dimer binding can be used as a reference state for trimers. This approach generally is not as useful, because many dimers bind too weakly to be studied. An important set of controls for all experiments is to use oligomers for which the complementary sequence does not exist anywhere on the RNA. These almost inevitably are found to bind much more weakly than any correctly complementary oligomers.

Results of oligonucleotide-binding studies

O. Uhlenbeck has carried out an extensive set of oligonucleotide-binding studies on *E. coli* tRNA^{Tyr}. Although the three-dimensional structure of this tRNA is not yet known, we summarize the results here anyway to demonstrate the power of this technique. Figure 24-17 shows the sequence of the tRNA folded as a cloverleaf. Uhlenbeck studied the binding of 63 of the 64 possible trinucleotides to tRNA^{Tyr}, and 24 of these showed binding constants above the level set by the background in

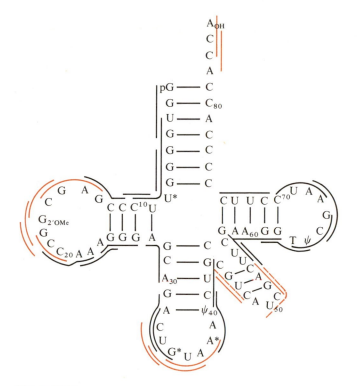

Figure 24-17

Oligonucleotide binding to tRNA. This diagram summarizes part of the results of Uhlenbeck's study of *E. coli* tRNA^Tyr. Solid colored lines show complementary oligomers that bind strongly to the native tRNA. Dashed bars represent intermediate binding strengths. Black bars denote oligomers that do not bind significantly.

the experiments. Then, some 70 selected tetramers out of 256 possible were tested. Only 13 of these 70 tetramers showed binding constants larger than twice the sum of the constants of their constituent trimers (Table 24-2).

Tetramer-binding enhancements ranged from just barely twice the constituent trimer constants to a 50-fold increase for the two tetramers UACA and UAUA complementary to the anticodon of the tRNA. Note that the strong binding of UAUA implies that the G–U pair predicted by the wobble hypothesis for the third position of the codon–anticodon complex may actually contribute to the stability of duplex formation.

When all the trimer and tetramer results are compared, the results are consistent with the availability for oligomer binding of only four discrete regions of the tRNA: the 3'-ACCA sequence, the end of the extra loop, the anticodon, and part of the dihydrouridine loop (Fig. 24-17). The few oligomer-binding data that do not fit this pattern can be rationalized. For example, trimers containing a GpG sequence

Table 24-2

Tetranucleotides that bind to tRNA^{Tyr} with $K_b \geqslant 2K_{tri}$ (see Note)

Tetranucleotide	K_b (M^{-1})	K_{tri} (M^{-1})	Control tetramers	Classification
UACA	100,000	1,800	UACG, UGCA	+
UAUA	37,000	600	UAUG, UGUA	+
UUAC	5,300	900	UUAA, UUAG	+
ACAG	3,800	1,900	ACAA, UCAG	I
GCUC	22,000	1,800	GCUA, GUCC	+
CCGC	12,500	3,900	CUGC	I
GCCG	16,600	4,500	——	I
GUCG	8,700	1,400	GUCC, UUCG	+
GAUG	4,700	600	UAUG, GAAG	+
GUCU	5,000	900	GUCC	+
GACG	4,700	2,000	UACG, GAAG	I
UCGA	3,300	1,400	UGCA	I
UGUG	4,200	1,500	——	I

NOTE: K_{tri} is the sum of the K_b values of adjacent constituent trimers of the given tetramer. The K_b values are given for those tetramers in which all four bases are judged to bind simultaneously to tRNA^{Tyr} (those with $K_b \geqslant 2K_{tri}$). The control tetramers are those of similar sequence and composition that do not bind to tRNA^{Tyr}. The binding tetramers are classified in the final column as strong binding (+) or intermediate binding (I).

SOURCE: After O. Uhlenbeck, *J. Mol. Biol.* 65:25 (1972).

appear to bind significantly to CC sequences, regardless of whether or not neighboring base pairs can form. This binding is reasonable in view of the fact that GG–CC is by far the strongest duplex interaction (Table 23-4).

Additional information is available from oligonucleotide competition binding. For example, suppose that a weakly binding, unlabeled trimer is added in a significant amount relative to the tRNA present, with total tRNA still in excess over a labeled oligonucleotide. Then the effect of the competitor is to complex part of the tRNA, depending on its binding constant,[§]

$$K_c = (\text{Compet} \cdot \text{tRNA})/(\text{Compet})(\text{tRNA}) \qquad (24\text{-}15)$$

This competitive binding reduces the available concentration of tRNA left to bind to the labeled oligomer by a factor of

$$(\text{tRNA})/[(\text{tRNA}) + (\text{Compet} \cdot \text{tRNA})] = [K_c(\text{Compet}) + 1]^{-1} \qquad (24\text{-}16)$$

[§] The complication that the labeled and unlabeled oligonucleotides might interact directly to form a complex is unimportant, so long as this complex is dialyzable.

Table 24-3

Competition between oligonucleotides that bind to tRNA^{Tyr}

Oligomer	Competitor	K_b (M^{-1})	K_{obs} (M^{-1})	K_c (M^{-1})
GGC	GGC	102,000	13,000	114,000
GGU	GGC	48,000	46,000	670
UGG	GGC	9,000	9,300	0
UACA	GGC	100,000	102,000	0
GUCG	GGC	8,700	8,700	0
GCCG	GGC	16,000	3,000	122,000
GCUC	GGC	22,000	15,000	21,000
GGC	GGU	102,000	97,000	500
GGU	GGU	48,000	11,700	51,800
UGG	GGU	9,000	2,000	58,400
UACA	GGU	100,000	102,000	0
GUCG	GGU	8,700	8,900	0
GCUC	GGU	22,000	22,000	0

NOTE: K_b is the measured binding constant between the oligomer and tRNA^{Tyr}. K_{obs} is the binding constant for the oligomer observed in the presence of the competitor; a reduction of the binding constant from K_b indicates competition. K_c is the binding constant of the competitor as computed by assuming equimolar competition for the same site. The experiments were conducted at 0°C.

SOURCE: After O. Uhlenbeck, *J. Mol. Biol.* 65:25 (1972).

Therefore, Equation 24-13 becomes

$$r_d = 1 + K_b(\text{tRNA})/[K_c(\text{Compet}) + 1] \qquad (24\text{-}17)$$

The constant K_c can be measured by determining K_b as a function of the concentration of the competing oligomer. Table 24-3 summarizes such experiments for tRNA^{Tyr}. In the simplest cases, the competition is a simple direct association with the binding site of the labeled oligomer. Then K_c is the same as the K_b that would be measured by directly studying the binding of the unlabeled competitor itself. Naturally, this is what is seen if unlabeled oligomer is used to compete with identical labeled oligomers.

Note from the results in Table 24-3 that GGU, which binds to the 3'-end of tRNA at A^{82}C^{83}C^{84}, competes against the binding of UGG, which binds to an overlapping site, C^{83}C^{84}A^{85}. Similarly, GGC (which binds to G^{18}C^{19}C^{20} in the dihydrouridine loop) effectively competes against the oligomer GCCG (which binds to the overlapping site C^{16}G^{17}G^{18}C^{19}). However, what is especially interesting is that GGC also shows competition with GCUC, which binds to the nonoverlapping site G^{13}A^{14}G^{15}C^{16}. The fact that K_c for GGC in this competition is less than its K_b

when measured alone suggests that both GGC and GCUC can bind simultaneously to the dihydrouridine loop, but that the binding of one weakens the affinity of the other. Neither GGC nor GGU affects the binding of any oligomers that associate with distant regions of the tRNA. This observation indicates that the binding is not triggering major conformational changes that are felt across the tRNA molecule, thus providing support for the validity of the technique.

Oligonucleotide-binding experiments have also been carried out on other tRNAs and on 5 S rRNA. For yeast tRNA^{Phe}, the results are not as extensive as those for *E. coli* tRNA^{Tyr}, but they are in general agreement. The results indicate that three regions are accessible for binding: the CCA end, the anticodon loop, and part of the dihydrouridine loop. The first two regions are in good agreement with expectations from the x-ray structure, which shows these regions as single strands. The third region is difficult to rationalize from the x-ray structure, but it is in reasonable agreement with the results obtained from slow tritium exchange (discussed earlier) and also from some chemical modification studies.

Cross-linking and energy transfer

None of the methods we have discussed give a clear picture of the macroscopic organization of the tertiary structure of tRNA. They indicate which residues may be affected by tertiary-structure interactions, but one would like to know specifically how the loops and arms of the tRNA cloverleaf are folded to make up the tertiary structure. The only direct approach to answering this kind of question is to determine the spatial proximity of fixed points in the tRNA structure. Intramolecular cross-linking can provide proof that two residues are close, assuming that forming the cross-link does not perturb the tertiary structure.

Many *E. coli* tRNAs contain a 4-thiouridine in position 8 of the CCA stem. Near-UV irradiation of these tRNAs produces a photoadduct between the 4-thiouridine and cytidine-13 in good yield (Fig. 24-18a). The cross-linked material shows no significant perturbation in physical properties and is biologically active in a wide variety of different tests of tRNA function. Therefore, one can conclude that the CCA and dihydro-U stems in the tertiary structure of tRNA are folded such that positions 8 and 13 are immediately adjacent. This prediction is in complete accord with the tertiary structure of yeast tRNA^{Phe} as revealed by x-ray crystallography (Fig. 24-18b). Note, however, that this tRNA species contains no 4-thiouridine, and so a direct test of the hypothesis is not possible. However, many lines of evidence indicate that almost all tRNAs have similar structures.

Cross-linking can yield information only about residues that are close to each other. Failure to observe a cross-link virtually never yields any useful information about the structure. A number of spectroscopic techniques are available for measuring proximity or remoteness, including various aspects of magnetic resonance and

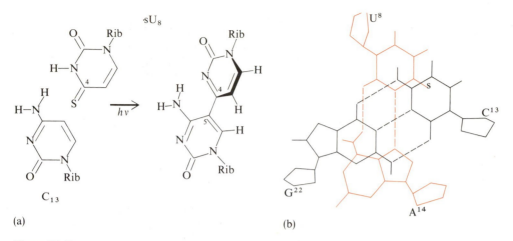

Figure 24-18

Cross-link formed by UV irradiation in many E. coli tRNAs. (**a**) Chemical structure of the cross-link between the 4-thiouridine in position 8 and the cytidine in position 13. [Courtesy of Nelson Leonard.] (**b**) Structure of adjacent A^{14}–U^8 and C^{13}–G^{22} base pairs in the yeast tRNA^{Phe} crystal structure, showing the proximity of residues 8 and 13. [After A. Rich and U.-L. RajBhandary, *Ann. Rev. Biochem.* 5:205 (1976).]

fluorescence-energy transfer. These techniques are discussed elsewhere in this book. Only the fluorescence technique has thus far been applied extensively to tRNA.

The first singlet energy-transfer experiments on tRNA by K. Beardsley and C. R. Cantor exploited the occurrence of the naturally fluorescent Y base adjacent to the anticodon of yeast tRNA^{Phe}. Several different fluorescent dyes were attached covalently to the 3'-terminus of tRNA and served as energy acceptors. Only small amounts of energy transfer were seen in all cases, indicating that the 3'-end and the anticodon must be more than 40 Å apart. The consistency of results obtained with several different dyes served as assurance that some extreme value of κ^2 (the geometric term in Eqn. 8-55) was not distorting the results.

Subsequently, D. Yang and D. Söll prepared a set of five samples of double-fluorescent labeled tRNAs. This was a much more difficult task, because both the donor and acceptor had to be put in by covalent chemical modification. Different tRNA species were used to construct different double-labeled samples, because the preparation of many of the covalent derivatives depended on the presence of particular unusual bases in the desired location and their absence at other locations. Figure 24-19 summarizes the results of energy-transfer measurements. If we assume that all tRNAs have identical tertiary structures (ignoring the "extra" loop), four of the five distance measurements available thus far agree fairly well with the crystal structure of yeast tRNA^{Phe}. This is evidence that, at least in broad outline, the tertiary structure of

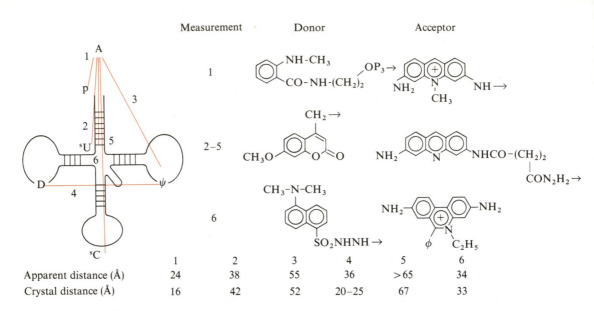

Measurement	1	2	3	4	5	6
Apparent distance (Å)	24	38	55	36	>65	34
Crystal distance (Å)	16	42	52	20–25	67	33

Figure 24-19

Singlet-singlet energy-transfer measurements compared with known distances within the yeast tRNA[Phe] crystal. Measurements 1 through 5 are from the work of Yang and Söll; measurement 6 is from Wells and Cantor. Most distances measured by energy transfer are in good agreement with the crystal distances. Because these measurements involved a number of different tRNAs, the results are a strong indication that all tRNAs can adopt very similar tertiary structures.

tRNA seen in the crystal must be the same as that occurring in aqueous solution. The energy-transfer results in Figure 24-19 are subject to some uncertainty, owing to lack of direct knowledge of κ^2 (the average value of 2/3 was used), but the good agreement is circumstantial evidence that extreme κ^2 values probably are not present.

Note that the results in Figure 24-19 are not detailed enough to allow an *a priori* determination of the geometric arrangement of the labeled regions. Dyes were placed in six positions on the tRNA: on the 3′- and 5′-ends, on the 4-thiouridine, and on bases in the anticodon, GTψC, and dihydrouridine loops. Energy-transfer measurements provide scalar distances and not vectors. To compute the geometric structure of N points in space can require as many as $4N - 10$ intramolecular distance ($N \geqslant 4$). For the six covalently labeled positions in tRNA, this means that 14 of the 15 possible pairwise distances would be measured to define their geometric orientation. Five distances are far from sufficient. Even when all 14 are available, the structure determined is uncertain with respect to reflection in a mirror. Scalar distances are invariant to a mirror reflection, but an actual asymmetric structure is not. However, in concert with other available structural information, a few quantitative distance measurements from energy transfer are quite informative, and they greatly restrict the range of plausible tertiary structures warranting further consideration.

Energy transfer also can be used to locate the sites of interaction of other molecules with tRNA. For example, in the presence of sufficient Mg^{2+} to ensure that tRNA is in the native tertiary structure, there is a single strong binding site for ethidium. This bound ethidium was used by B. Wells and Cantor as an energy acceptor for a dye attached at the 3'-end, yielding a distance of 34 Å between the dyes. An examination of the tRNA^{Phe} tertiary structure reveals that the only plausible double-strand binding site located this distance from the 3'-end is between residues 6 and 7 (Fig. 24-19). X-ray and NMR studies are consistent with this assignment, but the former yield the unexpected result that the bound ethidium is not intercalated.

A phase diagram for tRNA conformations

As environmental conditions are varied, tRNAs can assume a range of structures. Among the environmental variables thus far studied extensively are monovalent ion concentration, temperature, divalent ion concentration, and pH. One rationale for examining structural changes induced by alterations in conditions is that these may reflect possible conformational changes linked to tRNA function when it complexes with aminoacyl synthetases, protein synthesis factors, or the ribosome. It is far easier to examine the conformations of isolated tRNA than to study much larger nucleoprotein complexes. Studies of the tRNA conformational changes also can lead to increased knowledge about the forces that stabilize the native structure of tRNA. Furthermore, such studies can provide a framework for planning structural studies on more complex RNAs for which high-resolution x-ray or NMR data currently are unavailable.

We concentrate here on the effects of monovalent ions and temperature, because there are fairly comprehensive results for these two variables. A single tRNA, *E. coli* tRNA$_f^{Met}$, will be used to demonstrate how conformational changes have been resolved and analyzed. Available results for other tRNAs are qualitatively similar but differ in a number of important details. It is is conceivable that such differences are functionally important because tRNA$_f^{Met}$ is responsible for initiating protein synthesis but, unlike any other tRNA, it is not involved in protein chain elongation. However, the only obvious secondary-structure difference between tRNA$_f^{Met}$ and other tRNAs is one missing base pair at the end of the CCA stem (Fig. 3-7). From the known tertiary structure of yeast tRNA^{Phe}, it is not obvious that a broken base pair at this position should perturb any other region of the structure. L. Schulman has shown that chemical modifications allowing base-pair formation at the end of the CCA stem of *E. coli* tRNA$_f^{Met}$ render the tRNA able to participate in many of the protein-synthesis functions of normal tRNAs. Thus, it seems likely that the following results are not particularly related to the specialized role of tRNA$_f^{Met}$.

The first objective is to define the critical values of temperature and Na^+ concentration at which major conformational changes occur. The absorbance at 260 nm traditionally has been used to monitor RNA conformational changes. For *E. coli* $tRNA_f^{Met}$, it proved very useful to supplement such data with absorbance measurements at 335 nm. These data almost exclusively reflect the environment of the 4-thiouridine residue in position 8. Discrete conformational changes are easier to resolve at this wavelength, because thermal melting at this wavelength is clearly biphasic (Fig. 24-20). Data at 260 nm also show evidence of biphasic melting, but

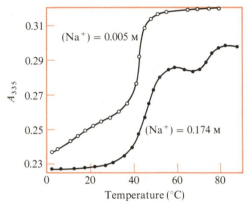

Figure 24-20

Absorbance melting of E. coli tRNA$_f^{Met}$ monitored at 335 nm and measured at two different salt concentrations. This wavelength essentially follows only 4-thiouridine absorbance and appears to be more sensitive than 260 nm to tertiary structure changes. [After P. E. Cole et al., *Biochemistry* 11:4358 (1972).]

it is more difficult to resolve because the transition with a lower T_m involves a rather small absorbance change. The 4-thioU is not involved in any cloverleaf duplex region, but it clearly is involved in tertiary structure. Thus, the very large 335 nm low-temperature absorption transition immediately suggests a tertiary-structure alteration.

From thermal-melting studies at a variety of ionic strengths, it is possible to construct a phase diagram for $tRNA_f^{Met}$ as described in Chapter 22 for poly $A \cdot U$ (Fig. 24-21). There is evidence for four phases in the absence of Mg^{2+}. The low-temperature, high-ionic-strength form (I) almost certainly is tRNA in its native tertiary structure. Addition of Mg^{2+}, known to stabilize the native structure, has very little effect on the absorption properties of tRNA in this region. The hypochromicity of the high-temperature, low-ionic-strength form (IV) is close to zero. Thus, this form is essentially a single-stranded coil. The exact position of the phase boundary between the other two forms (II and III) is not well defined by available experimental results. It is easy to see why; direct II $\leftrightarrow$ III interconversion is not

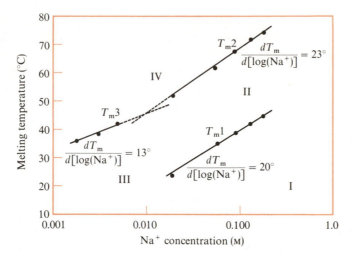

Figure 24-21

Phase diagram for E. coli tRNA$_f^{Met}$ at neutral pH. This diagram was constructed from the results of melting experiments like that shown in Figure 24-20 and is supported by the results of kinetic studies. See Figure 24-22 for a pictorial representation of the structures of tRNA in the four phases. [After P. E. Cole et al., *Biochemistry* 11:4358 (1972).]

resolved in thermal melting at constant salt concentration. In principle, one should be able to study the II ↔ III transition by varying the sodium ion concentration. However, the optical change for the II ↔ III transition is too small to work with conveniently, as you may be able to infer from Figure 24-21.

Relaxation kinetics of tRNA conformational changes

Several lines of evidence suggest that zones II and III in Figure 24-21 correspond to different tRNA conformations. The slopes of the II–IV and III–IV phase boundaries, $dT_m/d[\log (Na^+)]$, differ by almost a factor of two. The lower slope of the III–IV boundary suggests that the Na^+ binding is less important in stabilizing form III than form II. This observation is consistent with the fact that III is favored over II at low salt concentration and constant temperature. Stronger evidence is provided by kinetic studies. The process III → I has been monitored by rapidly increasing the Na^+ ion concentration and following absorbance changes. A very slow first-order rate is seen with a relaxation time of 1,200 sec at 13.3°C. The activation energy for this process is 61 kcal mole^{-1}, a very large value. This long relaxation effect helps to locate the I → III boundary at salt concentrations too low to resolve the entire I–III melting transition.

The rate of the I ↔ II transition must be studied by temperature-jump relaxation kinetics, because it is too fast for salt-jump techniques. At the lowest temperatures at which the process can be observed, the relaxation time τ for this transition is about 7 msec and is temperature-independent, implying a zero activation energy. At higher temperature, τ becomes markedly temperature-dependent, and an apparent activation energy of 50 kcal mole^{-1} can be derived from the data. The great difference in rates between the I $\leftrightarrow$ III and I $\leftrightarrow$ II transitions implies that states II and III are quite different entities. P. E. Cole and D. M. Crothers showed that the activation energies of various steps are particularly informative. For the I $\leftrightarrow$ II relaxation kinetics, the single τ observed must be $\tau^{-1} = k_{I \to II} + k_{II \to I}$. At temperatures well below the T_m for the transition, $k_{II \to I} \gg k_{I \to II}$ and, therefore, the zero activation energy observed must correspond to II $\to$ I. Conversely, $k_{I \to II}$ is completely dominant above T_m, and so its activation energy must be about 50 kcal mole^{-1}. Because form II can convert to I with no activation energy, no significant interactions within II are disrupted in order to produce I. Therefore, it is likely that form II has lost some of the tertiary or secondary interactions present in native tRNA I, but has not gained any new ones.

The rate of the I $\leftrightarrow$ II interconversion is orders of magnitude too slow for the melting of a simple hairpin, which we have shown to occur in microseconds. Therefore, some tertiary-structure changes must be involved. However, the large activation energy for $k_{I \to II}$ seems difficult to rationalize on the basis of tertiary structure alone, and so it seems reasonable to postulate that the conformational change involves additional loss of some secondary structure and thus large enthalpy changes. A tentative assignment suggests that the I $\to$ II transition involves loss of all tertiary structure and melting of the dihydrouridine loop (Fig. 24-22). From the energetics of RNA conformations discussed earlier, region II is predicted to be the least stable secondary-structure region.

The large activation energy of the III $\to$ I transition shows that this must be a process very different from II $\to$ I. Specifically, quite a few base–base interactions must be present in III that are not present in I, and these must be broken for the conformational change to occur. A reasonable guess is that form III is a hairpinlike structure (Fig. 24-22). This extended structure would have lower electrostatic free energy than the cloverleaf or native tRNA tertiary structure. Thus it could be more stable at low ionic strengths. Hydrodynamic data on tRNA at very low salt concentration are consistent with an elongated conformation.

The conformational change II $\to$ IV appears simple when observed by equilibrium thermal melting, but relaxation kinetic studies show otherwise. Three distinct relaxation processes are seen in the temperature range corresponding to this transition. These can be resolved because, although they take place in overlapping temperature ranges, they occur on quite different time scales. The results shown in Figure 24-23 demonstrate the power of kinetics to elucidate conformational changes. However, they also highlight a major limitation. The individual relaxation steps generally cannot be assigned to particular processes unless considerable additional information is available from higher-resolution structural techniques.

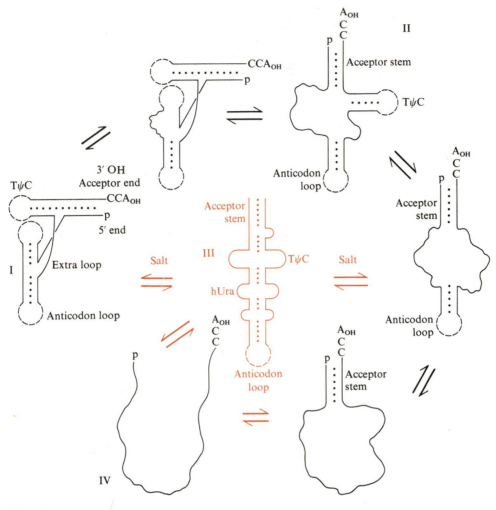

Figure 24-22

Schematic of thermal melting of E. coli tRNA$_f^{Met}$. The order in which various regions of structure are disrupted was established from the NMR and kinetic studies shown in Figure 24-23. Also indicated are the four states (I through IV) believed to correspond to regions of the phase diagram for this tRNA shown in Figure 24-21. The form of tRNA stable only at low salt and low temperature (III) is shown in color. [After D. M. Crothers et al., *J. Mol. Biol.* 87:63 (1974).]

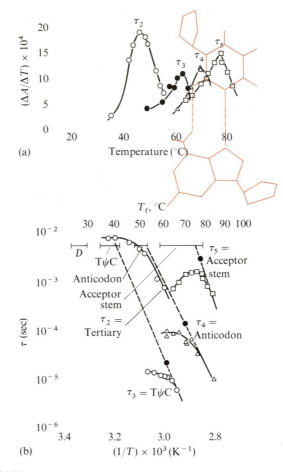

Figure 24-23

Thermal unfolding studies of E. coli tRNA$_f^{Met}$ in 0.17 M Na$^+$. **(a)** Differential thermal transition profiles for four discrete relaxation kinetic steps. $\Delta A/\Delta T$ is the amplitude of total absorbance change for each step, normalized by the size of the temperature jump employed. The different steps can be resolved so well because they occur on very different time scales, even though they overlap in temperature range. **(b)** Variation of four relaxation times with temperature. At the top are indicated the temperature ranges at which NMR melting of each region is observed. These ranges are plotted on the time scale at the characteristic NMR times for exchange broadening (see text). Also shown are the four relaxation processes seen by temperature-jump absorbance measurements. All points are directly measured, except for the dark ones, which were obtained from the fact that each $2\tau = 1/K_{-i}$ at $T = T_m$. Note the excellent extrapolation of the temperature-jump results to the NMR results. [After D. M. Crothers et al., *J. Mol. Biol.* 87:63 (1974).]

● **Using NMR measurements to assign relaxation values**

The proton NMR spectrum of $tRNA_f^{Met}$ was studied under the same sets of conditions used for relaxation kinetics. As the temperature of the tRNA sample is raised, selective broadening occurs of specific sets of low-field N–H···N proton resonances. For example, between 21° and 37°C, the integrated intensity of the NMR spectrum decreases by approximately four resonances. Two of these clearly are resolved single-proton peaks at -13.65 ppm and -11.45 ppm, and these can be assigned tentatively to A^{11}–U^{26} and G^{10}–C^{27}, respectively, as discussed earlier. Because both of these pairs are in the dihydrouridine stem, it is reasonable to conclude that this whole stem is melting concertedly, and the locations of the other two resonances that disappear are consistent with this. In a similar way, the melting regions of the $T\psi C$, anticodon, and acceptor stems can be assigned at successively higher temperatures (Fig. 24-23b).

To make the connection between an NMR melting temperature and an optical relaxation time, one must take into consideration the different time scales implicit in the two techniques. An NMR spectrum will broaden when exchange of the protons involved becomes comparable to NMR characteristic times. The situation is potentially quite complicated because two different processes can lead to broadening for the ith duplex region. The first process is duplex opening:

$$\text{H--Duplex} \underset{k_i}{\overset{k_{-i}}{\rightleftharpoons}} \text{H--Single strands} \qquad (24\text{-}18)$$

The second process is solvent exchange:

$$\text{H--Single strands} + H_2O \underset{k_{-ex}}{\overset{k_{ex}}{\rightleftharpoons}} \text{H--Single strands} + H_2O \qquad (24\text{-}19)$$

The rates seen optically are k_i and k_{-i}, whereas NMR could be monitoring these or k_{ex} and k_{-ex} or both sets of rates.

Some limiting cases allow a simple interpretation. Suppose that duplex opening is slow on the NMR time scale, but solvent exchange is fast once single strands are formed. Then it can be shown that k_{-i} determines the line broadening of the duplex resonances seen at -11 to -15 ppm. When duplex opening and solvent exchange both are slow, k_{-i} again dominates. If both processes are fast, then the relaxation time τ of the duplex transition becomes important: $\tau^{-1} = k_i + k_{-i}$. When $k_{ex}\tau \gg 1$, exchange with water occurs virtually every time the helix is open, and it can be shown that k_{-i} still dominates the line width. However, where $k_{ex}\tau \ll 1$, the solvent exchange dominates. We shall assume that this last case is not applicable to any of the tRNA data.

The measured optical relaxation data yield a value of k_{-i}. At high temperatures, this value must dominate the observed relaxation time $\tau_{opt}^{-1} = k_i + k_{-i}$ because, at the high-temperature part of each optical melting transition, $k_{-i} \gg k_i$. However, optical melting occurs on a time scale much faster than that required to cause NMR broadening. Hence, one must extrapolate the measured k_{-i} values back to the NMR time scale. It was found, from measurements on model oligonucleotide duplexes, that a choice of $\tau_{NMR} = 5$ msec brings melting curves measured by NMR and by optical techniques into coincidence. Figure 24-23b shows the NMR melting regions on this time scale. You can see that a rather long extrapolation of the optical data is required. One can construct an additional data reference point to increase reliability at lower temperatures by realizing that $k_i = k_{-i}$ at the optical T_m, and so each $\tau_{opt}^{-1} = 2k_i$ at T_m. The three high-temperature relaxation times extrapolate beautifully back to each one of the NMR melting transitions (Fig. 24-23b). Thus each optical transition is assigned.

Steps in the melting of tRNA

Figure 24-22 shows the detailed steps in the thermal melting of $tRNA_f^{Met}$, corresponding to processes I $\rightarrow$ II and II $\rightarrow$ IV of the phase diagram. Let us see how these steps can be rationalized. No separate optical melting was seen for the dihydrouridine helix or the tertiary structure. However, NMR clearly showed that the dihydrouridine-stem protons can exchange at a low temperature. Thus, it appears that a small degree of transient opening of this helix can occur prior to its actual thermal melting, which coincides with loss of tertiary structure. From the duplex interaction energies of Table 23-4, the TψC stem is predicted to be the next least stable helix after the dihydrouridine stem, and this is in accord with the assignments of Figure 24-23.

The highest-temperature optical melting transition is assigned to the acceptor-stem duplex. Note that this transition has the slowest relaxation time of any of the three, due solely to secondary structure. This is quite reasonable. Below its T_m, when the rate of helix formation dominates the relaxation kinetics, the relaxation time should be small, because the improbability of closing such a large loop will strongly inhibit the reaction. Note the unusual temperature dependence of this transition (Fig. 24-23). The relaxation time decreases below T_m as well as above T_m. The former effect probably is due to coupling between melting of the acceptor stem and of the anticodon stem.

We have been discussing each transition as though it were unaffected by all the others, but this clearly is an oversimplification. The acceptor-stem and anticodon-stem meltings overlap in temperature range. When the anticodon stem is partially duplex, it must speed the rate of formation of the acceptor stem, because the size of the loop required to close the acceptor stem becomes substantially reduced. It actually is surprising that more coupling effects are not seen. However, it also is fortunate, because severe coupling can lead to new relaxation times that are functions of several of the fundamental duplex times, and this effect can severely inhibit any attempt to assign and analyze each optical relaxation.

Note that the two duplex regions that turn out to be most stable in thermal melting are also the two postulated to remain intact when tRNA is converted into an elongated form at low salt concentration. In contrast, the relatively easy melting of the tertiary structure and of the dihydrouridine stem suggest that, in the various functional roles of tRNA, these two regions probably could undergo conformational changes. It is important to recognize from the studies of tRNA melting we have summarized how the different experimental techniques gain in strength when used in concert. NMR permitted indirect assignments of optical melting transitions to be confirmed, and the optical melting studies strengthened confidence in the assignments of particular resonances in the NMR spectrum. This symbiosis of various techniques is likely to be a necessity if the conformational equilibria and dynamics of some of the more complex RNAs ever are to be understood.

Secondary and tertiary structure of large RNAs

As an example of what the future has in store, consider the partial sequence of bacteriophage MS2 RNA in Figure 24-24. This sequence has been folded into a secondary structure locally consistent with the known duplex and loop interactions; but with such a large sequence (as with a typical protein), there is no practical way to examine all possible base-pairing schemes to be sure that a globally most-stable secondary structure has been found. Some aspects of the MS2 RNA secondary structure surely are important for function, and uncovering these aspects is a major challenge.

Techniques for the exploration of the structure of large RNAs are still in their infancy. One approach that appears to have considerable promise is intramolecular RNA cross-linking. For example, various psoralen derivatives bind specifically to double stranded regions and produce cross-linking efficiently upon UV irradiation. Other reagents appear able to cross-link specifically between two nearby single-stranded regions or between two nearby double-stranded regions. The approximate location of cross-links is found by examining the RNA in the electron microscope after vigorous denaturation. Molecules without cross-links appear as extended rods, whereas those with cross-links contain loops. Measuring the loop location reveals the approximate position of the cross-link. In ideal cases, an exact cross-link position should be accessible by rapid gel sequencing techniques.

A fundamental question about RNA structure is the relative potential for forming base pairs between residues nearby in the sequence (hairpins) versus those between residues far apart in the sequence (loops). The former are more favorable when the local thermodynamics are considered, but the cost in free energy of closing a large loop is really not that great and can easily be compensated for by a few G–C pairs. In preliminary studies on rRNA, numerous double-stranded interactions producing large cross-linked loops have been seen. However, the overall balance between hairpins and loops remains to be proven.

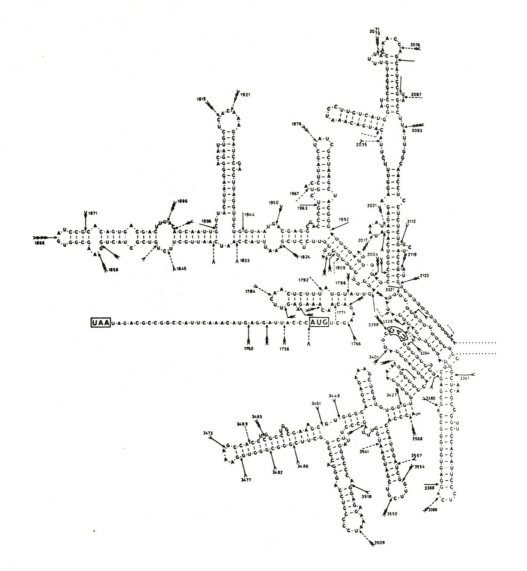

Figure 24-24

Part of the sequence of MS2 RNA, folded into a plausible secondary structure using (locally) the energy considerations of Tables 23-4 and 23-5. Arrows point to sites easily split by nucleases during partial digestion. [After W. Fiers et al., *Nature* 260:500 (1976).]

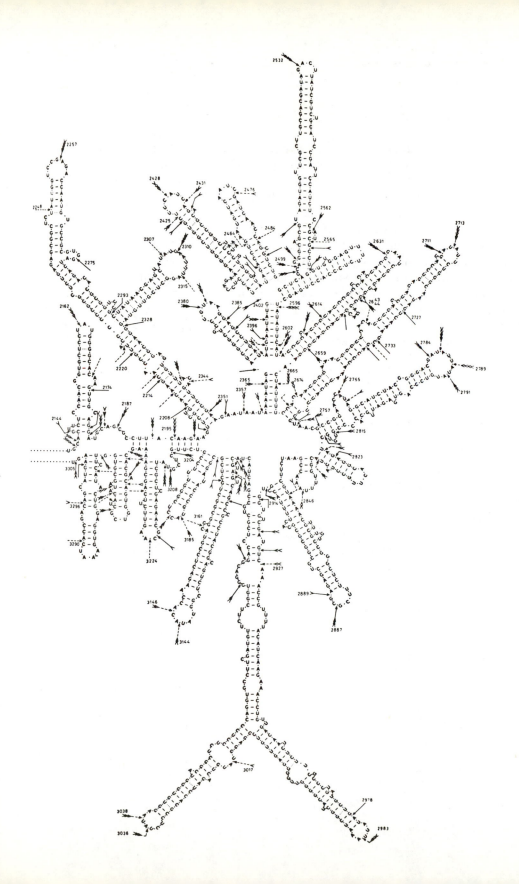

Summary

A topological constraint exists in closed circular duplex DNA. The linking number (the number of times one strand passes through the circle formed by the other) must remain constant. This leads to a coupling between the number of local helix turns (or the parameters characterizing a local helix) and the tendency of the DNA to coil into superhelices. Naturally occurring DNA closed circles appear to have negative toroidal superhelices. Adding an intercalating agent such as ethidium will progressively remove these superhelices to produce a relaxed DNA duplex circle. Further binding of ethidium leads to the incorporating of positive toroidal turns in the DNA. From a knowledge of the unwinding angle accompanying the binding of a single ethidium, it is possible to compute the average number of supercoils present in a DNA sample. DNAs with identical molecular weight but different supercoil number can be physically separated by gel electrophoresis. Supercoiling provides an extremely sensitive tool for monitoring local structural changes in DNA because it amplifies these into gross structural changes.

RNA tertiary structure in solution has been studied most extensively in tRNA. Spectroscopic techniques can provide estimates of the extent of base pairing. NMR appears capable in many cases of determining which particular residues are base-paired. Sequences that remain exposed or single-stranded can be identified by purine C^8-tritium exchange and by their ability to bind complementary oligonucleotides. The overall pattern of folding of individual helix stems and loops can be assessed by cross-linking and by singlet–singlet energy-transfer techniques. Various conformations become accessible to tRNAs as the salt concentration and temperature are raised. Combining equilibrium melting and kinetic studies allows construction of a phase diagram for tRNA structures. The detailed stages in tRNA melting can be analyzed by examining the temperature dependence of the NMR spectrum and then correlating the disappearance of sets of base-paired residues with relaxation times visible by temperature-jump kinetics.

Problems

24.1. Suppose that a DNA closed circular duplex of 6×10^6 mol wt is a relaxed structure containing the B helix in normal aqueous solution.
  a. Compute the number and type of supercoils expected when the DNA is converted into the A helix by the addition of alcohol.
  b. Suppose the DNA in part a is relaxed in alcohol by the nicking-and-closing enzyme. When the enzyme is removed and the DNA is returned to normal aqueous buffer, what happens?

24-2. How do the two structures shown in Figure 24-25 differ topologically from each other? from the structures shown in cases i and n of Box 24-2?

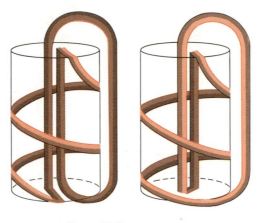

Figure 24-25

Structures for Problem 24-2.

24-3. A DNA is allowed to equilibrate with ethidium bromide. Nicking-and-closing enzyme then is added to relax the DNA completely.
  a. In one experiment, some ethidiums are found to be released from the DNA. What is the sign of the toroidal supercoils originally present in the DNA? If you knew how many ethidiums were released, could you calculate the number of supercoils originally present?
  b. In a second experiment, no ethidiums are found to be released. If 169 ethidiums were bound, what are the number and type of supercoils originally present in the DNA?

24-4. Demonstrate explicitly that $4N - 10$ intersite distances are sufficient to reconstruct the structure of N sites within a macromolecule. OPTIONAL: Suppose you know that a structure with eight sites has C_2 symmetry. We can write this schematically as ABCD·A'B'C'D'. Find the minimal number of distances that must be measured to reconstruct this structure. (Optional problem adapted from a suggestion by Francis Crick.)

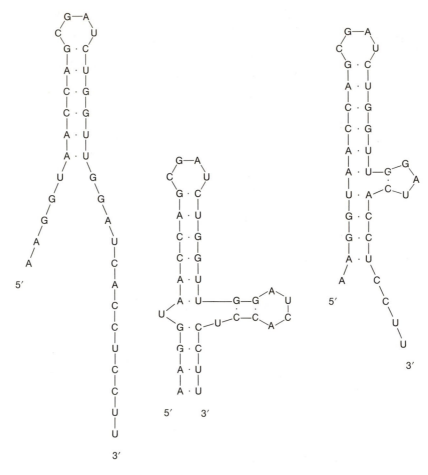

Figure 24-26

Proposed structures for Problem 24-5.

24-5. Figure 24-26 shows three structures that have been proposed for an RNA fragment.
   a. Estimate the NMR resonances expected between -14 and -15 ppm for each of these three structures. Will it be possible to distinguish among them using an NMR spectrum in this region?
   b. Name two tetranucleotides that could be used in binding studies to distinguish clearly among these structures.
   c. What single pancreatic RNase fragment would be most useful to analyze for C^8-H tritium-exchange studies?
   d. How could relaxation kinetics help to distinguish among the three structures?

References

GENERAL

Bauer, W. R. 1978. Structure and reactions of closed duplex DNA. *Ann. Rev. Biophys. Bioengin.* 7:287.

Bauer, W. R., and J. Vinograd, 1974. Circular DNA. In *Basic Principles in Nucleic Acid Chemistry*, vol. 2, ed. P. O. P. Ts'o (New York: Academic Press), p. 265.

Bloomfield, V., D. Crothers, and I. Tinoco, Jr. 1974. *Physical Chemistry of Nucleic Acids.* New York: Harper & Row.

Crothers, D. M., and P. E. Cole. 1978. Conformational changes of tRNA. In *Transfer RNA*, ed. S. Altman (Cambridge: MIT Press), p. 196.

SPECIFIC

Crick, F. H. C. 1976. Linking numbers and nucleosomes. *Proc. Natl. Acad. Sci. USA* 73:2639.

Crothers, D. M., P. E. Cole, C. W. Hilbers, and R. G. Shulman. 1974. The molecular mechanism of thermal unfolding of *Escherichia coli* formylmethionine transfer RNA. *J. Mol. Biol.* 87:63.

Dugaiczyk, A., H. W. Boyer, and H. M. Goodman. 1975. Ligation of *EcoRI* endonuclease-generated DNA fragments into linear and circular structures. *J. Mol. Biol.* 96:171.

Fuller, F. B. 1971. The writhing number of a space curve. *Proc. Natl. Acad. Sci. USA* 68:815.

Gamble, R. C., H. J. P. Schoemaker, E. Jekowsky, and P. R. Schimmel. 1976. Rate of tritium labeling of specific purines in relation to nucleic acid and particularly transfer RNA conformation. *Biochemistry* 15:2791.

Hilbers, C. W., G. T. Robillard, R. G. Shulman, R. D. Blake, P. K. Webb, R. Fresco, and D. Riesner. 1976. Thermal unfolding of yeast glycine transfer RNA. *Biochemistry* 15:1874.

Kearns, D. R. 1976. High resolution nuclear magnetic resonance investigations of the structure of tRNA in solution. In *Progress in Nucleic Acid Research and Molecular Biology*, vol. 18, W. E. Cohn (New York: Academic Press), p. 91.

Rhodes, D. 1977. Initial stages of the thermal unfolding of yeast phenylalanine transfer RNA as studied by chemical modification: the effect of magnesium. *Eur. J. Biochem.* 81:91.

Riesner, D., and R. Römer. 1973. Thermodynamics and kinetics of conformational transitions in oligonucleotides and tRNA. In *Physico-Chemical Properties of Nucleic Acids*, vol. 2, ed. J. Duchesne (London: Academic Press), p. 237.

Robillard, G. T., C. E. Tarr, F. Vosman, and B. R. Reid. 1977. A nucleic magnetic resonance study of secondary and tertiary structure in yeast tRNA[Phe]. *Biochemistry* 16:5261.

Sundaralingam, M., and S. T. Rao, 1975. *Structure and Conformation of Nucleic Acids and Protein-Nucleic Acid Interactions.* Baltimore: University Park Press.

Wang, J. C., and N. Davidson. 1966. On the probability of ring closure of lambda DNA. *J. Mol. Biol.* 19:469.

Introduction to membrane equilibria and to bilayers

25-1   MEMBRANE SYSTEMS

Every living cell is surrounded by an outer membrane (Chapter 4). This serves as the dividing partition between the cell's interior constituents and the extracellular environment. It also acts as the interface through which a cell communicates with the external medium and other cells. Membranes serve in other situations as well, such as a divider between the nucleus and the cytoplasm, and as the cablelike endoplasmic reticulum.

For many years there has been interest in membrane phenomena, and not solely because of the obvious importance of biological membranes. As an application of the thermodynamic principles that govern the distribution of molecules across semipermeable membranes, synthetic membranes (for example, fabricated from cellophane) are used in laboratory situations to gain useful information, for instance, on molecular weights of macromolecules and on small-molecule–macromolecule interactions. These principles apply as well to the more complex membranes encountered in biology, though in these cases additional special features must be accounted for as well.

In the present chapter, we first consider the fundamental thermodynamic principles that govern the distribution of small molecules across a simple semipermeable membrane, and we discuss how these principles enable us to study some interesting problems. These principles form a basic foundation for understanding essential features of all semipermeable barriers, including complex biological membranes. We then treat the properties of model biological membranes—phospholipid bilayers—with the aim of understanding the thermodynamic principles that dictate bilayer formation and of gaining insight into bilayer structure–function relationships. In this instance, we have made no attempt to cover even a reasonable fraction of the exciting work in this area; instead, we briefly summarize the essential ideas. This discussion of model membranes serves as a rational basis for understanding true biological membranes, which are briefly treated in Chapter 4.

25-2 MEMBRANE EQUILIBRIA

General comments on equilibria across a membrane

The most important thermodynamic principle governing the distribution of diffusible molecules across a membrane is the requirement that the chemical potential of a given component be the same on both sides of the membrane. Many interesting results can be derived from this principle.

To illustrate the basic idea, we consider a system in which a nonelectrolyte molecule D freely equilibrates across a membrane (Fig. 25-1). We use superscript (1) and (2) to denote side 1 and side 2 of the membrane. At equilibrium, it is required that

$$\mu_D^{(1)} = \mu_D^{0(1)}(T^{(1)}, P^{(1)}) + RT \ln a_D^{(1)} \tag{25-1a}$$

$$= \mu_D^{0(2)}(T^{(2)}, P^{(2)}) + RT \ln a_D^{(2)} = \mu_D^{(2)} \tag{25-1b}$$

where μ_D and a_D are the chemical potential and activity of D, respectively.[§] In general, equilibration occurs at constant T and P throughout the system, so that the standard chemical potentials $\mu_D^{0(1)}$ and $\mu_D^{0(2)}$ are identical on both sides of the membrane ($T^{(1)} = T^{(2)}$, and $P^{(1)} = P^{(2)}$). Consequently, we conclude that

$$a_D^{(1)} = a_D^{(2)} \tag{25-2}$$

Thus, D distributes itself equally across the membrane. If there are many different neutral molecules that freely equilibrate across the membrane, then equations identical to Equations 25-1 and 25-2 hold for each one.

Consider now a situation where a macromolecule is placed in solution on side 2, and where the membrane is not permeable to the macromolecule. For any diffusible neutral species, Equations 25-1 and 25-2 still hold. [Strictly speaking, the presence of

---

[§] We can write the chemical potential in different ways, depending upon our choice of the standard state. For example, ignoring activity-coefficient corrections, we can write $\mu_D = \mu_D^0 + RT \ln(D)$, where (D) is the concentration in moles per liter. In this case, the standard state is 1 mole per liter, and μ^0 is the partial molal free energy of a 1 M solution of D. On the other hand, if we write $\mu = \mu_D^0 + RT \ln \chi_D$, where χ_D is the mole fraction of D, then the standard state is unit mole fraction of D, and μ_D^0 is the partial molal free energy of D at unit mole fraction—i.e., pure D. The choice of how to express μ_D in terms of concentration of D depends on the problem under consideration.

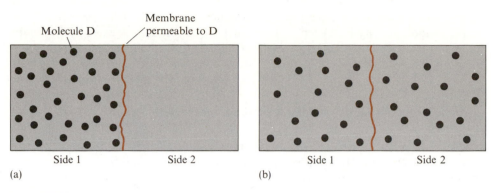

Figure 25-1

Equilibration across a semipermeable membrane. (**a**) Initial conditions, with D molecules on side 1. (**b**) After time for equilibration, there are equal numbers of D molecules on sides 1 and 2.

the macromolecule creates an osmotic pressure—discussed in the following sub-section—so that the pressure on side 2 is slightly greater than that on side 1. In this case, there is a small difference between $\mu^0(T, P^{(1)})$ and $\mu^0(T, P^{(2)})$. However, this difference usually is not great enough to affect significantly the result of Equation 25-2.] Equation 25-2 is particularly important because, in a situation where a species such as D binds to the macromolecule, it provides the basis for measuring the amount of bound D by equilibrium dialysis (Fig. 25-2). By this method, the freely diffusing ligand that binds is allowed to come to equilibrium, then measured. At equilibrium, the total concentration of D (bound and unbound) on side 2 minus the concentration on side 1 is the concentration of bound D:

$$(D)^{(2)} + (\text{Bound D})^{(2)} - (D)^{(1)} = (\text{Bound D})^{(2)} \qquad (25\text{-}3)$$

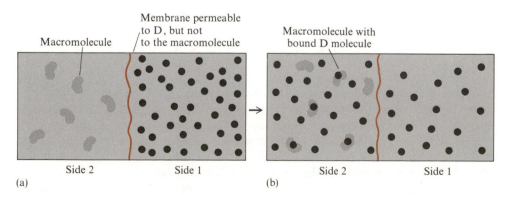

Figure 25-2

Equilibrium dialysis. (**a**) At the start, D molecules are on side 1 and macromolecules are on side 2. (**b**) After equilibration, equal numbers of free D molecules are found on each side of the membrane.

where, for simplicity, we have substituted concentrations for activities (clearly, we could include activity-coefficient corrections). Equation 25-3 is extremely valuable and has been widely applied to a variety of systems to study ligand interactions with macromolecules.

Osmotic pressure across a semipermeable membrane

Consider a system in which a macromolecule is constrained to one side of a semi-permeable membrane and is dissolved in a solvent that equilibrates freely across the membrane (Fig. 25-3). The chemical potential of the solvent is $\mu_s = \mu_s^0 + RT \ln$

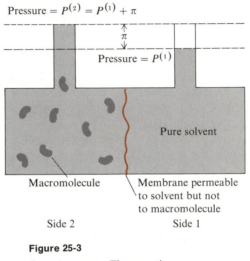

Pressure $= P^{(2)} = P^{(1)} + \pi$

π

Pressure $= P^{(1)}$

Pure solvent

Macromolecule

Membrane permeable to solvent but not to macromolecule

Side 2　　　　　　Side 1

Figure 25-3

Osmotic pressure. The osmotic pressure $\pi = P^{(2)} - P^{(1)}$, the difference in pressures on the two sides of the membrane.

χ_s, where μ_s^0 is the standard chemical potential of the solvent, and χ_s is the mole fraction of solvent. At equilibrium, the chemical potential of the solvent must be the same on both sides of the membrane. However, on the side of pure solvent (side 1), we have $\chi_s = 1$, whereas the presence of a second component (macromolecule) on side 2 means that $\chi_s < 1$ on side 2. No matter how much solvent flows from side 1 to side 2, the solvent mole fractions on the two sides of the membrane will not be equal. There-fore, in order to achieve equilibrium, the standard chemical potential of the solvent on the two sides of the membrane must differ by precisely the amount that is required to offset the difference in solvent activity on the two sides of the membrane.

In the experimental arrangement of Figure 25-3, equilibrium is achieved by solvent flowing from side 1 to side 2 until the pressure on the solution in side 2 becomes sufficiently greater than that on side 1 so as to achieve a balance in the chemical potential of solvent on both sides. The difference in pressure between the two sides is known as the *osmotic pressure*. As we shall see, the magnitude of the osmotic pressure depends on the number of macromolecule particles on side 2 of the membrane. Thus, osmotic pressure is a colligative property.

To obtain a relationship between the osmotic pressure and experimental variables, we work with expressions for the solvent chemical potential. Using again superscripts (1) and (2) to denote sides 1 and 2, and subscript s to denote solvent, we obtain the following expressions for the system at equilibrium and constant T:

$$\mu_s^{(1)} = \mu_s^{0(1)}(T, P^{(1)}) + RT \ln \chi_s^{(1)} \tag{25-4a}$$

$$\mu_s^{(2)} = \mu_s^{0(2)}(T, P^{(2)}) + RT \ln \chi_s^{(2)} \tag{25-4b}$$

and

$$\mu_s^{(1)} = \mu_s^{(2)} \tag{25-5}$$

It is clear that $\chi_s^{(1)} = 1$; therefore, from Equations 25-4 and 25-5, we obtain

$$\mu_s^{0(2)}(T, P^{(2)}) - \mu_s^{0(1)}(T, P^{(1)}) = -RT \ln \chi_s^{(2)} \tag{25-6}$$

Equation 25-6 establishes the basic relationship between the solvent activity on the macromolecule-containing side of the membrane (right-hand side of Eqn. 25-6) and the difference in solvent standard chemical potentials on the two sides of the membrane (left-hand side of Eqn. 25-6).

To convert Equation 25-6 into a more useful form, we consider dilute solutions and assume that the difference between $\mu_s^{0(2)}$ and $\mu_s^{0(1)}$ is due solely to the difference in pressure. Thus, from the basic thermodynamic relationship $(\partial \mu / \partial P)_T = \overline{v}$ (where $\overline{v}$ is the partial molar volume), we have

$$\mu_s^{0(2)}(T, P^{(2)}) - \mu_s^{0(1)}(T, P^{(1)}) = \int_{P^{(1)}}^{P^{(2)}} \overline{v}_s^{(2)} \, dP \tag{25-7}$$

Because we are dealing with dilute solutions, $\overline{v}_s^{(2)}$ can be replaced by the partial molar volume $\overline{v}_s$ of pure solvent. If we assume incompressibility and substitute $\overline{v}_s^{(2)} = \overline{v}_s$, Equations 25-6 and 25-7 yield

$$\pi \overline{v}_s = -RT \ln \chi_s^{(2)} \tag{25-8}$$

where π is the osmotic pressure: $\pi = P^{(2)} - P^{(1)}$.

Equation 25-8 can be converted into a more useful form by noting that $\chi_s^{(2)}$ is close to unity in dilute solutions, so that the logarithm may be expanded:

$$-\ln \chi_s^{(2)} \cong 1 - \chi_s^{(2)} = \chi_2^{(2)} \tag{25-9}$$

where subscript 2 denotes macromolecule. With $\hat{c}$ designating concentration in grams per liter and M designating molecular weight, and recognizing that $\hat{c}_s^{(2)}/M_s \gg c_2^{(2)}/M_2$, Equation 25-9 becomes

$$
\chi_2^{(2)} = \frac{\hat{c}_2^{(2)}/M_2}{\hat{c}_2^{(2)}/M_2 + \hat{c}_s^{(2)}/M_s}
$$

$$
\cong \frac{\hat{c}_2^{(2)}/M_2}{\hat{c}_s^{(2)}/M_s} \tag{25-10}
$$

From Equations 25-8, 25-9, and 25-10, we obtain

$$
\pi = \frac{RT}{\bar{v}_s} \left(\frac{\hat{c}_2^{(2)}/M_2}{\hat{c}_s^{(2)}/M_s} \right) \tag{25-11}
$$

Because we are dealing with a dilute solution, $\hat{c}_s^{(2)}/M_s \cong \bar{v}_s^{-1}$, so Equation 25-11 becomes

$$
\pi = (RT/M_2)\hat{c}_2 \tag{25-12}
$$

where we have dropped the superscript (2). Equation 25-12 is a useful result.

Before proceeding further, it is helpful to consider the magnitude of the osmotic pressure generated by a macromolecule solution. At room temperature, $RT \cong$ 24 atm liter mole^{-1}. If $M_2 = 10^4$, and $\hat{c}_2 = 10$ g l^{-1} ($= 10$ mg ml^{-1}), then $\pi \cong 0.024$ atm. In this case, the osmotic pressure is within a factor of 40 of atmospheric pressure. However, at the same value of $\hat{c}_2$ but with $M_2 = 10^6$, we have $\pi \cong 0.00024$ atm. In this case, the macromolecule exerts an osmotic pressure of only about 0.24 milliatmospheres. These calculations give a rough idea of the range of experimentally observed osmotic pressures; they also show the direct dependence of π on the number of solute molecules per unit volume.

In real solutions, Equation 25-12 is only approached as a limiting form. In general, the osmotic pressure may be written as a virial expansion in $\hat{c}_2$, so that

$$
\pi = (RT/M_2)\hat{c}_2 + B\hat{c}_2^2 + \cdots \tag{25-13}
$$

where B is the second virial coefficient. Clearly, Equation 25-13 reduces to Equation 25-12 as $\hat{c}_2 \to 0$.

In the event that the solute-containing side of the membrane contains several different macromolecular species 1, 2, 3, etc., then Equation 25-9 becomes

$$
-\ln \chi_s^{(2)} \cong \chi_1^{(2)} + \chi_2^{(2)} + \chi_3^{(2)} + \cdots \tag{25-14a}
$$

and it is easy to show then that Equation 25-12 becomes

$$
\pi = \sum_i \pi_i \tag{25-14b}
$$

where

$$\pi_i = (RT/M_i)\hat{c}_i \qquad (25\text{-}14c)$$

Thus, with a mixture, the total osmotic pressure is simply the sum of contributions by each species of macromolecule. This is to be expected for a property that depends only on the number of particles.

Using osmotic pressure to determine molecular weights

Equations 25-12 and 25-13 indicate that measurements of osmotic pressure as a function of $\hat{c}_2$ should yield molecular weight. Generally this is accomplished by constructing plots of $\pi/\hat{c}_2$ versus $\hat{c}_2$. In most cases, the plot is a straight line with a slope equal to the second virial coefficient B, and an intercept on the ordinate axis of RT/M_2.

Figure 25-4 shows such plots for four native proteins. In each case, the experimental points fall on a straight line that, over the concentration range investigated,

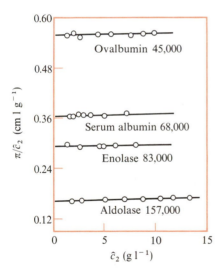

Figure 25-4

Osmotic pressure measurements on various proteins in their native states. Approximate molecular weights are indicated. [After F. J. Castellino and R. Barker, *Biochemistry* 7:2207 (1968).]

is rather flat. From the ordinate intercepts, these plots yield fairly precise values of the molecular weights of the proteins under investigation.

The question arises as to the meaning or interpretation of the apparent molecular weight $\bar{M}$ measured in a system where there is size heterogeneity with respect to the solute. This problem is equivalent to finding an expression for $\bar{M}$ in the equation

$$\pi/\hat{c} = RT/\bar{M} \qquad (25\text{-}15)$$

where $\hat{c}$ is total solute concentration (that is, $\hat{c} = \sum \hat{c}_i$). Dividing both sides of Equation 25-14b by $\hat{c}$, we obtain

$$\pi/\hat{c} = \left(1 \bigg/ \sum_i \hat{c}_i\right) \sum_i \pi_i = \left(RT \bigg/ \sum_i \hat{c}_i\right) \sum_i (\hat{c}_i/M_i) \qquad (25\text{-}16)$$

Comparing Equations 25-15 and 25-16, we see that

$$\bar{M} = \left(\sum_i \hat{c}_i\right) \bigg/ \sum_i (\hat{c}_i/M_i) \qquad (25\text{-}17)$$

Thus, $\bar{M}$ is a particular kind of average molecular weight. Multiplying each $\hat{c}_i$ in the numerator of Equation 25-17 by M_i/M_i enables us to see more clearly the meaning of $\bar{M}$:

$$\bar{M} = \left[\sum_i (\hat{c}_i/M_i)M_i\right] \bigg/ \sum_i (\hat{c}_i/M_i) \qquad (25\text{-}18)$$

Because $\hat{c}_i/M_i$ is the number of moles of the ith solute molecule per unit volume, it is clear that $\bar{M}$ is a *number-average* molecular weight ($\bar{M}_n$), and this is what is determined from an osmotic-pressure measurement on a solution containing a mixture of macromolecule species. In view of the fact that osmotic pressure depends on the number of solute particles, this result is not surprising.

The Donnan effect

In 1911, F. G. Donnan pointed out an important effect of charged macromolecules (limited to one side of a semipermeable membrane) on the distribution of small ions across the membrane. It is not difficult to understand how this Donnan effect comes into play.

Consider a system with a charged macromolecule M of charge z on one side (side 2) of a semipermeable membrane, and with a freely equilibrating, ionizable salt A^+B^- (Fig. 25-5). The salt ionizes according to the scheme $AB \rightleftarrows A^+ + B^-$ and, with the ionization reaction at equilibrium,

$$\mu_{AB} = \mu_{A^+} + \mu_{B^-} \qquad (25\text{-}19)$$

and

$$\mu_{AB} = \mu_{AB}^0 + RT \ln(AB) \qquad (25\text{-}20a)$$

$$= \mu_{A^+}^0 + \mu_{B^-}^0 + RT \ln(A^+) + RT \ln(B^-) \qquad (25\text{-}20b)$$

$$= \mu_{A^+}^0 + \mu_{B^-}^0 + RT \ln[(A^+)(B^-)] \qquad (25\text{-}20c)$$

where Equations 25-20b,c follow directly from Equations 25-19 and 25-20a. In these

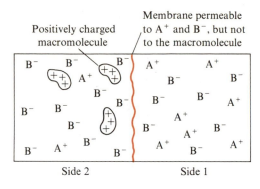

Figure 25-5

The Donnan effect. The ions A^+ and B^- are asymmetrically distributed across the membrane.

expressions, we have for simplicity used concentrations for activities; of course, the equations easily can be modified to include activity coefficients.

At equilibrium, we require that $\mu_{AB}^{(1)} = \mu_{AB}^{(2)}$. Assuming that the standard chemical potentials $\mu_{A^+}^0$ and $\mu_{B^-}^0$ are the same on both sides of the membrane, we conclude that

$$RT \ln[(A^+)^{(1)}(B^-)^{(1)}] = RT \ln[(A^+)^{(2)}(B^-)^{(2)}] \qquad (25\text{-}21a)$$

or

$$(A^+)^{(1)}/(A^+)^{(2)} = (B^-)^{(2)}/(B^-)^{(1)} = r_D \qquad (25\text{-}21b)$$

where r_D is the Donnan ratio. When a charged macromolecule M is present, this ratio may deviate significantly from unity. This deviation is a consequence of the condition for electrical neutrality,[§] which requires that

$$(A^+)^{(1)} = (B^-)^{(1)} \qquad (25\text{-}22a)$$

$$(A^+)^{(2)} + z(M) - (B^-)^{(2)} = 0 \qquad (25\text{-}22b)$$

Substituting Equation 25-21b into Equation 25-22b, we obtain

$$(A^+)^{(1)}/r_D + z(M) - r_D(B^-)^{(1)} = 0 \qquad (25\text{-}23)$$

Let $(s) = (A^+)^{(1)} = (B^-)^{(1)}$. [If the salt AB is completely dissociated, then (s) is the total salt concentration on side 1; otherwise, (s) is the concentration of each ionized

[§] It should be recognized that, when the macromolecule is introduced into the system, it is in a neutral form, so that it brings counterions with it. For example, if the charge z on M is positive, then the macromolecule could be added to the solution in the form of a salt with B^- counterions.

species on side 1.] Now we can rewrite Equation 25-23 as

$$r_D^2 - z(M)r_D/(s) - 1 = 0 \qquad (25\text{-}24)$$

Equation 25-24 is a quadratic equation in r_D with two solutions. Because $r_D > 0$, we choose the solution giving positive values of r_D as the physically significant one and thus obtain, by the familiar quadratic formula,

$$r_D = z(M)/2(s) + \{1 + [z(M)/2(s)]^2\}^{1/2} \qquad (25\text{-}25)$$

Equation 25-25 clearly shows that, in general, $r_D \neq 1$, so that A^+ and B^- do not equally distribute themselves across the membrane. This unequal distribution happens even though the system is at equilibrium, and even though we have assumed no binding of A^+ or B^- to M.

It is instructive to consider the magnitude of the Donnan effect. The important parameter is $z(M)/2(s)$, which is simply the ratio of the concentration of charge from the macromolecule to the total concentration of ionic species $[(A^+) + (B^-) = 2(s)]$ on side 1. Table 25-1 gives values of r_D for different values of $z(M)/2(s)$, for the case

Table 25-1

Values of the Donnan
ratio r_D for $z > 0$

$z(M)/2(s)$	r_D
10	20.05
1	2.41
0.1	1.10
0.01	1.01

that $z > 0$. For the case where $z(M)/2(s) = 10$, we have $r_D = 20$ so that, according to Equation 25-21b, there is a 20-fold excess of (A^+) on side 1 and a 20-fold excess of (B^-) on side 2 of the membrane. This situation would be achieved, for example, with a protein concentration of 10^{-4} M, a net charge on the protein of $+10$, and a mono-valent salt (such as NaCl) concentration, (s), of 50 μM on side 1. However, as (s) becomes equal to and greater than $z(M)$, the value of r_D drops toward unity. This effect is expected, because Equation 25-25 shows that $r_D \to 1$ as $z(M)/2(s) \to 0$. Thus, an effective way to suppress the Donnan effect is to raise the salt concentration to a sufficient level.

It should also be noted that, if $z < 0$, then each number in the first column of Table 25-1 is preceded by a minus sign. In this case, the reciprocal of each r_D value listed in the table applies. Thus, if $z(M)/2(s) = -10$, then $r_D = 0.05$. This example, together with those in Table 25-1, simply shows that the asymmetric distribution of ions is always such that the small ion bearing the same sign ($+$ or $-$) of charge as the macromolecular ion is in greater concentration on the side of the membrane excluding the macromolecule, whereas the opposite is true for the other small ion.

The Donnan effect clearly is a serious consideration in situations where the binding of an ion to a macromolecule is studied by means of equilibrium dialysis across a semipermeable membrane. Not to recognize this effect can lead to serious errors in calculating the amount of bound ion from the total amount of that ion present on each side of the membrane.

pH difference across the membrane

In the preceding treatment, we have ignored H^+ and OH^-; because they typically are present in relatively small amounts, this omission is not of serious consequence. However, it is of interest to note that the Donnan ratio r_D fixes the hydrogen and hydroxyl ion concentrations on each side of the membrane as well as the ratios of A^+ and B^-. To show this, we note that our system can form the acid HB, which dissociates to $H^+ + B^-$. At equilibrium,

$$\mu_{HB} = \mu_{H+} + \mu_{B-} \tag{25-26}$$

Equation 25-26 is exactly analogous to Equation 25-19. Proceeding as in the derivation of Equation 25-21b, we obtain

$$(H^+)^{(1)}/(H^+)^{(2)} = (B^-)^{(2)}/(B^-)^{(1)} = r_D \tag{25-27a}$$

Likewise, we can show that

$$(OH^-)^{(2)}/(OH^-)^{(1)} = (A^+)^{(1)}/(A^+)^{(2)} = r_D \tag{25-27b}$$

From Equations 25-27a,b, or by direct consideration of the species HOH (which dissociates to H^+ and OH^-), we have

$$(OH^-)^{(2)}/(OH^-)^{(1)} = (H^+)^{(1)}/(H^+)^{(2)} = r_D \tag{25-27c}$$

Thus, H^+ and OH^- establish a concentration difference across the membrane as a result of the presence of the charged macromolecule on one side of the membrane. From Equation 25-27a, it is clear that

$$\log(H^+)^{(1)} - \log(H^+)^{(2)} = \log r_D \tag{25-28a}$$

or, because $pH = -\log(H^+)$,

$$pH^{(2)} - pH^{(1)} = \log r_D \tag{25-28b}$$

From the values of r_D in Table 25-1, it is clear that, at very low salt concentration where $r_D = 20$, the difference in pH is greater than one pH unit. Also, it is evident that this pH difference can be eliminated by working at salt concentrations high enough that $r_D \cong 1$.

If a system contains many diffusible ions of comparable concentrations, then the equations for electrical neutrality (cf. Eqn. 25-22a,b) become more complex. However, it can be shown that, for any ionic species X_i with charge z_i, the ratio $(X_i)^{(1)}/(X_i)^{(2)}$ is a constant, with the same constant holding for each species. This ratio can be calculated readily in a system containing many species if one cation and one anion are in sufficient excess that the others may be ignored in the expression for electrical neutality.

Membrane polarization and membrane potential

It is well known from electrochemistry that an unequal distribution of ions produces a potential difference that can be measured as an electromotive force. However, if electrodes are placed on either side of the membrane in Figure 25-5, no potential difference will be measured. This effect is simply because the system is at thermodynamic equilibrium.

Although the net potential difference is zero, more careful analysis shows that this is true only because the membrane is polarized, as a result of the ion-concentration gradient. For simplicity, assume we are again dealing with the simple salt A^+B^- and the charged macromolecule M. Letting $\Delta\Phi = \Phi^{(1)} - \Phi^{(2)}$ be the potential difference between side 1 and side 2, we have

$$\Delta\Phi = 0 = \Phi_m + (RT/\mathscr{F}) \ln[(A^+)^{(1)}/(A^+)^{(2)}] \qquad (25\text{-}29\text{a})$$

$$= \Phi_m + (RT/\mathscr{F}) \ln r_D \qquad (25\text{-}29\text{b})$$

where Equation 25-29b follows from Equation 25-21b, and $\mathscr{F}$ is the Faraday constant (96, 487 coulomb mole^{-1}). The second term in Equation 25-29a is the familiar expression for the potential difference between solutions with different ion activities. The first term, Φ_m, is a junction potential; it is analogous to the liquid–liquid junction potential encountered when a saturated salt bridge is placed between two solutions. This potential is known in the present case as the membrane potential. From Equation 25-29b, it follows that

$$\Phi_m = (-RT/\mathscr{F}) \ln r_D \qquad (25\text{-}30)$$

Thus, the membrane potential is equal and opposite to that generated by ion gradients. This result shows that the membrane itself is polarized as a result of the ion gradients.

The potential difference $(RT/\mathscr{F}) \ln r_D$ actually can be measured by placing an impermeable barrier between the two solutions and connecting them with a salt bridge—for example, a saturated KCl bridge.

It is of interest to consider the magnitude of the membrane potential. Returning to the examples of Table 25-1, we calculate that for $r_D = 20$ the membrane potential $\Phi_m = 77$ mV, and for $r_D = 1.01$ the value of Φ_m is 0.26 mV. Thus, the membrane potential can be significant in magnitude, but sufficiently high salt concentrations suppress the Donnan effect to the point where the membrane potential is virtually eliminated.

Membrane equilibria and transport in biological membranes

Thus far, our discussion of semipermeable membranes has concentrated on simple types of membranes used in laboratory systems. Of course, the thermodynamic principles for these simple systems must hold also for all membranes, including the far more complex biological membranes that carry out a rich diversity of functions. However, because of the complexity of some membrane equilibria, the thermodynamic principles that always operate appear superficially to be obscured.

As a specific example, consider the fact that most animal cells maintain a sharp difference in concentrations of Na^+ and K^+ between the cell interior and the external medium. This fact may seem unsurprising in view of our discussion of the Donnan effect; because the cell interior contains many high-molecular-weight electrolytes, it is plausible that small cations might distribute themselves according to a Donnan ratio. However, for the case in point, the striking fact is that K^+ is more concentrated *inside* the cell membrane, whereas Na^+ is more concentrated *outside* the membrane. This observation is not in accord with simple Donnan predictions, which require that the ratio r_D of concentration inside to that outside be the same for each monovalent cation. Here is a situation where it appears superficially that the simple thermodynamic arguments do not hold.

The explanation lies in the fact that animal cell membranes have an active Na^+–K^+ pump, which concentrates K^+ inside the cell and pushes Na^+ to the outside. This pump is driven by the hydrolysis of ATP in a reaction catalyzed by a Na^+–K^+ ATPase. The hydrolysis of ATP is tightly linked to Na^+ and K^+ transport, so that the energy required to generate these unexpected ion-concentration gradients is completely dependent on ATP hydrolysis. It appears that, for every three Na^+ plus two K^+ transported, about one molecule of ATP is hydrolyzed.

The Na^+–K^+ pump is but one of many active transport systems present in biological membranes. Each of these systems is driven by an appropriate thermodynamically favorable reaction, permitting concentration gradients that by themselves (as isolated systems) would be thermodynamically disallowed. Thus, although the simple thermodynamic principles discussed earlier still hold, we must expand our perspective to include additional processes that are coupled to equilibration across membranes.

25-3   MICELLES

The essential framework of biological membranes is provided by lipid amphiphiles that aggregate to form bilayer vesicles encapsulating an internal cavity (see Fig. 4-13). The lipid amphiphiles are composed of polar head groups attached to long hydrocarbon tails (Chapter 4). Bilayers generally are constructed from double-chain amphiphiles—that is, molecules in which two hydrocarbon chains are attached to a single head group. On the other hand, single-chain amphiphiles do not make bilayers, but instead form micelles—globular aggregates with polar groups exposed to the surface and hydrocarbon portions clumped together (Fig. 4-13). This difference in behavior between single- and double-chain amphiphiles is explored in the remainder of this chapter, where we emphasize the thermodynamic principles that determine bilayer formation and the structure–function relationships of bilayers. (Chapter 4 discusses the general features of biological membrane structure and its relation to bilayers.)

Micelle formation

When an amphiphile such as dodecyl sulfate, $CH_3(CH_2)_{11}SO_4^{2-}$, is added in increasing concentrations to an aqueous solution, a critical point is reached at which the amphiphiles clump together and form aggregates. It is a cooperative process; most aggregates that form contain large numbers of amphiphiles, and no significant quantities form with only two or a few amphiphiles. Figure 25-6 qualitatively sketches this phenomenon as a plot of actual free-monomer amphiphile concentration in solution versus total added monomer concentration. The curve starts out as a straight line with a slope of unity; every added monomer becomes a free monomer in solution. At a critical concentration (known as the critical micelle concentration, or CMC), the concentration of free monomer in solution ceases to rise. A kind of phase separation

Figure 25-6

Distribution of amphiphile in monomer and micelle forms, as a function of total amphiphile concentration. [After C. Tanford, *The Hydrophobic Effect* (New York: Wiley, 1973).]

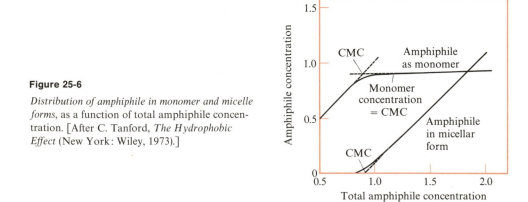

occurs, in which aggregates of the monomers in micellar structures are in equilibrium with an almost constant concentration of free monomer. As more monomer amphiphile is added, the amount of amphiphile in the micelle phase rises, while the free-monomer concentration increases very slowly. (Of course, if a true phase separation occurred, the monomer concentration would remain perfectly constant after the CMC.)

The micelles that form contain a large number of monomers, the exact number of which depends (among other factors) on the length of the hydrocarbon chain. For example, for n-alkyl betaines, $RN(CH_3)_2^+ CH_2COO^-$, the average number of monomers per micelle (under certain conditions) varies from 24 to 130 as saturated R groups vary from 8 to 15 carbon atoms.

The micelle interior

One might suspect that the interior of a micelle is chemically similar to a pure hydrocarbon solution. This hypothesis can be tested in many ways. One way is to examine the thermodynamics of transferring a hydrocarbon from water to the interior of a micelle, comparing it with transfer of the same hydrocarbon from water to the pure liquid-hydrocarbon state. Table 25-2 shows thermodynamic data for transfer of three probe hydrocarbons (ethane, n-propane, and n-butane) from water to the interior of a dodecyl sulfate micelle, and from water to the pure liquid hydrocarbon. Note that, for each kind of process, the nonaqueous phase is increasingly favored as the chain length of the probe hydrocarbon increases, and that the transfer is driven by a very favorable entropy change, which is sufficient to overcome the unfavorable enthalpy change. These effects are analogous to those discussed in Chapter 5, where free-energy-of-transfer data are considered in connection with the hydrophobic effect. Large entropy changes occur as the hydrocarbon moves from the aqueous to the nonpolar phase and water molecules are released from their relatively immobile

Table 25-2

Thermodynamic parameters for transfer of hydrocarbons from water to interior of dodecyl sulfate micelle and from water to pure liquid-hydrocarbon state

Hydrocarbon	Water → Micelle interior			Water → Pure liquid hydrocarbon		
	ΔG^0 (kcal mole^{-1})	ΔH^0 (kcal mole^{-1})	ΔS^0 (cal deg^{-1} mole^{-1})	ΔG^0 (kcal mole^{-1})	ΔH^0 (kcal mole^{-1})	ΔS^0 (cal deg^{-1} mole^{-1})
Ethane	-3.45	$+2.0$	$+18.3$	-3.9	$+2.5$	$+21$
Propane	-4.23	$+1.0$	$+17.5$	-4.9	$+1.7$	$+22$
Butane	-5.13	0.0	$+17.2$	-5.9	$+0.8$	$+23$

SOURCE: C. Tanford, *The Hydrophobic Effect* (New York: Wiley, 1973), pp. 18, 38.

positions in organized structures; at the same time, however, water–water hydrogen bonds are broken, so the enthalpy change is unfavorable.

It is interesting to note that the free energy change for C_2H_6 is more than half of that for C_4H_{10}. This observation is rationalized by assuming that the free energy change is proportional to the area of the hydrocarbon in contact with H_2O in the aqueous phase. Thus, both CH_3 groups in C_2H_6 are in contact with the water, whereas the intervening $-CH_2-$ groups in C_4H_{10} are partly shielded from the solvent. Furthermore, because the enthalpy changes become significantly less positive with increasing hydrocarbon length, it is clear that the shape of the hydration layer may be of consequence.

The key observation, however, is the similarity in the thermodynamics for the two kinds of processes. The magnitudes of changes in free energy, enthalpy, and entropy are somewhat smaller for the transfer into the dodecyl sulfate micelle, but the trends are the same for both kinds of transfer. The relatively small differences in thermodynamic parameters for the two transfers possibly are due to the constraints of the micelle structure. For example, hydrocarbons placed near the surface of the micelle may be somewhat immobilized (not as free as those near the center or as those in the pure liquid state); they also may be partly in contact with an organized water layer. These effects would indeed cause smaller entropy and enthalpy changes for transfer to the micelle interior, in agreement with the observed values. Regardless of the small differences, however, the data do indicate clearly that the interior of the micelle is similar to a liquid-hydrocarbon state. Other data also support this conclusion.

The principle of opposing forces in micelle formation

An informative thermodynamic parameter is $\mu_{mic}^0 - \mu_w^0$, where μ_{mic}^0 is the chemical potential of an amphiphile in the micelle, and μ_w^0 is that of the same amphiphile in the aqueous phase. For the n-alkyl betaines, this parameter varies with the number N_C of carbons in the amphiphile R chain according to

$$\mu_{mic}^0 - \mu_w^0 = 2,514 - 709\, N_C \tag{25-31}$$

If an amphiphile with a different head group is used—for example, $R–(OCH_2OCH_2)_6$ OH (alkyl hexaoxyethylene glycol monoethers)—the coefficient of N_C is virtually the same, although the other parameter is substantially smaller. Thus, whereas one of the parameters is sensitive to the head group, the other is not. It also is important to note that the variation of $\mu_{mic}^0 - \mu_w^0$ with N_C ($\cong -700\, N_C$) is somewhat less than the variation seen for transfer of variable-length hydrocarbons from water into the interior of a micelle (Table 25-2).

These and related data have been rationalized by C. Tanford (1973) in accordance with the *principle of opposing forces*. The concept is simple. Two forces dominate the determination of the structures and characteristics of micelles. One is the attractive hydrophobic interaction between hydrocarbon chains; the other is the repulsion between head groups. The repulsive force is obvious in amphiphiles that bear head groups with a net charge. In addition, however, there is the effect of solvation, which plays the dominant role in neutral head groups. The thick solvation layers around the head groups act to prevent their close association.

We can discuss these two contributions separately by expressing μ_{mic}^0 as

$$\mu_{mic}^0 = U_{mic}^0 + W_{mic}^0 \qquad (25\text{-}32)$$

where U_{mic}^0 is the attractive and W_{mic}^0 the repulsive component. Therefore,

$$\mu_{mic}^0 - \mu_w^0 = U_{mic}^0 + W_{mic}^0 - \mu_w^0 \qquad (25\text{-}33)$$

The contribution $U_{mic}^0 - \mu_w^0$ should be closely related to $\mu_{HC}^0 - \mu_w^0$, the free energy change in transferring a hydrocarbon from water to the pure liquid hydrocarbon. For a $-CH_2-$ group, $\mu_{HC}^0 - \mu_w^0$ is about -800 cal mole^{-1}; for a terminal $-CH_3$, it is $-2{,}000$ cal mole^{-1}. In a micelle, one or two $-CH_2-$ groups nearest the surface should contribute little or nothing to $U_{mic}^0 - \mu_w^0$, because they are in contact with the water around the polar head group. Moreover, $-CH_2-$ groups below (but still relatively near) the surface of the micelle are geometrically constrained somewhat and will contribute less to $U_{mic}^0 - \mu_w^0$ than will those deep within the micelle. Therefore, of the $N_C - 1$ $-CH_2-$ groups, effectively only $N_C - 3$ to $N_C - 5$ should be analogous to a hydrocarbon in the pure liquid state. The exact number will depend on the nature of the head group and its influence on the local geometry at the surface of the micelle. With this reasoning, we have

$$U_{mic}^0 - \mu_w^0 = (N_C - 3 \text{ to } N_C - 5)(-800) - 2{,}000$$
$$= (+400 \text{ to } +2{,}000) - 800\,N_C \qquad (25\text{-}34)$$

The repulsive electrostatic component W_{mic}^0 adds a positive contribution to $\mu_{mic}^0 - \mu_w^0$. Its effect on the coefficient of N_C is not hard to reason. Because micelle size increases as N_C increases, and head-group separation decreases as micelle size increases (see following subsection), we conclude that $dW_{mic}^0/dN_C > 0$. Therefore, $-800\,N_C$ becomes closer to $-700\,N_C$ (in the presence of some salt).

The principle of opposing forces provides a ready explanation for the increase in average micelle size as N_C increases. The larger N_C is, the greater the attractive component; this means that a larger repulsive component will be required to terminate micelle growth. The larger repulsive contribution comes from the increase in micelle size, which in turn brings about a closer crowding of head groups (see following subsection) that raises W_{mic}^0 to the point where further growth is discouraged.

Micelle shape and size

Small micelles have low intrinsic viscosities characteristic of roughly spherical particles. Large ones have sizable intrinsic viscosities, indicating that they are quite asymmetric. Thus, large sizes are accommodated by highly asymmetric shape, whereas small sizes fit into more spherical forms. This feature of micelle shape and size has been explained by Tanford (1973) on the basis of simple geometric considerations and the principle of opposing forces.

For a micelle with N_{ch} chains, the "core" volume V (in $Å^3$) is

$$V = (27.4 + 26.9 \, N'_C)N_{ch} \tag{25-35}$$

where N'_C is the number of carbon atoms of the amphiphile that are imbedded in the liquid hydrocarbon core and therefore are not influenced by surface effects; thus, $N'_C < N_C$. Equation 25-35 is based on volume measurements, but it also could be deduced (with essentially the same parameters) from published densities of liquid hydrocarbons. In addition to the volume relationship, another constraint on geometry and size is the maximal extension l_{max} of the amphiphile chain, which is an upper limit to one of the dimensions of the core. The value of l_{max} (in Å) is[§]

$$l_{max} = 1.5 + 1.26 \, N'_C \tag{25-36}$$

For a spherical micelle of radius l, it is clear that $l \leqslant l_{max}$; for an ellipsoidal micelle of semimajor axis a and semiminor axis b, we have $b \leqslant l_{max}$. If we assume that $l = l_{max}$, then the relationship for the volume V of a sphere of radius l_{max} ($V = 4\pi l_{max}^3/3$), together with Equations 25-35 and 25-36, enables us to determine uniquely the maximal value of N_{ch} for any given value of N'_C. Likewise, for an ellipsoid of a given axial ratio with $b = l_{max}$, we can determine the maximal N_{ch} for any N'_C.

Table 25-3 shows the results of such calculations for various values of N'_C, for a sphere and for prolate ellipsoids of axial ratios 1.25 and 2.0. It is clear that the number of amphiphiles per micelle increases as the dimension l_{max} increases with N'_C. This relationship is easy to understand. For the sphere, for example, the volume of the micelle increases as $(N'_C)^3$ ($\propto l_{max}^3$), whereas the volume per amphiphile (V/N_{ch}) is a linear function of N'_C (see Eqn. 25-35). Therefore, a spherical micellar shape demands a roughly quadratic change in N_{ch} for every change in N'_C. Moreover, for a given value of N'_C, more amphiphiles can be incorporated into an ellipsoidal micelle than into a spherical one, and the value of N_{ch} that can be accommodated is larger for a larger axial ratio (asymmetry) of the ellipsoid. These features are a simple reflection

---

[§] In the fully extended, planar zig-zag form, the distance between alternate carbon atoms is about 2.53 Å, or about 1.26 Å per carbon atom. Considering that the terminal methyl adds 2.1 Å, and counting half the bond length to the first carbon atom as not in the core, we obtain

$$l_{max} \cong 2.1 + 0.63 + (1.26)(N'_C - 1)$$
$$\cong 1.5 + 1.26N'_C$$

Table 25-3

Maximal micelle aggregation numbers (N_{ch})

Shape of micelle	N'_C				
	6	10	12	15	20
Sphere ($l = l_{max}$)[§]	17	40	56	84	143
Prolate ellipsoids ($b = l_{max}$):					
$a/b = 1.25$	21	50	70	105	178
$a/b = 2.0$	33	80	111	167	285

NOTE: For single-chain amphiphiles, N_{ch} equals the number N_{hg} of head groups; for double-chain amphiphiles, $N_{ch} = 2N_{hg}$.

[§] It is not physically possible to form a spherical micelle with $l = l_{max}$, so the N_{ch} values for physically possible, small spherical micelles are less than the values tabulated.

SOURCE: C. Tanford, *The Hydrophobic Effect* (New York: Wiley, 1973), p. 76.

of the larger volume of a prolate ellipsoid than of a sphere when the semiminor axis of the ellipsoid is equal to the radius of the sphere, and of the progressive increase in volume as an ellipsoid with a fixed b axis is elongated.

Ratio of micelle surface area to number of head groups

The size and shape a micelle actually assumes depend critically on the ratio of the surface area A_s (which contains the head groups) to the number N_{hg} of head groups in the micelle. (For single-chain amphiphiles, $N_{hg} = N_{ch}$.) This dependence is due to the importance of the repulsive component W^0_{mic} in regulating micelle growth; clearly, W^0_{mic} must be a function of the spacing between head groups. If a micelle is to grow beyond the size permitted by a spherical shape, it must adopt another shape. In this shape change, however, A_s/N_{hg} changes and, as we shall see, tends to become smaller. This change gives rise to an increase in W^0_{mic}, and that increase eventually can limit growth.

The decrease in A_s/N_{hg} with increasing micelle size is easy to demonstrate from approximate calculations for spherical micelles, cylindrical micelles, and planar bilayers (Fig. 25-7). For all three shapes, we assume that the volume $V \cong \beta N'_C N_{ch}$, and that the radius $l = \alpha N'_C$, where α and β are constants. (These relationships are analogous to Eqns. 25-35 and 25-36 when the N'_C-dependent term is sufficiently large to dominate over the constant term.) For a sphere, the relationships are the following:

$$A_s = 4\pi l^2 = 4\pi\alpha^2(N'_C)^2 \tag{25-37a}$$

$$V = \beta N'_C N_{ch} = (4/3)\pi\alpha^3(N'_C)^3 \tag{25-37b}$$

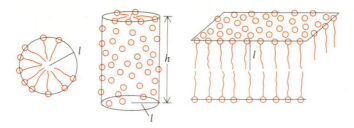

Figure 25-7

Spherical, cylindrical, and planar bilayer forms. In each case, the radius is l.

Therefore,

$$N_{ch} = (4/3)\pi\alpha^3(N_C')^2/\beta \qquad \text{(25-37c)}$$

and

$$A_s/N_{ch} = 3\beta/\alpha \qquad \text{(25-37d)}$$

For a long cylinder, where h is the height, and ignoring the area of the "ends,"

$$A_s = 2\pi l h = 2\pi\alpha N_C' h \qquad \text{(25-38a)}$$

$$V = \pi l^2 h = \pi\alpha^2(N_C')^2 h = \beta N_C' N_{ch} \qquad \text{(25-38b)}$$

Therefore,

$$N_{ch} = \pi\alpha^2 N_C' h/\beta \qquad \text{(25-38c)}$$

and

$$A_s/N_{ch} = 2\beta/\alpha \qquad \text{(25-38d)}$$

For a wide planar bilayer, and ignoring the area of the "edges,"

$$A_s = 2(V/2l) = \beta N_C' N_{ch}/\alpha N_C' \qquad \text{(25-39a)}$$

$$A_s/N_{ch} = \beta/\alpha \qquad \text{(25-39b)}$$

In each case, $A_s/N_{ch} = A_s/N_{hg}$ for single-chain amphiphiles, and $A_s/N_{ch} = 2A_s/N_{hg}$ for double-chain amphiphiles.

For a given value of l, the cylindrical micelle or the planar bilayer can accomodate any number of amphiphiles by simple growth on the ends. However, the preceding calculations show that, for a given l, there is a decrease in the available surface area per head group when the micelle adopts nonspherical shapes that can accommodate more amphiphiles. In general, deviations from the spherical shape produce a decrease in A_s/N_{hg} because, for a given radius, the sphere has the largest ratio of surface to volume.

Figure 25-8 graphically depicts the situation, showing A_s/N_{hg} as a function of N_{ch} for a single-chain amphiphile with $N'_C = 12$. The surface area was computed at a

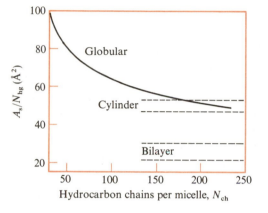

Figure 25-8

Surface area per head group versus number of hydrocarbon chains per micelle. [After C. Tanford, *J. Phys. Chem.* 76:3020 (1972).]

distance 2 Å beyond the surface of the core, in order to obtain a more realistic picture of the area actually available to the head groups. Note that, with a radius of l_{max}, the perfectly spherical shape has a volume of $(4/3)\pi l_{max}^3$. With fixed N'_C, this means that the spherical shape is adopted only at one particular value of N_{ch} ($N_{ch} = 56$; see Table 25-3). At larger values of N_{ch}, ellipsoidal shapes are adopted, as the micelle accommodates more and more hydrocarbon chains. Many of the other points used to construct this curve were obtained by assuming that the b axis of an ellipsoid is equal to l_{max}. Given this assumption, with any prescribed shape, it clearly is a simple matter to calculate A_s/N_{hg} for any value of N_{ch}.

Some additional points on the curve were generated by allowing spherical shapes with radii less than l_{max} and ellipsoids with the b axis less than l_{max}; for a sphere or for an ellipsoid of fixed axial ratio, a decrease in radius or b axis produces an increase in the surface-to-volume ratio and equivalently in A_s/N_{hg}.

In the cases of cylinders and bilayers, A_s/N_{hg} is independent of N_{ch} (ignoring end effects). Calculations are shown by horizontal dashed lines for cylinders of maximal radius (l_{max}, lower dashed line) and of somewhat smaller radius (upper dashed line); analogous calculations are shown for bilayers.

The figure shows that, at the smaller values of N_{ch}, globular shapes can be adopted readily. An average value of N_{ch} for globular micelles is about 100. As N_{ch} increases, A_s/N_{hg} decreases because the globular shapes become increasingly asymmetric. Eventually, the A_s/N_{hg} value for the cylindrical form is reached. This form now becomes preferred, because the micelles can continue to grow as cylinders without further change in A_s/N_{hg}. The value of A_s/N_{hg} for the planar bilayer falls well below that for the cylinder; this gives too much head-group crowding, so that no transition to the bilayer form occurs with single-chain amphiphiles. This generalization is consistent with experimental observations.

Bilayer formation

For double-chain amphiphiles (such as phospholipids), the situation is quite different from that we have described thus far for single-chain amphiphiles. For the double-chain amphiphiles, $N_{hg} = N_{ch}/2$. The A_s/N_{hg} values in Figure 25-8 are for single-chain amphiphiles; to obtain corresponding values for double-chain amphiphiles, we must multiply the values in the figure by two. The result is that the A_s/N_{hg} value for the bilayer of the double-chain species approach that of the cylinder for the single-chain species. Because this A_s/N_{hg} value is favorable, as we have seen for large micelles of single-chain amphiphiles, there is no reason to expect it to be unacceptable for the double-chain amphiphiles. Therefore, the double-chain amphiphiles form bilayers. Although cylindrical shapes would give even larger A_s/N_{hg} values for the double-chain amphiphiles, an excessively large value is not permitted because it allows penetration of water molecules into the hydrocarbon cores. Moreover, because $U_{mic}^0 - \mu_w^0$ for a double-chain amphiphile is about 60% more negative than for a single-chain amphiphile, the double-chain amphiphile can tolerate a closer packing of head groups than that tolerated by the single-chain amphiphile.

These then are the reasons that bilayers form in aqueous dispersions of double-chain amphiphiles. The bilayers often are in the form of closed vesicles, with the hydrocarbon-rich edges removed from the solvent. In the formation of such vesicles, there is only a small change in surface area from that of the planar form.

It should be recognized that cations will affect bilayer formation and structure, particularly because they can interact with negatively charged head groups. In the case of biological membranes, Ca^{2+} appears to play an important role. Evidence that it affects bilayer structure comes from studies showing that Ca^{2+} causes a condensation and restricts the motion of phospholipids.

25-4   BILAYER STRUCTURE AND FUNCTION

The preceding discussion has focused on macroscopic thermodynamic aspects of micelle and bilayer formation. To obtain a molecular picture of bilayer structure and its relationship to membrane function, we must use structural and kinetic methods that visualize the packing features and the dynamic aspects of individual hydrocarbon chains. For this purpose, x-ray diffraction, electron paramagnetic resonance (EPR), and NMR have been particularly useful techniques. The results obtained from these approaches have led to a rather detailed model for lipid bilayers that forms a basis for understanding natural membranes. (Chapter 4 gives structures of phospholipids discussed below.)

X-ray diffraction data

Maurice Wilkins and colleagues have obtained important structural data on oriented bilayers and multilayers. Particularly interesting is the scattering observed from multilayers of dipalmitoyl lecithin under various conditions of hydration. (See Chapters 13 and 14 for a discussion of scattering and diffraction.) At right angles to the lamellar diffraction spots, a sharp diffraction of 4.2 Å is observed, and this is not altered with the degree of hydration. This observation is reminiscent of the diffraction pattern obtained from an ordered array of hexagonally packed hydrocarbon chains (for example, in parafilm). The observation of a similar spacing in the phospholipid multilayer implies a similar ordered packing of chains.

Figure 25-9 shows the Fourier synthesis (of the diffraction data) across a bilayer in the wet and dry states. The electron density is computed relative to the reference

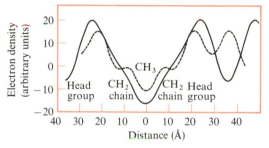

Figure 25-9

Fourier synthesis of electron density distribution across a lecithin bilayer. The solid curve is the spectrum for the wet state; the dashed curve is that for the dry state. [After Y. K. Levine et al., *Nature* 220:577 (1968).]

level of water. Assignment of particular groups is indicated. Note that the head-group regions move apart upon hydration. The negative trough in the electron density corresponds to the terminal $-CH_3$ group, which is known to have a negative electron density relative to water. The data show that the terminal methyl groups are located in the center of the bilayer. The $-CH_2-$ groups appear as a shoulder; this shoulder is more prominent in higher-resolution syntheses, where the $-CH_3$ well also becomes sharper. These observations collectively indicate that the hydrocarbon chains are not significantly interdigitated, and that they are oriented perpendicular to the bilayer surface. The perpendicular orientation means that the bilayer surface area per phospholipid molecule is not much greater than the cross-sectional area of two extended hydrocarbon chains.

More information has been obtained from oriented egg lecithin bilayers in the presence and absence of cholesterol (1:1 molar ratio). Figure 25-10 shows the diffraction pattern from the egg lecithin multilayers. The sharp meridonal arcs result from the lamellar spacing; at right angles, the widely spaced and diffuse equatorial band corresponds to a 4.6 Å hydrocarbon-chain spacing. The right-angle orientation of the chains with respect to the lamellar surfaces is clearly illustrated in this figure.

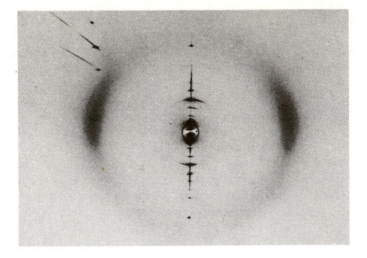

Figure 25-10

X-ray diffraction pattern of egg lecithin multilayers. Lamellar diffraction is seen as sharp meridional arcs; the 4.6 Å hydrocarbon-chain diffraction is oriented equatorially. [From Y. K. Levine and M. H. F. Wilkins, *Nature New Biology* 230:69 (1971).]

Effect of cholesterol on chain ordering

Figures 25-11 and 25-12 provide a more quantitative visualization of the hydrocarbon chains than does Figure 25-10. The figures show the angular variation of the intensity of the hydrocarbon-chain band from the equator (0°) to the meridian (90°) in the absence (Fig. 25-11) and presence (Fig. 25-12) of cholesterol. Each figure shows data for two different levels of water content. Notice that in all cases there is a substantial

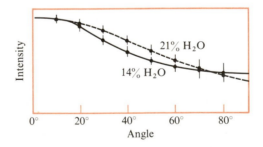

Figure 25-11

Angular variation of intensity of the hydrocarbon-chain band of egg lecithin bilayers. The radially integrated intensity of the 4.6 Å diffraction band is shown from the equator (0°) to the meridian (90°). [After Y. K. Levine and M. H. F. Wilkins, *Nature New Biology* 230:69 (1971).]

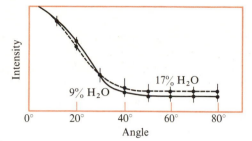

Figure 25-12

Angular variation of intensity of the hydrocarbon-chain band of egg lecithin bilayers in the presence of cholesterol. The radially integrated intensity of the 4.75 Å diffraction band is shown from the equator (0°) to the meridian (90°). [After Y. K. Levine and M. H. F. Wilkins, *Nature New Biology* 230:69 (1971).]

drop in intensity as the meridian is approached. Apparently the chain orientation is much more perpendicular to the surface than parallel to it.

Also, note that the intensity drop-off is more pronounced in the presence of cholesterol than in its absence, suggesting that cholesterol encourages a greater ordering of the chains. On the other hand, it is clear that, at least in the absence of cholesterol, increased hydration causes an increase in chain disarray.

The effects of cholesterol are clear also in the Fourier synthesis of the electron density in a direction perpendicular to the bilayer planes. Figures 25-13 and 25-14 show the syntheses in the absence and presence, respectively of cholesterol. Note that,

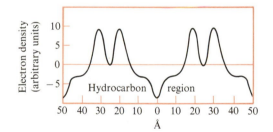

Figure 25-13

Fourier synthesis across egg lecithin bilayers at 57% relative humidity. [After Y. K. Levine and M. H. F. Wilkins, *Nature New Biology* 230:69 (1971).]

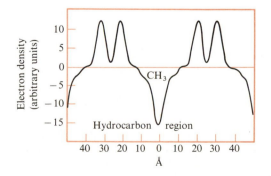

Figure 25-14

Fourier synthesis across egg lecithin/cholesterol bilayers at 57% relative humidity. [After Y. K. Levine and M. H. F. Wilkins, *Nature New Biology* 230:69 (1971).]

in the absence of cholesterol, the $-CH_3$ trough is broad, indicating considerable movement of these groups in the bilayer center; a much sharper trough occurs in the presence of cholesterol, indicating a substantial restriction of $-CH_3$ group positions. This observation also points to an ordering effect of cholesterol on the hydrocarbon chains.

In the data discussed, the hydrocarbon chains are in a "liquid-crystalline" state. Bilayers undergo a thermal transition between a lower-temperature, gellike, ordered state and the liquid-crystalline state at higher temperatures. (We discuss this transition in a subsequent subsection.) The ordering effects of cholesterol generally are seen in the liquid-crystalline state, whereas a disordering occurs when cholesterol is added to the gellike state. However, regardless of the exact behavior, it is clear that cholesterol influences the details of the bilayer structure.

EPR data

A considerable body of novel structural data has been gained from application of EPR spectroscopy (Chapter 9). An EPR signal is received only from an unpaired electron. For this reason, a phospholipid bilayer normally has no EPR signal. Useful information may be obtained, however, from spin-labeled probes inserted into bilayers. Such a probe molecule contains an unpaired electron, and its behavior within the bilayer is a sensitive monitor of structural and dynamic conditions.

The most commonly used paramagnetic probes are the nitroxide radicals. Figure 25-15 shows a typical radical. The spectrum of one of these radicals generally

Figure 25-15

A nitroxide radical.

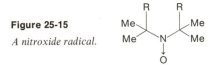

shows three lines (Fig. 9-29). These lines arise from the nuclear hyperfine interactions of the electron with the nitrogen nucleus. (The nuclear spin quantum number I of nitrogen is 1; there are $2I + 1$ lines produced by interaction of an electron with a nucleus having a spin quantum number of I.) The line shape and width are determined by anisotropies of the nitrogen nuclear hyperfine interaction and of the g factor for the nitroxide group (Chapter 9). The degree of anisotropy in these parameters is in turn sensitive to the molecular motion of the probe molecule.

EPR study of phospholipid movements between bilayer surfaces

An interesting question is whether phospholipids can move from one surface to the other in a bilayer and, if so, on what time scale. The measurement of phospholipid "flip-flop" in vesicle membranes was accomplished as follows. Phosphatidyl choline molecules were prepared with a nitroxide spin label. These labeled molecules were mixed with unlabeled phosphatidyl choline molecules, and the mixture was used to form bilayer-membrane vesicles. In Figure 25-16, which illustrates the experiment, the polar head groups of the paramagnetic and of the nonparamagnetic species are distinguished. When the bilayer is treated with ascorbate at 0°C, the paramagnetic

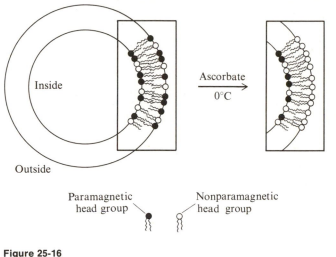

Figure 25-16

The effect of ascorbate treatment on paramagnetic molecules in a vesicle.

probe molecules on the outside of the bilayer are reduced to nonparamagnetic species. Those on the inside of the membrane, however, are unaffected by the ascorbate, and their paramagnetic signal persists after the treatment. Therefore, after all of the paramagnetic probes on the outside of the membrane have been reduced, the rate of loss of paramagnetism from the bilayer is a measure of the rate at which the phospholipids are flipping from the inside to the outside of the membrane.

Figure 25-17 shows an example of the data obtained from such an experiment. The spectrum of the untreated bilayer (Fig. 25-17a) is considerably higher in intensity than that of the bilayer after brief treatment with ascorbate at 0°C (Fig. 25-17b). Further treatment of the material with ascorbate produces no further change in signal intensity (Fig. 25-17c). This observation shows that the ascorbate is not penetrating the inside of the vesicle—and also, of course, that the rate of flip-flop at 0°C is extremely slow.

The inside–outside transition was measured by allowing flip-flop to occur at 30°C. The amount of flip-flop occurring in a unit of time was monitored at periodic intervals by chilling the system to 0°C, reducing with ascorbate, and taking an EPR spectrum to estimate the amount of flip-flop that had occurred during the interval at 30°C. Because the number of paramagnetic centers present is directly proportional to the integrated intensity of the EPR spectrum, it is a simple matter to measure the number of molecules on the inside of the bilayer. Using this approach, R. D. Kornberg and H. M. McConnell estimated the probability per unit time of the inside–outside

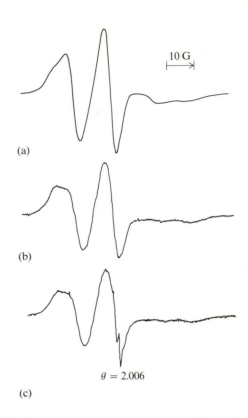

10 G

(a)

(b)

(c)

$g = 2.006$

Figure 25-17

EPR spectra of spin-labeled vesicles.
(**a**) Before ascorbate treatment; relative gain = 1.0. (**b**) After brief ascorbate treatment; relative gain = 6. (**c**) After further ascorbate treatment; relative gain = 6. [After R. D. Kornberg and H. M. McConnell, *Biochemistry* 10:1111 (1971).]

transition as 0.07 hr^{-1} at 30°C. However, this figure may represent an upper bound to the rate (see Rothman and Dawidowicz, 1975). In any event, flip-flop is a relatively slow process, which means that any phospholipid asymmetries in a membrane can persist for significant periods of time.

EPR studies of phospholipid lateral diffusion in membranes

Another aspect of membrane dynamics that has been investigated by the EPR method is the rate of lateral diffusion of phospholipids in vesicle membranes. One approach is to examine the NMR spectra of phosphatidyl choline (PC) vesicles in the presence of small proportions of spin-labeled PC. The idea is that rapid diffusion of the spin-labeled PC should have a significant effect on the line width of the proton NMR signal. In particular, line widths are broad as a result of interactions of nuclei with the unpaired electrons on the spin-labeled probes (Chapter 9).

Experiments have been done using the spin-labeled PC shown in Figure 25-18. Small proportions of this spin-labeled PC have a profound effect on the line widths of the proton resonance spectrum of PC vesicles (Fig. 25-19). Tracing **a** in the figure is the proton resonance spectrum of PC vesicles at 100 MHz, 35°C. Resonances

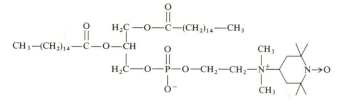

Figure 25-18

Spin-labeled phosphatidyl choline (PC).

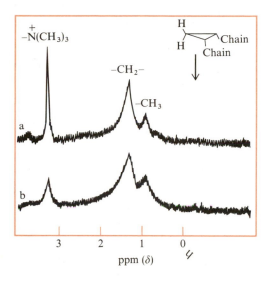

Figure 25-19

Effect of spin-labeled PC on 100 MHz proton magnetic resonance spectrum of PC vesicles at 35° C. **(a)** Pure didihydrosterculoyl PC vesicles; relative gain = 1.0. **(b)** About 1 mole percent spin-labeled didihydrosterculoyl PC vesicles; relative gain = 1.4. [After R. D. Kornberg and H. M. McConnell, *Proc. Natl. Acad. Sci. USA* 68:2564 (1971).]

associated with the protons of the choline head groups, the methylene groups, and the terminal methyl groups are clearly resolved. The addition of only one mole percent of the spin-labeled PC produces the spectrum shown in tracing **b**. Line broadening of all resonances occurs. When these line widths are analyzed in detail and other data are examined, it is clear that rapid lateral diffusion of the spin-labeled PC must occur. From these studies, the molecular frequency of the translational step in diffusion (corresponding to the rate of pairwise exchange of adjacent molecules) is estimated to be substantially greater than 3×10^3 sec^{-1} at 0°C.

A more accurate estimate of the rates of lateral diffusion may be obtained from a different kind of application of the spin-labeled PC molecules. Spin-labeled lipids are concentrated in one region of a bilayer (Fig. 25-20a). After a period of time, they diffuse laterally and eventually are distributed uniformly throughout the bilayer (Fig. 25-20b,c). When the labeled molecules are concentrated in one part of the bilayer,

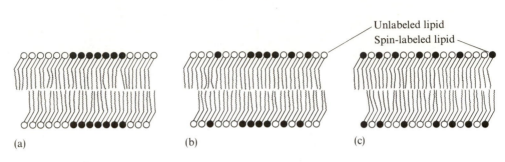

(a)　　　　　　　　　(b)　　　　　　　　　(c)

Figure 25-20

Measurement of lateral diffusion rates. [After P. Devaux and H. M. McConnell, *J. Am. Chem. Soc.* 94:4475 (1972).]

strong spin-exchange interactions occur between neighboring labeled molecules. The result is a rather diffuse, broad EPR spectrum. As the molecules separate by diffusion, the spin-exchange interactions decrease, and a quite different spectrum emerges. Finally, as the molecules reach a uniform distribution, the familiar three-line nitroxide spectrum reappears.

Figure 25-21 shows an example of the data obtained with this approach. Two sets of spectra were taken at various times after the onset of lateral diffusion of spin-labeled PC (concentrated initially in a patch) in bilayers of dihydrosterculoyl phosphatidyl choline. The spectra on the left were obtained with the applied field per-

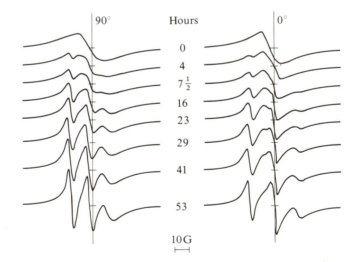

Figure 25-21

Paramagnetic resonance spectra of spin-labeled vesicles at various times after the onset of lateral diffusion. [From P. Devaux and H. M. McConnell, *J. Am. Chem. Soc.* 94:4475 (1972).]

pendicular (90°) to the plane of the bilayer; the spectra on the right were obtained with the applied field parallel (0°) to the plane of the bilayer. At any given time, the spectra in the two field directions are different. This result demonstrates that the spin-labeled molecules have a preferential anisotropic orientation. At the outset of the experiment ($t = 0$), the labeled PC molecules are clustered together, and a broad spectrum is obtained. As diffusion takes place, three lines begin to emerge, and the main features of the infinitely dilute spectra are clearly obtained in 40 to 50 hours.

These data may be used to obtain a rough estimate of the diffusion constant D. For diffusion, the root-mean-square distance traveled in time t is $(4Dt)^{1/2}$. Assume that the patch of spin-labeled PC initially has a radius of r_0. The infinitely dilute spectrum will emerge when the typical spin-labeled PC molecule has traveled a distance on the order of $2r_0$. With $r_0 \cong 0.8$ mm, the spectra of Figure 25-21 lead to an estimate for D on the order of 10^{-8} cm^2 sec^{-1}. A more exact analysis yields 1.8×10^{-8} cm^2 sec^{-1} at 25°C (Devaux and McConnell, 1972). For pairwise exchange of adjacent molecules, this corresponds to an exchange frequency of about 10^7 sec^{-1}.

The order–disorder transition in phospholipid bilayers

As we have mentioned, model membranes undergo a temperature-induced transition, in which the lipid molecules pass from an ordered to a disordered phase. A similar transition is observed in biological membranes. This transition has been studied by many techniques, including EPR. For example, E. Sackmann and H. Träuble studied the EPR signal of a spin-labeled steroid inserted into the lipid phase of a model monolayer membrane. Figure 25-22 shows the spin-labeled steroid, and Figure 25-23 shows the lipid monolayer vesicle containing an internal cavity of organic solvent.

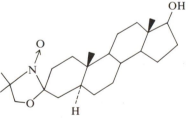

Figure 25-22

Structure of a spin-labeled steroid. The unpaired electron is on the nitrogen of the N→O group.

Figure 25-24 shows temperature-dependent spectra obtained with a label-to-lipid molar ratio of 0.01. The "phase transition" of this system occurs between 30° and 40°C. It is apparent from the figure that the spectra show significant changes in this temperature range. Particularly significant are the decreases in line widths as temperature

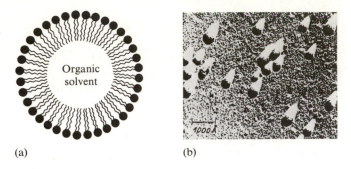

(a) (b)

Figure 25-23

Lipid monolayer vesicles. (**a**) Schematic diagram. (**b**) Electron micrograph of some monolayer vesicles. The average diameter is about 500 Å. [From E. Sackmann and H. Träuble, *J. Am. Chem. Soc.* 94:4482 (1972).]

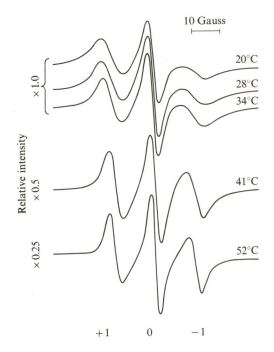

Figure 25-24

EPR spectra of spin-labeled steroids in monolayer vesicles at various temperatures. Magnetic quantum numbers ($+1$, 0, -1) of the three lines are indicated. [After E. Sackmann and H. Träuble, *J. Am. Chem. Soc.* 94:4482 (1972).]

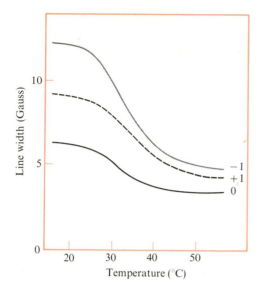

Figure 25-25

Line widths of the three lines in Figure 25-24 as a function of temperature. [After E. Sackmann and H. Träuble, *J. Am. Chem. Soc.* 94:4482 (1972).]

is raised. Figure 25-25 plots the effect of temperature on the line widths. It is apparent that the line widths can be used to follow the course of the transition.

The shapes of lines in the spectra of Figure 25-24 are determined in significant part by spin exchanges between labeled steroid molecules. It is possible, therefore, to calculate the frequency of the spin-exchange process. These calculations show that the exchange frequency *decreases* at the transition from low to high temperatures. Normally, however, one expects the rate of spin exchange to be proportional to T/η, where η is the viscosity of the solvent. Because the viscosity generally is a decreasing function of increasing temperature, the spin-exchange frequency normally increases with rising temperature. Observation of the opposite result in this case suggests that raising the temperature through the phase transition induces a substantial reorganization of the steroid spin label in the model membrane. Below the transition, the steroids probably are grouped into clusters in which substantial spin-exchange interactions can take place. Above the transition, these molecules are more uniformly distributed throughout the monolayer, so that spin-exchange interactions are greatly reduced.

In natural membranes, transitions tend to be broad. This broadening occurs because of the varied composition of lipid components, each of which in pure form can have a unique transition temperature. In general, the broad transitions of natural membranes fall in the range of 30° to 40°C—that is, within the range of physiological temperatures. The use of cells that are fatty acid auxotrophs enables an experimenter to engineer more precisely the lipid composition and thereby to narrow the thermal transition zone.

Bending of aliphatic chains in the bilayer

Abundant data indicate that the terminal methyl groups have considerably more freedom than do groups near the polar end of the molecule. Because of the greater rational freedom, one expects the methyl termini to occupy a larger effective volume than that of groups near the surface. This requires that the groups at the surface of the bilayer be more snugly packed than those near the methyl termini. However, this snug packing is impossible for parallel arrays of phospholipid chains. Figure 25-26 depicts one solution to this problem; here a systematic bending of the chains produces a smaller intrachain separation near the surface of the bilayer than near the terminal methyl groups.

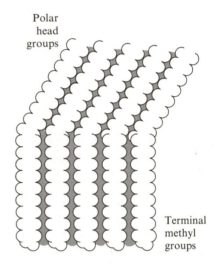

Polar head groups

Terminal methyl groups

Figure 25-26

Packing of fatty acid chains in one-half of a planar bilayer. [After B. G. McFarland and H. M. McConnell, *Proc. Natl. Acad. Sci. USA* 68:1274 (1971).]

This hypothesis has been tested in egg lecithin bilayers. For this purpose, different spin-labeled phospholipids were synthesized (Fig. 25-27). In these, the spin label is attached to different parts of the aliphatic chains of the phospholipids. The label positions are indexed by the variables m and n, indicating the number of methylene groups separating the label from the terminal methyl group and from the fatty acid carbonyl carbon atoms, respectively. By varying the position of the spin label on the aliphatic chain, one is able to examine the effect of position on the average label orientation.

Table 25-4 lists values of ϕ_{mp} for various values of m, n. The parameter ϕ_{mp} is the most probable angle between the plane of the paramagnetic oxazolidine ring and the plane of the bilayer. When the polymethylene chain is in its extended (all-*trans*) conformation, the plane of the oxazolidine ring is perpendicular to the hydrocarbon chain. Therefore, if the chain is perpendicular to the bilayer surface, then the plane of

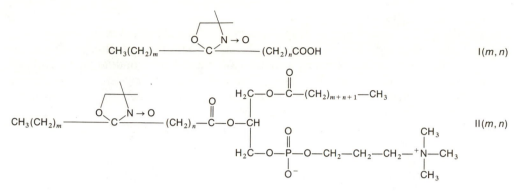

$CH_3(CH_2)_m$ — C — $(CH_2)_n COOH$ I(m, n)

$CH_3(CH_2)_m$ — C — $(CH_2)_n$ — $\overset{O}{\overset{\parallel}{C}}$ — O — CH   II(m, n)

H_2C — O — $\overset{O}{\overset{\parallel}{C}}$ — $(CH_2)_{m+n+1}$ — CH_3

H_2C — O — $\overset{O}{\overset{\parallel}{\underset{O^-}{P}}}$ — O — CH_2 — CH_2 — CH_2 — ^+N — CH_3 (with CH_3, CH_3)

Figure 25-27

Some spin-labeled phospholipids. The indices m and n can be varied.

Table 25-4

Values of ϕ_{mp} for phospholipid spin labels in egg lecithin bilayers

m	n	ϕ_{mp}
10	3	29° to 32°
7	6	28° to 31°
5	10	14° to 25°
1	14	5°

SOURCE: B. G. McFarland and H. M. McConnell, *Proc. Natl. Acad. Sci. USA* 68:1274 (1971).

the ring will be parallel to the surface ($\phi_{mp} = 0°$). The set of values of ϕ_{mp} is obtained by measuring the average angle between a principle axis of the probe and the perpendicular to the membrane surface. When the probe is close to the membrane surface, $(m, n) = (10, 3)$, then $\phi_{mp} = 29°$ to $32°$. The value of ϕ_{mp} decreases, however, as m decreases and n increases. Thus, with $(m, n) = (1, 14)$, we have $\phi_{mp} = 5°$. These data establish that the probe molecules have an average statistical bending near the surface of the membrane, whereas the chains are packed near the terminal methyl groups in an orientation perpendicular to the membrane surface. Moreover, the data show that the tilt in a given direction in the head-group region must have a lifetime greater than about 10^{-8} sec; if rapid interconversion between bent structures occurred with a frequency greater than 10^8 sec^{-1}, the measured average values of ϕ_{mp} would be about

0°. The tilt of about 30° near the head group apparently leads to a carbon-atom density about 12% larger near the surface of the bilayer than at the terminal methyl groups.

The following question arises: is the apparent tilting of phospholipids due primarily to perturbations arising from the paramagnetic oxazolidine ring introduced as a spin label? To answer this, A. Seelig and J. Seelig (1974) have studied deuterium-labeled phospholipid analogs, with the deuterium atoms located in specific positions (analogous to the approach used in the spin-label studies). The nuclear quadrapole splitting led to qualitative conclusions similar to those obtained in the spin-label studies. Quantitative differences between the two approaches might be due to the different time scales of the two kinds of measurements (McConnell, 1976). These findings indicate that the change in the order of the hydrocarbons with increasing penetration into the bilayer is an intrinsic property of the bilayer itself, and is not simply produced by the spin label.

Penetration of water below the bilayer surface

Although one visualizes the bilayer structure as devoid of water molecules, some penetration by water is expected. Hayes Griffith and colleagues have studied water penetration by making use of the small solvent effect on the EPR spectra of nitroxide labels. They measured the effect of solvents on the isotropic ^{14}N coupling constant. Although the solvent effects are small, the coupling constants can be measured with sufficient accuracy to obtain useful information. The observed ^{14}N coupling constants can be correlated reasonably well with solvent polarity. These measurements then provide a calibration for the splittings observed for spin labels within the bilayers. Measuring the spectra of spin labels attached to different positions along a fatty acid chain allows appraisal of the polarity of different parts of the bilayer.

Figure 25-28a shows results obtained from this kind of study. The figure is a plot of the polarity index (a measure of polarity) versus the distance from the center of a microsomal lipid bilayer. These data show that the limiting polarity of a pure hydrocarbon region is achieved in sections within 10 Å of the bilayer center. Outside of this region, the polarity increases sharply until the charged double-layer (head-group) region is reached at 30 Å from the center. A point at 40 Å corresponds to the polarity of the aqueous solution; here the polarity is only slightly less than that of the charged double-layer region.

These data show that some water molecules must penetrate significantly below the bilayer surface. Figure 25-28b shows a cross-sectional area of bilayer with some water molecules drawn to scale. The greater degree of random orientation of the terminal methyl groups also is indicated schematically.

The preceding discussion highlights some of the applications of the spin-label approach to membrane biophysics. It is clear that this approach has provided some of the most important insights presently available.

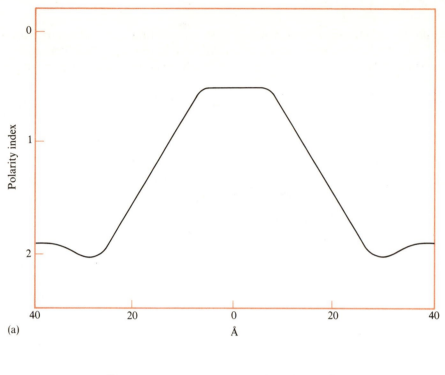

(a)

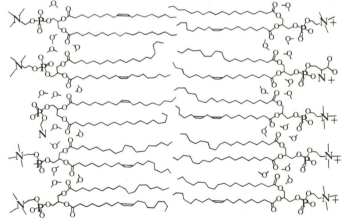

(b)

Figure 25-28

The hydrophobic barrier in a lipid bilayer. (a) Polarity index across the bilayer.
(b) A schematic view of the bilayer. [After O. H. Griffith et al., *J. Membrane Biol.* 15:159 (1974).]

Proton and ^{13}C NMR spectra

Proton and ^{13}C NMR spectra also have given information on bilayer structure. Figure 25-29 is a 220 MHz proton magnetic spectrum of sonicated dipalmitoyl L-α-lecithin. (See Chapter 9 for a discussion of NMR.) The sonication is performed

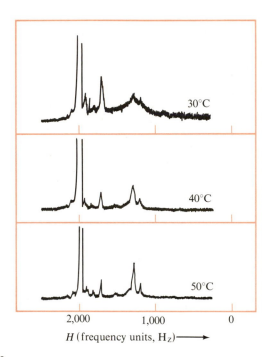

Figure 25-29

220 MHz proton magnetic resonance spectra of sonicated dipalmitoyl L-α-lecithin at three temperatures. [After A. F. Horwitz, in *Membrane Molecular Biology*, ed. C. F. Fox and A. D. Keith (Stamford, Conn.: Sinauer Associates, 1972, p. 164.]

to eliminate turbidity; presumably, aggregation of multilaminar vesicles is reduced, and the proportion of bilayer vesicles is increased. Spectra are shown for three different temperatures. At 50°C, the fatty acid methyl and methylene resonances (and the choline methyl group resonance) are clearly seen. At 30°C, however, the fatty acid methyl and methylene resonances are considerably broadened, but the head-group methyl proton bands show little change in line width. These spectra indicate a significant change in the bilayer between 30° and 50°C. This change is due to the sharp thermal transition, centered around 41°C, in which the phospholipids pass from an ordered crystallike state to one more disordered and mesomorphic.

Line broadening at 30°C of the fatty acid $-CH_3$ and $-CH_2-$ protons is expected if they pack into an ordered array. This broadening is caused by the shortening of the transverse relaxation time T_2. Because the width of the head-group methyl proton bands changes only slightly with temperature, this group's environment presumably is not greatly altered by the order–disorder transition.

Figure 25-30 shows ^{13}C NMR spectra of dipalmitoyl lecithin. These spectra show much greater resolution than do the proton spectra. At 64°C, the fatty acid

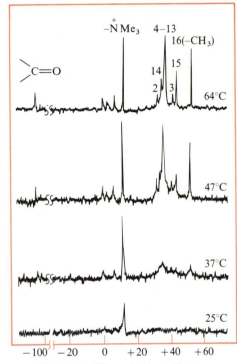

Figure 25-30

^{13}C NMR spectra of dipalmitoyl lecithin in D_2O at four temperatures. [After Y. K. Levine et al., *Biochemistry* 11:1416 (1972).]

carbonyl carbon (C-1), the carbons C-2, C-3, C-14, C-15, and C-16 (the terminal CH_3 group), and the head-group methyl protons are clearly resolved. Carbons C-4 through C-13 are combined into a single line. Except for the head-group resonance, all lines broaden progressively as the temperature is lowered; at 25°C, only the choline methyl groups are evident. This behavior is similar to the temperature dependence of the proton spectrum in Figure 25-29.

Differences in relative mobilities of specific segments in a bilayer

Spin lattice relaxation times T_1 have been monitored at 52°C for each of the discrete resonances in Figure 25-30. These measurements illustrate molecular motion at specific loci along the phospholipid. The relaxation time is sensitive to molecular tumbling of the molecule as a whole and to internal motion within the chain. If τ_m is the tumbling (or rotational) correlation time, and τ_I is the correlation time for internal motion, then $T_1 \propto \tau_m^{-1} + \tau_I^{-1}$. Fast molecular tumbling and internal motion give large T_1 values.[§]

Figure 25-31 shows spin lattice relaxation times at 52°C for the various carbon atoms in dipalmitoyl lecithin. These times are listed next to each carbon in the struc-

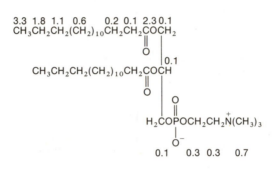

Figure 25-31

^{13}C *spin lattice* (T_1) *relaxation times* (in seconds) for carbon centers in dipalmitoyl lecithin in D_2O at 52° C. [After Y. K. Levine et al., *Biochemistry* 11:1416 (1972).]

ture; each number is an average for the two fatty acid chains. From the terminal methyl group, the T_1 values decrease as one moves to the center of the aliphatic chain. As the head-group region is approached, the methylene carbons reach their smallest T_1 values; the values rise again in the choline carbon atoms.

Of course, it must be kept in mind that these T_1 values (obtained at 52°C) are at a temperature well above the bilayer's thermal transition temperature (~ 40°C). Because of the extreme line broadening (Fig. 25-30), T_1 measurements for specific aliphatic-chain carbons cannot be made below the thermal transition.

Variation in the methylene carbon T_1 values indicates that the phospholipid must undergo internal degrees of motion. If rotation of the molecule as a whole accounted for the T_1 values, each methylene carbon would have the same spin lattice relaxation time.

[§] Recall that $1/T_1$ is the first-order rate constant for the rise of the longitudinal magnetization of the sample to its equilibrium value, after the magnetic field is turned on. T_1 is determined by the fluctuating magnetic interactions of the solution environment (lattice) with the precessing nuclei (see Chapter 9). Rapid molecular motion averages the local lattice fields, so that their net effect is diminished; this effect lengthens the time required to establish the equilibrium value of the sample magnetization.

The T_1 values increase sharply near the terminal methyl group. This means that chain motion is greater in the vicinity of the methyl terminus. The relatively long T_1 values of the choline methyl groups suggest a fair degree of mobility in this region as well. But the spin lattice relaxation times of the glycerol methylene carbons are among the shortest in the entire phospholipid molecule. This observation might mean that the greatest organization and rigidity near the bilayer surface are in the glycerol groups.

These data illustrate the valuable structural information that can be obtained from T_1 measurements on specific nuclei. Even though the bilayer is "melted," the systematic variation of T_1 values clearly indicates a significant structural organization for the bilayer at the elevated temperature. Thus, it must be kept firmly in mind that the "melted" bilayer is not in an amorphous state.

When the temperature is raised to 65°C, each T_1 value increases. This is expected if increased molecular motion occurs throughout the phospholipid. However, the relative magnitudes of the T_1 values for the various carbon atoms occur in the same order. This regularity is due to the persistence of some of the organized bilayer structure even at 65°C.

Microenvironments within the bilayer

Transverse relaxation times (T_2) also have been monitored in lecithin bilayers. These measurements have given a picture of the microenvironments of chain segments in the ordered form (below the thermal transition temperature). Figure 25-32 shows the proton free-induction decay of a lecithin bilayer sample at 30°C, as obtained by S. I. Chan and colleagues. The figure is a semilogarithmic plot of free-induction decay versus time.

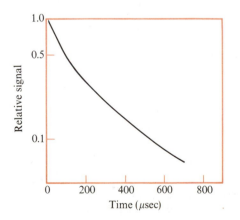

Figure 25-32

Proton free-induction decay of lecithin bilayers at 30°C, in a field of 14.1 kilogauss. [After S. I. Chan et al., *Nature* 231:110 (1971).]

For a single relaxation event, a straight line with a slope equal to T_2 should be obtained. The plot is markedly nonlinear, however. This suggests that a distribution of transverse relaxation times is present. The distribution is caused by the different environments of individual nuclei. This interpretation is strengthened by line-shape analysis on the 220 MHz spectrum and by the 100 MHz Fourier-transformed spectrum. These results confirm the expectation that, even in the ordered bilayer, the microenvironment varies considerably along the fatty acid chains. G. W. Feigenson and S. I. Chan (1974) report further studies, using lecithin multilayers, that take advantage of measurements of spin lattice relaxation times.

Some conclusions about bilayers

The studies described in this section elucidate many features of bilayers. It is clear that the bilayer arrangement of amphiphiles can be rationalized on the basis of elementary thermodynamic principles and some simple geometric considerations. X-ray diffraction studies have indicated, along with other structural features, a perpendicular orientation of the aliphatic chains with respect to the bilayer surface. Magnetic resonance investigations have elucidated characteristics and properties such as chain bending near the bilayer surface, phospholipid flip-flop, lateral diffusion of phospholipids in the bilayer, and segmental motion that varies along the length of the perpendicularly oriented aliphatic chains.

The study of bilayers forms a framework for a rational understanding of biological membranes. A typical biological membrane has a number of proteins imbedded within the bilayers. The proteins may be enzymes or receptors for specific molecules; some may play a role in active transport or in regulating the permeability of the membrane to specific substances. These proteins function within the framework of the structural and dynamic features of bilayers as described in this chapter. Chapter 4 discusses further the structure and function of biological membranes.

Summary

The equilibrium distribution (across a membrane) of diffusible species that cross the membrane is determined by the thermodynamic requirement that the chemical potential of a diffusible component must be the same on both sides of the membrane. This requirement in turn leads to many interesting results and applications. It explains the phenomenon of osmotic pressure in systems where a macromolecule is constrained to one side of a semipermeable membrane, and explains how osmotic pressure is related to the molecular weight of the macromolecule. It also accounts for the Donnan effect, whereby a charged macromolecule localized on one side of a membrane causes an unequal distribution of small ions that freely diffuse across the membrane. The Donnan effect also gives rise to a membrane potential. In biological membranes, active pumps (such as the $Na^+ - K^+$ pump) use energy from a source such as the hydrolysis of ATP to generate, across a membrane, ion-concentration gradients different from those predicted on the basis of simple thermodynamic arguments that do not take into account the energy source driving the "pump."

In biological membranes, the essential matrix is made up of lipid amphiphiles in a bilayer surrounding an aqueous cavity such that polar head groups are exposed to the aqueous medium, and the hydrocarbon chains of the lipids are together in the bilayer interior. The bilayer itself forms in accordance with the same principles that determine the formation of micelles. It can be shown that the ratio of micelle surface area to number of head groups influences micelle size and shape. Although single-chain amphiphiles form globular micelles, the ratio of aggregate surface area to the number of head groups favors formation of bilayer structures for double-chain amphiphiles (two nonpolar chains attached to a single polar head group).

Bilayer structure and function have been investigated by many physical techniques including x-ray diffraction, NMR, and EPR. The x-ray studies show that the hydrocarbon chains are at right angles to the bilayer surface, with an interchain spacing of 4 to 5 Å. It also has been found that lateral diffusion of amphiphiles in a bilayer occurs with a diffusion constant on the order of 10^{-8} cm^2 sec^{-1}. Other data suggest that the amphiphile flexibility, or segmental motion, varies from the head-group region to the center of the bilayer; it appears that segmental motion of an amphiphile is greatest near the bilayer interior and that the head-group region is more rigid. Thus, the bilayer structure is not static, but dynamic. Furthermore, it undergoes a transition from an ordered structure to a disordered, mesomorphic state. These and other results have provided a good foundation for understanding the behavior of biological membranes.

Problems

25-1. Assume that a macromolecule M_1 dimerizes according to the scheme $2\,M_1 \rightleftarrows M_2$ with a dissociation constant K defined as $K = (M_1)^2/(M_2)$, where (M_1) and (M_2) are concentrations in moles per liter. Let the total osmotic pressure π be given by Equation 25-14. And let $\hat{c}_1$ and $\hat{c}_2$ be concentrations in grams per liter of M_1 and M_2, respectively, and let $\hat{c}$ be the *total* concentration of monomeric macromolecule; that is $\hat{c} = \hat{c}_1 + 2\hat{c}_2$. Show that, at low concentrations $\hat{c}$, π can be expressed as a virial expansion in $\hat{c}$ such that

$$\pi = (RT/M)\hat{c} + B\hat{c}^2$$

where M is the monomer molecular weight. Express B in terms of R, T, M, and K.

25-2. Consider a charged macromolecule constrained to one side (side 2) of a semipermeable membrane. Its average charge is $+9.9$, and it is at a concentration of 10 μM. The macromolecule is in a dilute aqueous NaCl solution. It is found that, on side 1 of the membrane,

$(Na^+) = 10\ \mu M$. On side 2, it is found that $(Cl^-) = 50\ \mu M$. Are these values for the concentrations of ionic species consistent with the system being at thermodynamic equilibrium with respect to the distribution of Na^+ and Cl^- across the membrane? Why, or why not?

25-3. An investigator makes the following claim. "A long, wide planar bilayer is converted into a spherical bilayer vesicle. The outer surface area of the vesicle is essentially equal to that of one face of the planar bilayer, when the radius of the vesicle is much, much greater than the thickness of the bilayer." Do you agree? Check the investigator's claim by performing the appropriate calculation, assuming that the length of an amphiphile in each bilayer is l, that the volume of the bilayer itself is the same in both cases, and that the edges of the planer bilayer contribute a negligible amount to the total surface area.

25-4. An investigator studies the salt-concentration dependence of micelle formation with a single-chain amphiphile. He has techniques for studying the overall micelle shape. He observes that the shape changes dramatically as he alters the NaCl concentration in his solutions from 0.01 M to 0.5 M. Give an explanation for this shape change. Will the size also change as the salt concentration is increased? Explain your answer. What kind of shape changes (and size changes, if they occur) do you predict?

25-5. An isolated protein that normally occurs within a membrane bilayer is randomly labeled on its surface with a nitroxide spin label. The protein is inserted into the bilayer of a vesicle. It is known that the protein is cylindrically shaped, and that the long axis of the cylinder is parallel to the bilayer surface. The bilayer encapsulates ascorbic acid, and the decay of the EPR signal caused by reduction of the radical (by the ascorbic acid) is followed. After the initial reduction (at time zero) of paramagnetic centers that protrude into the internal cavity, it is found that the remaining decay can be described completely by two exponentials: a relatively fast phase in which the majority of the signal disappears, followed by a slower phase in which the rest is lost. Offer an explanation for these results, in terms of the position and motion of the protein in the bilayer.

References

GENERAL

Tanford, C. 1961. *Physical Chemistry of Macromolecules*. New York: Wiley. [Chap. 4 covers thermodynamics and takes up membrane equilibria and the subjects of osmotic pressure, Donnan effect, etc.]
————. 1973. *The Hydrophobic Effect*. New York: Wiley. [Includes a good treatment of the thermodynamics of micelles and bilayers.]
Weissmann, G., and R. Claiborne, eds. 1975. *Cell Membranes: Biochemistry, Cell Biology & Pathology*. New York: HP Pub. Co. [Very readable review articles on general membrane structure and function.]

SPECIFIC

Chan, S. I., H. Lindsey, K. E. Eigenberg, W. R. Croasman, and G. W. Campbell. 1979. In *NMR and Biochemistry*, ed. S. J. Opella and P. Lu (New York: Marcel Dekker), p. 249.

Donnan, F. G. 1911. Theorie der membrangleichgewichte und membranpotentiale bei vorhandensein von nicht dialysierenden electrolyten: Ein beitrag zur physikalisch-chemischen physiologie. *Z. electrochem.* 17:572.

Feigenson, G. W., and S. I. Chan. 1974. Nuclear magnetic relaxation behavior of lecithin multilayers. *J. Am. Chem. Soc.* 96:1312.

Fox, C. F. 1975. Phase transitions in model systems and membranes. In *Biochemistry of Cell Walls and Membranes*, ed. C. F. Fox (London: Butterworths, MTP International Review of Science, Biochemistry Series One), p. 279.

Keith, A. D., M. Sharnoff, and G. E. Cohn. 1973. A summary and evaluation of spin labels used as probes for biological membrane structure. *Biochem. Biophys. Acta* 300:379.

Levine, Y. K., and M. H. F. Wilkins. 1971. Structure of oriented lipid bilayers. *Nature New Biology* 230:69.

McConnell, H. M. 1976. Molecular motion in biological membranes. In *Spin Labeling*, ed. L. Berliner (New York: Academic Press), p. 525.

Nicolson, G. L., G. Poste, and T. H. Ji. 1977. The dynamics of cell membrane organization. In *Dynamic Aspects of Cell Surface Organization*, ed. G. Poste and G. L. Nicolson (Amsterdam: North Holland), p. 1.

Rothman, J. E., and E. A. Dawidowicz. 1975. Asymmetric exchange of vesicle phospholipids catalyzed by the phosphatidylcholine exchange protein: Measurement of inside–outside transitions. *Biochemistry* 14:2809.

Seelig, A., and J. Seelig. 1974. The dynamic structure of fatty acyl chains in a phospholipid bilayer measured by deuterium magnetic resonance. *Biochemistry* 13:4839.

Shipley, G. G. 1973. Recent x-ray diffraction studies of biological membranes and membrane components. In *Biological Membranes*, vol. 2, ed. D. Chapman and D. F. H. Wallach (London: Academic Press), p. 1.

A matrix is an array (or table) whose elements are numbers or symbols; for example,

$$\begin{pmatrix} 8 & 7 \\ 23 & 28 \end{pmatrix} \qquad \begin{pmatrix} a_{11} & a_{12} \\ a_{21} & a_{22} \end{pmatrix}$$

These are 2×2 matrices (having two rows and two columns); the symbol a_{12} denotes the element in the first row and the second column. In general, a_{ij} represents the element in the ith row and the jth column. For the matrix on the left, $a_{22} = 28$, $a_{21} = 23$, and so on. In this text, a matrix is denoted by a bold-faced letter with a "tilde" underscore. For example, we could represent the matrix at the right above as $\underset{\sim}{A}$, where

$$\underset{\sim}{A} = \begin{pmatrix} a_{11} & a_{12} \\ a_{21} & a_{22} \end{pmatrix}$$

A *row* matrix (also called a row vector) contains only one row:

$$\underset{\sim}{A}_r = (a_{11}, a_{12})$$

This matrix $\underset{\sim}{A}_r$ is a 1×2 row matrix (with one row and two columns). A *column* matrix (or column vector) contains only one column:

$$\underset{\sim}{A}_c = \begin{pmatrix} a_{11} \\ a_{21} \end{pmatrix}$$

The matrix $\underset{\sim}{A}_c$ is a 2×1 column matrix (with two rows and one column). In general, a matrix may be of any size $n \times m$ (with n rows and m columns).

MATRIX MULTIPLICATION

Two matrices may be multiplied together to give a third matrix:

$$\underset{\sim}{a}\underset{\sim}{b} = \underset{\sim}{c}$$

The element c_{ij} in matrix $\underset{\sim}{c}$ is defined as

$$c_{ij} = \sum_k a_{ik}b_{kj}$$

That is, the elements in the ith *row* of a are multiplied by counterparts in the jth *column* of b to produce c_{ij}. For example, if

$$\underset{\sim}{a} = (a_{11}, a_{12})$$

and

$$\underset{\sim}{b} = \begin{pmatrix} b_{11} & b_{12} \\ b_{21} & b_{22} \end{pmatrix}$$

then

$$\underset{\sim}{a}\underset{\sim}{b} = (a_{11}, a_{12})\begin{pmatrix} b_{11} & b_{12} \\ b_{21} & b_{22} \end{pmatrix}$$

$$= (a_{11}b_{11} + a_{12}b_{21}, a_{11}b_{12} + a_{12}b_{22})$$

Therefore,

$$\underset{\sim}{c} = (c_{11}, c_{12})$$

where

$$c_{11} = a_{11}b_{11} + a_{12}b_{21}$$

$$c_{12} = a_{11}b_{12} + a_{12}b_{22}$$

It is clear that multiplication is possible only when the number of columns of $\underset{\sim}{a}$ equals the number of rows of $\underset{\sim}{b}$. In general, an $n \times m$ matrix can multiply only an $m \times p$ matrix, where n and p are arbitrary; the resulting matrix is $n \times p$. Thus, in the example just given, $n = 1$ and $m = 2$ for $\underset{\sim}{a}$, and $m = 2$ and $p = 2$ for $\underset{\sim}{b}$; multiplication of $\underset{\sim}{a}$ times $\underset{\sim}{b}$ yields $\underset{\sim}{c}$, which is $n \times p$ or 1×2 (a row matrix).

As another example, let

$$\underset{\sim}{a} = \begin{pmatrix} a_{11} & a_{12} \\ a_{21} & a_{22} \end{pmatrix}$$

$$\underset{\sim}{b} = \begin{pmatrix} b_{11} & b_{12} \\ b_{21} & b_{22} \end{pmatrix}$$

$$\underset{\sim}{a} = \underset{\sim}{a}\underset{\sim}{b}$$

$$= \begin{pmatrix} a_{11}b_{11} + a_{12}b_{21} & a_{11}b_{12} + a_{12}b_{22} \\ a_{21}b_{11} + a_{22}b_{21} & a_{21}b_{12} + a_{22}b_{22} \end{pmatrix}$$

Thus, when a 2×2 matrix multiplies a 2×2 matrix, the product also is a 2×2 matrix.

When a 1×2 matrix multiplies a 2×1 matrix, the result is a 1×1 matrix that is just a number (a scalar) and is no longer considered a matrix. For example:

$$\underset{\sim}{a} = (a_{11}, a_{12})$$

$$\underset{\sim}{b} = \begin{pmatrix} b_{11} \\ b_{21} \end{pmatrix}$$

$$\underset{\sim}{a}\underset{\sim}{b} = a_{11}b_{11} + a_{12}b_{21}$$

If the matrix is square (number of rows equal to number of columns), then the matrix may be raised to any power whatever. For example, for a square matrix $\underset{\sim}{M}$,

$$\underset{\sim}{M}^{3} = \underset{\sim}{M}\underset{\sim}{M}\underset{\sim}{M}$$

$$\underset{\sim}{M}^{N} = \prod_{i=1}^{N} \underset{\sim}{M}$$

(which is just $\underset{\sim}{M}$ multiplied by itself N times). This procedure is possible only with a square matrix, because only in that case are the rules of matrix multiplication fulfilled. Note also that, if $\underset{\sim}{M}$ is $n \times n$, then $\underset{\sim}{M}^{N}$ also must be $n \times n$.

MATRIX INVERSION

For a square matrix $\underset{\sim}{a}$, you can find an inverse matrix $\underset{\sim}{a}^{-1}$ such that

$$\underset{\sim}{a}^{-1}\underset{\sim}{a} = \underset{\sim}{I}$$

where $I_{ij} = 0$ for $i \neq j$, and $I_{ij} = 1$ for $i = j$. The matrix $\underset{\sim}{I}$ is called the unit matrix.

For example, if $\underset{\sim}{\mathbf{a}}$ is a 2×2 matrix, then

$$\underset{\sim}{\mathbf{I}} = \begin{pmatrix} 1 & 0 \\ 0 & 1 \end{pmatrix}$$

Note that $\underset{\sim}{\mathbf{I}}$ has the property that any matrix multiplied by it is unchanged by the multiplication. Thus $\underset{\sim}{\mathbf{I}}$ is analogous to the number 1 in scalar algebra, so that $\underset{\sim}{\mathbf{a}}\underset{\sim}{\mathbf{I}} = \underset{\sim}{\mathbf{a}}$. The calculation of $\underset{\sim}{\mathbf{a}}^{-1}$ is relatively straightforward. For example, let

$$\underset{\sim}{\mathbf{a}} = \begin{pmatrix} a_{11} & a_{12} \\ a_{21} & a_{22} \end{pmatrix}$$

Then the rule for calculating $\underset{\sim}{\mathbf{a}}^{-1}$ is

$$\underset{\sim}{\mathbf{a}}^{-1} = \begin{pmatrix} a_{22}/\alpha & -a_{12}/\alpha \\ -a_{21}/\alpha & a_{11}/\alpha \end{pmatrix}$$

where

$$\alpha = a_{11}a_{22} - a_{21}a_{12}$$

which is the value of the determinant of $\underset{\sim}{\mathbf{a}}$. (Clearly, the determinant must be non-zero in order for $\underset{\sim}{\mathbf{a}}^{-1}$ to exist.)

MATRIX DIAGONALIZATION

For a 2×2 matrix **a**, there is a matrix $\underset{\sim}{\mathbf{T}}$ such that

$$\underset{\sim}{\mathbf{T}}^{-1}\underset{\sim}{\mathbf{a}}\underset{\sim}{\mathbf{T}} = \begin{pmatrix} \lambda_1 & 0 \\ 0 & \lambda_2 \end{pmatrix}$$

where λ_1 and λ_2 are the eigenvalues of **a**, and $\underset{\sim}{\mathbf{T}}$ is called a transformation matrix. The eigenvalues λ_1 and λ_2 are easy to find. Let

$$\underset{\sim}{\mathbf{a}} = \begin{pmatrix} a_{11} & a_{12} \\ a_{21} & a_{22} \end{pmatrix}$$

Then the eigenvalues of $\underset{\sim}{\mathbf{a}}$ are obtained by solving the determinant

$$0 = \begin{vmatrix} a_{11} - \lambda & a_{12} \\ a_{21} & a_{22} - \lambda \end{vmatrix}$$

$$= (a_{11} - \lambda)(a_{22} - \lambda) - a_{12}a_{21}$$

$$= \lambda^2 - (a_{11} + a_{22})\lambda + a_{11}a_{22} - a_{12}a_{21}$$

Using the quadratic formula to solve for λ, we obtain

$$\lambda = \{(a_{11} + a_{22}) \pm [(a_{11} + a_{22})^2 - 4(a_{11}a_{22} - a_{12}a_{21})]^{1/2}\}/2$$

There are two roots, λ_1 and λ_2; let λ_1 be associated with the $+$ sign, and λ_2 with the $-$ sign. Thus we take $\lambda_1 > \lambda_2$. The rule for calculating $\underset{\sim}{T}$ is simple. If

$$\underset{\sim}{a} = \begin{pmatrix} a_{11} & a_{12} \\ a_{21} & a_{22} \end{pmatrix}$$

then

$$\underset{\sim}{T} = \begin{pmatrix} (\lambda_1 - a_{22})/a_{21} & (\lambda_2 - a_{22})/a_{21} \\ 1 & 1 \end{pmatrix}$$

$$\underset{\sim}{T}^{-1} = \begin{pmatrix} a_{21}/(\lambda_1 - \lambda_2) & (a_{22} - \lambda_2)/(\lambda_1 - \lambda_2) \\ -a_{21}/(\lambda_1 - \lambda_2) & (\lambda_1 - a_{22})/(\lambda_1 - \lambda_2) \end{pmatrix}$$

where λ_1 and λ_2 are the eigenvalues of $\underset{\sim}{a}$. Two other useful facts for a 2×2 matrix $\underset{\sim}{a}$ are

$$\lambda_1 + \lambda_2 = a_{11} + a_{22}$$

$$\lambda_1\lambda_2 = a_{11}a_{22} - a_{12}a_{21}$$

These relationships, together with the expressions just given for $\underset{\sim}{T}$ and $\underset{\sim}{T}^{-1}$, suffice to derive the expression for the transformation matrix in Equation 20-53 and its inverse in Equation 20-54. It also is a simple matter to generalize these results to $n \times n$ matrices, where $n > 2$.

For a clear and useful introduction to matrix algebra, see Chapter 1 of F. B. Hildebrand, *Methods of Applied Mathematics* (Englewood Cliffs, N.J.: Prentice-Hall, 1965).

Chapter 15

15-1. Consider the equilibrium $M_{i-1} + L \rightleftarrows M_i$. There are $\Omega_{n,i-1}$ microspecies that make up M_{i-1} and $\Omega_{n,i}$ microspecies that make up M_i. Therefore, the microscopic dissociation constant is

$$k = [(M_{i-1})/\Omega_{n,i-1}](L)/[(M_i)/\Omega_{n,i}]$$

while the macroscopic dissociation constant is

$$K = (M_{i-1})(L)/(M_i)$$

Combining the two equations gives Equation 15-20.

15-2. If the sites are identical and independent with a microscopic ligand dissociation constant k, then $K_2 = (2/3)k$ and $K_3 = (3/2)k$ (Eqn. 15-37). Because it is observed that $K_2 = K_3$, then the sites are not independent and the interaction energy $\Delta G_{I,23}$ as calculated from Equation 15-39 is

$$\Delta G_{I,23} = -RT \ln K_2/K_3 + RT \ln[(\Omega_{4,1}/\Omega_{4,2})/(\Omega_{4,2}/\Omega_{4,3})]$$
$$= RT \ln[((2/3)/(3/2))] = RT \ln(4/9)$$

At $25°\,C$ ($298.16°\,K$), $\Delta G_{I,23} = -0.48$ kcal mol^{-1}.

15-3. The critic is correct because the K_i values check well with those predicted for identical, independent sites according to Equation 15-20. If k is the microscopic dissociation constant, then from Equation 15-20 we have $K_1 = (1/6)k$, $K_2 = (2/5)k$, $K_3 = (3/4)k$, $K_4 = (4/3)k$, $K_5 = (5/2)k$, and $K_6 = 6k$. From this we see that $K_5 = 15K_1$, and $K_4 = 8K_1$.

15-4. When half the L_1 and half the L_2 sites are filled, then

$$(L_2ML_1) + (ML_1) = (1/2)(M)_{Tot}$$

$$(L_2ML_1) + (L_2M) = (1/2)(M)_{Tot}$$

Dividing each of these equations by (M_{Tot}), we obtain

$$(ML_1)/(M)_{Tot} = (L_2M)/(M)_{Tot} = 1/2 - (L_2ML_1)/(M)_{Tot} = 1/2 - \bar{y}_{12}$$

From the definition of $(M)_{Tot}$ and the first two equations above, we have

$$(M)_{Tot} = (M) + (ML_1) + (L_2M) + (L_2ML_1)$$

$$= (M) + (1/2)(M)_{Tot} + (1/2)(M)_{Tot} - (L_2ML_1)$$

or

$$(M) = (L_2ML_1)$$

From the expression for K_{12} (Eqn. 15-89),

$$K_{12} = (L_2ML_1)(M)/(ML_1)(L_2M)$$

$$= (L_2ML_1)^2/(ML_1)(L_2M)$$

Using our expressions for $(ML_1)/(M)_{Tot}$ and $(L_2M)/(M)_{Tot}$,

$$K_{12} = (L_2ML_1)^2/[(1/2) - \bar{y}_{12}]^2(M)_{Tot}^2 = \bar{y}_{12}^2/(1/2 - \bar{y}_{12})^2$$

This is Equation 15-91. By taking the square root of this equation and solving for $\bar{y}_{12}$, we obtain Equation 15-92. From Equations 15-89 and 15-91, we have

$$\Delta G_{12}^0 = -RT \ln K_{12} = -2RT \ln(\bar{y}_{12}/[(1/2) - \bar{y}_{12}])$$

$$= -2RT \ln[(2\bar{y}_{12}/(1 - 2\bar{y}_{12})]$$

This is Equation 15-93.

15-5. From the definition of $\tilde{K}_1$, $\tilde{K}_2$, and $\tilde{K}_3$, we have

$$(L)/\tilde{K}_1 = (M_1)/(M_0)$$

$$(L)^2/\tilde{K}_2 = (M_2)/(M_0)$$

$$(L)^3/\tilde{K}_3 = (M_3)/(M_0)$$

From the definitions of the microscopic constants, we have

$$(M_1)/(M_0) = (L)/k_1 + (L)/k_2 + (L)/k_3$$

$$(M_2)/(M_0) = (L)^2/k_1k_2 + (L)^2/k_1k_3 + (L)^2/k_2k_3$$

$$(M_3)/(M_0) = (L)^3/k_1k_2k_3$$

Equating our two expressions for the $(M_i)/(M_0)$ quantities gives

$$(L)/\tilde{K}_1 + (L)^2/\tilde{K}_2 + (L)^3/\tilde{K}_3 = [1 + (L)/k_1][1 + (L)/k_2][1 + (L)/k_3] - 1$$

or

$$1 + (L)/\tilde{K}_1 + (L)^2/\tilde{K}_2 + (L)^3/\tilde{K}_3 = [1 + (L)/k_1][1 + (L)/k_2][1 + (L)/k_3]$$

This is identical to Equation 15-54, for the case $n = 3$. It is a simple expression because the product on the right-hand side, through multiplication, generates all the combinations of $(L)/k_i$ that multiply together to give the sum $1 + (M_1)/(M_0) + (M_2)/(M_0) + (M_3)/(M_0)$.

Chapter 16

16-1. For Equation A, the rate law is

$$V = k_2(X_1) + k_2'(X_2)$$

Solving for (X_1) and (X_2) in terms of $(E)_0$, (S), K_1, and K_2, we have

$$V = k_2(E)_0/[1 + K_1/(S) + K_1/K_2] + k_2'(E)_0/[1 + K_2/(S) + K_2/K_1]$$

We want to see whether it is possible to cast this equation into the form of Equation 16-37, which is the rate law for Equation B. To do this, first find the form for V_{max} by letting $(S) \to \infty$.

$$(V)_{S \to \infty} = V_{max} = k_2(E)_0/[1 + K_1/K_2] + k_2'(E)_0/[1 + K_2/K_1]$$
$$= (K_2 k_2(E)_0 + K_1 k_2'(E)_0)/[K_1 + K_2]$$

We can use this expression for V_{max} to alter the above expression for V to

$$V = V_{max}[1 + K_1 K_2/(K_1 + K_2)(S)]$$

This equation is identical in form to Equation 16-37, and therefore the bright student is correct.

16-2. By analogy with Equation 16-37, we have

$$v_i = V_S/[1 + K_S/(SM)]$$

It is simple to show that $(SM) = (S)_0/[1 + K_{SM}/(M)]$
Therefore, we can write

$$v_i = V_S/[1 + K_S(1 + K_{SM}/(M))/(S)_0]$$

and

$$1/v_i = 1/V_S + (K_S/V_S)[1 + K_{SM}/(M)][1/(S)_0]$$

Clearly, points a, b, and c are satisfied by this equation. However, point d is not, which means that a reaction can take place without M. Therefore, we need to add to the mechanism a simple step:

$$E + S \rightleftharpoons X \rightarrow E + P$$

16-3. For simplicity, we'll assume that the simple mechanism of Equation 16-48 and the equations that follow can be applied. Because V_S is pH-independent over the entire range of pH 7 to pH 8.5, the two pK values on the enzyme–substrate complex presumably are well separated. Therefore, the midpoints of the V_s versus pH curves should give pK_{aX} and pK_{bX}, and these are 5 and 10.1, respectively. Because K_S is pH-independent over the entire pH range, then we know from Equation 16-58 that $pK_{aE} = pK_{aX}$ and $pK_{bE} = pK_{bX}$.

16-4. The rate equation is

$$-d\Delta(A{=}{=}{=}U)/dt = -k_2 \Delta((A, U)) + k_{-2} \Delta(A{=}{=}{=}U)$$

The steady-state expression for $d\Delta((A, U))/dt$ can be simplified to

$$-d\Delta((A, U))/dt = (k_1((\bar{A}) + (\bar{U})) - k_{-2}) \Delta(A{=}{=}{=}U)$$
$$+ (k_{-1} + k_2) \Delta((A, U)) = 0$$

Solving for $\Delta((A, U))$ and substituting into the first equation gives the expression

$$-d\Delta(A{=}{=}{=}U)/dt = \Delta(A{=}{=}{=}U)/\tau$$

where $1/\tau$ is given in the problem. We expect $\Delta((A, U))$ to be short-lived because it has only one hydrogen bond. The concentration dependence of $1/\tau$ is identical in form to that for the simple one-step mechanism (see Eqn. 16-74), and the researcher is wrong.

16-5. To derive the expression for the relaxation time for the "very slow" step, we consider the fast steps always to adjust quickly while the very slow step relaxes. This means that each fast step can be considered to be at equilibrium, with respect to itself, while the very slow step relaxes. The derivation can proceed along lines similar to Equations 16-85 to 16-90. Our initial rate equation is

$$-d[(X_2) + (X_3)]/dt = -k_2(X_1) + k_{-2}(X_2)$$

The following relationships are useful.

$$(X_2) + (X_3) = (X_2)(1 + k_3/k_{-3})$$
$$\Delta[(X_2) + (X_3)] = \Delta(X_2)(1 + k_3/k_{-3})$$
$$\Delta(E) + \Delta(X_1) = -\Delta(X_2) - \Delta(X_3) = -\Delta(X_2)(1 + k_3/k_{-3})$$
$$\Delta(E) + \Delta(X_1) = \Delta(X_1)(1 + k_{-1}/\{k_1[(\bar{E}) + (\bar{S})]\}) \qquad \text{(see Eqn. 16-88)}$$

Solving for $\Delta(X_1)$ in terms of $\Delta(X_2)$, we obtain

$$\Delta(X_1) = -\Delta(X_2)(1 + k_3/k_{-3})/(1 + k_{-1}/\{k_1[(\bar{E}) + (\bar{S})]\})$$

Therefore,

$$-d[\Delta(X_2) + \Delta(X_3)]/dt = -(d\Delta(X_2)/dt)(1 + k_3/k_{-3})$$
$$= (k_2(1 + k_3/k_{-3})/\{1 + k_{-1}/k_1[(\bar{E}) + (\bar{S})]\} + k_{-2})\Delta(X_2)$$

and

$$-d\Delta(X_2)/dt = \Delta X_2/\tau$$

where

$$1/\tau = k_2/\{1 + k_{-1}/k_1[(\bar{E}) + (\bar{S})]\} + k_{-2}/(1 + k_3/k_{-3})$$

Chapter 17

17-1. Neither is correct. The lack of change in P when I binds, assuming an MWC model, would indicate that the enzyme starts out in the T state. If this is true, then S′ binding to the R state should give changes in P that lead, not parallel, the changes in $\bar{y}_{S'}$. See Equation 17–18 and the discussion that follows; see also Section 17-4.

17-2. Assume that I binds only to the T state and A binds only to the R state. If $L = 1$, then the data would be explained: almost hyperbolic binding of A and of I alone, but sigmoidal for A in the presence of I because I has pulled the enzyme into the T state; vice versa for I in the presence of A. But if $L \gg 1$, then A by itself should bind sigmoidally and not hyperbolically. Therefore, Dr. Alpha is wrong, and Dr. Omega is also wrong because he fails to realize that $L = 1$ is a solution.

17-3. Call the species along the top row I_0, I_1, I_2, and I_3, and those along the bottom row R_i, $i = 0$ to 4. The parameter L equals $(I_0)/(R_0)$. By analogy with Equations 17-9 to 17-12,

$$\sum_{i=0}^{4} (R_i) = (R_0)(1 + \alpha)^4$$

$$\sum_{i=0}^{3} (I_i) = (I_0)[(1 + c\alpha)^4 - c^4\alpha^4] = L(R_0)[(1 + c\alpha)^4 - c^4\alpha^4]$$

$$\sum_{i=0}^{4} i(R_i) = 4\alpha(R_0)(1 + \alpha)^3$$

$$\sum_{i=0}^{3} i(I_i) = 4Lc\alpha(R_0)[(1 + c\alpha)^3 - c^3\alpha^3]$$

$$\bar{y} = \left[\sum_{i=0}^{4} i(R_i) + \sum_{i=0}^{3} i(I_i)\right] \Big/ \left[4\left(\sum_{i=0}^{4} (R_i) + \sum_{i=0}^{3} (I_i)\right)\right]$$

$$= \{[\alpha(1 + \alpha)^3 + Lc\alpha[(1 + c\alpha)^3 - c^3\alpha^3]\}/[(1 + \alpha)^4 + L(1 + c\alpha)^4 - c^4\alpha^4]$$

This expression may be compared with Equation 17-13, which holds for the two-state MWC model. The two equations are quite similar, except that the one above subtracts out one term in the numerator and in the denominator. It would probably be hard to distinguish experimentally between the two expressions for $\bar{y}$.
The parameter r_i is

$$r_i = (I_i)/(R_i) = Lc^i$$

For $L = 10^3$, $c = 10^{-2}$, $r_i < 1$ for $i = 2, 3$. For $L = 10^4$, $c = 10^{-2}$, $r_i < 1$ for $i = 3$.

17-4. Let $\alpha = (L)/k$ and $\alpha' = (A)/k_A$.
Then

$$(PL_1) = 4\alpha(P)$$

$$(PL_2) = 6\alpha^2(P)$$

$$(PL_3) = 4\alpha^3(P)$$

$$(PL_4) = \alpha^4(P)$$

$$\bar{y} = \sum_{i=0}^{4} i(PL_i)/4\left[\sum_{i=0}^{4} (PL_i)\right]$$

Note that $(PL_0) = (P) + (PA)$; $(PA) = \alpha'(P)$. After simplifying,

$$\bar{y} = \alpha(1 + \alpha)^3/[(1 + \alpha)^4 + \alpha']$$

This equation is identical in form to Equation 17-16, where α' in the above equation plays the role of L in Equation 17-16. Therefore, increasing the concentration of (A) (greater α') will give a pronounced sigmoidicity to plots of $\bar{y}$ versus α in the same way that increasing L influences plots of $\bar{y}$ versus α in Equation 17-16.

For the K_i values, we have

$$K_1 = [(P) + (PA)](L)/(PL_1) = (k/4)(1 + \alpha')$$

$$K_2 = (2/3)k, K_3 = (3/2)k, \text{ and } K_4 = 4k$$

The only K_i affected by the concentration of A is K_1. As A increases, K_1 becomes steadily larger, so that

$$K_1/K_2 = 3/8, \text{ when } \alpha' = 0$$

$$= 3/4, \text{ when } \alpha' = 1$$

$$= 15/4, \text{ when } \alpha' = 9$$

Therefore, binding of the first L appears to become progressively weaker, as α' increases, relative to binding of the next molecule of L. As already mentioned, this gives the appearance of cooperativity to a system that has no site–site cooperative interactions.

17-5. If $L \gg 1, c = 0$ for A, and $c \gg 1$ for B, then the first two points will be satisfied. But binding of B alone should then be hyperbolic (protein starts out in the T state, which is the state to which B binds). Therefore, the data cannot be explained by the two-state MWC model.

Chapter 18

18-1. First calculate the denominator of Equation 18-23.

$$\int_0^\infty W(r)\, dr = (\beta/\pi^{1/2})^3 \int_0^\infty e^{-\beta^2 r^2} 4\pi r^2\, dr$$

$$\int_0^\infty e^{-\beta^2 r^2} r^2\, dr = (1/4\beta^2)(\pi^{1/2}/\beta) = (\pi^{1/2}/4\beta^3)$$

Therefore,

$$4\pi \int_0^\infty e^{-\beta^2 r^2} r^2 \, dr = (\pi^{3/2}/\beta^3)$$

and

$$(\beta/\pi^{1/2})^3 \int_0^\infty e^{-\beta^2 r^2} 4\pi r^2 \, dr = 1$$

Thus, the distribution function is normalized, as stated in the text.
For the numerator in Equation 18-23, we have

$$\int_0^\infty r^2 W(r) \, dr = (\beta/\pi^{1/2})^3 \int_0^\infty e^{-\beta^2 r^2} 4\pi r^4 \, dr$$

$$\int_0^\infty e^{-\beta^2 r^2} r^4 \, dr = (3/8\beta^4)(\pi^{1/2}/\beta) = (3\pi^{1/2}/8\beta^5)$$

Therefore,

$$4\pi \int_0^\infty e^{-\beta^2 r^2} r^4 \, dr = (3\pi^{3/2}/2\beta^5)$$

and

$$(\beta/\pi^{1/2})^3 \int_0^\infty e^{-\beta^2 r^2} 4\pi r^4 \, dr = (3/2\beta^2)$$

Because $\beta^{-1} = (2nl^2/3)^{1/2}$, then $(3/2\beta^2) = nl^2 = \langle r^2 \rangle_0$.

18-2. The idea of the statistical segment is that a chain may be described as equivalent to a random walk structure with N_e equivalent segments each of length l_e (see Section 18-9). The probability $p(N_e)$ that the chain ends are between 0 and l_e of each other is simply

$$p(N_e) = \int_0^{l_e} W(r) \, dr$$

where

$$W(r) = (\beta/\pi^{1/2})^3 e^{-\beta^2 r^2} 4\pi r^2 \, dr \qquad \text{(see Eqn. 18-22)}$$

and where

$$\beta = (3/2N_e l_e^2)^{1/2}$$

because $N_e > 300$, then $\beta^2 r^2 \approx 0$ for $r < l_e$, and $e^{-\beta^2 r^2} \approx 1$.

Therefore,

$$p(N_e) \cong (\beta/\pi^{1/2})^3 \int_0^{l_e} 4\pi r^2 \, dr = (\beta/\pi^{1/2})^3 4\pi l_e^3/3 = k\beta^3$$

where k is a constant.

Furthermore,

$$\beta^3 = k'N_e^{-3/2}$$

where k' is a constant, so that

$$p(N_e) = kk'N_e^{-3/2}$$

With $M \propto N_e$, then $p(N_e) = \text{const } M^{-3/2}$.

18-3. The general expression for calculation of $\langle r^2 \rangle$ is given by Equation 18-4. Virtual bond i runs from the α-carbon of residue $i-1$ to the α-carbon of residue i (see Section 18-6). The virtual bonds that run from residue k to residue p are numbered from $k+1$ to p. Therefore, to calculate $\langle r_{kp}^2 \rangle$, we need to construct the following summation:

$$\langle r_{kp}^2 \rangle = (p-k)l^2 + 2 \left\langle \sum_{j>i} (\hat{\mathbf{l}}_i \cdot \hat{\mathbf{l}}_j) \right\rangle$$

$$= (p-k)l^2 + 2 \left\langle \sum_{j=i+1}^{p} \sum_{i=k+1}^{p} (\hat{\mathbf{l}}_i \cdot \hat{\mathbf{l}}_j) \right\rangle$$

Thus, the professor is incorrect because his summation over i starts at k, rather than at $k+1$.

18-4. Yes. While the freely rotating chain tends to have relatively small chain dimensions, it is possible that, with restricted rotations, the chain would have even smaller dimensions. The key is whether the restricted rotations bias the chain toward highly compact structures at the expense of more extended ones. This could happen if attractive interactions between elements within the chain encourage adopting of internal rotation angles associated with less extended form. Thus, while the freely rotating chain embraces equally all values of the internal rotation angles, a chain with restricted rotations could, in principle, be biased toward internal rotation angles that give mostly compact structures.

18-5. With $n=2$ and $n=3$, there is not enough conformational flexibility for the charged ends to bias the conformation toward compact structures. This is because there are too few ϕ, ψ coordinates for $n=2$ and $n=3$, and steric constraints severely limit the accessible conformations. But with larger values of n, there are more conformations sterically available where the electrostatic attraction between chain ends can bias the chain into structures where the termini can interact. Therefore, not to account for the electrostatic attraction of chain ends leads to serious discrepancies between calculated and observed values of the end-to-end distances, for oligomers longer than $n=3$. For a treatment and discussion of this problem, see P. J. Flory and P. R. Schimmel, *J. Am. Chem. Soc.* 89: 6807 (1967).

Chapter 19

19-1. The sedimentation coefficient is given by Equation 19-22. If the equivalent sphere and the random coil have the same $\bar{V}_2$, then the identical value of s^0 can be obtained only if they have the identical frictional coefficients.

For the random coil, f is given by Equation 19-21, while Equation 19-20 gives the value of f for the equivalent sphere. Equating these expressions and solving for the radius r_e of the equivalent sphere, we have

$$r_e = (P_c \langle r^2 \rangle^{1/2})/6\pi = 27.1 \text{ Å}$$

19-2. Referring to Tables 19-1 and 19-2, we see that the solid sphere, the wide, thin disk, the long random coil, and the long random coil (unperturbed chain) can all give the result that $s^0[\eta]^{1/3}$ is proportional to $M^{2/3}$.

19-3. The mean square end-to-end distance is given by

$$\langle r^2 \rangle = \left(\int_0^\infty \exp(-\alpha r^2/kT) 4\pi r^4 \, dr \right) \Bigg/ \left(\int_0^\infty \exp(-\alpha r^2/kT) 4\pi r^2 \, dr \right)$$

After performing the integration and simplifying, we have

$$\langle r^2 \rangle = 3kT/2\alpha$$

From the intrinsic viscosity value, we can calculate $\langle r^2 \rangle^{1/2}$ from Equation 19-18. This gives $\langle r^2 \rangle^{1/2} = 350$ Å. The expression above can be used to calculate α from $\langle r^2 \rangle$ and kT. This gives, at 300°K, $\alpha = 50.7$ erg/Å^2.

19-4. Because $L_c \gg a$, we can use Equation 19-42 in lieu of Equation 19-41. To calculate $\langle r^2 \rangle_0^{1/2}$ in μm (1 μm $= 10^4$ Å),

$$\langle r^2 \rangle_0^{1/2} = (2aL_c)^{1/2} = [(2)(0.045)(60)]^{1/2} = 2.3 \ \mu\text{m}$$

19-5. From Equation 19-19 we can calculate $\langle r^2 \rangle$. This gives $\langle r^2 \rangle = 8.71 \times 10^{-10}$ cm or $\langle r^2 \rangle^{1/2} = 2.95 \times 10^{-5}$ cm $= 2950$ Å. Because $\langle r^2 \rangle^{1/2} \ll L_c$, we know that a must also be much less than L_c. Therefore, we may use Equation 19-42 to calculate a from $\langle r^2 \rangle$ and L_c. This gives $a = 145$Å. Note that this is only an estimate of a because, in Equation 19-42, we have used $\langle r^2 \rangle$ in place of $\langle r^2 \rangle_0$.

Chapter 20

20-1.
$$k_1 = (\Omega_{20}\sigma s^{20})/(\Omega_{50}\sigma s^{50})$$
$$k_2 = (\Omega_{25}\sigma s^{25})/(\Omega_{75}\sigma s^{75})$$

where Ω_k is given by Equation 20-27.
 With $\sigma = 10^{-4}$ and $s = 1$, we have

$$k_1 = \Omega_{20}/\Omega_{50} = 81/51$$
$$k_2 = \Omega_{25}/\Omega_{75} = 76/26$$

20-2. Let ΔG_1 be the free energy of helix 1 relative to the all-coil state and ΔG_2 be the free energy of helix 2 relative to the all-coil state. Therefore,

$$\Delta G_1 = -RT \ln \sigma s^{100} \text{ and } \Delta G_2 = -RT \ln \sigma' s'^{100}$$

Set $\Delta G_1 - \Delta G_2 = 0$, use the given values of σ and of s'/s, and solve for σ'. This gives $\sigma' = 10^{-2}$.

20-3. This is straightforward matrix multiplication.

20-4. This is straightforward differentiation.

20-5. Assign a matrix to each A residue and to each B residue. Refer to Equation 20-39 for the form of the matrix for a single residue. We have

$$q = (1, 0)\mathbf{M_A M_B M_A M_B} \cdots \begin{pmatrix} 1 \\ 1 \end{pmatrix}$$

where

$$\mathbf{M_A} = \begin{pmatrix} 1 & \sigma_A s_A \\ 1 & s_A \end{pmatrix}$$

$$\mathbf{M_B} = \begin{pmatrix} 1 & \sigma_B s_B \\ 1 & s_B \end{pmatrix}$$

We can also write q as

$$q = (1, 0)(\mathbf{M_A M_B})^x \begin{pmatrix} 1 \\ 1 \end{pmatrix}$$

where $x = n/2$ and $\mathbf{M_A M_B} = \mathbf{Q}$ is

$$\mathbf{Q} = \begin{pmatrix} 1 + \sigma_A s_A & \sigma_B s_B + \sigma_A s_A s_B \\ 1 + s_A & \sigma_B s_B + s_A s_B \end{pmatrix}$$

Chapter 21

21-1. Start with Equation 21-16 and use Equation 21-15 for K_{app}. This gives

$$d \ln K_{app}/dT = d \ln K_D/dT + d \ln\left(1 + \sum_i d_i K_i/K_D\right)\bigg/dT - d \ln\left(1 + \sum_i (1 - d_i)K_i\right)\bigg/dT$$

Carry out the differentiations and use the relations $d \ln K_D/dT = \Delta H_D/RT^2$ and $d \ln K_i/dT = \Delta H_i/RT^2$. After several steps of straightforward algebraic manipulations, Equation 21-17 is obtained.

21-2. With each physical property, we measure f_{app}. For each value of d_1, we can calculate what the f_{app} values should be, using Equation 21-10 and the values of K (to calculate f_x). We obtain the following three values of f_{app} with $d_1 = 0.1$ and with $d_1 = 0.2$ (shown in parentheses): 0.22(0.24), 0.52(0.54), and 0.81(0.82), corresponding to $f_D = 0.2$, 0.5, and 0.8, respectively. Because the f_{app} values at each f_D are within experimental error of each other, they cannot be distinguished experimentally. And because the two physical property measurements give, within experimental error, the same f_{app} values at three different points in the denaturation, we might erroneously conclude that no intermediates are present.

21-3. To calculate ΔH, ΔS, and ΔC_p, we can use the basic thermodynamic relations

$$\Delta H = -T^2[\partial(\Delta G/T)/\partial T]_p = a - cT^2$$

$$\Delta S = -(\partial \Delta G/\partial T)_p = -b - 2cT$$

$$\Delta C_p = (\partial \Delta H/\partial T)_p = T(\partial \Delta S/\partial T)_p = -2cT$$

To have a temperature of maximum stability requires that $(\partial \Delta G/\partial T)_p = 0$, and that $(\partial^2 \Delta G/\partial T^2)_p < 0$. (The latter condition is necessary for a maximum in ΔG, as opposed to a minimum. A maximum in ΔG means that the free energy of the denatured form is at its highest point relative to the native state; this gives the maximum stability to the native form.) By differentiation of the expression for ΔG, we have

$$(\partial \Delta G/\partial T)_p = b + 2cT$$

Setting $(\partial \Delta G/\partial T)_p = 0$ gives $T = -b/2c = 300°K \cong 27°$ C. The second derivative is

$$(\partial^2 \Delta G/\partial T^2)_p = 2c$$

Because c is negative, then $(\partial^2 \Delta G/\partial T^2)_p < 0$ and we are at a maximum in ΔG when $T = 27°$ C.

21-4. Because the kinetics as monitored at 280 nm are the same with or without the reporter group, it would seem unlikely that the reporter group seriously disturbs the structure. However, it is still difficult to rationalize results such as these. The problem is that, with the reporter group attached, biphasic kinetics are observed *in both directions* with the reporter group's absorption, but only monophasic kinetics are seen *in both directions* with the same molecule studied at 280 nm. However, the important data that are lacking are the rate parameters (λ_i values) for the kinetic phases. It would be useful to know the λ parameter for the 280 nm absorption kinetics and whether it corresponds to any of those monitored by the reporter's absorption.

21-5. Only four concentration variables, or fractional concentrations, are required to describe completely this system. The fifth variable is given as a linear combination of the other four; specifically,

$$\Delta f_{X_3} = -\Delta f_N - \Delta f_{X_1} - \Delta f_{X_2} - \Delta f_D$$

Therefore, four rate equations describe the system completely and this will give rise to four λ_i parameters (which are always equal in number to the number of independent concentration variables). When we study the reaction $N \rightarrow D$, we see only one phase, corresponding to the slow $N \rightleftarrows X_1$ transition. This is because the first step is slow and the following ones are much more rapid and are quickly equilibrated while the first one takes place. But as for the $D \rightarrow N$ case, there should be one or more rapid phases also observable as everything first goes as far as X_1, which is effectively the "bottleneck" while the slow $X_1 \rightleftarrows N$ equilibrium takes place. Therefore, the $D \rightarrow N$ case should reveal some rapid phases in addition to the slow one.

Chapter 22

22-1. a. Do an absorbance mixing curve, plotting mole fraction of A or U *residues*. An abrupt change in slope at 0.5 mole fraction will indicate saturation. Alternatively, one can ask whether the oligomer protects the polymer against chemical or enzymatic modifications known to be specific for single strands.

b. Compare the hypochromism of oligo A·poly U with that of poly A·poly U or compare their CD spectra.

c. Remember that the stability of a complex is always relative to that of the separated components. Although the structures of the complexes are the same, the interactions lost when the oligomers dissociate are different. In the oligo A case, strong stacking interactions between adjacent oligomers are lost; whereas in the oligo U case, the stacking interactions are much weaker. Thus the oligo A·poly U complex is much more stable.

22-2. Following the hint, write

$$C + H^+ + C \rightleftharpoons CH^+ {-} C \qquad \Delta H^0 = -7 \text{ kcal}$$
$$CH^+ {-} C + H^+ \rightleftharpoons 2CH^+ \qquad \Delta H^0 = +3 \text{ kcal}$$

Now use Equation 22-20b, which for H^+ as the variable becomes

$$\partial T_m / \partial \text{pH} = \Delta n_p R T_m^2 / \Delta H_p^0$$

The enthalpy changes per phosphate are half the values shown above. The change in bound protons per phosphate is $+1/2$ for each of the reactions. Thus $\partial T_m / \partial \text{pH}$ is negative for the top reaction and positive for the bottom. These two phase boundaries will meet at a pH that corresponds to the maximum possible T_m. Above this temperature there is no pH at which the acid double helix can be stable. There is no pH at which two melting transitions should be observed.

22-3. Actually this is not an easy problem. The best experiments would use an oligonucleotide of defined length. Then the concentration dependence of helix formation would give information on the strandedness. Alternatively, relaxation kinetics can be used. Length-dependent measurements can help sort out parallel and antiparallel structures. All parallel-stranded interactions (except in structures long enough to circularize) require intermolecular interactions, but antiparallel interactions can be formed by folding a strand into a hairpin loop. Such intramolecular complexes will show concentration-independent melting curves and relaxation kinetics.

22-4. There are three interactions:

$$\begin{array}{ccc} \text{A---T} & \text{T---A} & \text{A---A} \\ \cdot \quad \cdot & \cdot \quad \cdot & \cdot \quad \cdot \\ \text{T---A} & \text{A---T} & \text{T---T} \end{array}$$

One constraint can be generated because all runs of T on one strand must be bounded by A: $\chi_{AT} = \chi_{TA}$. Thus there are only two independent variables. Poly A·poly T and poly A—T·poly A—T contain all the required information to determine these variables.

22-5. One can speculate that the loop should roll along and lead to rapid tritium exchange. The kinetics of such a loop motion are unknown, but one can crudely estimate them to be similar to single-stranded branch migration (see Chapter 23). This is about 1,000 base pairs per second, which could lead to quite significant tritium exchange.

Chapter 23

23-1. Parts a and b are straightforward.

  c. This is done by evaluating each element one at a time. For example, $M_{12} = 0$ because, if sites 1 and 2 are both empty, one cannot allow a state for sites 2 and 3 in which site 2 is filled. Note that M_{42} is ambiguous. One could argue it should be 1 or 0, but it turns out this makes no difference since both resulting matrices have the same eigenvalues.

  d. One must solve the determinantal equation:

$$\begin{vmatrix} 1-\lambda & 0 & k(\mathrm{m}) & 0 \\ 1 & -\lambda & \zeta k(\mathrm{m}) & 0 \\ 0 & 1 & -\lambda & 0 \\ 0 & 1 & 0 & -\lambda \end{vmatrix} = 0$$

The result is $-\lambda^3 + \lambda^2 + \lambda\zeta k(\mathrm{m}) + (1-\zeta)k(\mathrm{m}) = 0$. In the limit $\zeta \to 1$, this equation becomes $-\lambda^3 + \lambda^2 + \lambda k(\mathrm{m}) = 0$, which rearranges to

$$\lambda^2 - \lambda = k(\mathrm{m})$$

To compare this with results in the text, start with Equation 23-101. Let the parameter ζ in that equation (which does not have the same meaning as the parameter ζ used above) equal zero. This produces the excluded-site model. The result is identical to the expression shown above.

23-2. The data in Tables 23-4 and 23-5 can be used to evaluate the following expressions. Free-energy change for forming the left-hand structure from a single strand:

$$\Delta G_{\mathrm{T}} = \Delta G_{\mathrm{GG}} + \Delta G_{\mathrm{GC}} + \Delta G_{\mathrm{CG}} + \Delta G_{\mathrm{loop}}(10) = -6.1 \text{ kcal}$$

For the right-hand structure:

$$\Delta G_{\mathrm{T}} = \Delta G_{\mathrm{GA}} + \Delta G_{\mathrm{AA}} + \Delta G_{\mathrm{AG}} + \Delta G_{\mathrm{AA}} + \Delta G_{\mathrm{GA}} + \Delta G_{\mathrm{GG}} + 2\Delta G_{\mathrm{loop}}(4) = -3.5 \text{ kcal}$$

Thus the free-energy change for forming the right-hand structure from the left-hand structure is: $\Delta G = +2.6$ kcal. The equilibrium constant for this reaction is just, at 25°C, $K = e^{-2.6/0.6} = 0.013$. This shows that the particular hydrogen-bonded sequences formed and the detailed loop sizes involved are much more important than the simple number of base pairs.

23-3. The repeating sequence is present in 5,000 copies and thus it will renature with a $(C_o t)_{1/2}$ of 1/5,000 that of the unique sequence DNA. If the DNA is cut into 60 base-pair fragments, the unique and repeated sequences will renature almost completely separately. With 300 base-pair fragments, most unique sequences will be on fragments containing some repeated DNA and will renature with the $(C_o t)_{1/2}$ of repeated sequence DNA. With longer fragments, virtually all the DNA will show fast renaturation kinetics.

23-4. At least three factors may be responsible for the extra stability: (1) stacking effects of neighboring bases in loops, presumably dependent on the particular sequence; (2) contact with other regions of the tRNA; (3) the mere presence of a loop rather than a single strand. To examine hypothesis (2), one could isolate hairpin loops from tRNA by partial nuclease cleavage and study their interactions. To examine (3), it would be useful to measure the strength of the interaction between a loop and a single strand to see if this is also stronger than expected from the known stability of simple duplexes. However, to critically evaluate (1) and (3), one would probably have to synthesize loops of 6, 7, or 8 residues with specific sequences or, alternatively, find mutant tRNAs with altered hypermodified bases or other anticodon-loop modifications. H. Grosjean, D. Crothers, and their coworkers are actively engaged in such studies.

23-5. The ΔG_{nuc} is given as $+4$ kcal. Equation 23-19b can be used for complementary oligo-nucleotide pairs to compute the total free energy of formation from the concentration needed to give a specific T_m:

$$\Delta G_T = -RT \ln q_c = -RT \ln(4/C_T)$$

Thus, from the data given, we can write for each complex:

$$\text{AATAA} \cdot \text{TTATT} \quad \Delta G_T = \Delta G_{nuc} + 2\Delta G_{AA} + \Delta G_{AT} + \Delta G_{TA} = -2.92 \text{ kcal}$$
$$\text{ATATA} \cdot \text{TATAT} \quad \Delta G_T = \Delta G_{nuc} + 2\Delta G_{AT} + 2\Delta G_{TA} = -3.59 \text{ kcal}$$
$$\text{AAAAT} \cdot \text{ATTTT} \quad \Delta G_T = \Delta G_{nuc} + 3\Delta G_{AA} + \Delta G_{AT} = -2.10 \text{ kcal}$$

These three equations in three unknowns can be solved simultaneously to yield: $\Delta G_{AA} = -1.56$ kcal; $\Delta G_{AT} = -1.42$ kcal; $\Delta G_{TA} = -2.37$ kcal. Using these values, one can calculate:

$$\text{TAAAA} \cdot \text{TTTTA} \quad \Delta G_T = \Delta G_{nuc} + 3\Delta G_{AA} + \Delta G_{TA} = -3.05 \text{ kcal}$$

Chapter 24

24-1. The DNA has approximately 9,090 base pairs. As the relaxed closed circular B helix, it has 909 turns. When it is converted into the A helix, only 826 turns are needed to accommodate this much DNA. Thus the remaining 83 twists originally present in the DNA must appear as 83 positive supercoils. If the A DNA is relaxed and then returned to the B-helical form, 83 additional helix turns will have to be introduced, and this will be accompanied by the appearance of 83 negative supercoils.

24-2. The right-hand structure is the same as case I, except that the supercoils are left-hand rather than right-hand. Thus $L = -2$ and $T = -2 \sin \theta$. The left-hand structure has the same twist as case N: $T = -2 \sin \theta$. However, the extra twistless loop introduces two more positive links, so that $L = +4$.

24-3. To solve this problem, it is useful to study Figure 24-8.
a. The fact that ethidium is released shows that the original DNA must have had negative supercoils and that the DNA has not bound enough ethidium to become fully relaxed. The number released will give information about how far from relaxed the structure is, but will not give information about the number of supercoils originally present.
b. Since there is no release, $\tilde{v} = \tilde{v}_c$. The DNA is fully relaxed. Using an unwinding angle of $-26°$ per ethidium, 169 ethidiums correspond to -12 supercoils.

24-4. Note that, in general, four points in space form a distorted tetrahedron, which is a rigid structure. This is completely defined by the six distances between the four points. To fix a fifth point with respect to these four requires four more distances. Three will locate it in space with respect to one face of the tetrahedron, but will not define on what side of that face the fifth point is located. The same argument applies to the sixth point and any additional ones. The answer to the optional problem is not known, at least to the authors.

24-5. a. Using the arguments given in the text, resonances between -14 and -15 ppm must be A—U pairs. These in isolation would appear at -14.8 ppm, but neighboring base pairs lead to upfield shifts, as shown in Table 24-1. When these shifts are considered, one can predict that the left-hand structure should show one resonance between -14 and -15 ppm, whereas the center structure should show two and the right-hand structure three. Thus, in principle, NMR will distinguish among the three structures.

b. AAGG will bind to the left-hand and right-hand structures, but not to the center one. CCUU, GUGG, CCUA, or others will bind to the left-hand structure, but not to the other two structures.

c. Pancreatic ribonuclease cuts after all pyrimidines. Among the fragments it generates, AAGGU would be particularly useful for tritium-exchange studies. In the left-hand structure, all four purines would be exposed; in the center structure, all four would be base-paired; in the right-hand structure, one A would be exposed.

d. The kinetics of melting of the left-hand structure should be a single exponential, whereas the right-hand structure would show two relaxation times and the center structure could have even three relaxation times.

Chapter 25

25-1. At low concentrations $\hat{c}$, there should be only a little dimerization, so we can estimate B by assuming that $\hat{c}_1 \cong \hat{c}$. Starting with Equation 25-14, we have

$$\pi = (RT/M)\hat{c}_1 + (RT/2M)\hat{c}_2$$

We can solve for $\hat{c}_2$ in terms of $\hat{c}_1$.

$$K = (M_1)^2/(M_2) = (\hat{c}_1/M)^2/(\hat{c}_2/2M)$$

and

$$\hat{c}_2 = 2\hat{c}_1{}^2/KM$$

With the approximation $\hat{c}_1 \cong \hat{c}$, we have

$$\pi = (RT/M)\hat{c} + (RT/KM^2)\hat{c}^2$$

so that $B = RT/KM^2$.

25-2. Refer to the section on the Donnan effect. On side 1 $(Na^+)^{(1)} = (Cl^-)^{(1)} = 10\ \mu M$. Because $(Cl^-)^{(2)} = 50\ \mu M$, we know $r_D = 5$ (see Eqn. 25-21b). Now calculate r_D from Equation 25-25, using $z(M) = 99\ \mu M$ and $s = 10\ \mu M$. From this equation we calculate $r_D \cong 10$. Therefore, the system is not at thermodynamic equilibrium because the observed value of r_D (5) is not in agreement with the one calculated for the system assuming it was at equilibrium.

25-3. Let r be the distance from the center of the spherical vesicle to the outer surface of the spherical vesicle. The volume of the bilayer portion of the spherical vesicle is

$$(4/3)\pi r^3 - (4/3)\pi(r - 2l)^3 = (4/3)\pi(6lr^2 - 12l^2r + 8l^3)$$
$$= 4\pi r^2[2l - 4l(l/r) + (8/3)l(l^2/r^2)]$$

The volume of the planar bilayer is $A_p \cdot 2l$, where A_p is the area of one face. Because the volumes of each bilayer (spherical and planar) are assumed equal, we have

$$A_p \cdot 2l = 4\pi r^2[2l - 4l(l/r) + (8/3)l(l^2/r^2)]$$
$$A_p = 4\pi r^2[1 - 2(l/r) + 4/3(l^2/r^2)]$$

The area of the outer face of the sphere is $4\pi r^2$. Clearly, if $l/r \ll 1$, then $A_p \cong 4\pi r^2$ and the investigator is correct.

25-4. See Section 25-3. Raising the NaCl concentration will permit closer head-group packing for charged, ionic head groups. This will, in turn, encourage micelle shapes where closer packing can occur, such as cylindrical shapes or planar bilayer structures. These shapes can grow larger without changing significantly the ratio of micelle surface area to number of head groups. Thus, at low ionic strengths, we expect there to be globular or spherical structures where head-group spacing is relatively large, and we expect a shape change to increasingly nonspherical structures as ionic strength is raised; these structures can grow considerably in size to give large, elongated micelles or bilayer structures.

25-5. The protein must rotate within the bilayer in two modes. One is around the long axis of the cylinder and should be rapid; as this phase occurs, it should expose most of the paramagnetic centers to the internal cavity containing ascorbate. This is the fast phase in the decay of the EPR signal. The other phase is due to slow end-over-end rotation, and exposes the remaining paramagnetic centers on the ends of the protein to the ascorbate.

Index to Parts I, II, and III